Library of
Davidson College

# BROUWER'S INTUITIONISM

# STUDIES IN THE HISTORY AND PHILOSOPHY OF MATHEMATICS

Volume 2

NORTH-HOLLAND
AMSTERDAM · NEW YORK · OXFORD · TOKYO

# BROUWER'S INTUITIONISM

## Walter P. van Stigt

*Wolfson College, Oxford, UK*

1990

NORTH-HOLLAND

AMSTERDAM · NEW YORK · OXFORD · TOKYO

ELSEVIER SCIENCE PUBLISHERS B.V.
Sara Burgerhartstraat 25
P.O. Box 211
1000 AE Amsterdam, The Netherlands

*Sole Distributors for the USA and Canada:*
ELSEVIER SCIENCE PUBLISHING COMPANY, INC.
655 Avenue of the Americas
New York, NY 10010, USA

ISBN 0 444 88384 3

**Library of Congress Cataloging-in-Publication Data**

```
Stigt, Walter P. van, 1927-
    Brouwer's intuitionism / Walter P. van Stigt.
    p.  cm.
    Includes bibliographical references (p.   ) and index.
    ISBN 0-444-88384-3
    1. Brouwer, L. E. J. (Luitzen Egbertus Jan), 1881-1966.
2. Mathematicians--Netherlands--Biography. 3. Mathematics-
-Philosophy.   I. Title.
QA29.B697S75  1990
510'.92--dc20
[B]                                                 90-7554
                                                       CIP
```

© Elsevier Science Publishers B.V., 1990

All rights reserved. No part of this publication may be reproduced, stored in a retrieval system, or transmitted, in any form or by any means, electronic, mechanical, photocopying, recording or otherwise, without the prior written permission of the Publisher, Elsevier Science Publishers B.V., P.O. Box 211, 1000 AE Amsterdam, The Netherlands.

Special regulations for readers in the USA – this publication has been registered with the Copyright Clearance Center Inc. (CCC), Salem, Massachusetts. Information can be obtained from the CCC about conditions under which photocopies of parts of this publication may be made in the USA. All other copyright questions, including photocopying outside of the USA, should be referred to the Publisher.

No responsibility is assumed by the Publisher for any injury and/or damage to persons or property as a matter of products liability, negligence or otherwise, or from any use or operation of any methods, products, instructions or ideas contained in the material herein.

Printed in The Netherlands

*To the memory of my parents*

# PREFACE

... und Brouwer, das ist die Revolution! (Hermann Weyl)

The loss of spatial intuition as the ultimate foundation of mathematics in the early nineteenth century had left the world of mathematical practitioners in a state of post-disaster shock and confusion. Many if not most, unaware or unwilling to face its implications continued the classical and Cartesian tradition, considering mathematics to be part of a universal search for the grand design of nature, the queen and hand-maiden of science. Others recognized the vacuum and the need for an alternative foundation of mathematics, independent of spatial perception and capable of supporting a rigorous treatment of analysis, algebra and number theory. Among the newly emerging class of pure mathematicians some simply welcomed their new freedom to pursue mathematics for its own sake in ever-increasing specialization. Those directly concerned with the fundamental issues such as the nature of mathematics and the ultimate source of mathematical truth almost universally sought its justification and rigour in some aspect of logic, the axiomatic deductive theory, which in the Aristotelian tradition had served merely as an 'organon', and instrument of disclosing the implications of known truths. The logicists Frege, Russell a.o. considered mathematics to be part of logic and all mathematical notions and theories to be reducible to logical ones; they identified mathematical truth with derivability from logical primitive notions by means of logical principles. Hilbert and the new formalists defined the object of mathematics as the formal structure of axiomatic systems, and mathematical truth as the consistency of the system, i.e. its guaranteed freedom from contradiction. Both schools agreed that the immediate object of mathematics, the elements of their systems, are symbols and symbol combinations, in the formalist view symbols without content or meaning. They confidently set about the task of putting their theories into practice and create the logical structures to serve as the new foundations of mathematics. Logicist confidence was badly shaken at the turn of the century by the discovery of the virus of contradiction within the logical structures they had created, and frantic efforts were made to remove the infected parts. At this moment of crisis Brouwer enters the foundational arena, challenging not just the soundness of some of the logicist and formalist systems but the fundamental assumptions on which the whole of their philosophy of mathematics was based.

Brouwer was a rebel. His doctoral thesis *On the Foundations of Mathematics* [1907] was the manifesto of an angry young man taking on the mathematical establishment on all fronts. His first and immediate attack was directed at the 'applied' philosophy, personified in his own Ph.D. supervisor D. J. Korteweg. Equally damning was his verdict on the logicist and formalist attemps to reduce mathematics to a mechanistic manipulation of symbols, a meaningless game. Even the neo-intuitionists' challenge of formalist trends did not escape his censure, their acceptance of a modified but essential role of language and logic in mathematics condemned as betrayal to the formalist cause. The *Foundations* was Brouwer's declaration of the independence of mathematics from both science and logic, and the launch of his alternative philosophy, in which the individual mind is the exclusive agent and seat of mathematical activity.

Brouwer was also an astute operator and a brilliant mathematician. He knew that his protestations would remain a cry in the wilderness until he had convinced his fellow-mathematicians of his mathematical mastery and he could expose mathematical errors resulting from his opponents' philosophy. During the following few years he concentrated his energies on the fundamental problems of topology, the youngest branch of mathematics, and he more than proved his mathematical mastery. The quantity and quality of his output were staggering: more than 40 major publications in less than five years in which he proved some of the fundamental theorems, forged new methods and determined the future course of topology. In this short time he established a world-wide reputation for himself; his genius and originality were acknowledged by the great mathematicians of the time such as Poincaré, Hilbert and Klein, and recognized at home by the award in 1912 of a professorship at his Alma Mater, the University of Amsterdam.

His authority established and position secured, Brouwer immediately returned to the cause of his major concern, the Foundations of Mathematics. His Inaugural Address, *Intuitionism and Formalism* [1912], is a renewed attack on the excesses of the formalist philosophy of mathematics as axiomatic theory, and a call for a full-scale reform of mathematical practice in line with his Intuitionist philosophy. It marks effectively the end of his topological work; his major energies for the rest of his creative life concentrate on his Intuitionist Programme: a further analysis of the implications of his fundamental theses and a complete re-construction of mathematics.

A start was made in 1918 with the re-construction of Set Theory. As in his approach to the problems of topology, he did not shrink from introducing wholly new fundamental concepts and methods as a basis of his constructive analysis of the continuum. Some, such as the Brouwer-set or 'spread' and 'species', were radical alternatives to traditional notions, others, such as the 'choice sequence', extended classical concepts. All were inspired by his 'Intuition of the Continuum' and his conception of mathematics as individual thought-construction.

Brouwer saw his Intuitionist re-construction as a programme of reform of classical analysis, flawed and discredited by its use of logical principles, in particular the Principle of the Excluded Middle, which had produced results that could not be effected by pure-mathematical construction. In this process of reform some fundamental theorems of classical mathematics had to be sacrificed; on the other hand the new Intuitionist concepts and methods held out the promise of a 'natural development' of a new Analysis.

The comprehensive re-construction of mathematics was a formidable undertaking, but perhaps even harder was the task of convincing fellow-mathematicians of the errors of their ways and bringing them round to what he considered to be the only true conception of mathematics. Searching for mathematical errors resulting from the use of classical logic he focused his attention on the Principle of the Excluded Middle and the concept of negation. Again his original approach led to some remarkable results, also in the field of logic; his strict interpretation of negation, 'absurdity', generated a new logical 'calculus of absurdities' and what became known as 'the Brouwer Logic'.

Brouwer's failure to gain support for his Intuitionist Programme and 'make school' was due partly to what Weyl described as 'the almost unbearable awkwardness resulting from the inapplicability of the simple laws of classical logic', partly to Brouwer himself. Ambition for international leadership and an almost pathological suspicion of fellow-mathematicians made cooperation impossible; academic debates turned into 'battles' and friends into foes. The Intuitionist-Formalist debate became a personal feud between the mathematical giants Brouwer and Hilbert, and ended in 1928 with the expulsion of Brouwer from the editorial board of the *Mathematische Annalen* by dictat of Hilbert. Forsaken, humiliated and disillusioned Brouwer abandoned his Intuitionist Programme and withdrew into silence just about the time when the Formalist Programme appeared to be fundamentally flawed and major opposition collapsed.

The intuitionist cause was kept alive by Arend Heyting, one of Brouwer's most loyal students. His Formalization of Intuitionist Logic of 1929 – in Brouwer's view 'an interesting but sterile exercise' – and his further publications made Intuitionism accessible to a wider audience, it also changed its course. The move towards 'separation of mathematics and logic' was gradually reversed and much of the Brouwerian philosophy of mathematics abandoned. Ironically, interest in Intuitionism has now become the almost exclusive preserve of logicians and historians; among the general mathematical public Brouwer is remembered mainly for his revolutionary challenge of the Principle of the Excluded Middle, his message of a new philosophy of mathematics as the foundation of mathematics largely forgotten.

Yet Brouwer's Intuitionism is primarily a philosophy of mathematics, the sole survivor from 'the crisis in the foundations of mathematics' and still relevant in the continuing debate about the nature of mathematical reality. In the broad divide of the historical debate his Intuitionism represents the last great attempt to seek the origin and seat of mathematics in the acting mind of man, challenging the Platonist tradition, which takes mathematics to be an objective, unique and timeless body of truths to be discovered by man. Others had professed their belief in the 'intuitive' nature of mathematics, but their 'intuition' was left vague – 'the guiding light' or the 'deeper reality' – and of little consequence to their practice. Brouwer searched for the genesis of mathematics within the deeper reality of the human thought-process, and having traced its ultimate source he made it the basis of his new philosophy and accepted the full consequences for mathematical practice.

His metaphysical investigation starts from and concentrates on the Kantian interpretation of mathematics as human knowledge which is synthetic and a-priori, i.e. created and yet independent from sense perception. His first and most fundamental thesis emerges as the radical individualist conclusion from the definition of mathematics as thought-construction. Thought and certainty are the monopoly of mind, which is essentially individual – 'there is no collective Mind' – and live, i.e. existing in time. The characteristics of mathematics are those of the live, individual Subject, the genitor and carrier of thought. Its distinctive 'synthetic' and 'a-priori' are determined by and originate in the time-bound nature of the live Subject, or as Brouwer impressed on Korteweg, 'must be rooted in life'. The synthetic construction of mathematics is building upwards in time with constructs 'previously acquired'. He recognized that such construction, like the system of deductive reasoning, ultimately requires primitive elements and construction methods. The a-priori as exclusion of the contingency of perceptional knowledge does not allow these to be determined by experience nor by the authority of arbitrary axioms. His solution to this most fundamental problem was found to be the a-priori as the necessary and sufficient condition for mathematics: the life of the Subject, time itself. In the thought-world of the Subject it is the becoming-aware of his time-existence which is the 'Primordial Happening' in which the first elements of mathematical construction are conceived. 'Intuition' is the Subject's direct grasp of Time in its two main and complementary aspects of discreteness and continuity and his power to create new constructs from these primordial elements, proceeding in time in 'free and unlimited unfolding'.

Brouwer's Intuition is more than an attempt to justify the natural numbers for a Kroneckerian construction of mathematics, it is the basis of a new Intuitionist philosophy and interpretation of mathematics: it defines the domain of mathematics and determines its principles and methods as well as its relation to other areas of 'reality' and human activity. Mathematics is the Subject's constructive thought-activity exclusively with the elements of the a-priori Intuition of Time. The interpretation of such fundamental notions as mathematical existence, truth, negation and proof follows from this definition, as do the role and function of representation in language, of logic and 'application'.

Rigorous adherence to Intuitionist principles imposes restrictions, felt by Brouwer's contemporaries to be excessive and destructive. On the other hand his Intuitionist philosophy is the inspiration and source of new developments; his new concepts 'Brouwer set' and choice sequence could only originate in a philosophy which gives central place to the individual mind, creating and sustaining a growing mathematical reality.

Brouwer, unfortunately, did not publish a systematic and comprehensive exposé of his philosophy of mathematics. His pronouncements on the major foundational issues are spread over a wide range of publications, often in rather cryptic form and as introductory remarks in mathematical papers. Commentators have dismissed his philosophy, altogether or in parts, as irrelevant and treat his Intuitionist mathematics in isolation.

This book tries to present an ordered account of Brouwer's views as can be established from all the evidence available and show that together they form a coherent and consistent philosophy and foundation of mathematics. It follows an earlier attempt, my dissertation of 1971 of the same title, when Brouwer's published papers were the only source available. Since then we have been able to trace and gather Brouwer's other writings, which even at the time of publication of his *Collected Works* by Heyting and Freudenthal were assumed to be lost; they include manuscripts of unfinished books, lecture notes, correspondence and other private notes. These 'other writings' proved very illuminating and relevant for an interpretation of Brouwer's Intuitionism. They also reveal a strong link between his foundational ideas, his Weltanschauung and his character, a link which in my dissertation was established on the evidence of such publications as *Life, Art and Mysticism*. Various aspects of his Intuitionist philosophy are clearly shown to have their roots in an egotistical and mystical outlook on life, which in turn is determined by his personality, his weaknesses and prejudices and by circumstances.

In the final composition of mathematical theories and philosophy there is no room for emotion, belief, opinion and personal experience, and Brouwer in his quest for the pure conception of mathematics was more radical than most in eliminating perceptional and other extraneous elements. However, the humble origin of ideas need not detract from their brilliance, indeed, it is an essential aspect of the product of human genius. In Brouwer's case, moreover, the human condition of the individual was the starting point in his analysis of the genesis of mathematics. His general method of investigating and interpreting human phenomena is, to use his own phrase, 'genetic': it searches for their origin and derives their nature and characteristics from the processes that brought them into being.

For an understanding of Brouwer's ideas the 'genetic' approach seemed most appropriate and natural and it prompted the overall design of this book: a geneaology of Brouwer's Intuitionism starting from the man Brouwer, followed by his philosophy of life, his conception of mathematics and its methodology and finally his Intuitionist mathematics.

*Chapters I and II* are a survey of Brouwer's life and work. A complete bibliography of his publications and other writings is included for the purpose of reference in this book and as a source for further reference by the reader. Chapter II attempts to place his works in their historical context; other biographical details are included as are relevant and helpful for an understanding of the man Brouwer and the development of his ideas.

*Chapter III* focuses on Brouwer's analysis and views on the main philosophical issues which bear directly on his interpretation of mathematics, such as the nature of man and mind, human knowledge, causality and the 'transcendental values of goodness, truth and beauty'.

*Chapter IV* analyses the Brouwerian conception of mathematics, its genesis in the Primordial Intuition, its characteristic constructions and its essential 'Subjectivity'.

The complete isolation of mathematics from the exterior world is the basis of the main tenets of Brouwer's philosophy of science: the ideal nature of all mathematical application and the non-contributary role of logic in mathematics. In mathematics as pure-thought construction language and logic do not play a role, but because of their dominant role in classical and formalist practice they figure prominently in Brouwer's campaign of Intuitionist reform.

*Chapter V* starts from Brouwer's interpretation of the origin of language and follows his development of a semantic theory and his critique of traditional use of logical principles in mathematics and of formalization.

*Chapter VI* introduces some of the fundamental parts of Brouwer's Intuitionist re-construction of mathematics, in particular those which deviate from the classical treatment and which clearly reveal their origin in Brouwer's Intuitionist philosophy: concepts such as the 'Brouwer Set' or spread, species and sequence, his construction of number systems and analysis of the continuum. They illustrate how Brouwer put his Intuitionist principles into practice. More generally, Brouwer's Intuitionism illustrates the dependence of mathematics on its philosophical foundations: how the adoption of a particular philosophy affects the content of mathematics, and that in turn a philosophical interpretation of reality is based on some ideological posit or commitment.

*The Appendices* are a selection of Brouwer's writings which are not accessible to a general readership either because they have not been published or have only been published in Dutch. They have been chosen for their relevance to the subject of this book: the development of Brouwer's ideas on the foundations of mathematics. Most of them were not intended for publication; they may lack the polish of Brouwer's published papers but in their direct and uninhibited form they often reveal more clearly his true intention and the origin of his ideas.

In my English translation of the Dutch papers I have tried to keep as close as possible to the original text; in some cases this meant leaving sentences incomplete and 'unpolished'. Well aware that every translation is a personal interpretation of the translator, I have included the Dutch original for further reference whenever the original is not available in published form.

The title *Brouwer's Intuitionism* reflects the scope and limits of my inquiry; it covers the whole of Brouwer's systematic speculation on the origin and nature of mathematics, his definition of norms and principles in accordance with this philosophy and his attempts to put these principles into practice.

The 'man and his work' are the main source of any such inquiry; they also provide a natural and practical limitation of its terms of reference. I have restricted myself to Brouwer's writings as my immediate and main source. Views and contributions of others are considered only in as far as they have or may have influenced the development of Brouwer's ideas. In no way do I hereby wish to discount the valuable contribution made by others to the development of Intuitionism and Intuitionist mathematics and logic, or indeed, to the understanding of Brouwer.

*Brouwer's Intuitionism* is written for the foundational specialist as well as for the 'general reader': all those who practise mathematics, are fascinated by it and may-be are concerned about modern interpretation of its status and function.

As a contribution to the specialist debate its claim is a modest one: it does not add to the treasure of Intuitionist mathematics and logic, but serves as a reminder of the roots of Intuitionism in the vision of its founder. His vision of the true nature of mathematics inspired his Intuitionist innovations, it should be the guide in our interpretation of his notions and principles and remain a continuing source of further intuitionist developments.

For the general reader, seeking a deeper meaning and purpose of mathematics, beyond mechanistic formalism and the isolation of specialisms, Brouwer and his Intuitionism may be a revelation and an education: here is a brilliant mathematician who was prepared to question the logicist and formalist conventions of his time and seek an interpretation of mathematics within the realm of human thought and in line with the constructive mode of thinking of the working mathematician; moreover, one who had the strength of his convictions. His foundational theory became the source of inspiration of his remarkable intuitionist contributions to mathematics and logic and has remained unchallenged to the present day.

I hope this work will help to keep alive the memory of one of the greatest mathematical philosophers of this century and may contribute to a wider public knowledge and better understanding of his ideas.

*Wolfson College*
Oxford, January 1990                                                         W.P.V.S.

## ACKNOWLEDGEMENTS

I wish to acknowledge my indebtedness to many friends and colleagues who have contributed to this work through their continued interest, encouragement and support.

First among them is Geoffrey Kneebone, who sparked off my interest in the philosophy of mathematics and Intuitionism, who guided me through my doctoral research and encouraged me to prepare the work for publication.

A research grant by ZWO enabled me to work with Dirk van Dalen at Utrecht University during the academic year 1976-77 and gather Brouwer biographical data. Our search led to the recovery of much of Brouwer's 'wetenschappelijke nalatenschap' and the setting-up of the Brouwer Archief. The 'new' material proved a rich source for further research; its wide use has undoubtedly added to the value of this book.

I like to use this opportunity to thank all those who made this year in Holland possible and enjoyable and contributed to its success, in particular the Utrecht 'Logic Group' under the leadership of Dirk van Dalen. I am grateful to Ir L.E.J. Brouwer, Brouwer's nephew and namesake and executor of his will, for his cooperation in the setting-up of the Brouwer Archief and for his kind permission to publish the Brouwer papers.

A very special debt of gratitude is due to Anne Troelstra, a long-time friend and adviser, for his interest in my work, his patience and loyalty. His critical comments on earlier working versions of the manuscript have been much appreciated and have contributed to the work in its present form.

Most of the work during the last few years was done at Oxford University and was made possible by the award of a Research Fellowship of the Leverhulme Trust. I am grateful to the Leverhulme Trustees for their confidence in me and their backing, to the President and Fellows of Wolfson College for electing me to join their august company and providing me with an academic home, and to the members of the Institutes of Mathematics and Philosophy for making me feel welcome.

I am particularly grateful for the intellectual companionship of the Oxford 'Foundations circle', for the encouragement and stimulus received in seminar and conversation. It is a pleasure to express my special thanks to Michael Dummett for his continued interest in the project, his comment and his encouragement at crucial times to persist and complete the work; and to Robin Gandy and Daniel Isaacson for their friendship, loyalty and and good advice.

Thanks also to many others who in their different ways have contributed to the development of the book: Brouwer's family and friends by their cooperation, his admirers such as Dale Johnson and Anthony Hill by their enthusiasm and encouragement, and all those participants in seminars and meetings over the past years who through their comment helped to crystallize my views.

Thanks to my wife Judith and my family for their patience with my long-time engagement in the project; special thanks to my son Nicholas for his expert advice and help with the technical problems of electronic type-setting.

Thanks are also due to North Holland - Elsevier Science Publishers for publishing the book and to its editorial staff, especially Arjen Sevenster and Jan Kastelein, for their helpfulness and competent contribution to its production.

W.P.V.S.

# REFERENCES

References to authors other than Brouwer are made in the usual way: the name of the author followed by the year of publication as given in the general bibliography at the end of this book: e.g. HEYTING[1956].

References to Brouwer's work are given in the codification of the BROUWER BIBLIOGRAPHY Chapter I, e.g. BR[1923A], or if no confusion can arise, as simply [1923A].

Most of Brouwer's papers were published in Dutch and later in a German or an English translation of Brouwer's own hand or approved by him.

For quotations and references in this monograph we have adopted the following practice:

Papers published by Brouwer in Dutch and English are quoted from the English version (as re-published in the Collected Works);

Quotations from papers published in Dutch and German are our English translation of the original Dutch version and reference is made to the relevant pages of the Dutch version;

Quotations from work published by Brouwer only in Dutch or in German are our English translation of the original, not necessarily the same as the translation published in the Collected Works. For reference to letters I have adopted the codification of the *Brouwer Archief*, which holds a collection of the Brouwer correspondence and photocopies of Brouwer's letters held elsewhere; e.g. (11.01.07 [DJK32]) was written on the 11th January 1907, the 32nd of the 142 letters of the Brouwer-Korteweg correspondence (originals held in the Universiteitsbibliotheek Amsterdam); (16.06.13 [DHI18] is Brouwer's letter to Hilbert of 16th June 1913, the 18th letter of the Hilbert-Brouwer correspondence (originals held in the Universitätsbibliothek Göttingen), etc.

W.P.V.S.

# CONTENTS

Preface
Acknowledgements
References

## Chapter I. The Brouwer Bibliography

| | | |
|---|---|---|
| 1.1 | Publications | 2 |
| 1.2 | Unpublished Papers | 16 |

## Chapter II. Brouwer, Life and Work

| | | |
|---|---|---|
| 2.1 | Childhood and Student-years | 21 |
| 2.2 | Post-graduate Study | 28 |
| 2.3 | Life, Art and Mysticism [1905] | 31 |
| 2.4 | The Foundations of Mathematics | 35 |
| 2.4.1 | Chapter One – The Construction of Mathematics | 37 |
| 2.4.2 | Chapter Two – Mathematics and Experience | 39 |
| 2.4.3 | Korteweg's Rejection | 40 |
| 2.4.4 | Chapter Three – Mathematics and Logic | 43 |
| 2.5 | Towards an Academic Career | 45 |
| 2.5.1 | The Unreliability of the Logical Principles [1908C] | 46 |
| 2.5.2 | The Rome Conference 1908 | 47 |
| 2.5.3 | Topological Activity 1908-1912 | 49 |
| 2.5.4 | The Nature of Geometry [1909A] | 49 |
| 2.5.5 | Zur Analysis Situs | 51 |
| 2.5.6 | The New Methods and the Invariance of Dimension | 53 |
| 2.5.7 | International Recognition | 55 |
| 2.5.8 | Professor L.E.J. Brouwer, 14 October 1912 | 57 |
| 2.6 | Intuitionism and Formalism | 61 |
| 2.6.1 | Intuitionist Set Theory | 62 |
| 2.7 | The Signific Interlude | 65 |
| 2.8 | The Second Act of Intuitionism | 71 |
| 2.8.1 | The Foundations of Set Theory Part I | 71 |
| 2.8.2 | The Foundations of Set Theory Part II | 75 |

| | | |
|---|---|---|
| 2.8.3 | The New Foundation Crisis | 75 |
| 2.9 | The Turbulent Years | 77 |
| 2.9.1 | Signific and Political Campaigns | 77 |
| 2.9.2 | Göttingen and Berlin Calls | 80 |
| 2.9.3 | The Intuitionist 'Putsch' | 82 |
| 2.10 | The Brouwer Logic | 85 |
| 2.10.1 | The Principle of the Excluded Middle – 1923 | 85 |
| 2.10.2 | The Calculus of Absurdities | 87 |
| 2.10.3 | La Logique Brouwerienne | 88 |
| 2.11 | The Reconstruction of Analysis | 90 |
| 2.11.1 | The Continuity Arguments | 91 |
| 2.12 | The Unfinished Books | 93 |
| 2.12.1 | Reelle Funktionen | 93 |
| 2.12.2 | Die Berliner Gastvorlesungen | 94 |
| 2.13 | The Brouwer-Hilbert Controversy and the Annalen Affair | 99 |
| 2.14 | The Silent Years | 103 |
| 2.15 | Post-war Lectures and Final Years | 107 |

## Chapter III. Brouwer's Philosophy

| | | |
|---|---|---|
| 3.1 | Introduction | 111 |
| 3.2 | Character and Early Background | 112 |
| 3.3 | Philosophy, Metaphysics and Ideology | 115 |
| 3.4 | Mysticism | 119 |
| 3.5 | God and the Moral Principle | 122 |
| 3.6 | Kant and the French Intuitionist Tradition | 127 |
| 3.7 | French Mathematical Intuitionism | 132 |
| 3.8 | The Self, Consciousness and Mind | 135 |
| 3.9 | Causal Attention, Will and Cunning Activity | 141 |
| 3.10 | The Phase of Social Acting and Language | 144 |

## Chapter IV. Brouwer's Philosophy of Mathematics

| | | |
|---|---|---|
| 4.1 | The Primordial Intuition of Time | 147 |
| 4.1.1 | Introduction | 147 |
| 4.1.2 | Genesis of the Concept and Definitions | 148 |
| 4.1.3 | Brouwer's Primordial Intuition of Mathematics | 150 |
| 4.1.4 | The Discrete and the Continuous in the Primordial Intuition | 153 |

| | | |
|---|---|---|
| 4.2 | Intuitive Thinking and Mathematics | 157 |
| 4.3 | The Nature of Mathematics | 159 |
| 4.4 | Reflection on the Nature of Mathematics | 162 |
| 4.5 | Constructiveness and Existence | 164 |
| 4.5.1 | Brouwer Construction | 164 |
| 4.5.2 | Classification of Constructions | 165 |
| 4.5.3 | Distinctive Features of Mathematical Constructive Thinking | 168 |
| 4.6 | The Subject and the Intersubjectivity of Mathematics | 171 |
| 4.6.1 | Brouwer's Idealism | 171 |
| 4.6.2 | The Idealized Subject | 172 |
| 4.6.3 | Mathematics, the Life of the Idealized Free Subject | 174 |
| 4.6.4 | Objectivity and Inter-subjectivity | 180 |
| 4.7 | The Application of mathematics | 182 |
| 4.7.1 | Brouwer's Critique of Scientific Practice | 184 |
| 4.7.2 | The Ideal Nature of Science | 185 |
| 4.7.3 | The Process of Application and its Mathematical Truth | 188 |

## Chapter V. Language and Logic

| | | |
|---|---|---|
| 5.1 | Society and Communication | 193 |
| 5.2 | The Purpose and Origin of Language | 196 |
| 5.3 | Language and Pure Thought | 199 |
| 5.4 | The Programme of Reform | 201 |
| 5.5 | Brouwerian Semantic Theory | 203 |
| 5.5.1 | The Privacy of Language | 203 |
| 5.5.2 | The Instability of Language | 204 |
| 5.5.3 | Words, Sentences and Truth | 211 |
| 5.6 | Brouwer's Analysis of Mathematical and Linguistic Activity | 213 |
| 5.6.1 | The Language of Mathematics | 215 |
| 5.6.2 | The Language of Logical Reasoning | 220 |
| 5.7 | Brouwers's Analysis of Logic | 224 |
| 5.7.1 | Logic, the Science of Reasoning | 224 |
| 5.7.2 | Pure Logic as Science and Abstraction | 225 |
| 5.7.3 | The Application of Logical Principles | 230 |
| 5.8 | Brouwer Negation | 238 |
| 5.8.1 | Brouwer's Constructive Notion of Negation | 239 |

| | | |
|---|---|---|
| 5.8.2 | Absurdity as Impossibility of 'Fitting-in' | 240 |
| 5.8.3 | Constructed Impossibility | 242 |
| 5.8.4 | Mathematical Absurdity and Other Notions of Negation | 244 |
| 5.9 | The Refutation of the PEM | 247 |
| 5.9.1 | The Unreliability of the PEM, 1908 | 248 |
| 5.9.2 | Mathematics Without the PEM, 1917 | 250 |
| 5.9.3 | Brouwer's Counterexamples | 252 |
| 5.9.4 | The Contradictority of the PEM | 255 |
| 5.10 | Logical Negation and the 'Brouwer Logic' | 257 |
| 5.10.1 | The Calculus of Absurdities | 257 |
| 5.10.2 | Logical Necessity and Truth | 260 |
| 5.10.3 | Logical Absurdity | 262 |
| 5.10.4 | Non-contradictority and Absurdity-of-absurdity | 265 |
| 5.11 | The Crisis in Brouwer's Intuitionism | 270 |
| 5.11.1 | The Use of Language and Logic in Brouwer's Reconstruction of Mathematics | 271 |
| 5.11.2 | The Pragmatism of Post-Brouwer Intuitionism | 274 |
| 5.12 | The Formalization of Mathematics | 279 |
| 5.12.1 | The 'Formalist' Schools and Formal Axiomatic Theory | 279 |
| 5.12.2 | The Formalization of Intuitionist Mathematics | 285 |
| 5.12.3 | Heyting's Formalization of Intuitionist Logic | 288 |
| | Selected Contributions to the Formalization of Intuitionist Logic and Mathematics | 293 |

## Chapter VI. Brouwer's Intuitionist Mathematics

| | | |
|---|---|---|
| 6.1 | Introduction | 295 |
| 6.2 | Brouwerian Separable Mathematics | 300 |
| 6.2.1 | The Primordial Intuition as Set-construction | 301 |
| 6.2.2 | The Fundamental Sequence | 305 |
| 6.2.2.1 | The Order-type $n$ | 306 |
| 6.2.2.2 | The Order-type $\omega$ | 309 |
| 6.2.2.3 | Brouwer's Generalization of the Concept Sequence | 309 |
| 6.2.3 | Cardinality | 312 |
| 6.2.4 | Systems and Operations of Separable Mathematics | 314 |
| 6.2.4.1 | The Operations on Ordinal Numbers | 314 |
| 6.2.4.2 | The Integers | 315 |

| | | |
|---|---|---|
| 6.2.4.3 | Rationals and the Order-type $\eta$ | 316 |
| 6.3 | The Brouwer Mathematical Continuum | 318 |
| 6.3.1 | Introduction | 318 |
| 6.3.2 | The Genesis and Nature of the Intuitive Continuum | 323 |
| 6.3.2.1 | The Construction of Intervals | 325 |
| 6.3.2.2 | Relations and Ordering of Intervals | 326 |
| 6.3.3 | Point-sets on the Continuum and the 'Made-measurable Continuum' (pre-1917) | 328 |
| 6.3.3.1 | The 'Made-measurable Continuum' | 328 |
| 6.3.3.2 | Point on the Continuum | 330 |
| 6.3.3.3 | Sets of Constructed Points | 332 |
| 6.3.3.4 | Denumerably-unfinished Systems and Definable Points | 333 |
| 6.3.4 | The New Set-theory 1917 | 334 |
| 6.3.5 | Mathematical Species | 335 |
| 6.3.6 | The Genesis of Species | 337 |
| 6.3.7 | The Hierarchy of Species | 340 |
| 6.3.8 | The Theory of Species | 345 |
| 6.3.8.1 | Union and Intersection of Species | 346 |
| 6.3.8.2 | The Empty Species | 347 |
| 6.3.8.3 | Relations between Species | 347 |
| 6.3.8.4 | Difference, Deviation and Apartness | 348 |
| 6.3.8.5 | Congruence and Agreement | 351 |
| 6.3.9 | The Complementarity of Species | 352 |
| 6.3.10 | Spreads and Choice-sequences | 357 |
| 6.3.11 | Real Elements of the Continuum | 361 |
| 6.3.11.1 | Interval | 362 |
| 6.3.11.2 | Real Point or Real Number | 362 |
| 6.3.11.3 | Law and Choice | 363 |
| 6.3.11.4 | Species of Coincident Points or Point-cores | 364 |
| 6.3.11.5 | Point-set or Point-spread | 366 |
| 6.3.11.6 | Fundamental Sequences of Intervals and Finite Point-spreads | 369 |
| 6.3.12 | The General Spread Concept | 370 |
| 6.3.13 | Continuity and Uniform Continuity of Spread Functions | 378 |
| 6.3.13.1 | Finite Spreads | 378 |
| 6.3.13.2 | The Brouwer Continuity Hypothesis or the Fundamental Hypothesis for Real Functions | 379 |

| | | |
|---|---|---|
| 6.3.13.3 | The Fundamental Theorem of Finite Spreads | 381 |
| 6.3.13.4 | The Uniform-Continuity Theorem | 384 |

**Appendices**

| | | |
|---|---|---|
| 1 | Profession of Faith of L.E.J. Brouwer [BMS 1A] (Dutch original and English translation) | 387 |
| 2 | Student Notebook [BMS 1B] (Dutch original and English translation) | 394 |
| 3 | The Rejected Parts of Brouwer's Dissertation [BMS 3B] | 405 |
| 4 | Intuitive Significs [BMS 22B] (Dutch original and English translation) | 416 |
| 5 | Will, Knowledge and Speech (English translation of BR[1933]) | 418 |
| 6 | Short Retrospective Notes on the Course "Intuitionist Mathematics" [BMS 52] | 432 |
| 7 | Disengagement of Mathematics from Logic [BMS 49] | 441 |
| 8 | Notes for a Lecture [BMS 47] (Dutch original and English translation) | 447 |
| 9 | Changes in the Relation between Classical Logic and Mathematics [BMS 59] | 453 |
| 10 | 'Second Lecture' [BMS 66] | 459 |
| 11 | Real Functions [BMS 37] | 469 |
| 12 | Intuitionist Mathematics or Die Berliner Gastvorlesungen [BMS 32] | 481 |
| 13 | The Brouwer-Korteweg Correspondence | 487 |

**Bibliography** 507

**Index** 517

# CHAPTER I
# THE BROUWER BIBLIOGRAPHY

This bibliography is a complete list of Brouwer's known writings; it includes his published as well as his unpublished work.
To avoid confusion we have for the published papers adopted the codification of *L.E.J. Brouwer Collected Works*, edited by Heyting and Freudenthal, although it is not always a correct indication of the chronological order of writing or publication.
Newspaper articles have been included in the list of unpublished work.
For the unpublished papers we have adopted the codification of manuscripts of the *Brouwer Archief*. Since the *Brouwer Archief* also holds the manuscripts of some published papers, our list of unpublished work is not the complete list of preserved Brouwer manuscripts.

*Abbreviations:*

| | |
|---|---|
| BA | : Brouwer Archief. |
| BMS | : Brouwer Manuscript (held in Brouwer Archief) |
| C.R., Paris | : Comptes Rendus Hebdomadaires des séance de l'Académie des Sciences, Paris. |
| CWI | : L.E.J. Brouwer, Collected Works I. Ed. A. Heyting, North-Holland P.C., 1975. |
| CWII | : L.E.J. Brouwer, Collected Works II. Ed. H. Freudenthal, North-Holland P.C., 1976. |
| Indag. Math. | : Indagationes Mathematicae ex actibus quibus titulus Proceedings KNAW. |
| Jahresbericht D.M.V. | : Jahresbericht der deutschen Mathematiker Vereinigung |
| KNAW Versl. | : Koninklijke Akademie van Wetenschappen te Amsterdam (Koninklijke Nederlandse Akademie van Wetenschappen), Verslag der gewone vergaderingen der wis- en natuurkundige afdelingen. |
| KNAW Verh. | : Koninklijke Akademie van Wetenschappen te Amsterdam, Verhandelingen. |
| KNAW Proc. | : Koninklijke Akademie van Wetenschappen te Am- |

sterdam, Proceedings of the Section of Sciences.
MA : Mathematische Analen.

## 1.1 Publications

1904 A1 Over de splitsing van de continue beweging om een vast punt 0 van $R_4$ in twee continue bewegingen om 0 van $R_3$'s. *KNAW Versl.* 12, pp. 819-839.

1904 A2 On a decomposition of a continuous motion about a fixed point 0 of $S_4$ into two continuous motions about 0 of $S_3$'s. *KNWA Proc.* 6, pp. 716-735; CWII, pp. 3-22.

1904 B1 Over symmetrische transformatie van $R_4$ in verband met $R_r$ en $R_l$. *KNAW Versl.* 12, pp. 926-928.

1904 B2 On symmetric transformation of $S_4$ in connection with $S_r$ and $S_l$. *KNAW Proc. 6,* pp. 785-787; CWII, pp. 25-27.

1904 C1 Algebraische afleiding van de splitsbaarheid der continue beweging om een vast punt van $R_4$ in die van twee $R_3$'s. *KNAW Versl.* 12, pp. 941-947.

1904 C2 Algebraic deduction of the decomposability of the continuous motion about a fixed point of $S_4$ into those of two $S_3$'s. *KNAW Proc.* 6, pp. 832-838; CWII, pp. 28-34.

1905 *Leven, Kunst en Mystiek* [Life, Art and Mysticism]. Waltman, Delft 1905, 99 pp.; selections in English translation, CWI, pp. 1-9.

1906 A1 Meerdimensionale vectordistributies. *KNAW Versl.* 15, pp. 14-26.

1906 A2 Polydimensional vectordistributions. *KNAW Proc.* 9, pp. 66-78; CWII, pp. 36-48.

1906 B1 Het krachtveld der niet-Euclidische negatief gekromde ruimten. *KNAW Versl.* 15, pp. 75-94.

1906 B2 The force field of the non-Euclidean spaces with negative curvature. *KNAW Proc.* 9, pp. 116-133; CWII, pp. 54-71.

1906 C1 Het krachtveld der niet-Euclidische positief gekromde ruimten. *KNAW Versl.* 15, pp. 293-310.

1906 C2 The force field of the non-Euclidean spaces with positive curvature. *KNAW Proc.* 9, pp. 250-266; CWII, pp. 73-89.

1907 *Over de grondslagen der wiskunde. Academisch proefschrift.* [On the foundations of mathematics. Dissertation.] Maas &

|  |  |
|---|---|
|  | van Suchtelen, Amsterdam 1907, 183 pp.; English translation, CWI, pp. 11-101. |
| 1908 A | Die mögliche Mächtigkeiten. In: *Atti IV Congr. Intern. Mat. Roma III,* pp. 569-571; CWI, pp. 102-104. |
| 1908 B | Over de grondslagen der wiskunde. [On the foundations of mathematics.] *Nieuw Archief voor Wiskunde* (II) 8, pp. 326-328; CWI, pp. 105-106. |
| 1908 C | De onbetrouwbaarheid der logische principes. [The unreliability of the logical principles.] *Tijdschrift voor Wijsbegeerte* 2, pp. 152-158. Also in [1919B]; CWI, pp. 107-111. |
| 1908 D1 | Over differentiequotiënten en differentiaalquotiënten. *KNAW Versl.* 17, pp. 38-45. |
| 1908 D2 | About difference quotients and differential quotients. *KNAW Proc.* 11, pp. 59-66; CWII, pp. 93-100. |
| 1909 A | *Het wezen der meetkunde. Openbare Les.* [The Nature of Geometry. Public Lecture.] (Clausen, Amsterdam); also in [1919B]; CWI, pp. 112-120. |
| 1909 B | Die Theorie der endlichen continuierlichen Gruppen, unabhängig von den Axiomen von Lie. In: *Atti IV Congr. Intern. Mat. Roma II,* pp. 296-303; CWII, pp. 109-116. |
| 1909 C | Die Theorie der endlichen kontinuierlichen Gruppen, unabhängig von den Axiomen von Lie, I. *MA* 67, pp. 246-267; CWII, pp.118-139. [Berichtigung, see [1910G].] |
| 1909 D | Over de niet-Euclidische meetkunde. [On non-Euclidean geometry.] *Nieuw Archief voor Wiskunde* (II) 9, pp. 72-74. [not in CW] |
| 1909 E | Karakteriseering der Euclidische en niet-Euclidische bewegingsgroepen in $R_n$. [Characterization of the Euclidean and non-Euclidean motion groups in $R_n$.] In: *Handelingen van het Nederlandsch Natuur- en Geneeskundig Congres 12*, pp. 189-199; CWII, pp. 185-192. |
| 1909 F1 | Over één-éénduidige, continue transformaties van oppervlakken in zichzelf, I. *KNAW Versl.* 17, pp. 741-752. |
| 1909 F2 | Continuous one-one transformations of surfaces in themselves, I. *KNAW Proc.* 11, pp. 788-798; CWII, pp. 195-206. |
| 1909 G1 | Over continue vectordistributies op oppervlakken, I. *KNAW Versl.*17, pp. 896-904. |
| 1909 G2 | On continuous vector distributions on surfaces I. *KNAW Proc.* 11, pp. 850-858; CWII, pp. 273-281. |

| | |
|---|---|
| 1909 H1 | Over één-éénduidige, continue transformaties van oppervlakken in zichzelf, II. *KNAW Versl.* 18, pp. 106-117. |
| 1909 H2 | Continuous one-one transformations of surfaces in themselves, II. *KNAW Proc.* 12, pp.286-297; CWII, pp.207-218. |
| 1910 A1 | Over continue vectordistributies op oppervlakken, II. *KNAW Versl.* 18, pp.702-721. |
| 1910 A2 | On continuous vectordistributions on surfaces, II. *KNAW Proc.* 12, pp. 716-734; CWII, pp. 283-302. |
| 1910 B1 | Over de structuur der perfecte puntverzamelingen, I. *KNAW Versl.* 18, pp. 833-842. |
| 1910 B2 | On the structure of perfect sets of points, I. *KNAW Proc.* 12, pp. 785-794; CWII, pp. 341-350. |
| 1910 C | Zur Analysis Situs. *MA* 68, pp.422-434; CWII, pp. 352-366. |
| 1910 D1 | Over continue vectordistributies op oppervlakken, III. *KNAW Versl.* 19, pp. 36-51. |
| 1910 D2 | On continuous vectordistributions on surfaces, III. *KNAW Proc.* 12, pp. 171-186; CWII, pp. 303-318. |
| 1910 E | Beweis de Jordanschen Kurvensatzes. *MA* 69, pp. 169-175; CWII, pp. 377-383. |
| 1910 F | Über eineindeutige, stetige Transformationen von Flächen in sich. *MA* 69, pp. 176-180; CWII, pp. 244-248. [Berichtigung, see [1910L] and [1919F]] |
| 1910 G2 | Berichtigung [referring to [1909C]]. *MA* 69, p.180; CWII, p.240. |
| 1910 H | Die Theorie der endlichen kontinuierlichen Gruppen, unabhängig von den Axiomen von Lie, II. *MA* 69, pp. 181-203; CWII, pp. 156-178. |
| 1910 J2 | Sur les continus irréductibles de M. Zoretti. *Annales Ec. Norm. Sup.* 27, pp. 565-566; CWII, pp.373-374. |
| 1910 L | Berichtigung [referring to [1910F]]. *MA* 69, p.592. |
| 1911 A | [Review of] G. Mannoury, Methologisches und Philosophisches zur Elementar-Mathematik, Haarlem 1909. *Nieuw Archief voor Wiskunde* (II) 9, pp. 199-201; CWI, pp. 121-122. |
| 1911 B1 | Over de structuur der perfecte puntverzamelingen, II. *KNAW Versl.* 19, pp. 1416-1426.. |
| 1911 B2 | On the structure of perfect sets of points, II. *KNAW Proc.* 14, pp. 137-147; CWII, pp. 384-394. |
| 1911 C | Beweis der Invarianz der Dimensionenzahl. *MA* 70, pp. 161-165; CWII, pp. 430-434. |

| | |
|---|---|
| 1911 D | Über Abbildung von Mannigfaltigkeiten. *MA* 71, pp.97-115; CWII, pp. 454-472. [Berichtigung, see [1911M] and [1921E].] |
| 1911 E | Beweis der Invarianz des *n*-dimensionalen Gebiets. *MA* 71, pp. 305-313; CWII, pp. 477-485. [Berichtigung, see [1921F].] |
| 1911 F | Beweis des Jordanschen Satzes für den *n*-dimensionalen Raum. *MA* 71, pp. 314-319; CWII, pp. 489-494. |
| 1911 G | Über Jordansche Mannigfaltigkeiten. *MA* 71, pp. 320-327; CWII, pp. 498-505. [Bemerkung, see [1911N].] |
| 1911 H1 | Over één-éénduidige, continue transformaties van oppervlakken in zichzelf, III. *KNAW Versl.* 19, pp. 737-747. |
| 1911 H2 | Continuous one-one transformations of surfaces in themselves, III. *KNAW Proc.* 13, pp. 767-777; CWII, pp. 221-231. |
| 1911 J1 | Over één-éénduidige, continue transformaties van oppervlakken in zich zelf, IV. *KNAW Versl.* 20, pp. 24-34. |
| 1911 J2 | Continuous one-one transformations of surfaces in themselves, IV. *KNAW Proc.* 14, pp. 300-310; CWII, pp. 233-243. |
| 1911 K | Sur une théorie de la mésure. A propos d'un article de M.G. Combebiac. *L'Enseignement Math.* 13, pp. 377-380; CWII, pp. 181-183. |
| 1911 L | Sur le théorème de M. Jordan dans l'espace à *n* dimensions. *C.R. Paris* 153, pp. 542-543; CWII, pp. 496-497. |
| 1911 M | Berichtigung [referring to [1911D]]. *MA* 71, p.598; CWII, p.474. |
| 1911 N | Bemerkung [referring to [1911G]]. *MA* 71, p.598; CWII, p.506. |
| 1912 A1 | *Intuitionisme en Formalisme; Inaugurale Rede.* (Clausen, Amsterdam ) [Also in: *Wiskundig Tijdschrift* 9, pp. 180-211 and in [1919B].] |
| 1912 A2 | Intuitionism and Formalism. *Bull. Am. Math. Soc.*, 20, pp. 81-96; CWI, pp. 123-138. |
| 1912 B | Beweis des ebenen Translationssatzes. *MA* 72, pp. 37-54; CWII, pp. 250-267. |
| 1912 C | Zur Invarianz des *n*-dimensionalen Gebiets. *MA* 72, pp. 55-56; CWII, pp. 509-510. |
| 1912 D | Über die topologischen Schwierigkeiten des Kontinuitätsbeweises der Existenztheoreme eindeutig umkehrbarer polymorpher Funktionen auf Riemannschen Flächen (Auszug aus einem Brief an R. Fricke). *Nachr. Göttingen* 1912, pp. 603-606; CWII, pp. 577-580. |

| | |
|---|---|
| 1912 E1 | Over omloopcoëfficiënten. *KNAW Versl.* 20, pp.1049-1057. |
| 1912 E2 | On looping coëfficients. *KNAW Proc.* 15, pp. 113-122; CWII, pp. 511-520. |
| 1912 F | Sur l'invariance de la courbe fermée. *C.R.Paris* 154, pp. 862-863; CWII, pp. 521-522. |
| 1912 G | Über die Singularitätenfreiheit der Modulmannigfaltigkeit. *Nachr. Göttingen* 1912, pp. 803-806; CWII, pp. 587-590. |
| 1912 H | Über die Kontinuitätsbeweis für das Fundamentaltheorem der automorphen Funktionen im Grenzkreisfalle. *Jahresb. D.M.V.* 21, pp. 154-157; CWII, pp. 568-571. |
| 1912 K1 | Over één-éénduidige, continue transformaties van oppervlakken in zichzelf, V. *KNAW Versl.* 21, pp. 300-309. |
| 1912 K2 | Continuous one-one transformations of surfaces in themselves, V. *KNAW Proc.* 15, pp. 352-360; CWII, pp. 527-536. [cf.[1920 G2]. |
| 1912 L | Beweis der Invarianz der geschlossenen Kurve. *MA* 72, pp. 422-425: CWII, pp. 523-526. |
| 1912 M | Sur la notion de 'classe' de transformations d'une multiplicité. In: *Proc. V Int. Congr. Math. Cambridge 1912*, II, pp. 9-10; CWII, pp. 538-539. |
| 1913 A | Über den natürlichen Dimensionsbegriff. *Journ. r. u. angew. Mat.* 142, pp. 146-152; CWII, pp. 540-546. [Berichtigung, see [1923D] and [1924M].] |
| 1913 B1 | Eenige opmerkingen over het samenhangstype $\eta$. *KNAW Versl.* 21, pp. 1412-1419. |
| 1913 B2 | Some remarks on the coherence type $\eta$. *KNAW Proc.* 15, pp. 1256-1263; CWII, pp. 396-403. |
| 1914 | [Review of] A.Schoenflies und H.Hahn, Die Entwickelung der Mengenlehre und ihrer Anwendungen, Leipzig und Berlin 1913. *Jahresb. D.M.V.* 23, pp. 78-83; CWI, pp. 139-144. |
| 1915 A1 | Opmerking over inwendige grensverzamelingen. *KNAW Versl.* 23, pp. 1325-1326. |
| 1915 A2 | Remark on inner limiting sets. *KNAW Proc.* 18, pp. 48-49; CWII, pp. 412-413. |
| 1915 B | Over de loodrechte trajectoriën der baankrommen eener vlakke eenledige projectieve groep [On the orthogonal trajectories of the orbits of a one-parameter plane projective group]. *Nieuw Archief voor Wiskunde* II, 11, pp. 265-290; English translation in CWII, pp. 319-337. |

| | |
|---|---|
| 1916 A | [Author's review of] L.E.J. Brouwer, Eenige opmerkingen over het samenhangstype $\eta$, *KNAW Versl.* 21, pp. 1412-1419. Jahrbuch über die Fortschritte der Mathematik 44, p.556; CWII, p.405. |
| 1916 B | [Review of] E.Borel, Les ensembles de mesure nulle. *Jahrbuch über die Fortschritte der Mathematik* 44, pp. 556-557; CWII, pp. 406-407. |
| 1916 C | [Review of] J.I. de Haan, Rechtskundige significa en haar toepassing op de begrippen: 'aansprakelijk', 'verantwoordelijk', 'toerekeningsvatbaar' [Legal significs and its application to the notions of 'liability', 'responsibility', 'accountability'] *Groot Nederland*, September 1916. [not in CW]. |
| 1917 A | Addenda en corrigenda over de grondslagen der wiskunde [Addenda and corrigenda to On the Foundations of Mathematics]. *KNAW Versl.* 25, pp. 1418-1423; also: *Nieuw Archief voor Wiskunde* II, 12, pp. 439-445; English translation in CWI, pp. 145-149. |
| 1917 B1 | Over lineaire inwendige verzamelingen. *KNAW Versl.* 25, pp. 1424-1426. |
| 1917 B2 | On linear inner limiting sets. *KNAW Proc.* 20, pp. 1192-1194; CWII, pp. 414-415. |
| 1918 A | [With H.P.J. Bloemers, H. Borel and F. van Eeden] Voorbereidend Manifest [Preparatory Manifesto, with French, German and English translations]. *Mededeelingen van het Internationaal Instituut voor Wijsbegeerte* I, pp. 3-12. [not in CW]. |
| 1918 B | Begründung der Mengenlehre unabhängig vom logischen Satz vom ausgeschlossenen Dritten. Erster Teil, Allgemeine Mengenlehre. *KNAW Verh.* 1e sectie, deel XII, no 5, pp. 1-43; CWI, pp. 151-190. |
| 1918 C | Über die Erweiterung des Definitionsbereich einer stetigen Funktion. *MA* 79, pp. 209-211; CWII, pp. 591-593. [Bemerkung, see [1919E]]. |
| 1918 D | Lebesguesches Mass und Analysis Situs. *MA* 79, pp. 212-222; CWII, pp. 595-605. |
| 1918 E | Beweging van een materieel punt op den bodem eener draaiende vaas onder den invloed der zwaartekracht [Motion of a particle on the bottom of a rotating vessel under the influence of gravity]. *Nieuw Archief voor Wiskunde* II, 12, pp. 407-419; English translation, CWII, pp. 665-675. |

| | |
|---|---|
| 1919 A | Begründung der Mengenlehre unabhängig vom logischen Satz vom ausgeschlossenen Dritten. Zweiter Teil, Theorie der Punktmengen. *KNAW Verh.* 1e sectie, deel XII, no 7, pp. 1-33; CWI, pp. 191-221. |
| 1919 B | *Wiskunde, Waarheid, Werkelijkheid.* [Mathematics, Truth, Reality]. (Noordhof, Groningen) [Re-print of [1908C], [1909A] and [1912A].] |
| 1919 C | [With G. Mannoury, H.Borel and F. van Eeden] Signifisch Taalonderzoek. Onderscheid der taaltrappen ten aanzien van de sociale verstandhouding. [With German and French translations] *Mededeelingen van het Internationaal Instituut voor Wijsbegeerte* II, pp. 5-32; the German translation, Signifische Sprachforschung. Vom Unterschied der Sprachstufen in Bezug auf die soziale Verständung, in CWI, pp. 222-229. |
| 1919 D1 | Intuitionistische Mengenlehre. *Jahresber. D.M.V.* 28, pp. 203-208; Also in *KNAW Proc.* 23 (1922), pp. 949-954, CW I, pp. 230-235. |
| 1919 D2 | Intuitionistische Verzamelingsleer. *KNAW Versl.* 29 (1921), pp. 797-802. |
| 1919 E | Nachträgliche Bemerkung über die Erweiterung des Definitionsbereiches einer stetigen Funktion. [referring to [1918C] *MA* 79, p.403; CWII, p.593. |
| 1919 F | Berichtigung. [corrections in [1910F]] *MA* 79, p.403; CWII, p.249. |
| 1919 G | Enumération des surfaces de Riemann regulières de genre un. *C.R. Paris* 168, pp. 677-678, 832; CWII, pp. 631-632. |
| 1919 H | Enumération des groupes finis de transformations topologiques du tore. *C.R. Paris* 168, pp. 845-848, 1168 and 169, p.552; CWII, pp. 634-638. |
| 1919 J | Sur les points invariants des transformations topologiques des surfaces. *C.R. Paris* 168, pp. 1042-1044; 169, p.552; CWII, pp. 639-641. |
| 1919 K | Sur la classification des ensembles fermés situés sur une surface. *C.R. Paris* 169, pp. 953-954 [not in CW]. |
| 1919 L1 | Over één-ééndudige continue transformaties van oppervlakken in zichzelf, VI. *KNAW Versl.* 27, pp. 609-612. |
| 1919 L2 | Über eineindeutige, stetige Transformationen von Flächen in sich, VI. *KNAW Proc.* 21, pp. 707-710; CWII, pp. 611-615. |

| | |
|---|---|
| 1919 M1 | Opmerking over de vlakke translatiestelling. *KNAW Versl.* 27, pp. 840-841. |
| 1919 M2 | Remark on the plane translation theorem. *KNAW Proc.* 21, pp. 935-936; CWII, pp. 269-270. |
| 1919 N1 | Over topologische involuties. *KNAW Versl.* 27, pp. 1201-1203. |
| 1919 N2 | Über topologischen Involutionen. *KNAW Proc.* 21, pp. 1143-1145; CWII, pp. 623-625. |
| 1919 P1 | Opsomming der periodieke transformaties van de torus. *KNAW Versl.* 27, pp. 1363-1367. |
| 1919 P2 | Aufzählung der periodischen Transformationen des Torus. *KNAW Proc.* 21, pp. 1352-1356; CWII, pp. 626-630. |
| 1919 Q1 | Opmerking over meervoudige integralen. *KNAW Versl.* 28, pp. 116-120. |
| 1919 Q2 | Remark on multiple integrals. *KNAW Proc.* 22, pp. 150-154; CWII, pp. 49-53. |
| 1919 R | Luchtvaart en photogrammetrie, I [Aviation and photographic survey]. *Nieuw Tijdschrift voor Wiskunde* 7, pp. 311-331. [not in CW]. |
| 1919 S | Über die periodischen Transformationen der Kugel. *MA* 80, pp. 39-41; CWII, pp. 620-622. |
| 1920 A1 | Over de structuur van perfecte puntverzamelingen, III. *KNAW Versl.* 28, pp. 373-375. |
| 1920 A2 | Über die Struktur der perfekten Punktmengen, III. *KNAW Proc.* 22, pp. 471-474, 974; CWII, pp. 643-646. |
| 1920 B1 | Over één-éénduidige, continue transformaties van oppervlakken in zichzelf, VII. [erroneously numbered VI.] *KNAW Versl.* 28, pp. 1186-1190. |
| 1920 B2 | Über eineindeutige, stetige Transformationen von Flächen in sich, VII. [erroneously numbered VI.] *KNAW Proc.* 22, pp. 811-814; CWII, pp. 647-651. |
| 1920 C | Enumération des classes de transformations du plan projectif. *C.R. Paris* 170, pp. 834-835, 1295. [not in CW]. |
| 1920 D | Enumération de classes de représentations d'une surface sur un autre surface. *C.R. Paris* 171, pp. 89-91, 830. [not in CW]. |
| 1920 E | Über die Minimalzahl der Fixpunkte bei den Klassen von eindeutigen stetigen Transformationen der Ringflächen. *MA* 82, pp. 94-96; CWII, pp. 652-654. |

1920 F     Luchtvaart en photogrammetrie, II [Aviation and photographic survey II]. *Nieuw Tijdschrift voor Wiskunde* 8, pp. 300-307. [not in CW].

1920 G1     Over één-éénduidige, continue transformaties van oppervlakken in zichzelf, VIII [erroneously numbered VII]. *KNAW Versl.* 29, pp. 640-642.

1920 G2     Über eineindeutige, stetige Transformationen von Flächen in sich, VIII [erroneously numbered VII] *KNAW Proc.* 23, pp. 232-234. [not in CW]; in part republished in [1912K].]

1921 A     Besitzt jede reelle Zahl eine Dezimalbruch-Entwickelung? *KNAW Versl.* 29, pp. 803-812; also in *KNAW Proc.* 23, pp. 955-964 and *MA* 83, pp. 201-210; CWI, pp. 236-245.

1921 C     Opmerking over de bepaling van alle complexe functies, voor welke $f(z) = \varphi(|z|)$ is. [Remark on the determination of all functions of a complex variable for which $f(z) = \varphi(|z|)$.] *Christiaan Huygens* 1, pp.352-354. [not in CW].

1921 D     Aufzählung der Abbildungsklassen endlichfach zusammenhängender Flächen. *MA* 82, pp. 280-286; CWII, pp. 655-661.

1921 E     Berichtigung [referring to [1911D]]. *MA* 82, p.286; CWII, p.474.

1921 F     Berichtigung [referring to [1911E]]. *MA* 82, p.286; CWII, p.488.

1923 A     Begründung der Funktionenlehre unabhängig vom logischen Satz vom ausgeschlossenen Dritten. Erster Teil, Stetigkeit, Messbarkeit, Derivierbarkeit. *KNAW Verh.* 1e sectie, deel XIII, no 2, pp. 1-24; CWI, pp. 246-267.

1923 B1     Over de rol van het principium tertii exclusi in the wiskunde, in het bijzonder in de functietheorie. *Wis-en Natuurkundig Tijdschrift* 2, pp. 1-7.

1923 B2     Über die Bedeuting des Satzes vom ausgeschlossenen Dritten in der Mathematik, insbesondere in der Funktionentheorie. *Journ. r.u.angew. Mat.* 154, pp. 1-8; CWI, pp. 268-274. [English translation, On the Significance of the principle of the excluded middle in mathematics, especially in function theory, in J. van Heyenoort, *From Frege to Gödel*, HUP, Cambridge Mass. 1967, pp.334-341].

1923 B3     Die Rolle des Satzes vom ausgeschlossenen Dritten in der mathematik. *Jahresbericht D.M.V.* 33, p.67.[Summary of [1923B2]]. [not in CW].

| | |
|---|---|
| 1923 C1 | Intuitionistische splitsing van mathematische grondbegrippen. *KNAW Versl.* 32, pp. 877-880. [not in CW]. |
| 1923 C2 | Intuitionistische Zerlegung mathematischer Grundbegriffe. *Jahresbericht D.M.V.* 33(1925), pp. 251-256; CWI, pp. 275-280. |
| 1923 D1 | Over het natuurlijk dimensiebegrip. *KNAW Versl.* 32, pp. 881-886. |
| 1923 D2 | Über den natürlichen Dimensionsbegriff. *KNAW Proc.* 26, pp. 795-800. [Revised edition of [1913A]]. [not in CW]. |
| 1924 A1 | Over de toelating van oneindige waarden voor het functiebegrip. *KNAW Versl.* 33, p.41. |
| 1924 A2 | Über die Zulassung unendlicher Werte für den Funktionsbegriff. *KNAW Proc.* 27, p.248; CWI, p.281. |
| 1924 B1 | Perfecte puntverzamelingen met positief irrationale afstanden. *KNAW Versl.* 33, p.81. |
| 1924 B2 | Perfect sets of points with positively-irrational distances. *KNAW Proc.* 27, p.487 CWI, p.282. |
| 1924 C1 | [With B. de Loor] Intuitionistisch bewijs van de hoofdstelling der algebra. *KNAW Versl.* 33, pp. 82-84. |
| 1924 C2 | Intuitionistischer Beweis des Fundamentalsatzes der Algebra. *KNAW Proc.* 27, pp. 186-188; CWI, pp. 283-285. |
| 1924 D1 | Bewijs dat iedere volle functie gelijkmatig continu is. *KNAW Versl.* 33, pp. 189-193. |
| 1924 D2 | Beweis dass jede volle Funktion gleichmässig stetig ist. *KNAW Proc.* 27, pp. 189-193; CWI, pp. 286-290. |
| 1924 E1 | Intuitionistische aanvulling van de hoofdstelling der algebra. *KNAW Versl.* 33, pp. 459-462. |
| 1924 E2 | Intuitionistische Ergänzung des Fundamentalsatzes der Algebra. *KNAW Proc.* 27, pp. 631-634; CWI, pp. 291-294. |
| 1924 F1 | Bewijs van de onafhankelijkheid van de onttrekkingsrelatie van de versmeltingsrelatie. *KNAW Versl.* 33, pp. 479-480. |
| 1924 F2 | Zur intuitionistischen Zerlegung mathematischer Grundbegriffe. *Jahresbericht D.M.V.* 36 (1927), pp. 127-129; CWI, pp. 295-297. |
| 1924 G1 | Opmerkingen aangaande het bewijs der gelijkmatige continuiteit van volle functies. *KNAW Versl.* 33, pp. 646-648. |
| 1924 G2 | Bemerkungen zum Beweise der gleichmässigen Stetigkeit voller Funktionen. *KNAW Proc.* 27, pp. 644-646; CWI, pp. 298-300. |

| | |
|---|---|
| 1924 H | Zuschrift an dem Herausgheber. *Jahresbericht D.M.V.* 33, p. 124. [not in CW]. |
| 1924 J1 | Opmerkingen over het natuurlijk dimensiebegrip. *KNAW Versl.* 33, pp. 476-478. |
| 1924 J2 | Bemerkungen zum natürlichen Dimensionsbegriff. *KNAW Proc.* 27, pp. 635-638; CWII, pp. 554-557. |
| 1924 K1 | Over de $n$-dimensionale simplex ster in $R_n$. *KNAW Versl.* 33, pp. 1008-1010. |
| 1924 K2 | On the $n$-dimensional simplex star in $R_n$. *KNAW Proc.* 27, pp. 778-780; CWII, pp. 606-608. |
| 1924 L | Zum natürlichen Dimensionsbegriff. *Mathematische Zeitschrift* 21, pp. 312-314. [not in CW]. |
| 1924 M | Berichtigung [referring to [1913A]]. *Journ. r. u. ang. Math.* 153, p. 253; CWII, p. 553. |
| 1925 A | Zur Begründung der intuitionistischen Mathematik I. *MA* 93, pp. 244-257; CWI, pp. 301-314. |
| 1925 B1 | Intuitionistischer Beweis des Jordanschen Kurvensatzes. *KNAW Proc.* 28, pp. 503-508; CWI, pp. 315-320. |
| 1925 B2 | Intuitionistisch bewijs van de krommenstelling van Jordan. *KNAW Versl.* 34, p. 657 [Summary of [1925B1]]. [not in CW]. |
| 1925 C | [With R. Weitzenböck and H. de Vries] Rapport over de verhandeling van P. Urysohn 'Mémoire sur les multiplicités cantoriennes. Deuxieme partie: les lignes cantoriennes'. *KNAW Versl.* 34, pp. 516-517. [not in CW]. |
| 1926 A | Zur Begründung der intuitionistischen Mathematik II. *MA* 95, pp. 453-472; CWI, pp. 321-340. |
| 1926 B1 | Intuitionistische invoering van het dimensiebegrip. *KNAW Versl.* 35, pp. 634-642. |
| 1926 B2 | Intuitionistische Einführung des Dimensionsbegriffes. *KNAW Proc.* 29, pp. 855-873; CWI, pp. 341-349. |
| 1926 C1 | De intuitionistische vorm van het theorema van Heine-Borel. *KNAW Versl.* 35, pp. 667-668. |
| 1926 C2 | Die intuitionistische Form des Heine-Borelschen Theorems. *KNAW Proc.* 29, pp. 866-867; CWI, pp. 350-351. |
| 1926 D1 | Over transformaties van projectieve ruimten. *KNAW Versl.* 35, pp. 643-644. |
| 1926 D2 | On transformations of projective spaces. *KNAW Proc.* 29, pp. 864-865; CWII, pp. 475-476. |
| 1926 E | [With R. Weitzenböck and H. de Vries] Rapport omtrent de |

|  |  |
|---|---|
|  | ter opneming in de werken der Akademie aangeboden verhandeling van P. Alexandroff en P. Urysohn 'Mémoire sur les espaces topologiques compacts'. *KNAW Versl.* 35 [not in CW]. |
| 1927 A | Zur Begründung der intuitionistischen Mathematik III. *MA* 96, pp. 451-488; CWI, pp. 352-389. |
| 1927 B | Über Definitionsbereiche von Funktionen. *MA* 97, pp. 60-75; CWI, pp. 390-405. [English translation 'On the domains of definition of functions' (§1-3) in J. van Heyenoort, *From Frege to Gödel,* Cambridge Mass. 1967, pp. 446-463.] |
| 1927 C | Virtuelle Ordnung und unerweiterbare Ordnung. *Journ. r. u. angew. Math.* 157, pp. 255-257; CWI, pp. 406-408. |
| 1928 A1 | Intuitionistische Betrachtungen über den Formalismus. *KNAW versl.* 36, p. 1189 [Summary of [1928A2]]. [not in CW]. |
| 1928 A2 | Intuitionistische Betrachtungen über den Formalismus. *KNAW Proc.* 31, pp. 374-379; also in: Sitzungsberichte der preuszischen Akademie der Wissenschaften zu Berlin 1928, pp. 48-52; CW1, pp. 409-414. [English translation of §1 in J. van Heyenoort, *From Frege to Gödel,* Cambridge Mass. 1967, pp. 490-492.] |
| 1928 B1 | Beweis dass jede Menge in einer individualisierten Menge enthalten ist. *KNAW Versl.* 36. p. 1189. [Summary of [1928B2]]. [not in CW]. |
| 1928 B2 | Beweiss dass jede menge in einer individualisierten Menge enthalten ist. *KNAW Proc.* 31, pp. 380-381; CWI, p. 415-416. |
| 1928 C1 | Zur Gechichtsschreibung der Dimensionstheorie. *KNAW Versl.* 37, p. 626. [Summary and announcement of [1928C2]]. [not in CW]. |
| 1928 C2 | Zur Geschichtsschreibung der Dimensionstheorie. *KNAW Proc.* 31, pp. 953-957; CWII, pp. 559-563. |
| 1929 | Mathematik, Wissenschaft und Sprache. *Monatshefte für Mathematik und Physik* 36, pp. 153-164; CWI, pp. 417-428 [In Sonderabdruck: 'Wissenschaft, Mathematik und Sprache'.] |
| 1930 A | *Die Struktur des Kontinuums.* Wien 1930 [Sonderabdruck]; CWI, pp. 429-440. |
| 1930 B | [Review of ] A. Fraenkel, Zehn Vorlesungen über die Grundlegung der Mengenlehre. *Jahresbericht D.M.V.* 39, pp. 10-11; CWI, pp. 441-442. |

| | |
|---|---|
| 1931 | Über freie Umschliessungen im Raume. *KNAW Proc.* 34, pp. 100-101; CWII, pp. 507-508. |
| 1933 | Willen, Weten, Spreken. *Euclides* 9, pp. 177-193; also in: *De uitdrukkingswijze der wetenschap, kennistheoretische voordrachten gehouden aan de Universiteit van Amsterdam.* Amsterdam, pp. 43-63. [English translation 'Will, Knowledge and Speech' in this monograph, appendix 5.] [not in CW; selected passages in CWI, pp. 443-446.] |
| 1937 | [With F. van Eeden, J.van Ginneken and G. Mannoury] Signifische dialogen [Signific Dialogues]. *Synthese* 2, pp. 168-174, pp. 261-268 and pp. 316-324; selections in CWI, pp. 447-452. |
| 1939 A | Zum Triangulationsproblem. *KNAW Proc.* 42, pp. 701-706; *Indag. Math.* I, pp. 248 -253; CWI, pp. 453-458. |
| 1939 B | Theodor Vahlen zum 70. Geburtstag. *Forschungen und Fortschritte* 15, pp. 238-239. [not in CW]. |
| 1941 | D.J.K.Korteweg [Obituary]. *KNAW Verh.*, pp. 266-267. [not in CW.] |
| 1942 A | Zum freien Werden von Mengen und Funktionen. *KNAW Proc.* 45, pp.322-323; *Indag. Math.* 4, pp. 107-108; CWI, pp. 459-460. |
| 1942 B | Die repräsentierende Menge der stetigen Funktionen des Einheitskontinuums. *KNAW Proc.* 45, p. 443; *Indag. Math.* 4, p.154; CWI, p.461. |
| 1942 C | Beweis dass der Begriff der Mengen höherer Ordnung nicht als Grundbegriff der intuitionistischen Mathematik in Betracht kommt. *KNAW Proc.*45, pp.791-793; *Indag. Math.* 4, pp. 274-276; CWI, pp. 462-464. |
| 1946 A | Synopsis of the Signific movement in the Netherlands. *Synthese* 5, pp. 201-208; CWI, pp. 465-471. |
| 1946 B | Address delivered on 16th September 1946, on the conferment upon Professor G. Mannoury of the honorary degree of Doctor of Science. *Jaarboek der Universiteit van Amsterdam 1946-1947 II*; CWI, pp. 472-476. |
| 1947 | Richtlijnen der intuitionistische wiskunde. *KNAW Proc.* 50, p.339; *Indag. Math.* 9, p.197. [English translation 'Guidelines of intuitionistic mathematics in CWI, p.477.] |
| 1948 A | Essentieel negatieve eigenschappen. *KNAW Proc.* 51, pp. 963-964; *Indag. Math.* 10, pp. 322-323. [English translation 'Essentially negative properties' in CWI, pp. 478-479]. |

| | |
|---|---|
| 1948 B | Opmerkingen over het beginsel van het uitgesloten derde en over negatieve asserties [Remarks on the principle of the excluded middle and on negative assertions]. *KNAW Proc.* 51, pp. 1239-1244; *Indag. Math.* 10, pp. 383-387. [not in CW]. |
| 1948 C | Consciousness, Philosophy and Mathematics. In: *Proceedings of the 10th International Congress of Philosophy, Amsterdam 1948 III*, pp. 1235-1249; CWI, pp. 480-494. |
| 1949 A | De non-aequivalentie van de constructieve en negatieve orderelatie in het continuum. *KNAW Proc.* 52, pp. 122-124; *Indag. Math.* 11, pp. 37-39. [English translation 'The non-equivalence of the constructive and negative order relation on the continuum' in CWI, pp. 495-496.] |
| 1949 B | Contradictoriteit der elementaire meetkunde. *KNAW Proc.* 52, pp. 315-316; *Indag. Math.* 11, pp. 89-90. [English translation 'Contradictority of elementary geometry' in CWI, pp. 497-498.] |
| 1950 A | Remarques sur la notion d'ordre. *C.R. Paris* 230, pp. 263-265; CWI pp. 499-500. |
| 1950 B | Sur la possibilité d'ordonner le continu. *C.R. Paris* 230, pp. 349-350; CWI, pp. 501-502. |
| 1950 C | Discours final. Les méthodes formelles en axiomatique. In: *Colloques Internationaux du C.N.R.S.*, Paris 1950, p.75; CWI, p.503. |
| 1951 | On order in the continuum, and the relation of truth to non-contradictority. *KNAW Proc.* Ser.A 54, pp. 357-358; *Indag. Math.* 13, pp. 357-358; CWI, pp. 504-505. |
| 1952 A | An intuitionist correction of the fixed-point theorem on the sphere. *Proceedings of the Royal Society London,* Ser. A 213, pp. 1-2; CWI, pp. 506-507. |
| 1952 B1 | Historical background, principles and methods of intuitionism. *South African Journal of Science* 49, pp. 139-146; CWI, pp. 508-515. |
| 1952 B2 | Voorgeskiedenis, beginsels en metodes van die intuitionisme. *Tydskrif vir Wetenskap en Kuns* 1956, pp. 1986-197. |
| 1952 C | Over accumulatiekernen van oneindige kernsoorten. *KNAW Proc.* Ser. A 55, pp. 439-441; *Indag. Math.* 14, pp. 439-441. [English translation 'On accummulation cores of infinite core species' in CWI, pp. 516-518.] |
| 1952 D | Door klassieke theorema's gesignaleerde pinkernen die on- |

vindbaar zijn. *KNAW Proc.* Ser. A 55, pp. 443-445; *Indag. Math.* 14, pp. 443-445. [English translation 'Fixed cores which cannot be found, though they are claimed to exist by classical theorems' in CWI, pp. 519-521.]

1954 A    Points and Spaces. *Canadian Journal for Mathematics* 6, pp. 1-17; CWI, pp. 522-540.

1954 B    Addenda en corrigenda over de rol van het principium tertii exclusi in de wiskunde. *KNAW Proc.* Ser. A 57, pp. 104-105; *Indag. Math.* 16, pp. 104-105. [English translation in J. van Heyenoort *From Frege to Gödel*, HUP, Cambridge Mass. 1967, pp. 341-342 and in CWI, pp. 539-540.]

1954 C    Nadere addenda en corrigenda over de rol van het principium tertii exclusi in de wiskunde. *KNAW Proc.* Ser. A 57, pp. 109-111; Indag. Math.16, pp. 109-111. [English translation 'Further addenda and corrigenda on the role of the principle of the excluded middle in mathematics' in: J. van Heyenoort *From Frege to Gödel* (HUP, Cambridge Mass. 1967), pp. 342-345 and in CWI, pp. 541-543.]

1954 D    Ordnungswechsel in Bezug auf eine coupierbare geschlossene stetige Kurve. *KNAW Proc.* Ser. A 57, pp. 112-113; *Indag. Math.* 16, pp. 112-113; CWI, pp. 544-545.

1954 E    Intuitionistische differentieerbaarheid. *KNAW Proc.* Ser. A 57, pp. 201-203; *Indag. Math.* 16, pp. 201-203. [English translation ' Intuitionistic differentiability' in CWI, pp. 546-548.]

1954 F    An example of contradictority in classical theory of functions. *KNAW Proc.* Ser. A 57, pp. 204-205; *Indag. Math.* 16, pp. 204-205; CWI, pp. 549-550.

1955    The effect of intuitionism on classical algebra of logic. *Proceedings of the Royal Irish Academy* Sect. A 57, pp. 113-116; CWI, pp. 551-554.

## 1.2 Unpublished Papers

BMS 1A    'Geloofsbelijdenis van L.E.J. Brouwer March 1898'. 7-page Dutch manuscript. [The original and an English translation 'Profession of Faith of L.E.J. Brouwer' published in this monograph as appendix 1.]

BMS 1B    Student Notebook; a collection of 'thoughts', original or quoted. 17-page Dutch manuscript. [The original and an English translation published in this monograph as appendix 2.]

| | |
|---|---|
| BMS 2 | Bollandiana. Brouwer's contributions to the Bolland-debate in the student magazine *Propria Cures* XVI(1904), under the pseudonym 'Lau van der Zee'; manuscripts and published articles [in Dutch]. |
| BMS 3A | Preparatory notes to the dissertation. 11-page foolscap manuscript of closely-written notes [in Dutch]. |
| BMS 3B | 'Over de Grondslagen der Wiskunde'. 22-page foolscap manuscript of part of the original version of the dissertation, consisting of the pages rejected by 'promotor' D.J. Korteweg. [The Dutch original published in D. van Dalen, *L.E.J. Brouwer Over de grondslagen de Wiskunde* (Mathematisch Centrum Amsterdam, 1981). An English translation published in W.P. van Stigt, The Rejected Parts of Brouwer's dissertation on the Foundations of Mathematics, *Historia Mathematica* 6 (1979), pp. 385-404; also in this monograph as appendix 3.] |
| BMS4 | 'Vlakke krommen en vlakke gebieden' [Plane curves and plane domains]. Lecture given to the 'Wiskundig Genootschap' in October 1908. 28-page Dutch manuscript. |
| BMS 5 | 'Die Unzuverlässigkeit der logische Principe'. [Incomplete translation of [1908C]. |
| BMS 6A | 'De invariantie van het aantal dimensies eener ruimte.' [The invariance of the number of dimensions of a space]. Lecture given to the Wiskundig Genootschap in October 1910. 4-page Dutch manuscript, published in R.H. de Jong, *Tweehonderd jaar onvermoeide arbeid*, tentoonstellingscatalogus Wiskundig Genootschap, Amsterdam 1978. |
| BMS 8 | 'Beweis dass die Modulmannigfaltigkeit der Riemannschen Flächen vom Geschlechte $p$ keinen singulären Punkt aufweist'. 2-page German manuscript. |
| BMS 10 | Karlsruhe-Lecture notes. [April 1912]. |
| BMS 14 | Corrections of Schoenflies' Entwicklung der Mengenlehre. [1913] |
| BMS 15A | A course on set theory. 50-page Dutch manuscript of a course on set theory given in Amsterdam between 1914 and 1916. |
| BMS 21 | 'Intuitionistische dimensie theorie.'[Intuitionist theory of dimensions.] Notes. |
| BMS 22B | 'Intuitieve Significa'. Brouwer introducing a lecture by Fr. van Eeden 13 March 1916. [The original and an English translation published in this monograph as appendix 4.] |

| | |
|---|---|
| BMS 23B | 'Signifische dialogen'. [Signific Dialogues]. The original version of [1937] in a 15-page manuscript, probably by Mannoury, circulated for approval with corrections by Brouwer dated 09.06.24. |
| BMS 32 | Berliner Gastvorlesungen - Intuitionisme. A course on Intuitionism given by Brouwer in Berlin in 1927 and intended for publication as a monograph; (B) an 80-page foolscap manuscript in German, and an incomplete version in Dutch. [English translation of extracts published in this monograph as appendix 12.] |
| BMS 35 | 'Definitionsbereiche von Funktionen', with preparatory studies by Alexandroff [cf. [1927B].] |
| BMS 37 | 'Reelle Funktionen'. An incomplete monograph on the intuitionist theory of real functions intended for publication; estimated date of writing 1924-27. 126-page foolscap manuscript in German. [English translation of Table of Content and extracts (pp. 1-7; and pp. 107-111; pp.112-114) published in this monograph as appendix 11.] |
| BMS 38 | 'Reële functies'. [Real functions.] Notes of a lecture probably given around 1930. 8-page Dutch manuscript. |
| BMS 40 | 'Zuschrift am Herausgeber'. Open letter to the Editor of the Jahresbericht D.M.V. concerning the Bologna Congress; printed and as a loose page inserted in the March edition of 1928. |
| BMS 41 | 'Wandlungen in den Beziehungen zwischen klassiker Logik und Mathematik'. A lecture probably delivered around 1930. 13-page type- and manuscript in German. |
| BMS 44 | Geneva Conference Course 1934. 6 recorded and cyclo-styled lectures given by Brouwer, with some corrections of Brouwer's hand. |
| BMS 45 | 'Discontinue intuitionistische functies eener reële veranderlijke'. [Discontinuous intuitionist functions of a real variable.] 3-page Dutch type-script of a lecture given to the Wiskundig Genootschap on 30 April 1938. |
| BMS 47 | Notes in Dutch giving the outline of a lecture on logic, given a second time in 1951. The original Dutch and an English translation are published in this monograph as appendix 8. |
| BMS 49 | 'Disengagement of Mathematics from Logic'. 5-page English manuscript of a lecture probably given around 1947. Published in this monograph as appendix 7. |

| | |
|---|---|
| BMS 50 | 'Program for a lecture in London on October 16th, 1947'. 3-page English manuscript. |
| BMS 51 | 'The Cambridge Lectures' or 'The Cambridge book'. A course on Intuitionisme given by Brouwer at Cambridge regularly between 1946 and 1951; the 80-page English typescript contains five of the six planned chapters of the monograph to be published by the CUP, together with preparatory notes for the final chapter. An annotated edition has been published by D. van Dalen, *Brouwer's Cambridge Lectures on Intuitionism* (CUP, Cambridge, 1981). |
| BMS 52 | 'Short retrospective notes on the course *Intuitionist Mathematics*', Easter term 1947. 7-page stencilled notes for those attending Brouwer's Cambridge course. [In this monograph published as appendix 6.] |
| BMS 54 | 'Bewustzijn, wijsbegeerte en wiskunde' [Consciousness, philosophy and mathematics], 12-page typescript with handwritten amendments of the first, Dutch version of [1948C]. |
| BMS 55 | 'Les Mathématiques Intuitionistes'. 30-page French type- and manuscript of lectures given in Paris in 1949. |
| BMS 59 | 'The influence of Intuitionist Mathematics on Logic'; [alternative title 'Changes in the relation between classical logic and mathematics']. 7-page English typescript of a lecture given in London on 2 November 1951. [published as an appendix to D. van Dalen, *Brouwer's Cambridge Lectures on Intuitionism* (CUP, Cambridge, 1981). An additional 7-page manuscript of the same title gives a slightly different version, which is published in this monograph as appendix 9.] |
| BMS 62 | Lecture notes. A 2-page English manuscript dd. 19 October 1953. |
| BMS 66 | 'Second Lecture', 9-page English manuscript dated May 1952 and 12.5.52 [University College, London University]. [In this monograph published as appendix 10.] |

# CHAPTER II
# BROUWER, LIFE AND WORK

## 2.1 Childhood and Student-years

Luitzen Egbertus Jan Brouwer, was born at Overschie – Rotterdam on the 27th February 1881, the eldest of three sons of Egbertus Luitzens Brouwer, a schoolmaster, and Henderika Poutsma. After their marriage in 1880 the Brouwers had left their native Friesland and settled in the more prosperous West of Holland, where Egbert Brouwer had been appointed headmaster, first at Overschie and in November 1881 of the 'Burgerschool' at Medemblik. There young Brouwer at the age of five received his first instruction from his father. From the very start he stood out above his contemporaries, physically as well as mentally. At the age of nine 'Bertus' had completed the primary programme, some two years ahead, and his father-headmaster decided that he would benefit from a more demanding programme and from competition. He was sent to the best secondary school within daily travelling distance, the H.B.S. at Hoorn.*
In 1892 the Brouwers moved to Haarlem, and Bertus joined the third year of the Haarlem H.B.S..

In spite of his age, young Brouwer set the pace of his class. His school-reports record top grades in all subjects of the curriculum: the sciences, mathematics as well as modern languages. The only grades below the 'excellent' grade-10 are for handwriting and 'effort' in mathematics.

The H.B.S. in these days, however, did not qualify for university entrance and Bertus' teachers agreed with his parents that only a university education would do full justice to his talents. A long academic tradition in the Poutsma family and a memory of hardship and poverty had established strong academic ambitions for their three sons, especially for their brilliant but frail eldest son Bertus. They decided not to wait until he had completed his H.B.S. studies and transferred him in the autumn of 1894 to the local Gymnasium. A special programme was designed which enabled him to complete his H.B.S. within the year and join the gymnasial courses in classics. He won a scholarship of the Frisian 'St. Jacobs op St

---
\* The H.B.S. - the 'Higher Burgher School'- was a five-year science-based secondary school.

Family Portrait ± 1895; 'Bertus' standing on the far right, his brother Isaac Alexander on the left and his younger brother Hendrik Albertus ('Aldert') sitting between his parents.

Jobsleen', which made him financially independent until the end of his university studies in 1907.

At the gymnasium at Haarlem Brouwer again excelled, even in the subjects he had recently started, years after his classmates. His love for the classics would remain with him all his life.

At the age of fourteen he took the final H.B.S. examination and passed with full honours. He remained at the gymnasium for another two years and in 1897 was awarded the Gymnasium Diplomas A and B.*

In the autumn of that year, at the age of sixteen Brouwer registered at the University of Amsterdam to read mathematics and science. He started his course with enthusiasm. His mathematical genius was soon recognized by his professors and fellow-students. He enjoyed the privilege of being invited as a first-year student to join various societies of the 'Student Corps'. Young Brouwer, however, did not wholly take to the sophistications of university life; he felt uncomfortable in the noisy company of fellow students and its atmosphere of adolescent affectation. He was not a socializer, a mixture of shyness and arrogance kept him away from the circle of his fellow mathematicians. He felt more at home in the company of his close poet friends, who shared his romantic pessimism and his quest for solitude and contemplation. Already then there was a deep-rooted lack of interest in or even a dislike of his fellow men, and a disregard for the opinion of others. He startled the congregation of the Haarlem Remonstrant Church when during the confirmation ceremony in March 1898 he read out his *Profession of Faith:*

> As to the love of my neighbour, I don't care twopence for most people... other human beings are the ugliest of my representations... I shall therefore never sacrifice myself for somebody else. [BMS1A], p. 5; app. 1).

Yet the *Profession of Faith* was not meant as a defiant declaration, mocking religion and religious feelings. Brouwer professed himself to be deeply religious; his religion, however, was a personal conviction of God's presence, resulting from his independent thinking and meditation. For himself he rejected 'organized religion' as worked out by theologians and imposed by a church:

> Of course, church domination and dogmatics are degenerate situations. I think such religious forms are right for the stupid masses, designed tot keep

---

* The Gymnasium-A and the Gymnasium-B course were each a full six-year programme; in the A-course the emphasis was on the Classics and the A-diploma qualified for university study in the Arts Faculty; in the Gymnasium-B course greater emphasis was placed on the sciences, the B-Diploma qualified for university study in Mathematics and Science.

them in respectful ignorance under the thumb of a power-thirsty church. (op.cit., p. 6).

*The Profession of Faith of L.E.J. Brouwer* [BMS1A] is the earliest record of Brouwer's thoughts. It reveals already at an early age the dual aspects of his egotism, which will determine and characterize his thinking, his philosophy and his attitude to others in later years: on the one hand an utter self-reliance and insistence on his own freedom and independence, on the other a complete lack of interest in other human beings and a disregard of other opinion. His misanthropy, however, was not due to a general lack of human emotions; Brouwer was sensitive and very much guided by his feelings. There is a growing need to be loved, admired and recognized; yet his dealings with others are dominated by a certain meanness, by personal ambition and suspicion of others. He was subject to moods of pessimism ands depression and could fly into violent tempers. During such spells of emotional upset work remained impossible.

In December 1900 Brouwer completed the first part of his studies and passed his 'Kandidaats' examination 'Cum Laude', and this in spite of frequent spells of illness and a short, unhappy period of national service as an army corporal. The 'pre-Kandidaat' course in mathematics and science had presented no difficulties to Brouwer's remarkable memory and sharp brain, but it left him utterly disillusioned.

Mathematics at Amsterdam University was then only just emerging as an independent subject within the Faculty of Science. Its development into a mathematics department and international centre of excellence was greatly due to the pioneering work of professors J.D. van der Waals, A.J. van Pesch and especially D.J. Korteweg.* The character and work of the

---

* *Diederik Johannes Korteweg* (1848-1941) studied at the Delft Polytechnic and worked as a secondary schoolteacher from 1869 until 1881. He was admitted to Utrecht University in 1876 and passed his 'Kandidaat's' examination in Mathematics and Natural Science in 1877. A year later he submitted his dissertation on 'The velocity of waves transmitted through elastic tubes' at Amsterdam University and was awarded the doctoral degree (the first doctorate after the Amsterdam Athenaeum had been granted the university charter). In 1881 a new chair in mathematics was created at Amsterdam University and Korteweg appointed Professor of 'Mathematics, Mechanics and Astronomy'. He delivered his inaugural lecture 'Mathematics as an ancillary science' on 10 October 1881. He vacated his chair in favour of Brouwer in 1912 and stayed on as extra-ordinarius until his retirement in 1918.

In spite of a heavy teaching load Korteweg was able to maintain a wide international correspondence and publish original work on a wide range of topics in algebra, geometry,

department at the turn of the century were determined by the personality and views of D.J.Korteweg. His inaugural address, *Mathematics as an Ancillary Science* (1881), was a statement of policy for the department at that stage of its evolution: a recognition of the right to practise mathematics for its own sake ('Man builds not only houses to live in, but also monuments'), but a personal preference for applicable mathematics. He loved and was inspired by the problems prompted by science; he promoted cooperation between scientists and mathematicians and was prepared to play the modest role of providing a mathematical service for his scientific colleagues; yet he remained aware of their separate identities and insisted on mutual understanding and respect for their disciplines and methods.

After many years and some tragic estrangement Brouwer would testify of his old master:

> He put his exceptional mathematical intelligence entirely at the service of physical science for two personal reasons: first, his psychological attention was so much directed towards the outside world that to him the value of mathematics was to be found exclusively in its practical use; secondly his character made him always do the work that in his view needed to be done, and indeed could not have been done better by anyone else. ([1941], p.266).

The Korteweg ideal failed to inspire the student Brouwer. The pre-Kandidaat mathematics syllabus with its emphasis on methods and applications left him bored, the subservience of mathematics to science evoked resentment and its 'use to conquer the world' felt as morally reprehensible. Years later he speaks of his undergraduate disillusionment:

> As happens so often, my academic studies started as a leap in the dark. After two or three years, however, full of admiration for my teachers I still could only see the figure of the mathematician as a servant of natural science or as a collector of truths. Truths, fascinating by their immovability but horrifying by their lifelessness, like stones from barren mountains of disconsolate infinity. And as far as I could see, there was in the mathematical field only room for talent and devotion but not for vocation and inspiration. I was filled with impatient desire for insight into the essence of the branch of work of my choice... ([1946B]).

The post-Kandidaat period was a time of instability, restlessness and

---

the theory of oscillation, electricity and sound, kinetic gas theory, hydronamics and probability, as well as edit and publish the works of Huygens and contribute to the International Catalogus of Scientific Publications.

doubts. Brouwer's extreme self-centredness and inability to love made human relations well-nigh impossible. In 1900 he became engaged to Elizabeth de Holl, separated from her first husband and twelve year older than Brouwer; she had started her pharmaceutical studies so that she could take over her father's pharmacies. Her 'Memlinck face' appealed to Brouwer's aesthetical sense. She seemed prepared to accept him without expecting too much in return and to provide him with the security he needed so much, to play the decorative and submissive role Brouwer had assigned to women, look after him and console him; above all, her two dispensaries would secure a steady income. The marriage took place in 1904.

During the early days of the engagement Brouwer turned more and more into himself. In search for solitude he moved from one appartment in Amsterdam to another. His only close friends were the young poets Carel Adema van Scheltema, Jan Lockhorst and Albert Plasschaert. Brouwer lacked the ability to express his feelings artistically, but he identified with their romantic moods of pessimism, despair of contemporary society, their 'Weltschmerz' and their rebellion against the establishment in all its forms. He appreciated their poetry and took a deep interest in every form of art.

On three occasions during these years he 'pilgrimaged' to Italy. Dressed in a large dark cloak, all on his own, he walked the long way to Florence, where he stayed with his friend Rudy Mauve, and then further to Rome. These trips whetted his appetite for foreign travel and fostered a cosmopolitan outlook. They allowed him to 'turn into himself' and think about the great issues: himself, life and human relations, truth and science. Of the philosophical views he had read, those of Schopenhauer were most in line with his own feelings.

At the depth of depression Brouwer began to think seriously of abandoning his university course. Family pressure, the need for financial security and his own conviction of physical unsuitability for any other career prevented an immediate break, but the thought of devoting his whole life to mathematical drudgery became more and more unbearable. He isolated himself and began to suffer from bouts of temper and nervous disease.

The turning point came when Brouwer met Gerrit Mannoury*, a bril-

---

\* *Gerrit Mannoury* (1867-1956) completed his training as a primary school teacher in 1885. Technically not qualified to register at a university, he pursued his interest in mathematics and philosophy through private, part-time study. His original publications and active participation in the Amsterdam Mathematical Society did not go unnoticed; he was

liant primary-school teacher and self-taught mathematician, whose eye for the essence of mathematics had not been obscured by the lumber of a traditional academic course. With gratitude Brouwer remembers this meeting:

> ... wanting to decide whether to stay or go, I began to attend the meetings of the Amsterdam Mathematical Society. There I saw a man apparently not much older than myself, who after lectures of the most diverse character debated with unselfconscious mastery and well-nigh playful repartee, sometimes elucidating the subject concerned in such a special way of his own that straight-away I was captivated. I had the sensation that for his mathematical thinking this man had access to sources still concealed to me, or had a deeper consciousness of the significance of mathematical thought than the majority of mathematicians... His articles had the same easy and sparkling style; and when I succeeded, not without difficulty, in understanding them, an unknown mood of joyful satisfaction possessed me, gradually passing into the realization that mathematics had acquired a new character for me. ([1946B]).

Gradually Brouwer's eyes were opened to the fundamental issues and controversies, the crisis that rocked the world of mathematics at the turn of the century. The 'crisis in the foundations of mathematics' had passed the Dutch traditional academe almost unnoticed. Mannoury was the first in Holland to be seriously concerned about the implications of Cantor's set theory and Frege's and Peano's logicist claims for mathematics and

---

licensed to teach at the University of Amsterdam as a 'Privaat Docent' in 'the logical foundations of mathematics'. He gave his first public lecture on 23 January 1903 'On the Significance of Mathematical Logic for Philosophy'.

Mannoury remained perhaps Brouwer's closest friend all his life. They shared a great many interests, although their views remained sufficiently divergent for Brouwer to describe their relation as that of 'dialectic partners'.

Mannoury's *Methodologisches und Philosophisches zur Elementar-Mathematik*, published in 1909, was based on his courses at Amsterdam University. He remained a (part-time) privaat-docent until 1917, when mainly on Brouwer's recommendation he was appointed 'extra-ordinarius' at the University of Amsterdam. A year later he was appointed Professor in Ordinary in Geometry, Mechanics and Philosophy of Mathematics.

Mannoury was an ardent socialist. He became with Brouwer one of the founders of the Signific Circle. His interest in the social aspects of language determined the later course of the Signific Movement in Holland.

Mannoury retired in 1937 and died on 30 January 1956.

A bibliography of Mannoury's many publications is given in *Synthese*, vol.Xa, pp. 418 - 422.

the nature of mathematical truth. After his appointment as 'privaat-docent' at Amsterdam University, Mannoury started a series of weekly lectures on logic and the foundations of mathematics. Brouwer was fascinated and attended every lecture of the course. The issues themselves concerned him deeply and Mannoury's personal stand appealed to him: there was a man who showed an understanding of recent publications in the field of foundations and who in all modesty dared to express his own personal opinion. The lectures focused his attention on such widely varied topics as: the role of symbolic logic in mathematics, the nature of mathematical truth, the role of mathematics in human sciety, the relation between language and logic and the relation between mathematics and human experience. Mannoury's approach invited a critical response and Brouwer readily accepted the challenge. He had found a new purpose: here was branch of his subject in which he could become wholly involved – the only way he was able to work – with all his energy and passion, relying on his own 'insight' and disregarding trodden paths.

With renewed energy Brouwer turned again to his course. Stimulated and encouraged he began to look afresh at some of the then current mathematical problems. Korteweg encouraged him to pursue an interest in the promising new field of topology, and on 27 February presented Brouwer's first publication to the Dutch Royal Academy, *On a decomposition of a continuous motion about a fixed point O of $S_4$ into two continuous motions about O in $S_3$* ([1904A]). There is already evidence of his search for geometric simplicity so characteristic of his later approach to the problems of topology. The paper also sparked off a sharp exchange with the Berlin-professor E. Jahnke, the first in a long series of controversies about 'authorship' and 'priority' which will mark Brouwer's dealings with fellow mathematicians.

## 2.2 Post-graduate Study

On 16th June 1904 Brouwer passed his final examinations with distinction. The phrase on his degree scroll '...et cum laude admissus est ut summos in Disciplina Matheseos honores petat...' would not remain an idle wish.

He had already decided to stay on at the university and work full-time on his doctoral dissertation, quite out of step with Dutch practice at the time, when doctoral theses were the result of years of private study of mature Gymnasium teachers, hoping to be promoted to a university career. The Governors of the St Jobsleen had agreed to extend their grant to Brouwer for another two years. His marriage in August that same year brought some administrative responsibilities for his wife's pharmacy but also provided security and a home.

On the subject of the dissertation Brouwer was in no doubt: he would devote these years of doctoral research to a study of the fundamental issues concerning mathematics, read and draw together his own thinking on the nature of mathematics, its relation to human experience and its value to human society and the individual. In the matter of 'promotor'*, however, there was no choice, the only professor in the faculty concerned with pure mathematics was Korteweg.

Behind Brouwer's impulsive and sometimes arrogant behaviour and his emotional arguments, Korteweg had recognized a great mathematical intellect. He trusted Brouwer's talent and was prepared to assist and guide him even if he had his reservations about his choice of subject; he would so much have liked to see his star pupil investigate a 'truly mathematical' subject. When Korteweg later expressed doubts about the mathematical content of of the first draft of the dissertation Brouwer reminded him of his early agreement:

> You know that when I chose the subject two years ago, it was not for lack of competence to tackle a more 'normal' topic, but only because I felt so strongly drawn towards this subject; it just originated and grew in me. You then agreed to it, provided there would be enough mathematics in it, probably suspecting that it would drive me strongly in the direction of philosophy, which indeed it has.' (Brouwer to Korteweg, 05.11.06 [DJK20]; app.13).

Work got off to a slow start. In the Autumn of 1904 Brouwer moved into the house over his wife's pharmacy on the Overtoom in Amsterdam, trying to adjust to married existence and to living together with his wife and her eleven-year old daughter of her first marriage. He did not find concentration on 'the deeper issues' very easy, he was quickly distracted and side-tracked into meddling in petty financial affairs and could not

---

* The 'promotor' at Dutch Universities has the general supervisory duties over a doctoral dissertation; he also carries the main responsibility for its academic standard.

resist becoming passionately involved in all kinds of causes of little relevance to his mathematical work.

With great interest he followed the rise and fall of G.J.P.J. Bolland, a self-made philosopher, who since his appoinment to the chair of philosophy at Leiden in 1896 had swept the country in a bid 'to bring philosophy to the educated citizens'. Philosophy in Holland since Descartes and Spinoza had made very little impact on international thinking. During the 19th century it was only a 'minor' subject of study at the universities, often taught by classicists or mathematicians. Philosophical trends were determined by movements abroad. Holland was violently roused from this slumber by the antics of Bolland, an ardent Hegelian and an eloquent but arrogant orator.

Brouwer was at first intrigued by Bolland, his violent attacks on materialism and hedonism appealed to him. He joined a student committee inviting Bolland to give a course of lectures in Amsterdam, and studied Bolland's *Pure Reason - a Book for the Friends of Wisdom*.

Bolland's triumphal march was halted in Amsterdam, his lecture course ended in November 1904 prematurely and in a riot. Brouwer, repelled by Bolland's arrogance and his worship of the human intellect and logical reasoning, became the leading spokesman of the the anti-Bolland campaign in the Amsterdam student journal *Propria Cures* and the general press. The impartiality of the inviting committee was preserved by Brouwer's use of the pseudonym 'Lau van der Zee'. The general tenor of the articles (*Troostgronden* [Grounds for Consolation], Propria Cures 31.12.04, etc.) is critical, and sneeringly critical of Bolland, yet there are clear indications of Brouwer's own thinking and moods at the time and his style: his 'essential difference' between the thinking of man and woman, his dislike of academic sophistry, his preference for solitude and for listening to the voice from within, etc.

When Bolland gave his lecture course at Delft, Brouwer was invited by the 'Society for Free Study' to speak in reply. He accepted the invitation and spent months preparing his answer, 'listening to the voice of his conscience', studying the works of the wise, who in their search for knowledge did not rely on their intellect but trusted their 'inner experience': the mediaeval mystics Jakob Böhme and Meister Eckehart, and the Baghavad Gîta.

Brouwer finished his 'Delft lectures' on 29 March 1905; the text was published in the form of a book, *Life, Art and Mysticism*.

## 2.3 Life, Art and Mysticism [1905]

*Life, Art and Mysticism* is the manifesto of an 'angry young man', rejecting and attacking all he sees at the surface of human society, with the vigour and in the style of the Dutch romantics at the turn of the century. But unlike these rebellious idealists, Brouwer lays the blame for all evil at the feet of humanity. With almost perverse relish he exposes human nature as the real villain: man polluting the world, driven and aided by his characteristic human ability to reason.

Chapter I, *The Sad World*, begins with a condemnation of the Dutch for building dykes and ends by comparing man to a bird 'arrogantly gulping up its own nest, interfering with mother earth, gnawing, mutilating her and making her creative power sterile until all life has been devoured.' (p. 10).

Chapter III, *Fall through the Intellect*, spells out Brouwer's anti-intellectualism in a fanatic attack on human science. Causality, which in his dissertation was to be identified with 'mathematical thinking', is condemned as fundamentally immoral:

> Intellect has done mankind a devil's service by linking the two phantasies of means and end. (p.19).

On medical science:

> The medical industry was with barbers and quacks in good hands; practised within the confines of the intellect, as medical science, it is far less effective... (p.27).

On science in general and the contemporary epistemological debate, and even the study of 'foundations':

> ...it does not stop with science serving industry: the means becomes an end in its own right and, horribile dictu, is even practised for its own sake...
> A scientific truth is no more than a certain infatuation of desire, living exclusively in the mind. Every branch of science gets into more trouble as it climbs higher. It even climbs so high that scientific truth is taken as something independent, living outside man.
> Some even start searching for the foundations of their science and that soon becomes a science in its own right and they practise a *theory of knowledge*. As they climb higher trouble increases and they go completely crazy.
> Some in the end quietly give up. Having thought for a long time about the elusive link between the intuiting consciousness and the world of phenomena (which again itself exists only through and in the form of the intuiting

consciousness, a confusion which originated in a sinful foundation of the world of intuition), they then plug the hole with the concept of the Ego, which was self-created with and at the same time as the phenomenal world. And then they say, 'Yes, of course, something must remain incomprehensible and that something is the Ego that comprehends.'
But there are others who do not know when to stop, who keep on and on until they go mad. They grow bald, short-sighted and fat, their stomachs stop working, and moaning with asthma and gastric trouble they fancy that in this way equilibrium is within reach and almost reached.
... So much for science, the last flower and ossification of culture. (pp.21-22).

*Life, Art and Mysticism* is much more than a reaction to Bollandian over-emphasis of the human intellect, it reveals a deep resentment in Brouwer against the two most characteristic human aspects: man's faculty of reason and his ability to communicate with fellow human beings. While the attack on the 'cunning Reason' is mainly directed at the male of the human species, 'social acting' – equally despicable – is more directly associated with women. There is a strange ambivalence in Brouwer's attitude towards women. He considers them intellectually inferior to men, a species of lower order:

> There is less difference between a woman in her innermost nature and an animal such as a lioness than between two twin brothers... (p.52).

In the context of Brouwer's anti-intellectualism this should have been a compliment and a redeeming feature. But in her whole being and sexuality, her directedness towards man and procreation, woman to him epitomizes the human condition in all its 'social' aspects. She becomes the focus of his anti-human feelings: she is a danger to man, through her touch even noble work turns unclean:

> The usurpation of any work by women will automatically and inexorably debase that work and make it ignoble.' (p.55).

In long drawn-out passages Brouwer spells out the servile role he has allotted to women, trying to justify his own lack of interest in other human beings and his inability to give himself and love others. One cannot help reflecting on his life at the time: a few months after his wedding, supported by his wife, who was herself working for a degree while running her pharmacy, especially when he writes:

> Two alarming features of modern times are: the debasement of universities

into places where wage-earners are trained in disagreeable, wretchedly necessary but degrading social work, and the appearance of women there...
That money for one's livelihood is usually earned by the man is of as little importance as money itself.This happens to be so at present, when earning money goes hand in hand with doing noble work. To the old Germanic tribes tilling the land was regarded as ignoble work and therefore done by women. When all productive labour has been made dull and ignoble by socialism it will be done exclusively by women. In the meantime men will occupy their time according to their ability and aptitude in sport, gymnastics, fighting, studying philosophy, gardening, wood-carving, travelling, training animals and anything that at the time is regarded as noble work, even gambling away what their wives have earned. For this really is much nobler than building bridges or digging mines. (pp. 55-56).

Another aspect of man as a social being, his ability to communicate his thoughts through *language*, comes in for equally harsh treatment. In line with the mystic tradition, Brouwer sees man as an essentially spiritual being, a soul or consciousness, imprisoned in an alien body. His 'consciousness' is a romantic mixture of the spirituality of the christian individual soul in the popularized version of Böhme and Eckehart and the universal soul of the Bhagavad Gîta in its exodus and return to Nirvana. Its characteristic vision in accordance with its spiritual nature is direct, timeless contemplation, 'Intuition' (Germ. 'Anschauung') in its purest form. Life is the journey of the exiled, spiritual soul travelling alone through this vale of tears. There is no certainty even as to the existence of other minds, and direct communication with other minds is impossible:

From life in the mind follows the impossibility to communicate directly or instinctively by gesture or looks, or even more spiritually, through all separation of distance. People then try and train their offspring in some form of mutual understanding by means of crude sounds, laboriously and helplessly, for never has anyone been able to communicate his soul by means of language. (p.37).

A whole chapter, chapter V, is devoted to human language, or rather to the impossibility of real communication and the deficiencies of language as a carrier of human thought. For communication between people to be at all possible, a 'harmony of will' must be presupposed:

Ridiculous is the use of language expressing fine nuances of human will if these are not part of one's own life, when so-called philosphers or metaphysicians speak together, people who don't even love one another, let alone share in a delicate communion of souls, people who sometimes don't even know each other... (p. 38).

The dominant positive theme of *Life, Art and Mysticism* is the excellence of inner visison, 'insight' or 'intuition'. By 'turning into himself' man finds real wisdom and release from the frustrations of this earthly existence. Brouwer's practical, romantic advice is to reject and shun all aspects of modern society and return to nature and the simple life. The apparent disagreement and conflict between his inner feelings and his practice as a mathematician is solved provisionally by a fatalistic acceptance of the inevitable.

There is a touch of insincerity about most of Brouwer's strong condemnations. He ridicules fashions and many of the human weaknesses which mark his own life, such as ambition, lust for power, jealousy and hypochondria. His condemnation of those seeking security by amassing capital rings rather hollow in a man whose life was so obsessed with money; his sarcasm when he deals with spiritism, theosophy and nudism must raise a smile in those who knew him and his obsession with health (he attended seances and theosophic meetings, was a vegetarian, visited spas and practised 'air and mud baths'). Somehow he seems able to detach his thinking from his own life, fatalistically accept his own idiosyncracies and wryly laugh at them:

> They emphasize the important role of skin-breathing for the human metabolism, the salutary effect of exposing one's skin to the pure air, and they become experts in hygiene and reformers... these inept people then take air-baths, and when they discover the effect of sunrays they take to light baths and sun baths, and finally to dusk baths, night baths, moon baths, star baths, forest baths and meadow baths as soon as somebody proclaims them to be healthy. (p.70).

Other instances, however, are true descriptions of his experience at the time; e.g. having advised the reader in Chapter VI, *The Freed Life*, to seek escape from the world into his solitary Self and ignore public opinion, he reflects:

> ...after this first escape he will not feel at home among other people, whom he begins to irritate through his eccentricities which unwittingly follow from this liberation. (p. 84).

Brouwer's escape from the world, away from the city and away from life with others, became reality when he found a plot of land in the forest and heath of 'Het Gooi', near the bohemian village of Laren, favoured by artists and progressive reformers. His friend Rudy Mauve designed a thatched cottage to Brouwer's specifications, 'de Hut' [the Cabin], which would remain his refuge all his life. Here he started work on his thesis and later produced his best topological work.

*Life, Art and Mysticism* cannot be written off as a rash, 'teen-age effort', of no relevance to Brouwer's views on the foundations of mathematics. He wrote it in the spring of 1905, after his doctoral examination, when he had already chosen the subject of his dissertation, and as he later wrote to his promotor, when the main themes of his dissertation were conceived. Far from disowning his 'booklet' as a youthful aberration, Brouwer backed it all his life. In 1916, four years after his appointment as Professor at Amsterdam University, he discussed it in detail in the 'Signific Circle', and Frederik van Eeden, a poet of national fame and friend of Brouwer, devoted a series of articles to it (*Een Machtig Brouwsel* [A Powerful Brew], De Groene Amsterdammer, 09.09.16; 16.09.16; 23.09.16 and 07.10.16). Brouwer also made attempts to have it re-published in 1927 and discussed the possibility of an English translation as late as 1964. It proudly features in every one of his entries for various biographical dictionaries as the first of his two books.

Most important, it is the clearest expression of his philosophy of life, which inspired his intuitionism. In spite of the triviality of much of its content it is a valuable source of information on Brouwer's strong feelings about science, language, logic and intuition at the time when he conceived his intuitionist philosophy of mathematics.

## 2.4 The Foundations of Mathematics

In the early summer of 1905 Brouwer moved into 'De Hut' and started work on his dissertation in earnest. The courses of his 'dialectical partner' Gerrit Mannoury had given him much to think about. They included courses on 'The logical foundations of geometry' and 'Geometry of $n$ dimensions' (1904), 'Mathematical logic', 'The theory of transfinite numbers', 'The philosophy of number' and 'The logical foundations of analysis situs' (1905) (cf. MANNOURY [1909]). Although Brouwer disagreed with Mannoury on most of these issues, particularly on the role of logic, Mannoury's treatment, linking these fundamental mathematical topics to the wider issues of human experience and society, appealed to him. Through his interests, erudition and familiarity with the contemporary mathematical-philosophical debate Mannoury showed Brouwer the way to the important writers and their publications.

Already in 1904 Brouwer had started studying Russell's *Principles* and his *Essay on the foundations of geometry*. The issues raised by Poincaré in *La science et l'hypothèse* were eagerly studied by both Brouwer and Mannoury and became a subject of frequent discussions between them. Brouwer followed with fascination the Poincaré-Hilbert debate in the *Revue de Métaphysique et de Morale*. Hilbert's work made a deep impression on him, Göttingen began to intrigue and attract him and Hilbert's Paris problems provided an immediate challenge. He used the good services of his promotor for an introduction to Hilbert and to obtain copies of Göttingen dissertations by Hamel, Tresdorf and Opitz on geometry and Gausz and Riemann space.

In his grand design of the dissertation Brouwer had planned to use *the Construction of Mathematics* only as an introduction, and he decided to publish separately some of his studies not directly relevant to his *Foundations*, much against the advice of Korteweg, who would have liked Brouwer to change the subject of the dissertation and move to 'firmer ground':

> I repeatedly told him 'I'd rather you reserved these studies for your dissertation', but he was so keen and insistent on the subject [of his dissertation] which in outline he had already in his mind that I thought it better not to press too hard. (Korteweg to St Jobsleen, 19.10.06 [DJK19]).

The publication of the first of these contributions, *Polydimensional vector distributions* [1906A], created some difficulties. Brouwer discovered some errors, which he wanted changed in the subsequent published English version, contrary to the editorial practice of the Proceedings of the Dutch Royal Academy. Korteweg intervened on his behalf:

> For somebody who is so conscientious it must be very hard to allow mistakes to remain in the text and we *must* be very careful with Brouwer. He has been doing much better the last few years, but not to such an extent that I am not rather worried when guiding him to his doctorate. But he is worth it, for we mathematicians, who know him so well from our meetings of the Mathematical Society, are convinced that he is an excellent mathematician. (Korteweg to the Editor of the Proceedings, 02.07.06 [DJK10]).

In September 1906 Brouwer started to order his notes and arrange them into chapters. From the notes, which he kept all his life, it appears that the main division he originally had in mind was:

i. The construction of mathematics [De opbouw zelf];
ii. Genetic and empiric foundations;
iii. Mathematics and the freedom of mind, and philosophical evaluation of various exact sciences;
iv. The so-called philosophical foundations of mathematics;
v. Axiomatic foundations of mathematics;
vi. Mathematics and society.

With some satisfaction Brouwer found that in spite of all his study of the last few years his main ideas had changed very little:

> I feel the more firm in my convictions since I now notice that I can still back all my notes of two years ago, even now after all the reading I have done. The only difference is that I can now support them better mathematically. (Brouwer to Korteweg, 07.09.06, [DJK13]).

The provisional division in chapters as sent to his promotor on 16 October 1906 is little different from the original plan:

I  The construction of mathematics
II  The genesis of mathematics related to experience
III  The philosophical significance of mathematics
IV  The foundations of mathematics on axioms
V  The value of mathematics for society
VI  The value of mathematics for the individual. (Brouwer to Korteweg, 16.10.06 [DJK15]).

The termination of his grant, with some difficulty extended to November 1906, made Brouwer abandon his grand design and reduce the planned last four chapters to one single chapter on Mathematics and Logic. It is interesting to note that twice in later life, when planning his magnum opus on Intuitionism, never to be published (*Die Berliner Gastvorlesungen* [BMS32] and *The Cambridge Lectures* [BMS51]) Brouwer also stuck to the magic number of six chapters.

Since Brouwer's views on the Foundations of Mathematics will be considered in detail in the following chapters, we shall concern ourselves here only with the historical background of the dissertation and briefly comment on some of its content.

### 2.4.1 Chapter One - The Construction of Mathematics

Chapter One seems to support the view of those 'constructive' mathematicians who, if not contemptuous of philosophy and foundational theory, maintain that these add precious little to their mathematical practice and that Brouwer's contribution to constructive mathematics can be seen as quite separate from his theory of mathematical foundations. Apart from some minor philosophical digressions on the union of the continuous and the discrete, and a preference for Helmholtz's philosophical analysis of space over Hilbert's naive treatment, Chapter I is an elementary-constructive treatment of some fundamental parts of arithmetic and geometry.

Without any speculation on the nature of number, Brouwer starts with the natural numbers as known and given, and proceeds to construct the order types $\omega$ and $\eta$ and the elementary propositions of algebra and geometry:

> One, two, three... we all know by heart the sequence of these sounds (spoken ordinal numbers) as a sequence without end, i.e. a sequence which will for ever proceed according to a fixed rule. Apart from this sequence of sounds, we know other intuitive sequences which proceed according to a certain rule, e.g. the sequence of written symbols 1, 2, 3, ... (written ordinal numbers). These are intuitively clear. ([1907], p.3).

In his correspondence with Korteweg Brouwer claimed that in Chapter I he had solved three of Hilbert's Paris Problems:

Cantor's problem of the power of the continuum:

> ...completely solved by my returning to the intuitive construction necessary for all mathematics;

the non-contradictority of the axioms of arithmetic:

> ...solved by characterizing the construction of arithmetic on the continuum as a two-parameter group for the operations of addition and multiplication;

elimination of the assumption of differentiability for Lie groups:

> ...I solved this for the simple case, i.e. linear groups. ([Brouwer to Korteweg, 05.11.06 [DJK20]).

However, the main object of Chapter I was:
1. to serve as 'a general survey of constructed mathematics to which I can refer in the following chapters', e.g. Hamel's investigation of the line as a minimal curve, which he needed in Chapter III to refute Russell;
2. 'to consider various recent investigations into the Foundations of Mathematics from one view-point: their significance for constructive mathematics';

3. to construct the group of operations on the continuum, 'I wanted to show the construction of groups independent of differentiability to be essential in the construction of mathematics';
4. 'to derive the non-euclidean arc-element by variation; I have not seen it done anywhere else, and it seems to me the only method: finding the arc-element for $n$ dimensions from what has been derived for 2 dimensions.' (Brouwer to Korteweg, 16.10.06 [DJK15]).

### 2.4.2 Chapter Two - Mathematics and Experience

After the more orthodox Chapter I as an introduction and as source for further reference, Chapter II was to be the manifesto of Brouwer's strongly held views, the result of his reflections and analysis of the nature of mathematics, its origin and its relation to human experience and the physical sciences.

The main theme of Chapter II in its original version is that the physical sciences originate in man's ability to enforce a mathematical interpretation on the world. This 'anthropomorphization' of the world is not only morally reprehensible but even misleading, since the 'mathematical interpretation' ignores many other factors and the 'observed regularity' is falsely credited with all the certainty of the mathematical system.

Like *Life, Art and Mysticism*, Chapter II, especially in its original version, reveals Brouwer's strong feelings of misanthropy and resentment of the applied emphasis in academic mathematics; it shows more clearly the direct relation between his ideology and his philosophy of mathematics. Its language is equally 'mystical' and fanatic. The original version of Chapter II opens with the following statement:

> All human life originated in a one-sided restriction of nature and has protracted its existence in an 'externalization', man impregnating nature with the human self and repressing other one-sided developments. This externalization of life and the holding off of death is seen by religion as a lack of wisdom and a lack of bond with the universe. Moreover, the will to destroy and rule immediately obstructs any nourishing of the heart by nature. Those who rule are already damned, and damned are those qualities that promote man's rule. ([BMS3B], p.1; app.3).

In the second half of the chapter Brouwer singles out aspects of the natural sciences to support his general thesis, and derives the a-priority of mathematics with respect to the sciences from his analysis of the mathematical nature of human observation of regularity in the world, i.e. 'seeing sequences of sequences'.

Even if there are frequent references to 'intuition' and 'intuitive' the passages dealing with pure mathematics are few and short. Almost hesitantly he introduces 'the primordial intuition of time', human awareness of time, as the basis for the conception of number and continuity and for the whole of mathematical construction. Some guilt seems to remain attached to the Primordial Intuition of Mathematics because of the monster it has fathered in the form of applied mathematics and science. But Brouwer holds out a glimmer of hope and optimism to the pure mathematician:

> But mathematics practised for its own sake can achieve all the harmony such as can be found in architecture and music. ([BMS3B], p.10).

### 2.4.3 Korteweg's Rejection

When Korteweg received the manuscript of Chapters II and III his worst fears were confirmed. He struck out the greater part of Chapter II and called Brouwer in immediately, on Sunday afternoon. In no mistaken terms he told him that this pessimistic and mystical philosophy had nothing to do with mathematics and that some of his comments on applications were 'absurd'. Brouwer's obvious contempt for mathematical applications and his remarks, such as 'the laws of astronomy are no more than the laws of our measuring instruments', were particularly painful to a man who had devoted his life to 'applications' and who had introduced astronomy in the undergraduate syllabus as the nearest to pure mathematics ('not-useful mathematics', 'mathematics for-its-own-sake').

After the initial shock Brouwer mounted his defence; in three long letters within a week he pleaded passionately with his promotor (5, 7 and 13 November 1906, [DJK20,23 and 26]). Characteristically, his counter-attack does not concern the main charge but concentrates on minor points and side-issues, blown-up out of all proportion in Brouwer's brooding mind:

> You remarked that Kant's name should not figure in a mathematical dissertation...
> You said on Sunday that you were not sure that I had studied Kant thoroughly enough to make a judgment...
> It is probably right that there are in my work traces of hurry and that there are here and there even inaccuracies, but I emphatically reject any charge

that the main thoughts are vague and that my preliminary study has been superficial... (Brouwer to Korteweg, 05.11.06 [DJK20]).

Korteweg's suggestion to remove the 'philosophical' first part and turn Chapter II into a survey of the physical sciences and a criticism of Russell is countered by Brouwer's insistence that he is no expert in the physical sciences or astronomy and that his digressions on mechanics and physics only serve as examples to support his leading thought:

> By themselves, wrenched from the context, some of these are even banal [Du.'onbenullig']...
> I don't wish this to turn into a bargaining session over what can remain in the dissertation and what must come out...
> If the chapter were reduced to a survey of the physical sciences and a criticism of Russell it would lack its framework; I'd rather scrap the whole chapter. (Brouwer to Korteweg, 07.11.06 [DJK23]).

There is no defense of the necessity of a philosophical base for the foundations of mathematics, only a repeated statement of the need for the main theme to be firmly expressed and a wish that he could bring his promotor round to sharing his convictions:

> I would so much like you to understand and recognize precisely this fundamental idea, the general rather than the details, that which can be read between the lines, even if your own philosophy is different and you consider mine absurd, because I am a product of my time and of a different generation...
> It is this general spirit which is the essential part of this work; that's why I wanted so much to send this into the world as a dissertation, which, I thought, traditionally has had the character of 'taking a stand'. Only if this spirit is valued by my promotor will my promotion give me satisfaction. (Brouwer to Korteweg, 05.11.06 [DJK20]).

Korteweg's initial reaction was a sharp reply, which, however, he did not send. (The original letter is marked 'not sent' and was kept by Korteweg [DJK21].) He was generous enough to overlook Brouwer's rudeness and take his difficult temper into account. But he remained adamant, and in his usual mild manner he writes on the 11th November:

> After receiving your letter I have again considered whether I could accept Chapter II as it stands, but honestly Brouwer, I cannot. I find it all interwoven with some kind of pessimism and mystical attitude to life which is not mathematics, nor has anything to do with the foundations of mathematics. In your mind it may well have grown together with mathematics,

but that is wholly subjective. One could totally disagree with you on this point and still share your ideas about the foundations of mathematics. I am convinced that any promotor, young or old, whether he shares this philosophy or not, would object to it being included in a mathematical dissertation. In my opinion yours will only improve by removing it. It now adds a rather bizarre flavour, which can only damage it. I have tried to indicate how it can be lifted out. Just consider it at your leisure and see if you can see your away to making something of it that *you* too think worth keeping. (Korteweg to Brouwer, 11.11.06 [DJK27]).

Brouwer had to agree to the removal of practically all of the rejected parts of Chapter II. He replaced them by a short, relatively sober introduction, leaving only indirect references to his moral valuation and rescuing his brief account of the primordial intuition of time:

> *The intellect and the jump of end to means.* The characteristic faculty of man which accompanies all his interactions with nature is his ability to view life mathematically, i.e. the ability to see in the world repetitions of sequences, or causal systems in time. The primordial phenomenon here is the simple intuition of time in which repetition in the form of 'thing in time and thing again' is made possible, and on the basis of which the moments of life break up into sequences of qualitatively different things... ([1907], p. 81).

Even if Brouwer tried to persuade Korteweg that 'whatever happens, the survey of the physical sciences and the critique of Russell must play a minor role' (Brouwer to Korteweg, 07.11.06 [DJK23]), the larger part of Chapter II in its final version is taken up by a refutation of Russell's views on the a-priority of space. Having rejected the a-priority of space, Brouwer in the last paragraph simply confirms the a-priority of time in the form of the 'mathematical Primordial Intuition'.

When some months later Korteweg made some critical comment on mathematical details of what was left of Chapter II, Brouwer, still resentful, blamed the general incoherence on the removal of his main theme, which these mathematical references only served as props:

> ...in my mind they were originally only incidental off-shoots of a fundamental idea which held them together and which is not any more to be found in the dissertation; they only had secondary importance.
> After their sudden appearance in the full limelight, substituting for their former leader, it was not possible to doll them up quickly in such a way that they together by themselves could save the entire performance.
> At least this is my impression when I now look at the chapter. (Brouwer to Korteweg, 11.01.07 [DJK32]).

## 2.4.4 *Chapter Three - Mathematics and Logic*

The immediate consequence of Brouwer's misanthropic philosophy and his denial of the possibility of any real communication was his unique stand on logic, which would determine his Intuitionism in its critical and perhaps most dramatic aspects. Whereas all other contemporary philosophies of mathematics relied in one way or another on linguistic forms and their structure, Brouwer started from an absolute distinction between human thinking and the linguistic medium, and from the necessary priority of mathematical thinking to its linguistic expression.

To show that 'mathematics is independent of logic, but logic is dependent on mathematics' Brouwer in Chapter III first stresses that the Aristotelian deductive form of reasoning, 'intuitive logical reasoning', or the 'language of logic', is not essential nor characteristic of human reasoning. Further, that a mathematical study of logical reasoning, 'theoretical logic', reveals that this form of reasoning is based on the very simple mathematical relation of *whole* and *part* ([1907], pp. 127 et ff.).

His most fundamental stand, however, is on the difference between mathematical constructive thinking and 'intuitive logical reasoning'. Even if traditionally mathematical reasoning is forced into the logical-deductive pattern, and expression of mathematical constructive thought is 'restricted by the poverty of language' and 'uses logical connectives', it has its own characteristic constructive methods determined by the mathematical nature of the human intellect. In a letter to Korteweg of 23 January 1907 Brouwer stressed that this warning was the main message of his Chapter III:

> It is precisely the main object of Chapter III to show that through the gullible use of logical instead of mathematical language, mathematics in some of its branches has been led astray. ([DJK35]).

The most important recent developments in Foundations research are then in turn submitted to a searching critique and their fundamental reliance on logic exposed:
1. The foundations of mathematics based on axiomatization (pp.133-142);
2. Cantor's theory of sets and transfinite numbers (pp.142-159);
3. The Peano-Russell attempt to found mathematics on logic (pp.159-169);
4. The logico-formal foundations of mathematics according to Hilbert (pp.169-175).

Only in the *Summary* of the original version does Brouwer return to his moralistic stance. As expected, these last two pages of the original manuscript are promptly crossed out by Korteweg with the remark:

> I would leave it at that; it is better to use the following passage elsewhere, not in your dissertation, if you so wish, and do more justice to it. ([BMS3], p. 33).

Chapter III, however, was substantially accepted by Korteweg, even if he had his reservations about some of Brouwer's pronouncements and his style, e.g.:

> It must be a rather strange race that does not reason logically! ([DJK35])... I would like to see some points expressed less bluntly; otherwise it could only rouse passion where it does not belong; and some statements should not be put so dogmatically. (Korteweg to Brouwer, 11.11.06 [DJK27]).

On the 19th February 1907 in a public 'Promotion' ceremony Brouwer defended his dissertation. His formal opponents were his friends Mannoury and Barrau. Mannoury concentrated his opposition on the intuitive mathematical nature of the 'and-so-on', whereas Barrau raised his objections to the intuitive nature of the continuum. He maintained that because of Brouwer's exclusively constructive view of mathematics there was no place for the 'continuous' in Brouwer's dissertation.

It is much to the credit of Korteweg's trust in Brouwer's genius and his total lack of vindictiveness that the doctorate was awarded to Brouwer with distinction, 'Cum Laude'.

As generally agreed by commentators, there is a certain immaturity in Brouwer's *Foundations*. The attempt of Chapter I at constructive mathematics has only historical interest. But many of the views and criticisms of Chapters II and III, even if later developed further and sometimes altered, have later been proved right. Practically all his intuitionist innovations can be traced back to the *Foundations:* his rejection of the principle of the excluded middle, the concept 'meta-mathematics' and even the later development of choice sequences.

In his later foundation papers Brouwer always refers to his dissertation and with some pride uses it to prove the 'priority' and authenticity of his views. He started negotiations for a re-print when in the late twenties it was out of print. *Addenda and Corrigenda* ([1917A]), published ten years later, is largely a list of references to further developments in later papers; only on two points does Brouwer want to withdraw statements made in the *Foundations*.

## 2.5 Towards an Academic Career

Academic life in the Netherlands at the turn of the century was run on a very modest and somewhat parochial scale; departments were small, as were the numbers of students, and 'scientific' output was rather low. But one of its characteristic, and redeeming, features was the strong link with secondary education and its emphasis on teaching. Professors (Dutch 'Hoogleraar', i.e. High-Teacher) regarded teaching as their prime function. In each of the mathematics departments of the four universities the teaching of the whole syllabus for the six-year course as well as the supervision of postgraduate research was shared between two professors, who could give as many as 16 lectures a week on topics ranging from analysis to astronomy. Many secondary teachers remained attached to universities as post-graduate students since a doctorate or evidence of 'working on' a dissertation was still a legal condition of service for gymnasium teachers. The usual route to a University career led through a period of secondary teaching, contacts with universiy departements as post-graduate students or through membership of professional bodies. In the case of mathematics, the 'Wiskundig Genootschap' (Mathematical Association) provided a meeting place where both universty staff and secondary teachers gave lectures and exchanged ideas. Aspiring teachers applied for a licence to teach part-time at the university as unsalaried 'privaat-docent'. Even the first appointment to the salaried staff was often made on a part-time basis as 'extra-ordinarius' and combined teaching at the university and a secondary school.

Even if during times of depression Brouwer considered alternative careers, he knew his own strength, and was quite determined to carve himself an academic career. He was equally convinced of his unsuitability for teaching in schools, and was not prepared to follow the traditional, slow route to academic high office. The unprecedented alternative, however, would involve him in some years of financial hardship and dependence. Above all, he would have to show his mathematical worth, make himself known in the mathematical world and draw the attention of those who 'mattered'.

The experience of his doctoral thesis, the difficulties he had encountered in convincing Korteweg of the mathematical value of his philosophical views, had taught him a lesson. His *Foundations* had been received with the customary suspicion of mathematicians of anything that cannot be proved mathematically; only those who already knew Brouwer's mathematical work in other fields were prepared to listen. If he wanted to achieve his immediate objective, an academic position, he would have to bury for the time being the subject dearest to his heart, concentrate on more patently mathematical subjects, publish and meet 'the great' in conferences and wherever possible.

During the next few years Brouwer retreated to his 'Hut' at Laren, continuing his early investigations of Lie Groups and Schoenflies-style topology. These years, especially the period 1909-1913, are regarded by topologists as the most fruitful of Brouwer's life. During that time he ventured into the field of Foundations on only two occasions.

### 2.5.1 The Unreliability of the Logical Principles [1908C]

Perhaps the most revolutionary conclusion of the general views on logic as expressed in the *Foundations* was drawn by Brouwer in an article *The Unreliability of the Logical Principles* submitted to the editors of the 'Tijdschrift voor Wijsbegeerte' (Journal for Philosophy) only a few months after its foundation. It is Brouwer's first attack on the validity of the Principle of the Excluded Middle (PEM). The article is rather verbose and obscure. The opening paragraphs illustrate the 'philosophical' origin of Brouwer's views on logic and mathematics, and his extraordinary use of language, which also in Dutch is difficult to grasp and even more difficult to translate:

> Science considers repetition in time of sequences, qualitatively different but supposably equal. This isolation of idea to perceptibility and as such to repeatability appears after the irreligious separation of subject and non-realized realizability which has become *something else* (a faculty which originates in the primordial sins of fear and lust, but which also returns without fear or lust etc... [1908C], p.152).

The significance of the article was not immediately clear. The use of the Principle of the Excluded Middle in mathematics is singled out for attack and its 'obvious' unreliability used to support a more general condemnation of the use of logic in science and mathematics. In the course of the general argument Brouwer claims:
1. the identification of the problem of the solvability of every mathematical problem and the PEM;
2. that the PEM is not a reliable principle in an infinite system;
3. that, since even improper use of the PEM does not lead to contradiction, non-contradictority is no guarantee of truth.

Also mentioned for the first time is Brouwer's definition of negation in terms of contradiction. The PEM, however, is not mentioned in the summing up:

> Summarizing:
> In Wisdom there is no logic.
> In science logic is often but not permanently effective;
> In mathematics: it is not certain whether or not logic is permissible; neither is it certain whether it can be decided that all logic is permissble or not. ([1908C], p.158).

That Brouwer was not fully aware of the implications of his rejection of the PEM and his strong definition of negation, or was not fully confident, is also shown by the fact that neither are mentioned in his inaugural address ([1912A].

The editorial board of the 'Tijdschrift voor Wijsbegeerte' accepted the article only after considerable discussion and opposition:

> Most members of the board frankly stated not to have understood a word of it... I only succeeded in overcoming their opposition by underlining what you already promised in your letter of 7 December 1907 to Bierens de Haan, that after publication of the article you would elaborate further on your point of view in a series of articles which will be more comprehensible to non-mathematicians. (Kohnstamm to Brouwer, 3.01.08).*.

### 2.5.2 The Rome Conference 1908

In April 1908 Brouwer attended the Fourth International Congress of Mathematicians, held in Rome, and contributed two papers: *Die mögliche Mächtigkeiten* (On possible powers) [1908A], and *Die Theorie der endlichen kontinuierlichen Gruppen, unabhängig von den Axiomen von Lie* (The theory of finite continuous groups independent of the axioms of Lie)

---

\* *Philip Abraham Kohnstamm* (1875-1951) studied mathematics and physics at the University of Amsterdam, but gradually moved in the direction of Philosophy and Pedagogy. He was awarded a doctorate on a dissertation concerning the philosophy of science, became a 'privaat-docent' at Amsterdam University in 1907 in 'logic and epistemology' and the following year extra-ordinarius in theoretical physics. In 1919 Kohnstamm was appointed to a special chair in Pedagogy at Amsterdam University, extra-ordinarius at Utrecht University in 1932 and again at Amsterdam University in 1938.
Brouwer had dealings, and disagreements with Kohnstamm on various occasions, especially concerning the issue of philosophy within the faculty of science.

[1909B]. Both papers were based on the treatment in his *Foundations* of Hilbert's Paris problems, resp. 1 and 5.

In the paper on continuous groups there is a clear development of the treatment of the elimination of differentiability from Lie's axioms as given in the *Foundations* (further extended in [1910H] to cover all plane transformation groups of no more than 2 parameters).

*On Possible Powers* simply reiterates the *Foundations* treatment of Cantor's continuum problem, which Brouwer had claimed to 'have solved completely'. The statement of the role of the Primordial Intuition is now given a prominent place in the first paragraph.

Brouwer wrote to Korteweg rather disappointedly about the reception his contributions received. But the journey was worth it, he was deeply impressed by 'those heroes of abstraction ... and the aura emanating from 500 honest thinkers'. His remarks about personal impressions and his reaction are interesting and revealing:

> Poincaré was a revelation to me; Darboux and Picard also made a great impression on me. In general, I recognized among these impressive faces almost all those whose work had filled me with admiration. Seeing them personally provides a real guide in choosing the authors I shall be reading. For example, having seen Mittag-Lefler's superficial and snobbish face, I shall never seek guidance in his work; but this I shall do in the case of Darboux, whose work I haven't read so far... (Brouwer to Korteweg, April 1908 [DJK39]).

He almost waxes lyrical when he describes the beau ideal of the mathematician to which the appearance of Poincaré had inspired him:

> To be able to raise oneself to a level and view from where one can produce a lecture such as Poincaré's *l'Avenir des Mathématiques*, whose truthfulness everyone experiences and accepts as a guide in his work, this to me seems to be the highest ideal for any mathematician! (ibid.).

In various letters to Korteweg Brouwer speaks with reverence and awe of the high office of professor. With this more concrete ambition in mind, and the experience of his foundational work and its reception so far, he set aside for the time being any hope of bringing these pragmatic mathematicians round to his Intuitionist point of view. He would not speak out openly again on Intuitionist foundations of mathematics until he had secured his appointment as professor in 1912. He now mustered all his energies to convince his fellow mathematicians of his mathematical genius.

## 2.5.3 Topological Activity 1908-1912

Hilbert's fifth Paris problem, solved for the one-dimensional variety in [1907] and [1908B], was an obvious topic for further research. Brouwer had realized that deeper topological work needed to be done and saw it as a golden opportunity to enter the international mathematical arena. His thorough study of Lie groups, approaching them through manifolds and one-to-one mappings ([1909C]), was the first work he considered good enough to offer to Hilbert for publication in the *Mathematische Annalen*. It set him firmly on the road to topology.

Yet Brouwer's choice of topology as a mathematical specialism was not entirely due to historical accident. The geometric, 'visual' approach was very much in line with his view of mathematics and his reliance on 'seeing with the inner eye'. Of all mathematical abstraction the geometric one is perhaps least grounded on a verbal base. In a review of *the Foundations* Mannoury had interpreted Brouwer's reluctance to follow formalist trends as conceptual conservatism (MANNOURY [1907B]); and indeed, one can feel a certain nostalgia in Brouwer's attempt to interpret mathematical entities ontologically, their essence being grasped by the mind. Rather than sitting at a desk, applying rules to symbols on paper, Brouwer preferred 'to think his mathematical problems through'; his favourite position when working on a mathematical problem was lying on his bed in the Hut, eyes closed, or sitting cross-legged on the heath. Such treatment is perhaps most appropriate to the geometric medium; it also demands and promotes a simplicity which is the hallmark of Brouwer's topological methods and solutions.

His views on geometry and topology and his programme of work are set out in his first public lecture at Amsterdam University on 'the Nature of Geometry'.

## 2.5.4 The Nature of Geometry [1909A]

Rather reluctantly Brouwer took up Korteweg's suggestion that he should apply for a licence to teach as 'privaat-docent' at the University of Amsterdam. Neither the unsalaried, non-recognized position, nor the thought of teaching ('to be dependent on the bon plaisir of my audience') appealed to him. He accepted it only as a necessary stepping stone to an academic career of research. The University Almanac records his request: 'L.E.J. Brouwer priv. doc. ven. imp. Geometriae projectivae partem and Geometriam non-Euclideam docebit die Mercurii hora I et die Lunae, hora X.' He gave his first public lecture on 12 October 1909.

*The Nature of Geometry* is important as a further extension of Brouwer's investigation into the foundations of mathematics, his views on geometry and especially the role of topology. It provides a link between the two areas of his interests usually seen in complete isolation.

Iterating the conclusion of his *Foundations*, that the only a-priori of mathematics is the intuition of time, he expresses his firm conviction that with the arithmetization of space the a-priori validity of geometry is not a mystery any more. The complete freedom of Geometry, earlier proclaimed by von Helmholtz, is based on the absence of any a-priori element other than the intuition of time. Brouwer surveys the contemporary position of geometry and sketches a programme of work still to be undertaken.

He suggests a definition of geometry as the invariant theory of a transformation groups, very much on the lines of Klein's Erlanger Programme of 1872:

> Geometry is concerned with the properties of spaces of one or more dimensions; in particular, it examines and classifies possible point sets, transformations and transformation groups in these spaces. (p.13).

His spaces, however, are continua and his simplices are paralellepipeds. Brouwer specifically excludes from geometry 'finite and denumerable sets of points as well as the abstract spaces of Veronese and Hilbert' (p.14). On the other hand he regards operations on real and complex numbers as part of geometry.

He stresses the problems of 'analysis situs' as the most topical and urgent, and advocates the use of topological methods in other branches of mathematics, as had been done by von Staudt in projective geometry; in particular he mentions the possibility of starting function theory from topology once a way has been found to eliminate the problem of the metric. *The Nature of Geometry* lists the various problems that were occupying Brouwer's mind at the time, such as the problem of dimension, the Jordan Theorem and the insufficiency of Schoenflies's notion of attainability, and contains suggestions as to his solutions.

It is, however, the visual-perceptive simplicity of geometric topology that Brouwer wanted to emphasize. In the final lines he equates 'topological' with 'geometric' and 'formula-less':

> Therefore, coordinates and formulae need not necessarily be banned from other theories even if one succeeds in founding them on analysis situs, but

Brouwer in the 'Hut'

the 'formula-less', the 'geometric' will be the starting point, while the analytic treatment becomes a dispensable expedient. It is mainly to the possibility and the desirability of the priority of the geometric treatment, also in parts of mathematics where this has not been done, that I have wanted to draw your attention. (p.20).

### 2.5.5 *Zur Analysis Situs*

Brouwer often expressed his indebtedness to Schoenflies. Schoenflies's specifically topological approach to set theory, his work on the 'closed curve' had guided Brouwer's investigations already in his undergraduate days and was the basis of his early mathematical work. His first lecture at the 'Wiskundig Genootschap', *On plane curves and plane domains* (October 1908 [BMS4]) was still firmly based on Schoenflies's results.

It was, however, his discovery of flaws in Schoenflies's work during the winter of 1908/09 that gave him the break he had been waiting for, an opportunity to conduct a searching mathematical investigation in the international limelight and under the eyes of 'the great'. Early in 1909 he writes to Hilbert:

When during this winter I had the second part of my *Finite continuous groups* ready to send off for publication in the Mathematische Annalen, I suddenly noticed that Schoenflies's investigations into the Analysis Situs of the plane, on which I had so entirely based my work, cannot in all its parts be sustained. This calls into questions also my own group-theoretical results. To clear his matter up, it was necessary to work thoroughly through the relevant parts of Schoenflies's theory and determine precisely on which results we can rely in full confidence. That's how the enclosed work originated. (Brouwer to Hilbert, 14.05.09 [DHI1]).

Brouwer enclosed the original, rather aggressive draft of his *Zur Analysis Situs* [1910C] with a request for publication in the Mathematische Annalen. Another copy he sent to Schoenflies. Hilbert's reply was a simple acceptance, his comment only concerned the size of the illustrations. But it was Schoenflies's reaction that pleased Brouwer most. Immediately he writes to Korteweg:

At last some fish has taken the bait... I am so glad at last to receive something more than just a polite postcard about my work. Schoenflies has gone into my paper in considerable detail, but I did have to put the thumb-screws on rather hard. (Brouwer to Korteweg, 18.06.09 [DJK58]).

*Zur Analysis Situs* marks the beginning of a series of many papers of Brouwer's hand during the period 1909-1912 which completely changed the course and role of topology. There are already the characteristic, startling counterexamples, which completely upset current views (in topology as well as foundation study) and which opened the way to new possibilities and methods. The immediate aim of the paper was to show that the Schoenflies theory of the general closed curve and general domain was insufficient and that the proofs of some of his theorems were 'completely wrong'. Schoenflies accepted Brouwer's suggestion to publish a reply, but there were difficulties about an agreed text both of *Zur Analysis Situs* and Schoenflies's reply. Schoenflies did not wholly understand the subtlety of Brouwer's arguments and objected to his abrasive style. Brouwer called in Hilbert's help to make sure that Schoenflies's reply did not in any way hint at triviality in Brouwer's work, and he more or less demanded Schoenflies's public recognition of his superior arguments.*

Brouwer's hard and aggressive treatment of Schoenflies is partly due to the practical, financial need to prove his mathematical superiority and so achieve his academic ambition, partly also to a lack of tact or even a

---

* Cf. VAN STIGT[1979]

deliberate disregard of feelings of others and his 'love of a fight'. His energies and mathematical talent seemed to thrive on controversy.

Brouwer's study of the Theory of Groups also led to a series of papers on *Transformations of surfaces in themselves*. He considered this work as a routine exercise and had it published in the *Proceedings* of the Dutch Royal Academy. He was rather unpleasantly surprised when he learned that Hilbert had read these papers and writes to Korteweg of his 'mixed feelings':

> Hilbert has seen my latest contributions in the *Proceedings*; I consider this work to be of much lower calibre than my other work. Anyone whose attention happened to have wandered in that direction could have found these results. Doesn't the value of a mathematical composition, as of any work of art, lie in its penetration and not in some surprising, but easily-grasped result, no more than the value of a Dutch painting lies in the little mill. (Brouwer to Korteweg, 22.05.09 [DJK 50]).

Hilbert did not share Brouwer's low opinion of this work and asked him to write a summary of the results for the Mathematische Annalen. Later, when Brouwer had shaken off the Cantor-Schoenflies pure point-set approach, this line of investigation led to his *Plane translation theorem* [1912B].

### 2.5.6 The New Methods and the Invariance of Dimension

Brouwer's dissatisfaction with this early work reflects his growing doubts not only about some of Schoenflies's results but about the fundamental approach to topological problems. The combination of a sharp, original mind, an eye for simplicity and a preparedness to question accepted methods and rigid adherence to specialization, led to a successful search for new methods and tools in topology for which Brouwer will always be remembered.

During the summer of 1909 he submitted to Hilbert a new, elegant *Proof of the Jordan Theorem* [1910E]. His new approach was announced in the introductory lines, where he simply states:

> Veblen was the first to give a proof which instead of considering the curve as a limit of polygons, concentrates on the curve itself, and uses its inner properties. We shall also proceed in this direct way, but achieve our result by a completely different, and I believe, a considerably shorter route. ([1910E], p.169).

At the time of writing his *Foundations* Brouwer was still confident that the two-dimensional case is usually a good guide for generalization to $n$ dimensions (cf. Brouwer to Korteweg, 16.10.06 [DJK15], quoted above). But one of the first surprising results of his study of Schoenflies was the discovery that simple generalization of some plane results even to three dimensions can lead to false conclusions. There is evidence that already in 1908 Brouwer began to concern himself seriously with the problem of dimension. His 1908 lecture for the Wiskundig Genootschap (*On plane curves and plane domains* [BMS4]) surveyed the then current situation. He reduced the difference of dimension to the impossibility of a homeomorphism, in Brouwer's terminology 'appliceerbaarheid' ('applicability'):

> It can be shown that it is precisely this applicability which is lacking between a line and a space of higher dimension, in particular between a line and a plane... Also the impossibility of 'applying' the plane to a space of higher dimension is now clear; perhaps the same can even be said of $R_3$; but one cannot say anything with certainty about $R_4$, so that one may not rule out the possibility that between spaces of four and higher dimensions there will be no clear difference. ([BMS4], p.3).

In that same month, in his first public lecture [1909A], he still speaks of some of the dimension problems as 'extremely hard to prove, probably remaining unsolved for some time to come' (p.20), but at Christmas of that year he made a dramatic discovery. While staying with his brother Aldert in Paris, he hit upon two major, original ingredients for his proof of one of the most fundamental notions of topology: the 'degree of mapping' and the method of simplicial division and approximation. During the spring of 1910 he completed his proof that 'the Euclidean $m$-space and the Euclidean $n$-space are not homeomorphic unless $m = n$'. Almost casually he writes to Hilbert:

> I am writing a paper for submission to the editorial board of the Mathematische Annalen in which I solve the problem of invariance of dimension... (Brouwer to Hilbert, 18.03.10 [DHI7]).

The paper *Proof of the invariance of the dimension number* [1911C] was submitted in June 1910 and published in February 1911. In October 1910, in a lecture to the Wiskundig Genootschap ([BMS6A]) he had revealed his remarkable proof 'that $R_n$ cannot contain a one-one continuous image of a domain of higher dimension', defining his revolutionary method of 'simplicial approximation' and using the degree of a mapping.

In a series of articles during the next two years Brouwer developed his notion of degree of mapping and his theory of what he described as 'the natural concept of dimension'. By defining the degree of dimension as an integral-valued function and proving that the 'degree of dimension' of Euclidean $n$-space is $n$, he suceeded in giving a natural foundation to Dimension Theory. The significance of Brouwer's work in dimension theory is summed up by Alexandroff and Hopf:

> Brouwer has been the first to prove the invariance of the dimension number. His proof is the beginning of a new era in topology: his method of simplicial approximation and degree of a mapping is powerful enough to overcome all difficulties in the construction of 'Polyhedral Topology', first of all to give the necessary proof of invariance. The effects of these methods, however, have reached far beyond the proof of invariance; they are the kind of creation which have made Brouwer, with Cantor and Poincaré, one of the founders of modern topology. (ALEXANDROFF/HOPF[1935]).

## 2.5.7 International Recognition

With the publication of his Proof of the Invariance of dimension number in February 1911 Brouwer had established himself on the international scene as one of the great innovators. But again his path to fame and recognition was not without its heated controversy. Some months before the publication Blumenthal, the editorial manager of the *Mathematische Annalen*, had mentioned the proof to Lebesgue, who then claimed to have had various proofs for years. Lebesgue then sent Blumenthal a proof using his famous 'paving principle', which Blumenthal liked even better than Brouwer's proof. It was also published in the February issue of the *Mathematische Annalen* after Brouwer's proof. Brouwer, however, regarded it as a challenge to his 'priority' and immediately went into action. In a rather aggressive short note, *Remarks on Mr Lebesgue's proof of the invariance of dimension* (March 1911 [HLE2]) Brouwer showed that the proof depends on what Lebesgue calls 'faits bien évidents', which form the essence of the proof. Blumenthal held up the publication of Brouwer's *Remarks* to give Lebesgue the opportunity to reply and provide the proofs of the 'faits bien évidents'. When Brouwer received Lebesgue's 'defective' proof he challenged him to publish it, which Lebesgue declined. Brouwer in the meantime had found a proof of Lebesgue's 'faits bien évidents' and sent it to his friend Hadamard:

> You see that this proof of Lebesgue is based on a 'théorème évident' which is far more difficult to prove. It has no other merit than to have reduced the

problem of invariance of dimension to a more difficult problem. (Brouwer to Hadamard, 02.07.11).*

Brouwer had more than proved his mathematical mastery. In a frenzy of activity he now used his newly found tools to prove his famous fixed point theorems for the $n$-dimensional element and the $n$-dimensional sphere, the invariance of the $n$-dimensional domain and the Jordan Theorem for $n$-dimensional space, to name only some results which have become classics in topology.

His researches spread over a wide range of topics in topology; in the short period 1909-1912 he published over 40 papers in Dutch, German and French journals.**

At the age of thirty Brouwer had achieved his first great ambition: he had received international recognition and was in regular close contact with other famous mathematicians. His contacts with Hadamard of the 1908 Rome Congress had developed into a personal friendship. In the summer of 1909 he met Hilbert in the Hague for the first time, and the relation Hilbert-Brouwer until 1920 can only be described as friendship and mutual admiration. Hilbert wrote of Brouwer in 1911:

> I consider Brouwer to be a scholar of singular talent, of the most wide-ranging and richest knowledge and of an exceptionally penetrating mind. His special field of work is the theory of point sets, a theory which affects almost all branches of mathematics and whose development is one of the most important tasks still to be done.
> It is part of his character never to be satisfied with easy results which the usual research offers; he always starts on specially hard and deep problems and does not stop until he has succeeded in finding a solution which completely satisfies him. I think in particular of his solution of the problem of

---

\* Brouwer's resentment is clear from a private note in which he gives his 'pedagogic' reasons for not divulging his proof to Lebesgue: 'I know from experience that if one wants to find out from someone who makes errors how deep his fault lies, one should never tell him anything positive because he will then immediately pretend he had known this all along.' ([HLE], F).

\*\* Hans Freudenthal in his editorial notes of *Brouwer's Collected Works Part II* gives a well-documented historical introduction and analysis of various aspects of Brouwer's topological work.
A detailed analysis of Brouwer's contribution to dimension theory is given by Dale M. Johnson in JOHNSON[1979], which also includes transcripts and copies of relevant documents.

finite continuous groups and his wonderful theorem of the Jordan Curve Theorem. (Hilbert to Korteweg, 06.02.11 [DJK79]).

In July 1911 Brouwer visited Göttingen and he would remain a regular visitor of the Göttinger Kreis for a long time.

In the controversy with Koebe, following Brouwer's Karlsruhe lecture in September 1911, 'Über den Poincaréschen Kontinuata' (published as [1912H]), he received the full support of Hilbert, Klein, Fricke, Bernstein, Bieberbach and Rosenthal.

There was friendship and frequent contact too with the younger generation of promising mathematicians, especially Hermann Weyl, and Constantin Carathéodory. His international reputation for rigour is evident from a letter of Carathéodory to Hilbert:

> The length of the manuscript is due to my trying to be absolutely rigorous (you know how much people sin against this in this part of mathematics, with the exception of course of Brouwer). (Carathéodory to Hilbert, 05.05.12, Nieders. Staatsbibl. Göttingen).

Perhaps the greatest satisfaction Brouwer found in the recognition by Poincaré, who shortly before his death wrote to Brouwer:

> 'Je suis heureux d'avoir cette occasion d'entrer en rapport avec un homme de votre valeur.' (Poincaré to Brouwer).

### 2.5.8 Professor L.E.J.Brouwer, 14 October 1912

The fulfilment of Brouwer's second ambition, a professorship with its financial security, was greatly due to the relentless efforts of his promotor. Korteweg was generous enough to overlook the occasional rudeness of his protegé and, what's more, his lack of experience and interest in teaching\*, which he himself considered to be 'the primary function of a professor'. He used all his influence with the civic and academic authorities to

---

\* Brouwer had not been a great success as a teacher. Students' comment in the University Almanac refers to 'the rather low attendance at lectures of Dr Brouwer, an indication that his lectures were too difficult for first-year students; perhaps this is also due to the high speed at which Dr Brouwer delivers his lectures, which makes it very difficult for us to gain understanding in this subject [projective geometry Ed.] which is new to us.' (Amsterdamse Studenten Almanak 1909, vol.81).
Again in 1911: 'Its seems to us that Dr Brouwer's audience would have benefited if his lectures had been delivered at a slower speed.' (ASA 1911).

Professor L.E.J. Brouwer – 1912

secure Brouwer a professorship at Amsterdam University.

When in 1909 Korteweg failed in his first attempt to create a lectureship for Brouwer at Amsterdam university, he impressed on him 'to let no opportunity pass to show your worth as a teacher'. (Korteweg to Brouwer, 16.02.09 [DJK43]). Brouwer's reaction to the City Council's refusal was angry resentment. To emphasize his threat to seek alternative employment he asked Korteweg to support his application for the Directorship of the Teyler Museum (Brouwer to Korteweg, July 1909 [DJK54]).

In spite of his disagreement with Brouwer on the teaching and research aspects of a university post, Korteweg continued his campaign to retain him for mathematics and the University of Amsterdam. His second attempt in 1910 is a well-documented request, based on the needs of the Mathematics department and the need to secure such a brilliant mathematician for the University; even the controversial dissertation is quoted in support:

> After completing his studies at the University of Amsterdam with truly brilliant results and crowning his studies with a dissertation whose originality and importance no one will deny, Dr Brouwer has continued to deploy his mathematical talents in papers published in the proceedings of the Royal Academy and in the best mathematical journal of Germany, the Mathematische Annalen. We are absolutely convinced that through these publications he has placed himself on a level with the best mathematicians of our time. (Korteweg to the Council of Amsterdam, 06.10.10 [DJK72]).

When also this attempt failed, Korteweg tried again, seeking the support of Hilbert, Poincaré, Klein and Borel. He also lobbied his fellow members of the Royal Academy in support of Brouwer's nomination for fellowship, to strengthen his candidature for a professorship.

After much work behind the scenes Korteweg finally succeeded: a vacant lectureship provided the opportunity to create an extraordinary professorship, 'much more in keeping with Brouwer's rising fame as an outstanding mathematician both in Holland and abroad' (Korteweg to the Council of Amsterdam, April 1912 [DJK111].

In April 1912 the Faculty and the University Curators approved Brouwer's appointment to 'Professor in Mathematics'. His main responsibility was to be the 'Theory of Functions', and with Korteweg he would share the teaching of Analytic Geometry. On the 14th October 1912 Brouwer delivered his Inaugural Address, *Intuitionism and Formalism*.

Only half a year later the death of Schouten created a vacancy at Groningen University, and Brouwer was offered the Ordinary Chair at Groningen. Again Korteweg started a campaign, this time for an ordinary chair for Brouwer at Amsterdam. In a most unselfish gesture he offered to vacate his own chair in favour of Brouwer, which ironically would make Brouwer responsible for applied mathematics. In a letter to Hilbert Brouwer speaks of his dilemma and asks for advice:

Lieber Herr Geheimrat,

I come to you for advice. I can now become ordinarius at Groningen, where on the one hand I shall be completely free in my work, but where on the other hand I shall find a narrow-minded provincial town and probably much less sympathetic colleagues. I also have an offer of an ordinary chair in Amsterdam, a lively metropolitan city with which my life has been so intimately connected, near my home in Blaricum and near the dunes, and where I feel at home in the faculty but where I shall be mainly concerned with the teaching of mechanics. If my duties at these universities had been identical I would, of course, not hesitate and opt for Amsterdam, but I cannot possibly see how a committed involvement in applied mathematics would not fail to divert my research from its natural path, and so more or less paralyse it. What do you think? Dare I stay where I feel at home and trust that the harmony between my thinking and mechanics will automatically be maintained? You know me and you have such a wealth of experience as a scholar. I don't know anyone whose advice I can trust more in this most difficult struggle to come to a decision. Social life at Groningen with its pettiness and straight-laced conventions will be awful and scenic beauty is non-existent... (Brouwer to Hilbert, 16.06.13 [DHI18]).

Hilbert's reply reassured Brouwer on the matter of mechanics and recommended Amsterdam, the choice Brouwer in the meantime had already made.

The 'applied' aspects of mathematics were no longer the bogey, representing a moral threat to the purity of mathematics. Neither did Brouwer's rebellious spirit any more identify the mathematical establishment with the applied emphasis of Amsterdam and Korteweg. His rebellion now shifted away from 'application of mathematics' and centred on the formalistic tendencies and the supremacy of the international mathematical establishment.

In June 1913 Korteweg retired as 'ordinarius' and Brouwer took on the full responsibility for applied mathematics and projective geometry.

The offer of a chair at Leiden in 1916, engineered by Brouwer and Lorentz, and Brouwer's 'resistance to the Sirensong' (Minutes of the Faculty of Science and Mathematics AU, 06.02.16), were used to strengthen his position at Amsterdam. He successfully campaigned for an improvement in the staffing of his Mathematics Department in 1917 and a professorship for his friend Mannoury. In the re-organization of the department he was now able to off-load mechanics on Mannoury and establish for himself the chair in 'Set Theory, Function Theory and Axiomatics'.

## 2.6 Intuitionism and Formalism

On the 14th October 1912 Brouwer broke his long, calculated silence. His inaugural adress *Intuitionism and formalism* [1912A] signals his return to the battlefield of the Foundations of Mathematics. His professorship had given him the financial security and the freedom to speak his own mind. He could now also speak confidently and authoritatively, knowing that his voice would be heard and command attention. There is an element of defiance in his opposition to the established policy of world-famous mathematicians, whom he now knew personally, perhaps even an element of revenge in his repeating in public and in the presence of Korteweg views that had been dismissed by his promotor and removed from his dissertation.

In many respects *Intuitionism and Formalism* re-states Brouwer's position of 1907. There is still some detachment in his description of the differences between the Intuitionist and the Formalist viewpoint, he does not yet identify the Intuitionist cause with his own 'intervention'. The 'Primordial Intuition' is still described as the direct source of the linear continuum and the immediate foundation of geometry as well as arithmetic.

But there is a shift of emphasis. The introductory reflection on science no longer carries a direct moral condemnation. There is also a clear dissociation from the French Intuitionist School. Brouwer's own analysis of the 'fundamental phenomenon', the *Intuition of Time*, becomes the prominent philosophical basis of what he now describes as 'neo-intuitionism'. No sharp distinction is any more made between logicists, axiomatici and formalists, they are all bracketed together as:

> Formalists, who maintain that human reason does not have at its disposal exact images either of straight lines or of numbers larger than ten... for whom mathematical exactness consists merely in the methods of developing meaningless series of relations which have mathematical existence only when they have been represented in spoken or written language. (p.8).

There is no long diatribe against the inadequacy of language and logic; the Principle of the Excluded Middle is mentioned only indirectly in an added footnote to the English translation [1912A2].

Most significant, however, is the shift of emphasis to Set Theory. The greater part of the offensive is now directed at the constructive inadequacy, or rather the meaninglessness of Zermelo's axiomatic basis of Set

Theory (ZERMELO[1908]). Singled out are two of Zermelo's axioms, 'Auswahl' and 'Aussonderung', to which Brouwer refers as 'the Axiom of Selection' and 'the Axiom of Inclusion' ([1912A2]). He shows that in particular the Axiom of Inclusion leads to contradiction (the Burali-Forti paradox), and he claims that even in its modified form, which avoids known paradoxes, it forms the basis of a totally meaningless and unacceptable theory of potencies of sets, of number and the continuum. Later, in [1919D], Brouwer would claim that the rejection of the Principle of the Excluded Middle and the Axiom of Inclusion had been the guiding principle in his systematic construction of an Intuitionist set theory.

## 2.6.1 Intuitionist Set Theory

The shift of emphasis and Brouwer's pre-occupation with set theory in 1912 and the following years was sparked off by Schoenflies's work. It is remarkable that Brouwer's involvement in topology had been occasioned by his critical analysis of Schoenflies's work and that now, in 1912, his move away from topology to Set Theory and back to Intuitionism was again partly due to his involvement in the work of Schoenflies, this time the re-edition of his standard work on point sets (SCHOENFLIES-HAHN[1914]). On both occasions the direction of Brouwer's work during the following years was determined by it.

Towards the end of 1911 Brouwer's attention began to centre on Set Theory. He describes the reason for this interest in a letter to Hilbert:

> You probably know that for some time I have been busy with the re-edition of Schoenflies's *Bericht über die Mengenlehre*. What happened is this: I was pressed repeatedly from various quarters to write a book on the theory of sets, since the present textbooks and contributions in encyclopediae are inadequate and superficial. When I was in Göttingen in the summer of 1911 I was asked again, and at the same time I heard that Schoenflies was preparing a new edition of his book. I thought it would solve the problem if I could control the edition during the printing stage and, where necessary, improve it and complement it. The difficulty of persuading Schoenflies to allow me the control was soon solved when Fricke, who knows Schoenflies personally, offered to mediate. Schoenflies was very pleased with my proposal. (Brouwer to Hilbert, 16.01.13 [DHI16]).

In Brouwer's imagination his collaboration had grown into a right of control (cf. VAN STIGT[1979]). His correspondence with Hilbert at that time is dominated by his obsession with the Schoenflies edition; he repeatedly pleads with Hilbert to put pressure on Schoenflies to allow him 'full control' ('I beg you to send him a telegram immediately and tell him that he must give in to me' etc. (Brouwer to Hilbert, 05.09.13 [DHI21])). He does not hide from Hilbert his conviction that he considered himself to be *the* authority on Set Theory at the time:

> My concern is a good book on the Theory of Sets... and at the moment I am the most competent person... (Brouwer to Hilbert, 23.04.13 [DHI17]).

His passionate concentration on Set Theory at that time made any serious topological work impossible and Schoenflies is given the blame:

> Schoenflies is getting worse. If there is not a complete change I shall have to give up the work on which I have spent more than eight months... I have not done any work of my own... (Brouwer to Hilbert, 04.07.13 [DHI20]).

The complaint is echoed by Mrs Brouwer in a letter to her sister-in-law, in which she specifically blames Schoenflies for Brouwer losing the contest in proving Poincaré's last problem:

> Bertus thinks he cannot stand it any longer. It has also made him lose the priority in solving Poincaré's famous last problem... (12.04.13).

From the Brouwer-Korteweg correspondence it appears that during the summer of 1912 Brouwer had been working on a solution of Poincaré's problem. Just before his inauguration he writes:

> As to my solution of Poincaré's problem, it will take some more weeks. Please, don't speak to anyone about it. When I have a fully worked-out version I shall present it to the Academy. (Brouwer to Korteweg, 30.09.12 [DJK116].

Brouwer's 'loss of interest in topology' at that time is due to a variety of factors. There was increasing pressure of work; mastering the areas of his new responsibility demanded more time and energy than he had expected. Moreover, he could not resist the temptation and promise of power and admiration associated with positions of leadership, whether in international mathematical organizations or in campaigns for marginal causes. In 1914 Klein invited him to join the editorial board of the *Mathematische Annalen*. Brouwer enthusiastically accepted 'this great honour for himself and for the Dutch universities' (Brouwer to Klein 10.07.14 [FKL3]). He would give his editorial duties first priority.

Above all, the newly found security of a permanent university post had removed the pressure to prove himself and had given him the freedom to to devote his energies to the mathematical cause nearest to his heart, the foundations of mathematics.

His involvement with Schoenflies's work on set theory had made him more and more aware of the deficiencies of set theory and in particular its fundamental presuppositions. In his *Review of Schoenflies's Bericht* [1914], Brouwer comments on its usefulness as a survey of work on set theory to-date, but he castigates its unquestioning acceptance of philosophical principles, some of which are mutually exclusive and contradictory:

> For the Intuitionist it contains too much ...much that is superfluous..., for the Formalist it constains too little...

But his main criticism concerns Schoenflies's refusal to adopt a consistent and coherent philosophy of mathematics:

> Schoenflies avoids difficulties by steering clear of any particular philosophical commitment, or rather by adopting the most generous possible axiomatic foundation, which does not concern itself with non-contradictority nor with the question of existence outside the axiomatic system. In this way he manages to fit-in all published developments in Set Theory. (p.78).

Brouwer's involvement in Schoenflies's work had made him recognize the need for a coherent philosophy of set theory which questioned fundamental assumptions, a systematic fundamental theory of sets in accordance with rigorous constructive demands.

Central to all his thinking during these years is the problem of the continuum, the growing realization of the need to accommodate the real number constructively in the geometric continuum. In [1912A] Brouwer still states the position of the 'Intuitionist School', to which he would later refer as the 'Pre-intuitionists', which only recognized real numbers generated by a law in a finite number of operations. He refers to 'the Intuitionist to whom the real number of the Formalist determined by an elementary sequence of freely selected digits is devoid of all meaning'. But already then he mentions the possibility of 'admitting, on the basis of the linear continuum, an elementary sequence of free selections as an element of construction'. (p.23).

In his *Review of Schoenflies* [1914] Brouwer has clearly moved to a more independent position. He introduces his notion of 'well-constructed set', which includes 'elements each determined by a sequence of choices from elements of a finite set or fundamental sequence' (p.79). Some basic properties of the 'wel-constructed set' are listed, but the immediate implications for the reals and the continuum would not be drawn until Brouwer in 1917 presented his thorough and revolutionary *Foundations of Set-Theory independent of the logical Principle of the Excluded Middle*.

## 2.7 The Signific Interlude

The early war years mark a period in Brouwer's life of little mathematical activity beyond his regular lecturing commitments. International communication was disrupted and academic life in particular at the great German centres almost completely halted. Restless and ambitious, Brouwer now begins to stretch his search for recognition, admiration and power beyond the frontiers of the academic world of mathematics. Apart from some minor campaigns, his energies in these years concentrate on three main issues: the Signific Movement, the setting up of an national mathematical research centre at Amsterdam University under his leadership, and national as well as international politics.

The term 'Significs' was used for the first time by Lady Victoria Welby in *What is Meaning?* (WELBY[1903]). Her call for a critical re-appraisal of all aspects of human language and for linguistic reform was enthusiastically answered by Frederik van Eeden, a Dutch medical doctor, a poet of national standing and a prominent figure in the literary revival and romantic social reform of the 'eighteen-eighties'. His writings and the foundation of his commune in 1889 were inspired by a romantic idealism and his naive confidence that society and the lot of the poor masses can be changed by the benign interference of a few wise men. He became the hero of the young intellectuals of Amsterdam, students and artists, and was elected 'Rector of the Student Corps of Amsterdam University'.

The personal friendship between van Eeden and Brouwer dates back to 1915. Van Eeden recognized in Brouwer a man of genius, 'the greatest genius I have ever met', who could provide the intellectual power in the realization of his romantic projects. After his first meeting he describes Brouwer as:

> A very engaging man with the pleasant childlike manners of a man of genius. It is surprising how often mathematical talent runs hand in hand with freedom of thought and noble character. An intelligent, spiritual face, clean-shaven, with fine-cut features and early appearing lines. He is only 34. Dressed in a white cotton suit. During the conversation he sometimes sits down on the floor, deep in thought. (*Diary*, 22 October 1915).

Brouwer was flattered by the admiration of the famous poet, and he shared van Eeden's ambition for a leading role in a campaign of social reform. Initially they sought a platform for the launching of their programme in the 'Amersfoort Academy'. As chairman of the newly formed circle Brouwer promulgated 'a programme of revaluation of the fundamentals of life for the individual and society'. When by a democratic vote, 'a majority of mediocrities', Brouwer was forced to resign the chairmanship, a new circle was formed of friends of Brouwer, prepared to accept his leadership in a *Signific Movement*.

The early circle consisted of Brouwer, van Eeden, H.P.J.Bloemers – a sciologist and burghomaster – and H. Borel, a journalist and sinologist and for many years one of Brouwer's closest friends. They were soon joined by the geologist L.S.Ornstein and by J.I.de Haen, a poet who had taken on the respectability of a lawyer after disgracing himself earlier as the author of homosexual novels, and later also by Mannoury. Their leader, the driving force behind the early Signific Movement was Brouwer, and his ideas formed the basis of the initial programme. It accepts the fundamental role of language in all collective human activity, from the single interaction between two human beings in love to the sophistication of scientific cooperation, and it signals the need for systematic study of all aspects of language and the need for some wise men to influence and redirect its historic course. To Brouwer and the early Signific enthusiasts, however, Significs was more than a sideline or hobby of linguistic reform and purification, it was an alternative discipline, akin to philosophy, embracing all the branches of science and the arts and all forms of culture, mathematics as well as poetry.

The general need for reform arose mainly from Brouwer's pessimistic view of society and man. During the first of the frequent meetings of the circle during the year 1916 Brouwer lectured on his *Life, Art and Mysticism*. These lectures were a revelation to van Eeden, who enthusiastic and over-awed devoted five long articles in a national paper to 'this powerful brew':

> These hundred pages of Dutch prose are in my opinion the most powerful but also the most horrifying published this century. They are beautiful and deep and full of truth. But they are violently revolutionary and completely hostile to the whole of our society. (VAN EEDEN[1916], I).

If *Life, Art and Mysticism* was recognized as an early clarion call for general Signific reform, Brouwer's *Foundations of Mathematics* was hailed as the first attempt of putting the Signific programme into practice. Jacob Israel de Haen followed him in the legal field; in his doctoral dissertation, *Legal Significs*, Brouwer's *Foundations* is frequently quoted and referred to as 'Brouwer's Mathematical Significs'.

For a shortwhile Brouwer's Intuitionist campaign becomes part of his grand design of Signific reform. How much his intuitionist and signific diagnoses and ideals are interwoven is evident from his introduction to a lecture by van Eeden where Brouwer speaks of 'Intuitive Significs'. Misuse of existing language is the common diagnosis. The new Signific ideal, however, reflects a shift of emphasis and a new mood of optimism in Brouwer: the inadequacy of language is due to its origin as an instrument of power, but an instrument which is capable of improvement:

> The words of our present language are nothing but commando-signs of social/labour regulations...
> *Intuitive Significs* concerns itself with the creation of *new words* forming a code of the basic means of mutual understanding for the systematic activities of a new and holier society. ([BMS22B], App.4).

The *Preparatory Manifesto* ([1918A]) of the Circle, now constituted as 'The International Academy of Practical Philosophy and Sociology' reflects the new obsession of its principal author, Brouwer. 'The coining of words of spiritual value for the languages of western nations' is described as its primary task. Brouwer's earlier solipsistic despair of the possibility of human communication has made place for a new trust in the collective ability of 'a group of pure-feeling and independent thinkers':

> The failure of earlier attempts to purify language was inevitable and due to the individual character of the work of such philosophers; their words were only useful for recalling the expressed thoughts in the mind of the writer and isolated readers, they never found a place in the mutual understanding of the masses...
> But if now this task is undertaken collectively, by a circle of pure-feeling and independent thinkers then their insights, formed in mutual understanding, will acquire a linguistic accompaniment which can find a place in the mutual understanding of the masses. As to the realization of these insights, one should bear in mind that a thought as the embryo of action has far greater chance of development and realization if it is the collective, intimate conviction of *a group of human beings* than if it belongs to a single individual, however courageous he may be and however large the crowd of half-understanding followers surrounding him. ([1918A], pp.4-5).

While other members of the Academy tried to promote the Signific movement nationally by the creation of a university chair in Significs to be held by Jacob Israel de Haen, Brouwer's dream and grandiose design reached beyond national frontiers. The Circle was to be no less than an International Academy whose members all had an established international reputation and together 'covered almost every domain of human mental activity'. *The Manifesto* and invitations to join were sent a.o. to Norlind, Bjerre, Romain Roland, Landauer, Buber, Carus, Ehrlich, Mauthner, Tagore and Peano ('the creator of mathematical token language (pasigraphy) which had proved not only to be a very useful means for the investigation of the role of language in mathematics but to be susceptible of generalization in other domains of signific research as well.' (ibid.)).

Mannoury's move to Amsterdam allowed Brouwer to leave the general cooordinating task to his unambitious and trusted friend, while he himself could concentrate his energies on the organization of Signific reform in the field of mathematics. His dream and ambition was to create an international centre of mathematical research in Amsterdam under his leadership, which would undertake the giant task of re-constructing the whole of mathematics, a 'second Göttingen' and ultimately the world centre of mathematics.

Brouwer's newly found trust in collective ability, however, extended only to fellow-kings in other fields; 'cooperation' in the mathematical field was that of the beehive, where the queen has the monopoly of creative activity. His self-confidence and his conviction that he alone, single-handedly could undertake the task of the intuitionist re-construction of mathematics spurred him into renewed activity, boldly identifying himself with Intuitionism and pioneering new concepts and their linguistic representation. His inability to cooperate at a level with fellow-mathematicians, his jealous mistrust even of his own students and friends, were the main reason for his failure to surround himself with an able team and ultimately for his tragic failure to 'create an Intuitionist School'.

In order to achieve his great ambition Brouwer first set out to secure himself a position of power within his department and faculty at the University of Amsterdam, and then proceeded to campaign for his department to become a national and international centre. The offer of a chair at Leiden in 1916 and his refusal strengthened his bargaining power with the governing board of the city fathers. It is typical of Brouwer's love of intrigue and a certain inconsistency between theory and practice that he could use his personal influence to secure the appointment of his friend

Mannoury to a chair in mathematics (and that of his own brother to the chair of geology) and almost at the same time lead a campaign in the faculty and – much to the dismay of his fellow-professors – in the national press for an open system of academic appointments on the sole ground of academic excellence and against the tradition of secrecy and personal preference.

At the same time he used his influence as president of the Wiskundig Genootschap to press for an expansion of his department and for the setting up of a mathematical centre. In an open letter to the Amsterdam City Council and the Minister of Home Affairs* a case was made for 'the urgent expansion of the department of mathematics of at least one University'. The classification of areas of responsibility and the proposed team of professors are an interesting reflection of Brouwer's views on mathematics:

1. 'Undergraduate mathematics (including mechanics and astronomy);
2. Epistemological mathematics (theory of sets, analysis situs, axiomatics, the mathematical foundations of natural philosophy);
3. Mathematics of finite and denumerable systems;
4. Mathematics of systems of the power of the continuum (real functions etc.);
5. The theory of geometric transformations and relations;
6. The theory of analytic functions of one or more variables.'(op.cit.).

Korteweg's retirement and Mannoury's appointment allowed Brouwer to re-shuffle 'his' department, off-load mechanics and take on the responsibility for 'epistemological mathematics'. To secure subordination in the department he launched the idea of 'associate professorship' which would allow expansion of the department without loss of his authority**.

By the end of World War I most of Brouwer's dreams and ambitions had come true. He had acquired a strong position of leadership in his

---

\* During Brouwer's chairmanship of the Wiskundig Genootschap in 1917 Brouwer's friends de Vries and Ornstein proposed that a committee be set up to inquire into the development of mathematics in the Netherlands; 'from the floor' it was suggested that Brouwer should head this inquiry. At the meeting of the Wiskundig Genootschap of 18 July 1918 Brouwer submitted the result in the form of a draft letter to the Council of Amsterdam and the Minister of Home Affairs. The letter was accepted and sent. (Verslag van de Algemeene Vergadering van het Wiskundig Genootschap 18 July 1918)

\*\* Another of Brouwer's campaigns by letter and in the press; cf.his letter to the Faculty of Mathematics and Natural Science A.U., 20 November 1918.

University and academic life in Holland. The Signific Movement had given him the opportunity to assume leadership at an international level. Within the international mathematical community he had forged strong links with the established great, especially of Germany. The cool working relation with Schoenflies had turned into a warm friendship, and Brouwer carefully nurtured a personal friendship with Blumenthal, the managing editor of the Mathematische Annalen. He was delighted to receive Hilbert's and Klein's accolade in the form of membership of the Göttingen Club in August 1918 and writes:

> Lieber Herr Hilbert,
> 
> It was an extremely pleasant surprise to receive today the news of my election as member of the Gesellschaft der Wisschenschaft of Göttingen. I could not possibly confine my thanks on this occasion to an official thank-you to the Secretary. I shall never forget that all my many personal and official contacts in Göttingen have been the results of our personal relationship ever since we met nine years ago in the Hague, and even more that the wealth of knowledge and inspiration which I gained from your work have played a decisive role in my mathematical development... I shall also write a special letter of thanks to Klein... (Brouwer to Hilbert, 28.08.18 [DHI23]).

There is no doubt that Brouwer's non-mathematical interests and campaigns during these years were time-consuming and wasteful of energy and detracted from pure-mathematicial work. They also created a climate in which he thrived and which helped him to develop the new elements of the second phase of his intuitionist campaign. The Signific ideals had temporarily and partly lifted him from his solipsistic isolationism and his pessimistic despair of human communication. Total rejection of language has now made room for a new trust at least in his own ability to create a new language and be understood by 'the masses'. Dialogue between the Signific Founders remained restricted to mutual encouragement of a naive self-confidence and a conviction of their superiority, each in his own field. Brouwer in particular accepted the ensuing responsibility to create for his own domain a new language, a truly mathematical language freed from the straitjacket in which traditional logic had placed it, and faithfully reflecting the mental process of mathematical construction.

The 'First Act of Intuitionism' had been the exposure of existing language as the villain, totally inadequate as a carrier of mathematical thought and guilty of distorting the mathematical reality. The Signific task of creating a new mathematical language, however, could not be separated from the task of re-constructing mathematics. As 'accompaniment of mathematical activity' a new language could only be created together with mathematics itself. This, Brouwer recognized, was the task that faced the world of mathematics. New hope, self-confidence and ambition made him undertake this task single-handedly. In later historical surveys he describes his attempt dramatically as 'an intervention in the course of history... the second act of Intuitionism'.

## 2.8 The Second Act of Intuitionism

> Und Brouwer, das ist die Revolution! Brouwer ist es, dem wir die neue Lösung des Kontinuumproblems verdanken, dessen provisorische Lösung durch Galilei und die Begründer der Differenzial- und Integralrechnung der geschichtliche Prozess von innen heraus wieder zerstört hatte.' WEYL [1921].

### 2.8.1 The Foundations of Set Theory Part I

In 1917 Brouwer presented to the Royal Academy Part I of his *Begründung der Mengenlehre unabhängig vom logischen Satz vom ausgeschlossenen Dritte* [1918B] (The Foundations of Set Theory independent of the logical principle of the excluded middle). It is the first part of his reconstruction of mathematics on Intuitionist lines. In spite of its title it is not an attack on logic or an exercise in re-writing Cantorian set theory without the use of the Principle of the Excluded Middle; it is a positive, non-polemical attempt to establish an alternative constructive basis for analysis. The 'Brouwer Revolution' is the radical proposal to replace the traditional notion of 'real number' and Cantorian set theory by new concepts, derived naturally from the continuum and the Subject's power of abstraction.

In his dissertation Brouwer had stressed the dimensional gulf between the discrete and the continuous, point and space. The starting point of his dimension theory had been his unwillingness to accept the cantorian set-element relation as the natural characterization of two subsequent dimensions. Underlying both was his deep conviction that discrete points cannot be bound into a continuum by logical comprehension nor the continuum be broken up into atomic parts in the set-theoretical sense.

The dissertation describes the Continuum as 'intuitively given', its most characteristic feature being its infinite divisibility, 'the intuition of the between that is never exhausted by division'. In Brouwer's new analysis

Brouwer at work ± 1924

the continuum remains the intuitive, primitive concept and its characteristic divisibility in intervals becomes the basis of a new, analytical set construction, generating 'real elements of the continuum' as convergent sequences of nested intervals. The primitive element of construction is the interval, of the same nature and dimension as the continuum, relating to the continuum as part or as subset to set.

Already in his *Review of Schoenflies* [1914] Brouwer had proposed a new set-construction, the 'well-constructed set', the 'free act' of the Subject generating an infinite sequence of nested intervals whose measures converge positively to zero. Analysing the notion of a sequence of nested intervals Brouwer recognized that although at any stage of the constructive process such a sequence is unfinished and defines an interval, a sequence defined by an effective procedure can be thought-of as 'completed' and be identified with its limiting point in the traditional sense. No such 'limitation' adheres to the sequence generated by 'free acts'. Being wholly dependent on human choice for each of its terms the *Free-Choice infinite sequence* can never be thought of as 'complete'. The essential incompleteness of the infinite sequence generated by free choice represents the characteristic and 'chaotic' infinite divisibility of the continuum and its elements and safeguards their measurability. Incompleteness, however, seemed incompatible with the requirement of mathematical existence as 'having been constructed', and as late as 1912 Brouwer dismissed the free-choice sequence as 'non-constructive'.

[1914] introduces 'free-choice' as a constructive generator, but does not resolve the apparent conflict between its 'incompleteness' and mathematical existence. A solution is first hinted-at in a marginal note to his 1916 course on set theory ([BMS15A]) and is based on a distinction between the infinite sequence and its finite 'initial segments'. By a remarkable stroke of genius, his 'second insight', Brouwer recognized that while the essentialy unfinished and undetermined choice-sequence retains all the characteristics essential of an element of the continuum, the 'completeness' of the construction of the finite initial segment gives it the legitimacy of a mathematical entity for the purpose of constructive analysis and function theory:

> With such a sequence one can work very well as long as, at every stage [of the function assignment ED.], one can work with a suitable intitial segment. ([BMS15A], p.1).

The 'Second Act of Intuitionism' is later described as an act of 'admitting' new concepts as legitimate mathematical entities, in particular the widening of the concept 'infinite sequence' to include those generated by free choice. The new *Brouwer Set* or *Spread* (cf. 6.3.12) is identified with the procedure generating convergent sequences, the generator being the free Subject and the self-imposed restrictions of this freedom defining the individuality of the Set. It is inspired by Brouwer's new concept of real number, and is a generalization of 'Point-Set', the procedure for generating 'real points of the continuum'. Much of the alleged incomprehensibility of Brouwer's treatment and definition of Set is due to his attempt to grasp in one definition – and in one sentence – every aspect of his complex notion of set and its elements, and this without reference to its main purpose. Without guiding the reader along the route of his own thinking and relating the new concept to the continuum Brouwer in [1918B] immediately launches his general Set concept in what he later admits is 'unfortunately but unavoidably a rather long-winded definition' ([1925A], p.245).

Starting from a denumerable sequence of labels or 'indices' he defines Set as a 'law', consisting of 1) a procedure for generating essentially unfinished, indefinitely proceeding sequences of indices by free choice within certain restrictions, and 2) a law assigning to each index 'a mathematical entity previously acquired', such as a constructed interval. (see further 6.3.12).

The 'Brouwer Set' provides a constructive interpretation of the concepts convergent infinite sequence and real number; it is a constructed procedure by means of which the Subject generates elements of the set, unlike the Cantorian predicative definition of set, which presumes the existence of a Platonic universe partitioned by a property into a set and its complement.

Reflective analysis of completed constructions, however, and the needs of Analysis and Function Theory made Brouwer realize that some form of 'comprehension' by property is needed, indeed is an indispensable element of such fundamental relations as 'equivalence'. Again his constructive alternative, the 'Species', is first mentioned in [BMS15A], and referred-to as 'soort' (Engl. 'sort' or 'species'). In [1918B] the concept 'species' is introduced in one line and identified with 'property of mathematical entities', in later definitions invariably specified as 'mathematical entities previously acquired'. Unlike the Brouwer-Set, the Species does not generate its elements, its potential elements are previously constructed mathematical entities, which only become elements of the species when their element-hood is proved. (see further 6.3.5).

During the following years Brouwer becomes increasingly more aware of the fundamental role played by 'species' in the construction of mathematics, 'at every stage of its construction' and generating a hierarchy of species (6.3.7). In [BMS32] and his post-war papers the emergence of the concept spread as well species is described as due to an intuition, an 'insight' into the nature of mathematical construction:

> ...The Second Act of Intuitionism which recognizes in the self-unfolding of the Primordial Intuition not only the construction of natural numbers but also the spread construction and the foundation of Analysis on properties of spreads... ([BMS32], p.7; app.12).

In a wave of Signific enthusiasm Brouwer did not hesitate to introduce a host of new notions and to coin new words, some as vital ingredients of his new set theory, others as a result of further analysis of the consequences of his refusal to accept the Axiom of Choice and the Principle of the

Excluded Middle. His distinction between the various degrees of equipotency ([1918B] p.10-12), to which he attached great importance at the time, is just one example of his critical analysis of concepts which in traditional set language were covered by just one word. In some cases the implications of this rejection were not immediately seen; in particular its implications for negation and the logical calculus of negation were not fully recognized until 1923.

### 2.8.2 The Foundations of Set Theory Part II

With *The Foundations of Set Theory, Part II, Theory of Point Sets* [1919A], submitted to the Royal Academy in October 1918, a beginning was made with the construction of an Intuitionist point-set topology and analysis on the basis of the Brouwer Set and Species. Accepting the possibility of similar treatment of point-set of other dimensions, Brouwer singles out plane point sets for treatment throughout. A theory of point sets is developed within the restrictions imposed by avoidance of the use of the Principle of the Excluded Middle and the Axiom of Comprehension.

The starting point of Brouwer's Set Theory had been his constructivist rejection of Cantorian set theory and Zermelo's axiomatization as well as his refusal to accept any contribution to mathematics by the use of logical principles. It was not surprising that in the new Set Theory there was no room for the spectacular new notions of 'Cantor's Paradise'. More alarming was the realization that mathematics would have to abandon some classical and some newly established results which had become fundamental tools of Analysis; some results had to be given up completely, such as the Well-ordering Theorem and the Bolzano-Weierstrass Theorem while others, such as the Heine-Borel Covering Theorem, would need considerable adjustment.

### 2.8.3 The New Foundation Crisis

The full implications of Brouwer's set Theory were not immediately recognized by the mathematical world of 1917/18. International communication had been disrupted during the war years and remained confused for some time. The first to study Brouwer's paper and realize its full import was Hermann Weyl. Weyl's admiration for Brouwer's topological work, mutual appreciation and a shared interest in the fundamental problems of mathematics had led to a real friendship. In a letter to Hilbert in 1911 Brouwer mentions with delight his meetings with Weyl during his visit to

Göttingen. In 1913, the year they were both appointed professor, they became deeply involved in the theory of the Riemann Plane. Much of their time during the war years was spent on the problem of the continuum and the creation of an adequate mathematical language. Weyl describes how 'during a short vacation spent together he full under the spell of Brouwer's personality and became an apostle of Intuitionism'. He dramatically renounced his just published work, *Das Kontinuum*, and hailed Brouwer as 'die Revolution'. In a series of lectures at Zürich in 1920 *On the new foundations of mathematics* Weyl expounded Brouwer's views on the continuum and set theory. It is significant that, while preparing these lectures for publication, Weyl changed the title of the manuscript to *On the new crisis in the foundations of mathematics*.

It almost seemed that in Weyl Brouwer had found the partner he so desperately needed in his programme of constructing an Intuitionist Mathematics. Weyl shared Brouwer's interest and enthusiasm for mathematical reform, he accepted the main tenets of his philosophy and had the ability to make a positive contribution to Brouwer's programme. Temperamentally he was perhaps the only mathematician capable of working with Brouwer on equal terms. Brouwer was keen to secure Weyl for his departement, as part of his grand design to create an international centre of mathematics. Under his pressure the University of Amsterdam and the City Fathers had 'promised' an expansion of his department, and in 1921 agreed to offer Weyl a professorship. Weyl seriously considered accepting the offer – as well as a similar 'Call' from Göttingen – and discussed with Brouwer the practical implications of a permanent move to Holland such as housing, the education of his children and the language problem in his work at Amsterdam university. However, the Brouwer-Weyl partnership which might have changed the historical course of Intuitionism, did not materialize. Weyl declined the call from both Amsterdam and Göttingen and did not make further contributions to Intuitionist Mathematics. He would remain faithful to the Brouwer ideal and conception of mathematics, but became more and more disillusioned by the difficulties of putting intuitionist philosophy into practice. Many years later he wrote:

> Mathematics with Brouwer gains its highest intuitive clarity. He succeeds in developing the beginnings of analysis in a natural manner, all the time preserving the contact with intuition much more closely than had been done before. It cannot be denied, however, that in advancing to higher and more general theories the inapplicability of the simple laws of classical logic

eventually results in an almost unbearable awkwardness. And the mathematician watches with pain the larger part of his towering edifice which he believed to be built of concrete blocks dissolve into mist before his eyes. (WEYL[1949], p.54).

## 2.9 The Turbulent Years

The decade following the publication of *The Foundations of Set Theory* is the most productive stage of Brouwer's Intuitionist campaign and also the most critical time in his career. His many publications of these years almost exclusively concern 'foundational' issues and the Intuitionist reconstruction of mathematics. They became the subject of international debate and controversy. It is a time when Brouwer took the centre stage as one of the chief contenders in the great Foundational Debate; a time of heated controversy which often turned into personal feuds and led to alienation from friends and fellow mathematicians.

A concerted movement of mathematical reform on Intuitionist lines failed to gather momentum. Brouwer's solitary efforts led to some remarkable results, but much of his energy and time was wasted in a search for power and leadership in all kinds of marginal campaigns. The relative failure of his Intuitionist campaign and the frustration of his other ambitions would leave him bitter and would drive him into isolation.

### 2.9.1 Signific and Political Campaigns

The first in a sequence of shattered dreams was the Signific Movement. The response to the *Signific Manifesto*, Brouwer's invitation to the world-famous to join him in an International Academy for linguistic reform was lukewarm if not wholly negative. Buber mockingly rejected the idea of 'creating words of spiritual value for the languages of the western world' and wrote:

> However high I value the possibility of a community of like-minded men and their creative power, this to me seems in principle to go beyond the function of such a community. The creation of words to me is one of the most mysterious processes of spiritual life ... a mystery enacted in the enlightened universal soul of man... (Buber to the Secretary of The International Institute of Philosophy at Amsterdam, 18.03.17).

In his reply Brouwer blames the absence of words of exclusively spiritual value in the languages of the western world on over-population ('people living too closely together') and he demands 'immediate control of human propagation by law':

> Only then will the introduction of words of spiritual value be possible when the 'mystery of becoming' of these words has been enacted not in the individual but in mutual understanding of a community of men of clear perception and sharp intelligence, men who do not live physically too close together. (Brouwer to Buber, 04.02.18, published in *The Communication of the International Institute for Philosophy at Amsterdam* I, March 1918).

Similar doubts were expressed by Gutkind and others. Peano was a little more encouraging; he mentioned 'difficulties created by the multitude of languages' but offered his conditional cooperation:

> I have already sent you some of my publications on an international language. If you want further explanation and consider adopting an international language, I shall be at your disposal. (Peano to Brouwer, 25.10.19).

The International Institute of Philosophy and its own journal in four languages was set up with all legal formalities. The journal, however did not survive its second issue. The 'International Academy' remained a talking club of the few founder members. It was re-named the 'Signific Circle' in 1922. Brouwer remained an interested member, but its course was gradually determined by Mannoury.*

'Signific' ambitions to play a leading role in social and political reform, especially in the turbulent years following the end of World War I, met with similar fate. The problem of world peace was the subject of serious debate among the founder members especially Brouwer and van Eeden, who were convinced that the only lasting solution could be found by great minds such as themselves. In anticipation of peace negotiations in The Hague van Eeden and Brouwer went to see the American Ambassador and suggested a 'conference of all great scholars'. In his diary on 1st August 1918 van Eeden describes two of Brouwer's proposed measures: '1. the elimination of the anarchy of marriage and human reproduction through control and approval by the community; 2. the institution of

---

\* The Signific Circle consisting of Brouwer, van Eeden, van Ginneken and Mannoury remained active until 1925. Mannoury revived the Signific Movement after World War II and founded in 1948 the 'International Society for Significs' and the journal 'Synthese'.

temporary money. Banknotes should have a fixed value for a definite period of time, and be cancelled at the end of the period'.

There is gradual disillusionment among the Signific members with the peace efforts of the Allies, especially with French and Belgian intransigence, which imposed humiliating conditions on the defeated Centre countries and even denied and renounced any German contribution to Western civilization and scholarship. Brouwer began to look more and more towards Germany as a model of academic and mathematical tradition, particularly in times of frustration and disillusionment at home. He had felt badly let down when he was not appointed a member of the Royal Commission for Defence. His publications on aerial survey, *Aviation and Photographic Survey* ([1919R] and [1920F], he claimed, were important contributions to the country's defence. Mrs Brouwer, 'without her husband's knowledge', wrote to Lorentz, the chairman of the Royal Commission, to tell him how bitterly upset her husband was, 'so upset that he is incapable of any work'. Brouwer turned down Lorentz's offer of membership of a subcommittee and, forgetting his moral disapproval of applied mathematics, claimed that 'the setting-up of a national laboratory for photogrammetry has been my life's ambition' (Brouwer to Lorentz, 16.02.18); and he did not hide his suspicion of a conspiracy of Zeeman, Lorentz and his old master Korteweg. Private negotiations during the next two years with Fokker and the Ministry of War to establish such a laboratory did not lead anywhere and Brouwer did not pursue his interest in 'Photogrammetry' any further.

But on other fronts battles continued: in his Faculty, the Royal Academy and the national press Brouwer waged war against the inclusion of Philosophy in science-degree courses, campaigned for the introduction of associate-professorships and, against the agreed policy of the Mathematics and Science Section, for an Academy-Fellowship of his own preferred candidate Denjoy etc.. Brouwer's friends despaired at the sight of so much waste of time and energy. His colleague, friend and admirer Hendrik de Vries writes in a touching letter:

> Consider your obligations towards Mathesis. You are one of those who should devote all their energies to Mathematics and who do not have the right to waste their time on affairs of no scientific importance. What Gausz, Riemann, Poincaré and Steiner have achieved for mathematics, that only matters, not whether and for what cause they have been quarrelling. (de Vries to Brouwer, 03.03.20).

## 2.9.2 Göttingen and Berlin Calls

Most unsettling to Brouwer, however, were the offers of University Chairs in Germany. In December 1919 he received almost simultaneously 'Calls' from Göttingen and from Berlin. Brouwer later claimed that Hilbert offered him the Klein Chair, which had fallen vacant when Erich Hecke left Göttingen in the autumn of 1919. In a letter to Schoenflies he writes:

> I have received at the same time a Call from Göttingen and from Berlin, or to be more precise, I have been proposed by both faculties as the number-one candidate on the short-list. I am at the moment in Berlin to discuss the financial implications with the Minister. It is a difficult decision to make. I'd love to work in Germany; the universities of Göttingen and Berlin are infinitely better than Amsterdam; I can expect greater incentives and a better atmosphere. But in the current situation, I am afraid, it would be a considerable financial loss. It seems that a university professor in Germany without private capital can hardly support a family. I therefore may for the time being have to abandon the idea. (Brouwer to Schoenflies, 29.12.19).

Brouwer never considered Hilbert's offer seriously. Even if Hilbert was approaching 60, his presence dominated the Göttingen mathematics department, which was united in admiration for and unswerving loyalty to Hilbert.

Until recently Brouwer's diverging views on the foundations of mathematics, although known to Hilbert, had not presented a personal, frontal attack on Hilbert's formalism; Brouwer had so far always spoken of the the Intuitionists in the third person. But when at last in 1918-19 he presented his long-heralded re-construction of Set Theory – even if the words 'Intuitionist' or 'Intuitionism' are nowhere mentioned in the text – Hilbert could not remain blind to its implied threat. Brouwer at this stage still wanted to avoid an open confrontation and was anxious to declare his loyalty to Hilbert. The last item in the Brouwer-Hilbert correspondence is a postcard dated 25th September 1919; it bears the message:

> From two admirers of the scholar and friends of the man - L.E.J. Brouwer, C. Caratheodory.

Hilbert became alarmed and irritated when Hermann Weyl, one of his most promising students, fell under Brouwer's spell and openly declared Brouwer as 'the greatest revolution since Galileo'. Brouwer's and Weyl's appointment to a chair in Göttingen might secure their loyalty to Hilbert and keep 'the crisis in Foundations' in the family.

Brouwer's refusal of the Göttingen chair and his serious consideration of the Berlin offer were felt by Hilbert as a personal snub. At the height of the row about the Mathematische Annalen in 1928 Brouwer would remind his co-editors that one of the real reasons of Hilbert's displeasure was his 'turning-down the Call' in 1919. Hilbert knew Brouwer well, he had recognized and admired his mathematical genius and was well aware of his pride and his passionate involvement in whatever he undertook. Whatever Hilbert had known of Brouwer's earlier interest in Intuitionism had been dismissed as youthful flirtation with philosophy, his recent publications on Set Theory presented a serious challenge. The thought of Brouwer in Berlin leading the opposition against the formalist and cantorist establishment could not but provoke fears of a revival of old vendettas. Brouwer's challenge, described by Weyl as 'a revolution', looked uncannily like Kronecker's 'Putsch', a threat to the hard-won supremacy of Göttingen. Above all, Brouwer's programme threatened to invalidate much of the mathematics he, Hilbert, had created and practised.

As to Brouwer, Berlin appealed to him. His programme and especially his newly undertaken task of re-constructing the foundations of Analysis could hardly be described as a continuation of Kronecker's programme but he felt at home in Berlin University where some of the strict constructivist tradition and a scepticism towards Cantorism had remained alive, and there was a spirit of rivalry with Göttingen in what was once the centre of the mathematical world. Carathéodory's recent appointment had re-inforced the band of admiring colleagues. Berlin seemed to offer the long-dreamed-of opportunity for academic research and form a natural home-base to take on the mantle of world leadership from an ageing Hilbert.

The choice between Göttingen and Berlin presented Brouwer with the opportunity to set himself free from an allegiance to Hilbert. His open preference for Berlin caused a breach between him and Hilbert which was almost necessary in the next stage of his Intuitionist programme. Ironically, in his last letter to Hilbert Brouwer had compared the professional detachment of scholars to the quarrelsomeness of politicians:

> In the end we scholars are in this respect most fortunate since most of our thinking is so completely free of the mischievous strife of politics. (Brouwer to Hilbert, 28.06.19, DHI25).

Neither Brouwer nor Hilbert were temperamentally capable of keeping mathematical controversy at the level of a detached professional debate. Brouwer in particular needed the stimulus of a personal challenge to stir him into action; he was a fighter, who needed a personal enemy on whom to concentrate his attack. Even if Hilbert's and Brouwer's views on mathematics could hardly be described as antipodal in every respect, the foundational debate now became polarized into a battle between Intuitionism and Formalism with international leadership as the prize.

From now on Brouwer completely identified with Intuitionism. When describing the controversy between Intuitionism and Formalism he no longer plays the role of an observer and interested outsider: Intuitionism now is identical with his philosophy of mathematics and the programme he had set himself. Others, like Poincaré, Borel and Kronecker become 'fore-runners of Intuitionism' and 'Pre-intuitionists'. He took exception when Fraenkel in his manuscript of *Foundations of Set Theory* bracketed them together with Brouwer as 'Intuitionists':

> Please, don't continue the expropriation of which German literature has been guilty by letting me share with Poincaré, Kronecker and Weyl what is my exclusive spirtual property. I am partly myself to blame, since in past accounts of Intuitionism I wrongly included my predecessors, with whom I a share the battle against Formalism. (Brouwer to Fraenkel, 28.01.27, AFR8).

### 2.9.3 The Intuitionist 'Putsch'

*Intuitionistische verzamelingsleer* [1919D] (Intuitionist Set Theory), published simultaneously in Dutch and German, is the first publication in which Brouwer entitles his own work as 'Intuitionist'.

He describes it as 'a report introducing my recent publication *Foundations of Set Theory*'. Nothing new is added. It is a publicity pamphlet, a public notice drawing attention to his earlier papers, which because of their sober technical appearance in a restricted publication had gone unnoticed; an open declaration of his true intuitionist colours, quoting at large from his publications as far back as 1907 in evidence of the independence and originality of his views. The emphasis is on the principal differences between Classical and Intuitionist Set Theory and on the radical consequences of the changes he had introduced. As his two guiding principles Brouwer quotes his rejection of the Axiom of Comprehension and his rejection of the Principle of the Excluded Middle, now identified with Hilbert's 'axiom of the solvability of all mathematical Problems'.

Brouwer's open challenge of Formalism irritated and increasingly alarmed Hilbert. He had recognized the validity of some of Brouwer's criticism, in particular of the need to distinguish between 'mathematics and mathematical language of the first order' on the one hand and 'the mathematical system and language of the second order'. Hilbert, however, was a pragmatist; unlike Brouwer, a mathematician and philosopher who was prepared to accept the consequences of his philosophy and who with the missionary zeal of a prophet of doom preached the hard and unpleasant truth, Hilbert was a practical mathematician concerned with 'more results'. Irritated he left his research in theoretical physics 'to solve the problem of foundations once and for all'. In lectures in Hamburg and Kopenhagen in 1922 he launched his attack and started the defense of the Principle of the Excluded Middle:

> What Weyl and Brouwer are doing is precisely what Kronecker did. They seek to save mathematics by throwing overboard all that is troublesome... If we would follow the kind of reform they suggest we would risk losing a great part of our most valued treasures. (HILBERT[1922], p.157).

During the next few years Hilbert would concentrate on the distinction to which Brouwer had drawn attention between 'formal theories' and 'contentual mathematics'. But whereas Brouwer maintained that 'real mathematics' is 'intuitive' mathematics, i.e. thought-constructions on the basis of the Primordial Intuition − 'mathematics of the first order' in which there is no room for the use of extended logical principles, and dismissed formal theories as meaningless linguistic structures − Hilbert would claim the title 'mathematics' for these formal theories while accepting the need for 'contentual' mathematics as 'meta-mathematics', a system necessary to prove the consistency of his formal theories. He would accept the limitations of contentual mathematics and even restrict it to finitary methods.

The Brouwer-Hilbert controversy, however, was by no means just a matter of words; certainly for Brouwer it was a fundamental, philosophical issue that was at stake. His Intuitionist programme during the next few years was to re-construct mathematics and in particular Analysis in accordance with his vision of the nature of mathematics, and to expose the weaknesses and falacies resulting from the alternative 'formalist' philosophy.

*The Foundations of Set Theory* represented the first stage of this re-construction, a constructive basis for the continuum. The next stage was the re-construction of measure- and function theory.

Brouwer was well aware of the disastrous consequences of his intuitionist restrictions for classical mathematics, which he describes as: 'at first destructive and emasculating or sterilizing', but he felt confident that 'the Second Act of Intuitionism will produce ample opportunity to make up for these losses and lead to new developments and new discoveries'.

Contributions to the new, Intuitionist Mathematics followed in a steady flow of publications during the remainder of his creative life: re-definitions of fundamental notions of Analysis and Algebra ([1921B], [1924B], [1924K], [1926B]), re-constructions of proofs of classical results ([1924C], [1924E], [1925A], [1925B], [1926A], [1926C], [1927A], [1927C], [1928B]) and the development of a new function theory, which indeed led to new results ([1923A], [1923B], [1924A], [1924D], [1924G], [1927B]).

However, the handicaps in Brouwer's undertaking were formidable, and personal ambition made him incapable of accepting and welcoming contributions from others. His almost pathological suspicion and jealousy turned friends into enemies and even his own students into rivals. The list of former friends turned enemies, Korteweg, Kohnstamm, Lebesgue, Lorentz, Hilbert, Wijdenes would sadly grow longer and include Denjoy, Blumenthal, Alexandroff and Fraenkel as well as his own students Menger, de Loor, Schogt, Haalmeijer, van der Waerden, Heyting and Freudenthal. And again, his aspirations for world-leadership would consume a great deal of time and energy during these, Brouwer's best years.

After an extended stay in Berlin as the guest of his friend Gutkind, Brouwer returned to Amsterdam in January 1920. He had declined the Berlin 'Call', but remained determined to lead the international mathematical community ultimately from Berlin where he had bought himself a house and would remain a regular visitor.

Amsterdam was a convenient launching pad for his campaign: his authority in his department was almost absolute and neutral Holland seemed a natural base for his peace-making efforts in the post-war bitterness between French and German mathematicians. 'Loyalty to Amsterdam' was again used with the University governors to strengthen his department and establish a Mathematics Institute. In an effort to gather an international mathematical community around himself Brouwer pressed for the appointment of foreign staff in his department and provided personal hospitality to many distinguished mathematicians such as Carathéodory, Weyl, Urysohn, Schoenflies and Blumenthal. His cosmopolitan outlook was not shared by the University authorities, who could not understand 'that there was not sufficient local talent available', but who had to agree first to the appointment of Weyl and later of the Austrian Roland Weitzenböck.

He set himself up as the champion of the depressed German interests in scientific international cooperation, campaigning against the 'Conseil International de Reserches' and the 'Union Mathématique Internationale'. Both organizations had been founded in 1919 as branches of the League of Nations, and included in their statute a humiliating boycott of German scholarship. Brouwer made gallant attempts to end the boycott, by campaigning with his colleagues Mannoury, de Vries and Weitzenböck for disaffiliation of the Dutch Royal Academy from the 'Conseil' and for severance of links between the Wiskundig Genootschap and the Union Mathématique. His efforts on behalf of Germany earned him many friends, a.o. Sommerfeld, Planck, Bieberbach, Hölder, Dyck and Schmidt. Ironically, they were not appreciated by Hilbert, who interpreted them as 'interference in German affairs', to spite him and keep him, Hilbert, out of the international mathematical community.

## 2.10 The Brouwer logic

Growing opposition to Brouwer's Intuitionist claims forced his attention back to the Principle of the Excluded Middle (PEM). His re-construction of Set Theory had avoided the use of the PEM, his arguments against its validity had not progressed since [1908C]. He then had simply dismissed the PEM as based on the assumption that every mathematical hypothesis can either be proved to be true or be proved to be false, concluding that 'therefore the Principle of the Excluded Middle is unreliable when applied in an infinite system', but conceding that 'its unjustified use will never lead to a contradiction and so show-up the fundamental flaws in the argument.' ([1908C], p.6).

*2.10.1 The Principle of the Excluded Middle – 1923*

In 1923 there was a marked change in Brouwer's attitude towards the PEM. There is a new conviction that contradictions can be shown to arise from the unjustified use of the PEM in mathematics. Further reflection made him recognize the full implications for logic of his strong definition of negation and of his rejection of the PEM.

In August 1923 Brouwer delivered a paper to the Antwerp Congress for Natural and Medical Science on *The Role of the Principle of the Excluded Middle in Mathematics, especially in the Theory of Functions* [1923B]. After almost literally repeating the arguments of [1908C] against the reliability of the PEM, Brouwer proceeds to give counterexamples, disproving the validity of two fundamental theorems in the theory of functions of the Paris School, both based on the PEM, viz. that the points of the continuum form an ordered point species and that every mathematical species is either finite or infinite. Counterexamples are further given, refuting classical results or proofs: the existence of a maximum of every continuous function in a closed interval, The Heine-Borel Covering Theorem and the Bolzano-Weierstrass Theorem.

In passing Brouwer also mentions, for the first time, the Principle of Reciprocity of Complementary Species:

> ...a special case of the Principle of the Excluded Middle... the principle which asserts that in every system the correctness of a property follows from the imposssibility of the impossibility of this property. ([1923B], p.1).

The concept 'species' had been introduced in [1918B] as 'a property of previously constructed mathematical entities' to be used as a selective procedure resulting in one collection, unlike the Cantor Set, which partitions the 'Universe' into two complementary sets. Difference relations were generated by 'mutually exclusive properties', or 'properties which are mutually contradictory' or 'properties which are mutually exclusive a-priori', each constructively defined. Double negation, 'impossibility of impossibility' or 'absurdity of absurdity' had so far not been seriously considered as a mathematical construction.

[1923B] dismisses 'absurdity-of-absurdity' and its equivalence to truth as a logical notion and 'a logical principle', applied to the 'language of mathematics'. The chief criticism is directed at the Formalists, who are concerned with the elimination of contradictions in the language of mathematics:

> ...the Formalist critique subjects the *language accompanying mathematics* to a mathematical examination as a result of which the laws of theoretical logic appear as operators acting on primitive formulae or axioms. Their aim is to reform the axioms in such a way that the linguistic effect of these operators, which remain essentially unchanged, can no longer be disturbed by the linguistic figure of contradiction. One need not at all despair of

achieving this goal one day, but nothing of value will be achieved; an incorrect theory remains incorrect even if it cannot be disproved by contradiction, in the same way that a criminal policy remains criminal even if it cannot be condemned and stopped by any legal process. ([1923B], pp.2-3).

## 2.10.2 The Calculus of Absurdities

*Intuitionistische splitsing van mathematische grondbegrippen* [1923C] (Intuitionist splitting of fundamental mathematical concepts), presented to the Dutch Royal Academy on 24 November 1923, departs dramatically from the wholly negative attitiude to logic which had characterized Brouwer's thinking so far. It re-states the Intuitionist rejection of the classical Principle of Reciprocity as a special case of the PEM. The rejection is 'made plausible' by counterexample but is no longer based on an outright rejection of logic and logical principles. Indeed, Brouwer now enters the logicians' domain himself: he considers the 'predicates' absurdity and absurdity-of-absurdity and proposes an intuitionist alternative to the classical logic of negation and the Principle of Reciprocity, his calculus of absurdities.

Some underlying logical axioms are stated and a theorem is proved within the alternative logic.

An operation of cancelling double negation is introduced, applied to sequences of absurdity predicates and referred-to as 'crossing-out':

> The Intuitionist view allows such cancellations provided the last absurdity predicate is not included in the cancellation. ([1923C], p.878).

The theorem 'Absurdity-of-the-absurdity-of-the-absurdity is equivalent with absurdity' is proved in some seven lines on the basis of a few axioms and modus ponens. The proof will be further discussed and is quoted in full in 5.10.1. It illustrates Brouwer's preference for 'words' and his reluctance to use logical symbols. Yet even in this long-worded form his general theory of absurdity or calculus of absurdities occupies only half a page, most of the eight-page paper concerns the generation of relations beween points and point-species by means of the absurdity operator, and the relation between these relations. The calculus of absurdities is clearly intended as an introduction, clarifying the principles by means of which relations can be generated, although the lines of demarcation between mathematical species construction and logical operation become somewhat blurred in the process. The distinction between a property and the absurdity-of-the-absurdity of that property leads to a range of new relations which will play a role in Brouwer's re-construction of Analysis, such as e.g. the relation between two points coined as 'deviation', which is distinct from both relations of 'coincidence' and 'apartness' (see further 6.3.8.4).

Application of the absurdity operator also led to a proliferation of relations and distinctions in [1923C] and other papers, many of which did not find a place in Brouwer's constructive analysis. They were intended as examples, showing how 'absurdity' can be used as an operator; and as Brouwer re-assured Heyting – then one of his Ph.D. students – the given list is not complete:

> My paper 'Intuitionist Splitting' is by no means intended to give a complete system of all possible relations of 'envelopement' and 'fusion' (it is questionable whether such a system would not have to be infinite), but only to show relations that can be generated from some definite fundamental concepts by means of the operation of stating its absurdity...
> The number of ways of extending the given tables are legion. (Brouwer to Heyting, 26.06.24).

The significance of *Intuitionist Splitting* is that it states the immediate consequence of the Brouwer negation and his rejection of the PEM, and that it sets down the principles of a new, Intuitionist Logic. Like all strokes of genius, Brouwer's discovery of an alternative logic may in retrospect seem simple and obvious. A combination of consistent thinking and preparedness to question age-held beliefs enabled him to take the simple step which would shake the world of logic.

### 2.10.3 La Logique Brouwerienne

*Intuitionist Splitting* [1923C] and its German version, published in 1925, sparked off an international debate on the 'Brouwer Logic', especially in Belgium where Brouwer had first stated his rejection of the Principle of Reciprocity of Complementary Species. Its full import was perhaps best grasped by Kolmogorov, who in 1925 published a complete formalization of the Intuitionist Propositional Calculus and compared the status of mathematical conclusions in a transfinite domain resulting from the use of Intuitionist logic with those resulting from the use of classical logic ('the mathematics of pseudo-truths').Kolmogorov's paper (KOLMOGOROV[1925]), published in Russian, however, seems to have gone unnoticed for some time. Robin Wavre was the first in the West to proclaim Brouwer's Calculus of Absurdities as the basis for a 'logique Brouwerienne'. His paper *Sur le principe du tiers exclu*, WAVRE[1926], started a debate on the Brouwer Logic, mainly conducted in the *Bulletin de l'Académie Royale de Belgique*. (For details of contributions see the Bibliography under AVSITIDISKY[1927], BARZIN AND ERRERA[1927], BOREL[1927], GLIVENKO[1928] and [1929], KHINTCHINE[1928], LEVY[1926A], [1926B] and [1927], WAVRE[1924], [1926A] and [1926B].) Some contributions showed complete lack of understanding of Brouwer's views on negation and the PEM; Barzin and Errera in *Sur la Logique de M.Brouwer* introduced a third truth-value and, using classical negation side by side with the Brouwer negation, arrived at a 'Principle of the Excluded Fourth'.

The debate ended with the publication of two papers in 1928 and 1929 by another Russian, V. Glivenko. In these papers Glivenko introduced his formalization of the Brouwer Logic and proved that the introduction of as third truth value is as illegitimate in the Brouwer Logic as it is in classical logic; moreover, that 'in the Brouwer logic 'the proposition $p \vee \sim p$ is absurd' is absurd in the Brouwer sense, or $\vdash \sim(\sim(\sim p \vee p))$, which amounts to a proof of the universal non-contradictority of the PEM.

Brouwer did not take part in the debate himself in spite of repeated requests by M. de Donder, the Editor of the *Bulletin*.* A man so much-concerned about public recognition must certainly have been flattered, being the centre of international controversy and being heralded by some as the innovator of logic of the century; but he remained true to his Intuitionist principle that logic was no more than a post-factum analysis of the language of mathematics with no relevance to mathematics itself. His intervention in the domain of logic never became one of the 'Acts of Intuitionism' and is never mentioned in his later Historical Surveys.**

As to the formalization of Intuitionist Logic, Brouwer maintained an

---

\* Having promised on various occasions to round-off 'the debate', Brouwer finally referred de Donder to HEYTING[1930C] as the authoritative account of Intuitionist Logic: 'which clarifies in a masterly way all the points I had wanted to clear up... nothing more remains to be said on the questions under discussion.' (Brouwer to de Donder, 09.10.30)

\*\* The nearest to a mention is [1954A], p.3, where Intuitionist Mathematics is claimed to have 'its general introspective theory of mathematical assertions, a theory which with some right may be called *Intuitionist Logic* and to which belongs a theory of the *Principle of the Excluded Middle*'.

attitude of benign indifference. Heyting's prize-winning entry in a competition organized by Mannoury in 1928, *Stones for Bread*, had Brouwer's full approval since it expressly – and in the title – endorsed Brouwer's reservations as to the function of logic and formalization. He encouraged Heyting to publish it and arranged for its publication (*Die formalen Regeln der intuitionistischen Logik*, HEYTING[1930A]).

In his appraisal of its value for mathematics, however, Brouwer remained consistently negative and disparaging. His verdict, repeatedly expressed to Heyting was: 'an interesting but irrelevant and sterile exercise' (see further section 5.12.3 Heyting's Formalization of Intuitionist Logic).

Heyting's publication, more than those of Kolmogorov and Glivenko, caused a stir in the world of logic and led to many further investigations. It made Intuitionism accessible to a wider mathematical audience and drew attention to aspects of Intuitionism which Brouwer tended to ignore or undervalue. He changed the direction of Intuitionism at a critical time of its history and kept it alive when Brouwer retired into silence in the thirties.

## 2.11 The Reconstruction of Analysis

Brouwer's embargo on the PEM undoubtedly presented the most dramatic threat to established mathematical practice, a handicap likened by Hilbert to 'denying the boxer the use of his fists'. It also contributed to the alleged 'unbearable awkwardness' in the construction of an alternative Intuitionist Analysis. However, Brouwer's ultimate failure to develop a universally acceptable and relatively simple alternative, is probably due more to complications and problems arising from his new fundamental concepts, in particular from his new interpretation of the continuum. The 'natural' characterization of 'points of the continuum' as convergent sequences of nested intervals generated by choice seemed to offer a simple solution to the problem of 'power of the continuum' and the problem of measure. But the indeterminate nature of the points-of-the continuum created new problems as to the 'order' of points, the definition of functions and continuity.

An alternative system of 'ordering', sufficient for the purposes of Analysis, was developed in [1926A], [1927A], [1927C], [1930A] and [BMS32].*

---

* Also [1950A], [1950B], [1954A] and [BMS51] chapter 3.

The solution of the problem of definition of functions on an indeterminate, unfinished domain proved more elusive.

*2.11.1 The Continuity Arguments*

Brouwer launched his new theory of the continuum in 1918, confident that the solution to the apparent conflict between the effective determination required of functions and the indeterminacy of points of the continuum was within his grasp. At this stage it is an 'insight', a conviction based on his conception of constructive existence of infinite sequences and on the concept of function as an effective procedure assigning a definite value to each element of its domain: If a function is to assign a definite value to an element generated by a convergent infinite seque in its most general form, the assignment must be wholly determined by a completed, finite 'initial segment' of the sequence.

Combined with Brouwer's characterization of real points of the continuum it constitutes what we may call *the Fundamental Hypothesis of Brouwer's function theory*. It asserts that for all functions defined on the continuum the function value of each point is wholly determined by a definite segment of the sequences generating the point.

The Fundamental Hypothesis is first stated in [1918B], where it is used to prove the non-denumerability of the continuum. The immediate and most sensational implication, however, is the continuity and uniform continuity of all functions defined on any closed continuum interval. Intuitive arguments for continuity and uniform continuity of all real functions so defined are already mentioned in the Brouwer-Weyl correspondence of 1920, and Brouwer claims in [1927B] that 'this immediate consequence of my intuitionist view has been frequently mentioned in my lectures ever since 1918' (p.62). He was in no doubt about the importance of his con-

---

Brouwer's view on the impossibility of a complete and natural ordering of the intuitive continuum is perhaps best summed-up in [1930A]: 'not only is such a natural ordering out of the question, but one can also show that any other ordering on the intuitive continuum has no hope of success, a conclusion which squarely contradicts the until recently generally held belief that the continuum not only can be ordered in the most varied ways, but even that it can be well-ordered'. (p.8)

A 'pseudo-ordering' was suggested in HEYTING[1925] and a 'virtual ordering' introduced in BR[1926A]. (see also [BMS51], p.52).

Zermelo's well-ordering theorem had already been rejected in [1907]; the classical theory of well-ordering is replaced in [1918B] and [1927A] etc. by an Intuitionist alternative with which its has little in common save its name.

jectured Uniform Continuity Principle: it would provide the much needed simplification in his re-construction of function theory. But to be universally acceptable, it required a proof of the Fundamental Hypothesis on which it was based, a proof which derives the conjectured property from the definition of the Brouwer-Set or Spread which generates the elements of the continuum.

The proof of the Fundamental Hypothesis presented a challenge to Brouwer throughout the twenties. He made several attempts between 1923 and 1927, but a constructive, 'simple' proof that fully satisfied him, ultimately eluded him.

The Uniform Continuity of all 'full' functions, i.e. functions defined on a closed unit interval, is first proclaimed in *The Foundations of the Theory of Functions independent of the logical Principle of the Excluded Middle, Part I* [1923A]. Only a sketch of proof is given at this stage, but elements of later proofs are already mentioned, in particular the requirement of effective computability of functions values in a finite number of steps and the restriction of the function domain to points of intervals generated by 'Finite Spreads'.

*Proof that every full function is uniformly continuous* [1924D] shows clearly the dependence of the Uniform Continuity Theorem on the Fundamental Hypothesis and makes a first attempt at proving the hypothesis.

The proof of this theorem, now called *The Fundamental Theorem*, is simplified by being restricted to integral valued functions of 'Finite Spreads' (also referred to as 'the Fundamental Theorem of Finite Spreads' (e.g. [BMS32]), 'the Bunch Theorem' (e.g. [BMS51]) or 'the Fan Theorem' (e.g.[1952B]). It is based in turn on an auxiliary theorem, later referred-to as *the Bar Theorem*, which asserts that an integral-valued function 'splits the domain $M$ into a well-ordered species of subsets, $M_\alpha$, each determined by a finite segment of choices'. ([1924D], p.191 et ff.).

The proof of the Bar Theorem marks an important shift in Brouwer's view on what in 1907 he had branded as 'mathematics of higher order'. It is based on reflective analysis of the structure of possible proofs and on a new principle of regressive induction. (see further 6.3.13).

Again, these innovations were hailed subsequently as an important logical breakthrough, introducing in proof theory a new constructive tool of higher type. To Brouwer the Fundamental Theorem was the real breakthrough, leading directly to the basic theorems of an alternative Analysis, such as 'the Intuitionist Covering Theorem', 'the Impossibility of Splitting the Continuum' and 'the Uniform Continuity Theorem'. The Bar Theorem remained an auxiliary theorem and perhaps the weakest link in the chain. His feelings about the Fundamental Theorem and his unsuccessful search for a more direct, constructive proof are summed-up in a lecture in 1952:

> For Fans a truly wonderful theorem holds whose importance would justify to call it the Fundamental Theorem of Intuitionism, but whose absolutely rigorous proof till now has not been sufficiently simplified...([BMS66], p.6; app.10).

It remains an open question whether dissatisfaction with his proof of the Bar Theorem, his failure to find a 'simple', constructive alternative is the main cause of the demise of Brouwer's programme of re-constructing mathematics. It is a fact, however, that all his attempts at a systematic construction of an Intuitionist Analysis abruptly end at this point.

His mathematical publications of this period do not extend beyond Set Theory and the beginnings of Function Theory. *Towards a Foundation of Intuitionist Mathematics* ([1925A], [1926A] and [1927A]) is a revised version of [1918B] and [1919A], incorporating new developments in the theory of species, relations, ordering and well-ordering. The sequel to *The Foundations of Function Theory, Part I* ([1923B]) never materialized.

The abrupt ending of Brouwer's re-construction programme at this stage is even more evident in his unfinished, unpublished books.

## 2.12 The Unfinished Books

Among Brouwer's personal papers which I was fortunate to trace in 1977 there were manuscrips of three unfinished books, *Reelle Funktionen* [BMS37], *'Die Berliner Gastvorlesungen'* [BMS32] and *'The Cambridge Lectures'* [BMS51]. Each of these manuscripts was clearly intended as a comprehensive exposé of Intuitionist Mathematics in book-form, and each attempt was abandoned after the introduction of the Fundamental Theorem.

### 2.12.1 Reelle Funktionen

*Reele Funktionen* (Real Functions) is a 124-page typescript in German. It was started probably soon after the publication of [1923B]. From the completed parts it appears to have been planned as a comprehensive survey of Intuitionist Analysis.

Part I is complete. It is entitled 'Die Grundlagen aus der Theorie der Punktmengen' (Foundations from the Theory of Point-sets) and consists of 101 pages together with a detailed table of content and an alphabetical glossary. A note on the cover gives the date of completion of 'the final draft' as 24 February 1924. Part I is further divided into 4 chapters headed 'Point-species and point-spreads', 'Boundary Points, Cataloguizing', 'The Genetic Concept of Content' and 'The General Notion of Content'. (For 'the Table of Content' see Appendix 11, which also gives a translation of parts of the manuscript.)

Part II is entitled 'Hauptbegriffe über reelle Funktionen einer Veränderliche' (Fundamental notions in the theory of functions of one variable). Only the first Chapter 'Stetigkeit, Extreme' (Continuity, Extremes) has been completed; it covers the notions of function, continuity and uniform continuity and includes a proof of the Uniform Continuity Theorem along the lines of [1924D]. And here the manuscript ends.

It might be argued that Brouwer tired of the project or considered the further construction of theory a routine matter, less vital than the fundamental notions and principles. There is, however, sufficient evidence of the importance he attached to Intuitionist Function Theory and of the considerable divergence he expected from the classical treatment as a result of his Fundamental Theorem, in particular in the treatment of measurable functions and integral. This is clear e.g. from [BMS32]:

> ... a fourth application of the Fundamental Theorem of Finite Sets: if an unlimited sequence of full functions on the continuum converges to a limit function this convergence is uniform. This leads in the Intuitionist Theory of Real Functions to the collapse of the theory of Baire function classes which fuse into the species (not split into classes) of measurable functions... The Intuitionist Theory of measurable functions, which is founded on the theory of measurable point-species and forms the basis of a construction of a theory of integration in which the notion of Integral embraces the Riemann Integral as a special case, differs in a great many respects from the classical theory of measurable functions. (pp.64-65).

*2.12.2 Die Berliner Gastvorlesungen*

In the spring of 1927 Brouwer gave a series of lectures on Intuitionism at the University of Berlin. The enthusiasm which these lectures and Brouwer's presence generated was a natural response to positive leadership in an ailing constructive tradition. The age-old Berlin-Göttingen rivalry revived and Brouwer was hailed as the champion challenger of Hilbert. The 'Putschist' cause was hotly debated in senior common rooms and fashionable drawing rooms. There were feature articles in popular papers and even poems in university Journals.* Admirers and publishers pressed Brouwer for a book on Intuitionism, 'which the public will tear out of your hands' (Dr Lily Herzberg – publ. Walter de Gruyter – to Brouwer 08.11.27).

As late as 1919 Brouwer had tried to find support for a revised publication of his *Foundations of Mathematics* [1907]. He now realized that after the developments of the past ten years a re-print would be out-of-date. He accepted the invitation of Walter de Gruyter to publish his Berliner lectures in the form of a book on Intuitionism and Intuitionist Mathematics.

In the summer of 1928 the book was nearing completion; Brouwer

---

\* Mathematische Schnaderhüpfl

...

Und wird mir das ganze
Getu hier zu trist,
Dann kauf ich mir'ne Kanone
Und werde Putschist*)
(* In Berlin gebräuchlicher Spitzname der Intuitionisten.)

Ja klassich da schlieszen s'
Mit falschem Genie:
Sie mag net die andern,
Drum mag's also mi.

Wir Putschisten aber sagen,
Des stimmt net deswegen,
Denn sie braucht nämlich leider
Überhaupt kein net z'mögen!

Ach, zwischen 'Sie liebt mich'
Und 'Sie liebt mich nicht',
Da gibt's noch ein Drittes,
Die alte Geschicht.

Und grade das Dritte,
Des gibt mir an Risz:
Man kann's net entscheiden,
Man weisz 's halt net g'wisz!

*Häufungspunkte* Mathematischer Reimsalat für die studierende Jugend von 3 bis 17 Semestern, Mathematischer Verein an der Universität Berlin, 1927.

writes to Heyting:

> The Berlin Lectures will shortly appear in print. There may be some delay...
> (Brouwer to Heyting 17.07.28).

In this letter Brouwer also refers to 'improvements needed in my *Towards a Foundation of Intuitionist Mathematics*' and introduced in the Berlin Lectures, in particular in the treatment of species. (see further 6.3.7 The Hierarchy of Species).

The *Berlin Lectures* were never published. The manuscript, however, is preserved and is of considerable historical interest: it is Brouwer's last comprehensive statement on Intuitionism and Intuitionist Mathematics before his departure into 'silence' and it reflects the unfinished state of his re-construction program.

The manuscript of *The Berlin Lectures* consists of a 77-page German version ([BMS32B]), 75 pages typed and 2 pages written, and 24-page Dutch version ([BMS32A]), 12 pages typed and 12 pages written. The Dutch version appears to be the original 'working version'; it contains many alterations, some of them made after the German translation was finalized (see app.12 and illustration).

Like the orginal plan of the dissertation and the *Cambridge Lectures*, *The Berlin Lectures* is divided into 6 Chapters headed:

I    Historical background of Intuitionism;
II   The object of Intuitionist Mathematics: species, points and spaces, the continuum;
III  Ordering;
IV   Precision analysis of the continuum;
V    The Fundamental Theorem of Finite Sets;
VI   Intuitionist criticism of some elementary theorems.

*Chapter One* is a historical survey, Brouwer's standard interpretation as given in his post-war public lectures. It names the 'Two Acts of Intuitionism', claimed to have changed the course of history: 1. 'the uncompromising separation of mathematics from mathematical language and thereby from the linguistic phenomena as described by theoretical logic'; 2. 'the admission, as a form of self-unfolding of the Primordial Intuition of mathematics, not only of finite sequences of mathematical systems and of lawlike indefinitely proceeding sequences of mathematical systems already formed by induction, but also of sequences of mathematical systems proceeding indefinitely in complete freedom or in freedom subject to possibly changing restrictions ... and the admission at every stage of this construction of mathematics of properties which can be supposed for mathematical entities already constructed, as new mathematical thought-constructions named species'.

*Chapters Two, Three and Four* are a summary of Brouwer's earlier publications on set theory, the continuum and ordering.

*Chapter Five* is entirely devoted to the Fundamental Theorem and its 'applications'. The proof of the Fundamental theorem is a modified version of the proof given in [1927B].

The first and 'most important application' is the Intuitionist Covering Theorem: 'If a law $\omega$ assigns to every element $e$ of a located compact pointcore-species $Q$ a neighbourhood of $e$ then a finite number of these neighbourhoods can be indicated such that for any arbitrary pointcore of Q at least one of these is a neighbourhood.'

Unlike the Heine-Borel Theorem, the Brouwer Covering Theorem requires that the pointcore-species is 'catalogized' or 'located', i.e. that its distance from any element of the 'space frame' can be approximated with any degree of accuracy.

The second application is the theorem that 'the continuum cannot be split', i.e. whenever the continuum is divided into two subspecies, one of these subspecies will be identical with the continuum. This application then leads to two important corollaries: 1. the proposition 'any infinite binary fraction is either rational or irrational' is contradictory; 2. any ordering of the continuum is necessarily identical with the naive, i.e. intuitive 'virtual ordering'.

The Theorem of Uniform Continuity of all full functions on the closed unit interval is now relegated to third place in the applications of the Fundamental Theorem. The same argument is used as in [1927B]. There is, however, an interesting indication of Brouwer's earlier hunch of the truth of the theorem: it is 'based on the from the intuitionist viewpoint immediately evident truth of the somewhat weaker theorem of the *apparent continuity* of all full functions on the unit-continuum'.

The fourth and final application of the Fundamental theorem leads to the uniform convergence of an unlimited sequence of full functions on the unit-continuum converging to a limiting function.

The final paragraphs of Chapter Five are perhaps the clearest indication that at this point in the re-construction of function theory Brouwer had met an obstacle which seemed insurmountable. He briefly refers to the Intuitionist theory of measurable functions based on a theory of measurable point-species, 'on which a theory of integration can be constructed...which in many repects differs from the classical theory of measurable functions'. But then he suddenly declares: 'Since we cannot enter into further discussion of the Intuitionist theory of real functions we shall use our final chapter to show by means of some examples, how radical a re-construction Intuitionism must demand, even of the beginnings of the various mathematical disciplines.' The German word 'verzichten', which Brouwer uses here, expresses a certain resignation and implies a forced abandon of any further attempts.

The *Berlin Lectures* is a true reflection and a complete summary of Brouwer's Intuitionism: there is a statement of his fundamental philosophical views on the nature of mathematics, the limited role of logic and the implications of these views for the practice of mathematics; there is a critical analysis of classical practice; the development of a constructive set theory consistent with the nature of natural numbers as well as of the continuum; and finally, there are the beginnings of a re-construction of analysis and the creation of a whole armoury of tools still unused. It leaves a feeling of unfulfilled expectation; staying within Brouwer's favourite analogy, it gives the impression of a partly completed and abandoned building, the scaffolding and building materials a silent testimony of the hopes and vision of its architect. Dissatisfaction with his own efforts was probably the main reason why Brouwer never published his *Berlin Lectures*.

The *Berlin Lectures* were used by him as the basis of his courses on Intuitionism at Amsterdam University in 1927/28, Geneva in 1934, his post-war lecture courses in Leuven and Paris, and especially his annual series of lectures at Cambridge between 1947 and 1951. Plans to publish the lectures were revived, this time in an English version, 'The Cambridge Book' or *The Cambridge Lectures*, but they met with a similar fate. From the correspondence with A.Steen, who consulted Brouwer on the English translation, it appears that Brouwer again planned a book of six chapters. Only the first five chapters were written and have been preserved (manuscript [BMS51]), Chapter Five consisting of a proof and 'applications' of the Fundamental Theorem now called the 'Bunch Theorem'. And again, the *Cambridge Lectures* were never published by Brouwer. (An annotated edition has now been published by Dirk van Dalen; see bibliography.)

The *CambridgeLectures* confirm that Brouwer's programme of re-constructing mathematics had come to its final stop in 1928: nothing substantially different is added to the content of the *Berlin Lectures*. Even his later publications (1949-1954) in which classical theorems are refuted and replaced by 'Intuitionist versions' are no more than up-dated versions of the seven points of Intuitionist criticism which make up the final chapter of the *Berlin Lectures*.

## 2.13 The Brouwer-Hilbert Controversy and the Annalen Affair

Around the year 1926 Brouwer's home in Laren had become an international meeting place. There was a constant stream of mathematicians who became regular visitors or stayed as temporary assistants; some, like Alexandroff, Carathéodory, Hopf, Noether, Hurwicz, Wilson, Newman, Vietoris and Gawehn, in search of inspiration from the presence of the 'Altmeister der Topologie', others, like Freudenthal and Menger, intrigued and inspired by the Intuitionist claims of the great topologist.

These halcyon days, however, were short-lived. By the end of 1927 the centre of gravity had moved from Laren to Göttingen. The building of a new Mathematics Institute at Göttingen was symbolic of its re-establishment as the world centre of mathematics. An increasing number of young, gifted mathematics collaborated in various aspects of mathematical research under the undisputed leadership of Hilbert. Pure mathematics flourished in regular contact and interchange with the physical sciences. Hilbert remained involved in the work of assistants and colleagues of his 'Göttingen Circle' and maintained a lively interest in mathematical physics. His chief concern, however, was the final solution of the problem of the foundations of mathematics, the consistency of Arithmetic. He had secured the assistance of Ackermann and later von Neumann and Bernays.

In Hilbert's early sketch of an independent proof of the consistency of Arithmetic (HILBERT[1904]) the confusion of logic and arithmetic in a direct axiomatic approach, 'a partly simultaneous development on the laws of logic and arithmetic', had obscured the distinction between conceptual mathematics and its symbolized recording. As early as 1909 Brouwer had pointed out to Hilbert that, although the symbolized recording could be analysed mathematically, the resulting mathematical system was a form of applied mathematics, 'mathematics of second order', and could not serve to justify or form a basis of 'first-order mathematics'.*

When Brouwer's Intuitionist challenge forced Hilbert in 1922 again to face the problem of the consistency of Arithmetic, Hilbert sought a solution based on this very distinction. In his paper *Neubegründung der Mathematik* he launched the term 'meta-mathematics', 'the contentual theory of formalized proofs', a theory or system which, as analysis of mathematical theory, deals with formulae and formula-construction and is finite in character. Brouwer's observation that the mathematical structure of such analysis was a simple and basic one was used by Hilbert to seek the justification of mathematics in the finiteness of his meta-mathematics and turn it into the missing link in the somewhat circular chain of his consistency proofs.

By the end of 1927 Brouwer's fortunes seemed to be at their lowest ebb. His programme of re-constructing mathematics had run aground. His dreams of a Mathematics Institute in Amsterdam, the recognized centre of the mathematical world, were shattered. Almost all international support had fallen away. Hilbert's renewed and public attack on Brouwer and, even more, his annexation of meta-mathematics 'without mention of autorship', turning it into a weapon against Intuitionism, was the last straw. returning the attack in an emotional outburst, *Intuitinistische Betrachtung über den Formalismus* [1928A], Brouwer first proved from his own and from Hilbert's work his paternity claim of the notion 'meta-mathematics' and then listed the improvements in the Formalist Programme due to his questioning of the validity of the Principle of the Excluded Middle:

> Formalism has received nothing but benefits from Intuitionism and can expect further benefits. The Formalist School should therefore show due recognition instead of war-mongering against Intuitionism in sneering tones, never once making proper reference to autorship. Moreover, Formalism should remember that in the Formalist structure so-far nothing mathematical has been achieved (we are still waiting for a proof of the non-contradictoriness of its axiom system), whereas Intuitionism on the basis of its constructive definition of sets and the Fundamental Property of Finite Sets has already erected new structures in real mathematics of unshakable certainty. ([1928A], p.4).

---

\* In his letters to Hilbert of 28.10.09 and 28.08.18 Brouwer refers to Hilbert's and Mrs Hilbert's visit and their meeting in the Hague in the autumn of 1909, 'their walk in the dunes... and their exciting conversation'. In [1928A], footnote 2 Brouwer claims that on that occasion he spoke to Hilbert about his distinction between 'mathematics' and 'mathematics of second order'.

The last stages in the Brouwer-Hilbert battle were fought on a semi-political front and concerned the International Congress of Mathematicians at Bologna in September 1928.

The Bologna Congress was organized by the Union Mathématique Internationale under the auspices of the Conseil International de Recherches. Brouwer's public call in 1928 on all German mathematicians to boycott the Congress followed years of his campaigning for the admission of German mathematicians as full members of the Union and the removal from its constitution of the humiliating references to German scholarship (cf. also 2.9.3). Hilbert, however, interpreted Brouwer's action as interference in German affairs, a new obsession of Brouwer, disguising a deliberate attempt to prevent him from attending the Congress as head of the German delegation.

Hilbert's address to the Bologna Congress was his last public appearance before retirement; for a moment his leadership seemed undisputed and his Programme almost complete. Brouwer stayed away and so did many leading German mathematicians.

For Hilbert, however, the matter was not closed; it had shown up Brouwer's ambitions and personal influence. Hilbert felt confident of the victory of his Formalist programme and the loyalty of his Göttingen Circle, but he feared Brouwer's strong personality and his influence on the Editorial Board of the *Mathematische Annalen*, which he considered as almost his personal property. On his return from Bologna in a curt note he dismissed Brouwer from the Editorial Board, without even first obtaining the permission of Einstein and Carathéodory, who with Hilbert formed the chief editorial board. The only reason given was: 'the incompatibility of our views on Foundations, which makes it impossible for me to work with you' (Hilbert to Brouwer, 25.10.28).

Hilbert's friends and co-editors were taken by surprise, but they were anxious to avoid any unpleasantness. Schmidt sent a telegram to Brouwer; Carathéodory, one of Brouwer's few remaining friends and admirers, was dispatched to Laren to persuade Brouwer not to take immediate action, because 'Hilbert is desperately ill and will regret his steps in a few weeks'. Brouwer, however, soon found out that his co-editors' first and only concerns were the wishes of their master and that they were prepared to sacrifice Brouwer in the process. To him the dismissal from the Editorial Board was an injustice and a blow to his career, which he was not prepared to accept without a struggle. The frantic correspondence between Blumenthal, Courant, Carathéodory, Springer, Einstein, Bohr and Brouwer gives a day-to-day account of the 'Annalenstreit' and the almost conspiratory efforts of the inner circle to legitimize Hilbert's rash action. When legal advice was sought it became clear that Hilbert had exceeded his authority, dismissal could only be effected by a unanimous decision of the chief editors; both Carathéodory and Einstein remained unwilling to sign the dismissal notice and a considerable number of assistant-editors were opposed to Brouwer's removal. Legal niceties had to be bent; the only solution was to disband the present company and start a new journal with Hilbert as the sole, supreme authority to appoint editors. Carathéodory and Einstein declined a seat on the new editorial board in spite of repeated pleas of Blumenthal and Courant. Carathéodory felt so strongly about this 'dishonourable affair' that he considered turning his back on Europe and accepting a chair at Stamford. Einstein, relieved at the end of the 'frog and mouse battle', refused the 'glamour of his name' on the cover page of the new Annalen.

In all these dealings Brouwer's continued membership of the editorial board was never at issue; Hilbert's decision, however unconstitutional was accepted by all except a few Berlin dissidents among board members. The debate among the senior board-members was about the justification of Brouwer's dismissal, no more than a post-mortem analysis of Hilbert's motives. Blumenthal in his circular letter to his fellow-editors had to admit that Brouwer's work as editor could not be faulted: 'Brouwer has been a very conscientious and capable editor' (Blumenthal to all editors of the MA 16.11.28). Carathéodory reacted angrily when Hilbert referred to the Bologna affair as the only reason for his decision: 'Brouwer has insulted me' (Carathéodory to Courant, 23.12.28). Blumenthal and Courant were more concerned about Hilbert's reputation and tried to persuade Carathéodory to accept and spread their agreed version:

> Hilbert feared that Brouwer's personality might be damaging and dangerous for the future of the Annalen. It is not 'an interpretation construed after the event' if one emphasizes this factual motive, even if Hilbert's action at first might give a different impression. For Hilbert's sake we cannot allow a version of his reasons to become public which does not do him full justice. If you already accept such a version what can we expect from the public at large? (Courant to Carathéodory, 23.12.28).

Brouwer with his wife and Cor Jongejan, his secretary and life-long companion – 1927

## 2.14 The Silent Years

The Annalen Affair left Brouwer bitter and disillusioned and marks the end of his creative life. It came at the end of a growing crisis of confidence in the success of his mission. His programme of diverting the world of mathematics away from formalistic trends, leading it towards a new, constructive practice, seemed to have failed. His two Vienna Lectures in March 1928 *Wissenschaft, Mathematik und Sprache* and *Die Struktur des Kontinuum* were to be his last public appearances for many years.

*The Structure of the Continuum* [1930A] is a summary of Brouwer's interpretation of the continuum and its properties as given and amended in *Intuitionist Mathematics* and the *Berlin Lectures* \*; it uses a special and simple form of spread-construction, 'the pure $n$-finite set' to generate and characterize the continuum. (see further 6.3.12).

*Mathematics, Science and Language* [1929]\*\* reflects Brouwer's mood at the time: there is a return to the despair of the possibility of human communication of his student days and a mistrust in the goodwill of his fellow-men. New is the Schopenhauerian emphasis on the Will as the agent of human activity, in particular of mathematical activity, and his analysis of human activity into three stages: 1. causal acting, 2. the pure-mathematical abstraction and construction and 3. the social acting of organized society in which language plays a fundamental, instrumental role. In line with Aristotelian tradition, pure mathematics is described as resulting from mathematical abstraction at a higher stage of civilization. Seen in historical perspective, causal acting, although based on the mathematical concepts of repetition and order, precedes pure mathematical activity. There are again the familiar strong images and words of moral disapproval:

> The so-called causal cohesion of the world is no more than a mental force projected onto the world, serving the dark function of the will of man, who uses it to subdue the world and render it powerless, like a snake paralyzing its prey by its hypnotic stare or the inkfish by spraying its venom. ([1929], p.154).

Language is no more than a dubious instrument used by man to impose his will in the organization of social groupings:

---

\* Brouwer's amendments to [1925A] 'entered in writing in 1929 in all his copies' are given by Heyting in his *Notes* to *Brouwer's Collected Works* I, pp. 590-592.

\*\* Brouwer's lecture on 10 March 1928 was announced as *Wisschenschaft, Mathematik und Sprache*; this is also the title printed on the cover of off-prints of [1929] and referred-to by Brouwer in his letter to Heyting 01.07.34. It was published in *Monatshefte für Mathematik und Physik* as *Mathematik, Wisschenschaft und Sprache*.

It is neither exact nor certain... and this applies no less when the transfer of will relates to the construction of pure-mathematical systems. There is no certain language in pure mathematics. (op.cit. p.157).

The experiences of 1928 had killed what little trust Brouwer had in his fellow men and human nature. The Annalen affair remained a 'conspiracy' of his fellow editors against him personally; Carathéodory's however reluctant agreement to accept Hilbert's decision was felt as betrayal by the last of his mathematical friends.

Brouwer's 'silence' during the coming years, however, was more than a defiant resolve not to waste pearls on a mathematical fraternity unwilling to accept his leadership or recognize his mathematical contribution. There was a loss of self-confidence during these years, a growing doubt in his own ability to continue his programme of re-construction. Only a few trusted friends were aware of his despair and depression. In a letter to a friend he writes: 'All my life's work has been wrested from me and I am left in fear, shame and mistrust, and suffering the persecution of my baiting torturers.' (Brouwer to Pogany, 03.06.32). Mrs Brouwer consoles him in one of his bouts of depression: 'Please, don't grieve over your mathematics. I am sure you can still do it as you did in the past, once peace has returned. And even if you could not do it all again, you certainly have already done more than your share.' (Mrs Brouwer to Brouwer, 07.09.32).

The offer of a chair at Göttingen again in 1934 came too late. By then Brouwer's energy and fighting spirit were channelled into non-mathematical controversies. He had become deeply involved in local government, school organization and petty property litigation. He put himself up as the champion of the duped shareholders of a Hungarian spa-enterprise, 'Sodalitas', in which he had made a small investment (after shares had plumeted). He seemed to look for an excuse that made any further mathematical research impossible. His presidency of the Sodalitas shareholders association for more than 14 years required lengthy absences abroad and an extensive correspondence. It is sad to read Brouwer's own report on his activities during these years in his expense claim to the association:

> Journeys undertaken by professor L.E.J. Brouwer in respect of Sodalitas Medicorum:
> 1930 in October, November and December,
> 1931 in February, March, April, June, July, August and November,
> etc....

After listing some 36 periods of absence during the years 1930-1939 he states:

> My Sodalitas files, or rather what has been saved of them from the fire on 3rd November 1941, contain, apart from contracts and press-cuttings, first of all:originals and copies of legal reports, documents, memoranda, circular letters and requests, covering some 1000 pages of foolscap text; apart from those received, they have practically all been drafted by me or with my help. Further: originals and copies of about 2000 letters concerning the Sodalitas affair, many of them very detailed, received or sent by me. The preserved files now weigh some 67 kg.
> The number of working hours devoted by me to Sodalitas runs into many thousands. (Report on my work relating to Sodalitas, during the years 1930-1944; BA).

One of the positive outcomes of the 'Annalen Affair' was the foundation of *Compositio Mathematica*, an international mathematical journal intended by Brouwer as a rival of the Mathematische Annalen. Brouwer invited and appointed all the members of the first editorial board, but never since its first edition of 25 January 1934 did he make a mathematical contribution. The day-to-day editorial management was wholly left to his assistant Hans Freudenthal.

The only 'Intuitionist' publication during these years is *Willen, Weten and Spreken* [1933] (Wil, Knowledge and Speech). Although substantial parts are a re-statement of [1929], it was considered by Brouwer to break some new ground. When Heyting omitted it from the bibliography of a planned 'book' Brouwer wrote:

> I would appreciate it if you would include in your bibliography my article 'Will, Knowledge and Speech' in the ninth annual edition of Euclides, which covers much but certainly not all of 'Science, Mathematics and Language'...' (Brouwer to Heyting, 01.07.34).*

There is an important shift of emphasis and the introduction of the new notion of the 'idealized mathematician', no doubt grown from Brouwer's awareness of the limitations of the human mind in any individual's lifetime. Still unwilling to accept the complete reliability of mathematics as a

---

\* In my Review of *Brouwers Collected Works I* in *Philosophia*, vol.9 no 1 I pointed at the considerable differences between [1929] and [1933] and expressed regret at the omission of a full translation of [1933] in the CW.

Some months after writing the Review I received from Heyting photocopies of his correspondence with Brouwer, including the quoted letter, which gives added weight to this criticism.

To make the text of [1933] more widely available, I have attached my English translation as appendix 5.

collective effort based on human communication, Brouwer now makes the mental activity of the hypothetical human mind endowed with unrestricted power of memory and the ideal communication between such idealized mathematicians the starting point of his thinking on mathematical exactness and logic. In this way the human parentage of mathematics is preserved and the presumption of an existing Platonic real world avoided. Although in some of his post-war papers the 'creative subject' is mentioned and used, the concept itself and its philosophical implications were not further developed by Brouwer.

## 2.15 Post-war Lectures and Final Years

At the start of World War II Brouwer was approaching sixty. The German occupation seriously affected academic life in Holland during the war years. Brouwer's pragmatic attitude to politics, his concern for the continuance of academic life and his fear of 'becoming involved' laid him open to petty accusations of the many enemies he had made in local government and at the University. In the post-war hysteria these were blown up into serious crimes before a kangeroo court of Amsterdam University. He was reprimanded and suspended from his duties (Du.'gestaakt') for nine months.

The setting-up of a mathematical research centre in Amsterdam by the Dutch Governement from which Brouwer was deliberately excluded, came as the final blow. Once again he felt robbed of the rightful parentage of a long-fostered idea.

He now turned his back on the Dutch mathematical community and his university. Once again there was a short episode when he became deeply involved in the post-war re-organization of international cooperation especially in the field of Philosophy of Science. He was one of the founder members of the 'Académie Internationale des Sciences' and the 'Union Internationale de Philosophie des Sciences' and its first president.

He was now one of the grand old men of mathematics, a celebrated guest speaker at conferences all over the world. His carefully prepared addresses at conferences in Leuven, Paris, London, Cambridge, Cape Town, Toronto and Dublin drew large audiences, curious to see the man who once rocked the foundations of the mathematical world. They all started with a historical survey and his interpretation of the crisis in the Foundations of Mathematics.

Brouwer lecturing at Cambridge – 1949

*Consciousness, Philosophy and Mathematics* [1948C], his address to the 10-th International Congress of Philosophy at Cambridge in 1948, evidences his continued concern with the moral aspects of mathematics and its roots in philosophy and mysticism.

*Historical Background, Principles and Methods of Intuitionism* [1952B] and *Points and Spaces* [1954A] state further developments of earlier ideas but held few surprises.

Of Brouwer's post-war publications perhaps the most significant are those on 'essentially negative properties' ([1948A] and [1948B]) in which he dissociates Intuitionism completely from the radical attempts of van Dantzig and especially Griss to eliminate negation from constructive mathematics. (see relevant entries in the Bibliography).

Brouwer's post-war appearances and publications aroused a renewed interest in the fundamental issues raised by his Intuitionist challenge. The collapse of Hibert's Formalist programme had left a vacuum in the realm of Mathematical Foundations. The battle of foundational philosophies had left Intuitionism as the sole surviving contestant. Brouwer's profound vision of pure mathematics and uncompromising adherence to its principles intrigued a new generation of mathematicians. The longest and most comprehensive of Brouwer's post-war courses were those given at Cambridge between 1947 and 1951. His annual Trinity-term at Christ College was a welcome retreat from the in-fighting at home and gave him the opportunity to work on his 'comprehensive Introduction to Intuitionism'. As mentioned above (2.12.2), the 'Cambridge Book' or *The Cambridge Lectures* was never completed and not published in Brouwer's life-time.

Other courses or single lectures were given a.o. at St Andrew's Edinburgh, University College London and Dublin University. (Manuscripts of his texts have been preserved, a selection of them attached as appendices 6-10 to this monograph).

In 1948 Heyting was appointed Professor of Mathematics at Amsterdam University with Brouwer's full support. In his inaugural address he acknowledged his indebtedness to Brouwer and promised him 'to pass the wealth of ideas on the foundations of mathematics, which we owe to you, to a younger generation'. His publications and his encouragement of doctoral research by his most promising students into aspects of Intuitionism secured a continued interest in Intuititionism, albeit with a considerable change of emphasis (see further 5.9.3 and 5.10.2).

In 1951 Brouwer reached retirement age. Relations with his University colleagues had been rather strained during the past few years. The post-war revival of the *Compositio Mathematica* under new editorial management was felt as yet another snub. He refused to accept the 'Festschrift' in his honour on his seventieth birthday and the tributes of his old friends and students, and stopped Heyting's first attempt to prepare a publication of 'Brouwer's Collected Works'. Confiding in his old friend Mannoury, he spoke of his bitterness and his feelings of loneliness and isolation from the Dutch mathematical community:

> The last straw has been the theft of my journal by means of the most incredible deceit and fraud. I know that for the sake of self-survival I should try to forget, but I find it impossible to accept. I am sure it is this experience that during the last few years has made me frightened of human company within the borders of this country.
> It is strange how little one's own philosophy affords protection against psychosomatic reflexes and reactions. (Brouwer to Mannoury, 08.05.53).

*The effect of Intuitionism on classical algebra of logic* [1955] was Brouwer's last public lecture, a centenary tribute to George Boole, the 'originator of the formal image of the laws of common-sensical thought'. Almost fifty years after his revolutionary attack on the identification of mathematics

Final sitting for bust by Wertheim (now in the Mathematics Institute Amsterdam University)

and logic, Brouwer could confidently re-state his conviction of their separate identity and autonomy. The mildness of old age made him recognize 'the formal image of common-sensical thinking as a thing of exceptional beauty and harmony' ([1955], p.116).

The last ten years of Brouwer's life were spent in sad isolation. After the death of his wife in 1959 his only comfort was the patient support of Cor Jongejan, who since 1916 had been his devoted secretary and companion. Life was dominated by – totally unfounded – financial worries and a paranoid fear of bankruptcy, persecution and illness.

Brouwer died at the age of eighty-five. Death came sudden and unexpected; on the evening of 2 December 1966 he was accidentally killed when crossing the road in front of his house.

The funeral on the 6th December was attended by only a few relatives and friends. The only mathematician present was Max Euwe, Master of chess, one of Brouwer's former students and faithful friend. He paid the last tribute to 'the greatest of Holland's mathematicians since Huygens'.

# CHAPTER III
# BROUWER'S PHILOSOPHY

## 3.1 Introduction

If philosophy is defined as the sustained process of reflection directed towards ultimate understanding, Brouwer can claim to be a philosopher. His concern with the fundamental questions of human understanding goes back to his adolescent years. The *Profession of Faith* [BMS1A], written in 1898, puts a full stop to the apprentice stage of Brouwer's thinking and learning; it dispenses with external authority and factual school-learning and with unusual intellectual maturity probes the mysteries of reality, the human self, knowledge and morality. These questions would remain central to his thinking during his student years and throughout his life.

A study of Brouwer's life leaves no doubt that his chief and continued interest has been the Foundations of Mathematics as philosophical reflection on the nature of mathematics, its origin, its fundamental principles and concepts, its purpose and its relation to other human activities. Unlike the maturing mathematician, gradually attracted towards the deeper issues of his discipline and from there moving-on to the wider problems of philosophy, Brouwer was moved into the direction of Mathematical Foundations by his innate pre-occupation with the fundamental questions of philosophy.

Brouwer's philosophical thoughts have been only fragmentarily reported and they, naturally, tended to concentrate around mathematics, the activity he was professionally engaged in. One may be tempted to disregard the general philosophical views of a mathematician as irrelevant and also consider such subjective factors as personal inclination, temperament and religion as out of place in a philosophical enquiry. Professional detachment is expected in practitioners of either discipline and blatant prejudice is usually avoided. Detachment, however, was not one of Brouwer's virtues, his involvement in any cause was emotional and passionate, and reverberated in every aspect of his thinking. The philosophical speculation in his mathematical work has been a source of irritation to mathematicians from Korteweg to Bishop, and to some extent even Heyting. His philosophy, in turn is a clear illustration of Fichte's tenet that ultimate philosophical commitments are pre-rational and often made on temperamental grounds.

Brouwer's search was for a philosophy which on the one hand would provide a reasoned justification of his mystical beliefs and the ideas to which he was committed by temperament and prejudice; on the other hand, a philosophy which could accommodate a conception of mathematical activity, morally acceptable and superior to the mercenary role of science and human communication. His search led to a revolutionary and original view of mathematics consistent with- and rooted in his general philosphy of life. In this peculiar conception of mathematics originate most, if not all of Brouwer's revolutionary ideas; it was the source of inspiration in his attack on current mathematical practice and the key in identifying its shortcomings and in conceiving his alternatives.

Analysing Brouwer's philosophical views one meets striking resemblances to Fichte's extreme solipsistic interpretation of Kant's subjectivism, to Schopenhauer's pessimism and Bergson's intuitionism. Yet Brouwer's philosophy is unique; by assimilating certain parts of various forms of idealism and by pressing for a predominant role for mathematics in the general process of human thinking he created a philosophy which is strictly his own.

We shall not attempt to reconstruct a complete and coherent philosophy which could be distilled from Brouwer's writings, nor shall we try to give a full analysis of his temperamental prejudices and the whole gamut of circumstances which conditioned his thinking. Our main concern is Brouwer's philosophy of mathematics, which we shall discuss in Chapter 4.

In this chapter, Brouwer's Philosophy, we shall restrict ourselves to the main themes of Brouwer's general philosphy which dominate his writing and are directly relevant to his philosophy of mathematics, and we shall briefly comment on those aspects of the contemporary and personal milieu which are indispensable for an understanding and interpretation of Brouwer's views.

## 3.2 Character and Early Background

An investigation into Brouwer's philosophical views justifiably focuses on the early years. Certain ideas developed and matured during later years, but there is remarkably little change in his fundamental ideology. There are no major U-turns, apart perhaps from a temporary flirtation with the 'Signific' ideal. Moreover, Brouwer spoke out more unreservedly and passionately during his student years than in his later publications. *Life, Art and Mysticism* and *The Foundations of Mathematics*, especially in its first version, are clear statements of his commitments and his motivation.

Searching for any external sources of Brouwer's ideas among the philosophical trends in the Netherlands during his formative years, one is struck by the absence of any direct links of Brouwer's ideas with an academic philosophical tradition. In the University of Amsterdam there were unusually close contacts between the departments of mathematics and philosphy* but there is no evidence that Brouwer attended lectures in philosophy during his student days or had any other contacts with the philosophy department.

Full credit for arousing his interest in the philosophy of mathematics was given to Mannoury. Yet, apart from a shared interest in topics such as the role of language and logic in mathematics, Brouwer's views differed in almost every respect from those of his 'dialectical partner'. Their opposition to Hegel's rationalism as popularized by Bolland was in matters of philosophy perhaps the only common denominator. Both reacted against Bolland's arrogance and sophistry, but for different reasons and with different emphases. Mannoury's universal love and respect for his fellow-men, his harmonious appreciation of science and philosophy and his social convictions were deeply offended by Bolland's intellectual elitism. Brouwer's resentment stemmed partly from pride hurt by Bolland's brash claims of personal superiority, partly from a romantic anti-intellectualism, fashionable among the bourgeois student circles of Amsterdam.

Bollandian rationalism was the immediate and express object of Brouwer's attack in his articles in the student magazine (cf. 2.2) and *Life, Art and Mysticism*. He had also given the impression to his friends that his

---

\* Staff 'change of department' seems to be common-place, considering the small size of the University at the time:
C. Bellaert-Spruyt (1842-1901), a mathematician-turned-philosopher and perhaps the most eminent philosopher in Holland during that period. His inaugural address reflects his special interest: 'The relation between philosophy and the various branches of science'. J.D. van der Waals Jr, one of Bellaert-Spruyt's students, moved from philosophy to the department of mathematics and physical science.
Cf. also Kohnstamm, footnote p. 47.

dissertation was aimed at Bolland. Mannoury in his Review of Brouwer's *Foundations* writes:

> The man who caused this stir in the Netherlands is not actually named in Brouwer's book, but his ideas and pronouncements are so often directly referred-to that many readers will be surprised not to find the name of the 'Giant of Leiden' under the letter B in the register of names at the end of the book. (MANNOURY[1907], p.175).

Surveying the scene of religious and quasi-philosophical trends in contemporary Holland, one meets many elements which feature in Brouwer's writings. Among the intelligentsia there is a cynical attitude towards Dutch society dominated by the established christian religion, a reaction against its formalism and conformity. There is a growing interest in Eastern, mystic and pantheistic ideology with its emphasis on personal meditative experience. The Theosophic Society, founded in 1875, promoted a non-dogmatic and syncretic approach to religion; it favoured the Hindu tradition, the mediaeval mysticism of Boehme and Eckehart, as well as the philosophies of Plato, Schopenhauer and Bergson. There is no doubt about Brouwer's familiarity with the ideas of the Theosophic movement, nor about his involvement and his sympathy with their ideals, even if with typical inconsistency he ridiculed the movement in *Life, Art and Mysticism*.

We have already referred to a romantic, though male-dominated fashion among the Amsterdam student body at that time, and their almost modern obsession with industrial pollution.

Recognizing many familiar aspects of Brouwer's ideas among contemporary trends and available ideas, one is tempted to trace the origin of Brouwer's views amongst them. Brouwer, however, was not a blind follower of any master, school or trend. Like any other innovator or original thinker, he was conditioned by contemporary thinking and built upon past achievements. From the multitude of ideas and trends available he rejected some and adopted others, using, transforming and adding to them.

The only real key to the source of his views and philosophy is the person of Brouwer himself: the mixture of intellectual brilliance, up-bringing, complex character and personality, which determined his particular selection and his assimilation of current ideas into a coherent and original philosophy.

Analysing Brouwer's character and personality, one finds a peculiar blend of the usual human qualities and vices, perhaps a predominance of the less attractive properties. For many around him his pride, arrogance, jealousy and meanness were so dominant that they fully characterized and typified him. Such impressions were magnified by Brouwer's obsessive involvement in whatever he undertook, a certain fanaticism with which he pursued his causes at the cost of an inordinate amount of time and energy. They are, unfortunately, the kind of vice which lack the redeeming aspects of other, sympathy-provoking weaknesses.

The underlying and perhaps most basic of Brouwer's personal characteristics is his complete self-centredness and almost total lack of interest in other human beings. His rather negative appreciation of human company, social intercourse and language are not the detached views of an academic philosopher, neither is his declaration in the *Profession of Faith* 'as to the love of my neighbour, I don't care twopence about most people' just a brash boast of a teenager wishing to shock his elders. His inability to love resulted on the one hand in a natural shyness and unease in the company of others and in an unwillingness to listen in sympathy and be open to their views; on the other hand, it resulted in a feeling of inadequacy and insecurity and an aggressive determination to prove himself. From his inability to relate to his fellow human beings stems his low opinion of the physical aspect of the human make-up, which even more than the mind is orientated towards fellow beings and emphasizes human interdependence, epitomized in sexuality and in language. This self-reliance also made him less susceptible to outside influences; it accounts in part for the originality of his views and for his search for the true nature of mathematics in the confines of the Self.

## 3.3 Philosophy, Metaphysics and Ideology

There is some confusion as to Brouwer's attitude towards Philosophy; it arises partly from a characteristic inconsistency between what he said and what he did, seemingly scorning and ridiculing what he himself practised, partly from the notion of philosophy itself. Leaving aside the notion of philosophy as a general outlook on life ('Weltanschauung'), one can distinguish between philosophy as an academic discipline, a systematic study conducted by the professional philosopher on the basis of reasoning alone (Academic Philosophy), and the general human activity of reflective thinking on the nature of things (Reflective or Intuitive Philosophy).

Brouwer's attitude towards philosophy as a profession, at least in his early years, is one of almost contemptuous dismissal. His campaigns against professional philosophers such as Bolland, Reiman and Kohnstamm and against the validation of philosophy as a subject within his Faculty are partly the result of personal grudges; they also reflect his doubts about philosophy as a genuine academic discipline. More generally, the art of the the professional philosopher, the skilful activity of making subtle distinctions and arguing by chains of reasoning in non-mathematical matters, is rejected outright and philosophy condemned as 'perverse trickery'. In the 1906 *Notes* ([BMS3A] philosophy in this sense is referred-to as 'reasoning philosophy' and branded as 'estrangement compounded' (C.V.25) . Such philosophical reasoning, especially when expressed in words, is totally alien to the nature of Philosophy as reflective thinking: 'Saying or writing philosophy is contrary to the nature of philosophy' (op.cit., C.V.26).

Reflective or Intuitive Philosphy is direct insight and lacks structure:

> In Philosophy structures and systems are useless (one wants to be struck by direct insight). Systems have value only when applied in the struggle with an enemy; philosophy *should not* be applied. Philosophy cannot work mathematically. (Op.Cit., C.VIII.73).

Philosophy in this sense is a kind of natural wisdom, the result of meditative reflection by the mind, still innocent and not yet polluted by long experience or by study of the work of others. Almost nostalgically the 25-year old Brouwer writes:

> The peak of one's mathematical power is reached in old age; the ability to do philosophy is lost at 18 (apart from the knack of applying it to some things). ([BMS3A], A.I.19),

a typical Brouwerian exaggeration, but a clear endorsement of his *Profession of Faith*, written at the age of 17.

In parallel with Brouwer's view of the nature of mathematics, philosophy is the meditative activity itself, 'philosophizing', reflecting on the deepest human issues. It is an interpretation that suited his preferred style of study, which was not that of the scholar at his desk, pain-stakingly consulting references but lying on his bed in the solitude of his Hut, or sitting cross-legged on the heath or in a tree, concentrating his mind, 'thinking-through' the problems of individuality, consciousness, the intellect, God, the universe and human society.

These metaphysical problems had been the concern of academic philosphers too; indeed, the main pre-occupation especially of nineteenth century philosophers was the foundation of philosophical thinking, the principles of knowledge, morality and the whole of western culture.

Kant had initiated a critical questioning of the assumptions of the rationalist philosophers before him and of Reason itself. Traditional Aristotelian philosophy and in particular the Cartesian School maintained that human reason can transcend the boundaries of phenomena and apprehend the essence of spirit and matter, the permanent and necessary, grasp even Reason itself and determine its own criteria of validity. Kant concluded that the reality of things in themselves, although not in doubt, cannot be the object of human knowledge; the *forms of understanding*, inherent in all human percepion and reasoning, safeguard the universality and objectivity – or rather inter-subjectivity – of human concepts.

Kant's successors followed his critical lead, but some carried his rejection of the dogmatic authority of Reason to its extreme idealist conclusion and built their philosophy on the primacy of the human will (Fichte and Schopenhauer).

Some considered reality, including human nature, to be in constant flux and made 'historical conscience' the primary arbiter of the changing reality. There were others who rejected any philosophical speculation beyond empirical science, materialists for whom physical matter is the only reality, empiricists who claimed that all knowledge should be tested by experience and positivists who only accepted sensations themselves as elements of human knowledge and recognized the scientific method as the only legitimate method of philosophical inquiry.

They all have in common that their philosophical system is built ultimately on one or more axiomatic presumptions or some ideological commitment; after the heuristic preliminaries the structure is then continued rationally and logically.

Brouwer's bias against 'academic' philosophy during his early years was partly a reaction against sophistry and in particular against Bolland's glorification of 'Pure Reason', his arrogant claims for the superiority and power of the Intellect. There is, however, evidence of his more fundamental suspicion of the arbitrariness of the basis of any philosophical system and the circularity or paradox in a critique of reason by reason itself. In *Life, Art and Mysticism* for example he ridicules epistemology and points at the gap between human thinking about reality and reality itself:

> They practise so-called 'theory of knowledge'...some after thinking for a long time about the elusive link between the perceiving consciousness, which as life itself develops from the world of perception, and the world of perception, which itself only exists through and in the form of the perceiving consciousness; they then plug the hole with the 'Ego', which was also self-created at the same time as the world of perception, and say 'Yes, of course, there must in the end be something that cannot be understood... ([1905], p.22).

In his search for truth Brouwer was not prepared to accept authoritatively given first principles or build on solutions given by others. Hilbert testified to his preference for 'difficult and deep problems ...not giving up until he has found a solution that fully satisfies him' (Hilbert to Korteweg, 06.02.11 DJK80). Korteweg warned him 'not to walk near the abyss' and remarked 'I find it difficult to follow you where you dig so deep down to fundamentals' (Korteweg to Brouwer 13.05.05, DJK2).

Rather than accept some fundamental adage justifying reason and build a rational system on it, Brouwer looked beyond the boundaries of reason and searched for the source of truth and certainty within the wider complex of mind-activity. *Life, Art and Mysticism* describes Reason, the Intellect as just one of the faculties, almost as an organ of the human soul, an organ of cognition an ratiocination. Within the make-up of the Soul he acknowledged other faculties, operating differently, with higher authority and of greater (moral) value. The distinction between perception, emotion and volition is more than a subtle differentiation within the unique and indivisible psyche. Moreover, knowledge is not the monopoly of the reasoning Intellect, which forms concepts by abstraction and sense perception, 'a child of time and space' (p.19). He recognizes an ability to 'withdraw from passion, from time and space and from the whole of the perceptional world' (p.13) and grasp truth directly.

Analysing knowledge, certainty and truth Brouwer recognized that certainty is inner conviction and that the only authority and arbiter of truth is to be sought in the individual conscious Self. His point of departure is the individual mind in its widest sense, the Aristotelian 'Soul', and the existence of a supra-sensual reality which is the Soul's natural world, which presents itself directly to the Soul and is by its very nature impossible to capture in words. Occasionally it reveals itself indirectly, through the veils of the physical world e.g. through works of art. The intellect can grasp aspects of this reality through abstraction; these intellectual images, however, are earthbound and man-made.

Most philosophers restrict their attention to the intellect and its abstracted images, excluding all that is 'unsagbar', i.e. cannot be expressed in words (Wittgenstein), or that cannot be verified by experiment (neopositivists). Truth to them is 'adequatio rei and intellectus', agreement between the images and that to which they relate, or agreement and coherence with other images and ultimately with the axiomatic posits which form the basis of their systems. Any knowledge that transcends the bounds of the intellect is rejected as 'mysticism' and 'religion', lying outside the domain of their philosophies. Brouwer accepted the distinction between the intellectual and the mystical, but he embraced both within his philosophy and gave the mystical a prominent and essential place. He does not hide his sympathies with the thoughts and writings of 'mystics' and openly referred to Mysticism in the title of his first book and in his later writings.

## 3.4 Mysticism

The word 'Mysticism' is often loosely used to describe any belief in the supernatural, the occult and magical, or any theory which accepts the authority of mysterious agencies for which no rational account can be given. Associated with self-delusion and dreamy confusion of thought, it is ridiculed and rejected as dangerous and false. Mysticism is sometimes identified with parapsychology and described as scientific investigation of occult phenomena. Many philosophers use the word mysticism to describe those aspects of human awareness and knowledge which transcend the the physical reality and the boundaries of reason. Science and mysticism are then seen as the two poles of human understanding, science: discursive, analytic knowledge produced by the senses and reason, mysticism: understanding founded on direct insight, intuition, creative and embracing the emotional, artistic and moral aspects of human life. Bertrand Russell in his *Mysticism and Logic* writes:

> The greatest men who have been philosophers have felt the need both of science and mysticism: the attempt to harmonize the two was what made their life, and what always must, for all its arduous uncertainty, make philosophy, to some minds, a greater thing than either science or religion. (RUSSELL[1918], p.9).

In the religious tradition of the West and the East, 'mysticism' is the personal and direct experience of communion of man with the divine through ecstatic contemplation. Distinction is usually made between various types of mysticism according to the nature of the divine, ranging from a personal, transcendent God to pantheistic conception of the godhead, the Absolute, or even the Buddhist state of Nirvana. Most of the great mystics have been faithful sons and daughters of established religions; yet the emphasis on individual experience as the ultimate source of inspiration and authority made them often appear as critics of the formalism, the external aspects of their religion, and of church authority, and they were often branded as 'spiritual anarchists'.

Although mystics like Eckehart, Boehme and Flaubert have written about their experience, the language they use is allegorical. No words can be adequate to describe their purely spiritual experience, descriptions have to be suggestive rather than literal or precise. They speak of life as a journey of the Exiled Soul, the imprisoned, sense-fed consciousness attempting to shed its fetters, only sometimes 'reaching Home' in ecstatic contemplation. This mystic experience of union of the Soul with the Divine is achieved in 'stillness'. The inner vision of the 'triple star of Goodness, Truth and Beauty' is more than observation of something else, it is the fusion of the knowing and loving capacities of the Soul with the Divine.

The main characteristics of religious mysticism are described by Evelyn Underhill in her classical analysis of mysticism (UNDERHILL [1911], p.81 ff.):

1. True mysticism is active, in the jargon of psychologists – and Brouwer –, a 'conative', dynamic movement of love, not a passive, theoretical knowing or intellectual cognition.
2. Its aims are wholly transcendental and spiritual. It is in no way concerned with adding-to, exploring, re-arranging or improving anything in the visible universe; the 'mystic heart is always set upon the changeless One'.
3. This 'One' is for the mystic not merely the reality of all that is, it is also a living and personal object of love. Mysticism is not self-seeking.
4. The living union with the 'One' is achieved by a process of liberation, transmutation of the Self to higher levels.

The atheistic variant of mysticism, or rather mystic philosophy, shares with religious mysticism an acceptance of the spiritual nature of man and a reality beyond the grasp of the senses and the intellect. Russell characterizes mystical philosophy by:

1. a belief in insight, intuition as a way of knowing, distinct from the

discursive, analytic processes of the intellect;
2. its belief in unity and its refusal to admit opposition or division;
3. the denial of the reality of time;
4. its belief that all evil is appearance.

Brouwer could certainly not be called a faithful son of any particular religion or church; he condemned what he describes as 'organized religion... good for the stupid masses'. Yet he was convinced of a religious reality, deeply moved by religious emotions and attracted to religious mystique. When he used the word 'mysticism' he refers to the tradition of the great mystics such as Boehme, Eckehart and the authors of the Bhagavad Gîta, not the philosophic interpretation of Russell or Mannoury.

Brouwer is not a mystic himself, nor is *Life, Art and Mysticism* an account of his mystic experience; it is an attempt to describe and promote the mystic way of life. He lacked the ability and will to surrender and love even the Divine Being. Religious mysticism appealed to him as a tradition in which the inner experience of the individual is supreme and where the authority of society and hierarchy are absent. Moreover, mysticism accepts the possibility of direct knowledge, independent of Reason and dialectic proof, providing him with an independent criterion, a 'metarational' justification of Reason.

Brouwer's interest in mysticism is more prominent in times of depression, but it is not a youthful, passing phase, as is obvious from his writings (in particular [BMS1A], [1905], [BMS3],[1907], [1929], [1933], [1948C]). Nor was it a part-time pre-occupation, 'a Sunday-suit for special occasions': elements of mysticism became a fundamental part of his 'Weltanschauung' and permeated even into his philosophy of mathematics. In his philosophical accounts of the foundations of mathematics we frequently meet words, references, ideas and attitudes which have their parallel in the mystic tradition or Brouwer's understanding of it, and which originate there; words such as 'intuition', 'insight' or 'introspection', 'stillness', 'externalization'; and 'attitudes' such as a preference for spiritual and moral values over Reason and over the 'external', physical aspects of human life. These words, ideas and attitudes in Brouwer's philosophy of mathematics are often more easily understood and interpreted when traced to their 'mystical' origin or parallel in Brouwer's thinking, where they seem more 'natural' or appropriate. An example is his view of language and logic. Speaking in [1905] about mystic experience, he reluctantly accepts the use of language as necessary but inappropriate and incapable of capturing mystic experience:

> ...mystic experience will most frequently be represented in words, the nearest to the human curse, the Intellect... (p.75).

He praises the absence of logic in mystical accounts:

> Nowhere in mysticism is there a thread or appropriate sequence; every sentence stands by itself and does not need another to precede or follow it. (p.76).

### 3.5 God and the Moral Principle

In *The Profession of Faith* ([BMS1A]) Brouwer gives perhaps his most direct and detailed account of his 'universe': the Self, Consciousness, God and the physical world, and of the respective forms of knowing these aspects of the universe. Later references conform to the views expressed in this, Brouwer's first preserved paper.

He distinguishes between 1. knowing (Du.'kennen'), 2. believing (Du. 'geloof') and 3. accepting the result of deductive reasoning, to which one could refer as 'intellection'. All three seem to be forms of awareness of being or being-true. Their difference, though not precisely defined, is more than a matter of degree in the strength of conviction. Both 'knowing' and 'believing' are direct and spontaneous acts of consciousness, while 'intellection' is indirect, achieved through the reasoning activity of the Intellect by means of perceptual images. (In his later work Brouwer moves the motive power and seat of intellectual activity in the direction of the Will.)

The difference between 'knowing' and 'believing' is less clear. In [BMS1A] Brouwer reserves the word 'know' exclusively for awareness of the Self, 'the conviction of my Ego'. The words 'belief' and 'believing' are used to express awareness and acceptance of God and of the mind's images. To emphasize this difference the words are underlined in the manuscript:

> My life is my *conviction* of my Ego and my *belief* in my images. (p.1).

The Kantian problem of knowing the 'thing-in-itself' is solved by confining the 'knowable', real world to the Self and accepting the images as real only in as far as they are part of the Self:

> There is no such thing as a second reality independent of my Ego to which these images would correspond. (ibid.).

His knowledge of God, though a 'belief', is no less strong a conviction. He refers to it as a 'direct spontaneous feeling in me', not a reasoned result of a metaphysical proof along the lines of the Aristotelian or Thomistic five ways:

> In the many words I use one should not read a rational deduction of the existence of God. Belief in God is the deepest ground for deductions, but cannot itself be deduced. (ibid.)

Belief in God is stronger and more fundamental than belief in his images:

> My Weltanschauung accepts my Self and my God as the only living beings, of which I *Know* my Self and *feel* my God, my Master. (ibid)

He accepts God as supreme and as the 'originator', but again only in relation to the Self, not the perfect, infinite, spiritual personalized being, the Creator of the whole Universe. He speaks of 'my God', 'my Master', 'the Origin of my Ego, who gives me images, who lives and stands above me... that is my God'.

Also: 'my God stands under my life, i.e. is the originator of my life'. Unlike the theologians of Immanence, who see God 'dwelling in the world and in the Self', Brouwer describes his God as independent even from the Self. The existence of his God arises from the need for an independent origin of the Self:

> ...something independent of me, therefore something which, like me, lives and which stands above me; that is my God... my God stands under me, i.e. is the originator of my life, and outside my life... (ibid)

Brouwer does not engage in theology, metaphysical speculation on the nature of God, nor in the mystical tradition does he attempt to describe his God in allegory. His religion is what he describes as 'my feeling of my God'. Attempts to describe or even form a conceptual image of God are bound to fail:

> Language is too clumsy an instrument . The feeling of God, trust in God, is not a conscious thought, an image, for this would remain outside God; but it is something which transcends human thought, that cannot be thought and even less described. It is something bound to the unconscious Ego as it becomes conscious, or as it receives its images, but separate from these images. (ibid.).

It is worth noting Brouwer's obsession, already at the age of 17, with the 'primordial happening', the dawn of consciousness, seeking an explanation of the Divine – as later of 'number' – in the human thought-process prior to and independent from perceptional images; an obsession also with independence, the need to seek justification of any human function or system outside the realm of that system, and the impossibility of describing the system from within:

> My life is an accomplished fact on which I cannot express an opinion; to do so I would need to be able to look upon it from the outside, which is impossible. My life for me is the only 'It' to which I cannot ascribe any qualities. (ibid.)

Brouwer's conception of God provided him with an independent origin of the Self. It also gave him an a-priori moral sense of values and a moral mission:

> My God has given me the incentive to make my life, i.e. my images, as beautiful as possible. This has caused me to be shocked by the loathsomeness and ugliness in the world which surrounds me and is part of me, and to endeavour to remove them, also where it affects human society. (ibid.).

The *Profession of Faith* is not very explicit about the nature of Conscience. *Life, Art and Mysticism* frequently refers to 'conscience' in the traditional popular sense of a 'voice from within', counselling the Soul as to good and evil. In other instances Conscience is almost identified with Consciousness, the knowing Self, in moments when less distracted by images and homesick it longs for 'the paradise lost':

> ...when conscience manages to break through the shackles of this sad world, which to many happens round the age of eighteen, taking the form of a purely central, not merely artistic admiration for dreamers, monks and hermits. ([1905], pp.27-28).

Surprisingly, neither in the *Profession of Faith* nor in *Life, Art and Mysticism* nor in his later writings is there any reference to or defense of conscience in the sense of a direct intuition of moral principles, in pre-Kantian ethics often referred to as 'synderesis'. However, of Brouwer's acceptance of conscience in his sense there can be no doubt. The main theme of *Life, Art and Mysticism* is mystic conversion, a plea for a return to a state of original goodness by 'introspection', away from the distractions of a sensual world and from the machinations of the Intellect. The physical world, spoilt by the interference of human reason, is 'sad' and 'wicked', and the 'human Intellect' is exposed as the source of all evil, sin personified. The title of Chapter II, 'Fall through the Intellect', suggests a Brouwerian interpretation of original sin, leading to knowledge of good and evil. Good, in Brouwer's terminology identified with 'beautiful', and evil are no more than attributes referring respectively to the original pure state of consciousness – and of the physical world – and its corruption by Reason. Conscience is consciousness itself knowing its 'home' and recognizing the enemy within; 'intuition', a-priori in the sense that it precedes sensual perception and reason.

This principle of morality differs in every respect from the Kantian categorical imperative based on practical reason. The moral duty it imposed on Brouwer bears no relation to the accepted moral codes of society and church; it leads to an apparent contradiction between the moral stance in his philosophy of mathematics and science and his supercilious attitude to social morality, as evident e.g. in *Bollandiana* [BMS2]:

> Self-liberation for the woman will lead her in sweeping moves of flight to acts against the laws of the land, against current morality. It will move her to acts against religious morality... She will scatter these acts on her path, unabashed, making her way with wonderful brilliance between the laws and public opnion, protected by noble disgust from hurting herself. Her way will lead her first to murder, theft, lies and adultery – while the world has no hold over her – and ultimately to suicide...(*Propria Cures*, jg. 16, no.10, p.110).

In Chapter VIII of *Life, Art and Mysticism*, 'The Liberated Life', breaking the accepted moral code is described as inevitable and excusable:

> ...he stands outside the world, where he has no further obligation, he cannot sin any more... his attention moves in better spheres and 'L'apostasie est permise quand le coeur est pure' (Flaubert)...([1905], p.86).

Laws and moral codes are imposed by the State and Church-organizations to enforce their will on individuals. The *Profession of Faith* scathingly calls them 'good for the stupid masses, to keep them in respectful ignorance under the thumb of a power-thirsty Church'. [1933] exposes the attempts by society to justify laws and moral codes in a moral theory:

> To secure the highest possible degree of stability in the feelings of loyalty and contentedness within the organized human groupings, the organization, well-aware of the limited means of horse-training [Du. 'dressuurmiddelen'] at its disposal, resorts to the propaganda of moral theories, i.e. mathematical views which base the suitability and necessity of the existing organization not only on common aims and emergencies which are acceptable to the egoistic mind, but also on so-called moral values in their attitude to life, i.e. values which depart from egoistic notions. Classic examples are religious commandments, and notions such as fatherland, property, family, solidarity, class-consciousness, honour and duty.' ([1933], p.50).

The intuition of God plays no further role in Brouwer's general philosophy, nor in his philosophy of mathematics. While others resorted to God to justify existence, infinity or even the natural numbers, Brouwer's world of mathematics, though in the realm of Mind, is wholly man-made; its elements have a time-existence, are constructed and 'becoming', qualities which are the reverse of those characteristic of the Divine.

Brouwer's moral intuition, his moral conscience, however, did affect his philosophy of mathematics. Indeed his moral prejudice inspired his Intuitionist campaign. In its early stages the attack was focused mainly on the application of mathematics, man's cunningness, his mathematical ability, using and abusing his power of reason. The critical analysis of the role of mathematics in the physical sciences in his *Foundations of Mathematics* is strictly based on scientific and philosophical argument, but he does not hide his underlying resentment and moral disapproval:

> Man has the faculty, accompanying all his interactions with Nature, of objectifying the world, of seeing in the world causal systems in time...This externalization by man, the will to destroy and rule, immediately obstructs any nourishing of the heart by nature. Those who rule are already damned, and damned are those qualities that promote man's rule. ([BMS3B], p.1).

When the focus of attack increasingly shifts to language and logic, the deep underlying motive remains the same. The opening lines of *The Unreliability of the Principles of Logic* [1908C] refer to the abstracting reason as 'a faculty originating in the primordial sin of fear or desire' (p.5). Quite apart from its close association with reasoning ('the word, nearest to the human curse, the Intellect' [1905], p.75)), language as communication with other human beings carries the stigma of evil, detracting consciousness from the pure, naive state of stillness and self-concentration. The evil move-away, 'the exodus from its deepest home', is described in 1928 and 1948 as a 'deteriorative process'. Worse even than the 'causal phase', when causal attention moves consciousness on an almost irreversible path away, the 'social phase' of cooperation with others makes a return impossible, and 'naivety in a receding distance vanishes for ever' ([1948C], p.1242).

## 3.6 Kant and the French Intuitionist Tradition

Brouwer likes to trace the lineage of his Intuitionism right back to Kant: 'In Kant we find an old form of intuitionism' ([1912A], p.8; cf. also [1907], p.94, [1909A], p.5, [1912A], p.10 ff, [1930A], p.6, etc.). He claims that with some adjustments Kant's philosophy of the a-priori and of synthetic judgments could be accepted as the ultimate basis for the validity of mathematics:

> However weak the position of intuitionism seemed to be after this period of mathematical development, it has recovered by abandoning Kant's a-priority of space but adhering the more resolutely to the a-priority of time. ([1912A], pp.11-12).

The greater part of Chapter II of *The Foundations* is devoted to an examination of the adjustments needed in Kant's philosophy to accommodate the non-euclidean geometries. Commentators have not failed to characterize Brouwer's philosophy as essentially Kantian, although Heyting qualifies this by saying that 'this interpretation of Kant's theory of knowledge was rather special' (HEYTING [1958],p.101).

The Romantic revolt against the Age of Reason had to some extent been initiated by Kant; he had placed the basis of the validity of mathematics firmly in the human mind itself, which he saw as the active, synthesizing agent in the formation of concepts. In this respect Brouwer's philosophy may be called Kantian, as can indeed those of most nineteenth-century philosophers. But apart from this most general basis of agreement and an emphasis on the a-priori intuition of time as the foundation of mathematics, Brouwer's philosophy can only be related to that of Kant as the extreme consequence of some of Kant's doctrines, such as the subjective tendencies in the process of human cognition and the primacy of the human will.

When considering the fundamental attitudes to knowledge and the central theme of Kant's philosophy, one might even describe Brouwer's views as in almost complete opposition to those of Kant. Kant's *Critique* placed an a-priori restriction on the limits of all human knowledge, including knowledge of the Self; the whole of being is exhausted in the knowledge that Reason can gain of it.

As to knowledge of the 'exterior world', 'things', including such animate things as other human beings, Brouwer not only denies the possibility of knowing things-in-themselves, he even rejects the hypothesis of an existing reality and most, if not all of Kant's Categories, the elementary a-priori concepts which make objective reference possible. Causality to him is no more than a uniform sequence, and other things only 'iterative complexes of sensations whose elements are permutable in point of time' ([1948C], p.1235). As to knowledge of the Self, Brouwer lays claim to a special kind of insight outside the sphere of perception, a knowledge which is direct and transcends Reason. Even when professing his allegiance to Kantian doctrine and using his terminology Brouwer strays far from the interpretation given by Kant. These differences of interpretation and 'reference to Kant' were among the contentious points in the Korteweg-Brouwer conflict. Repudiating Korteweg's doubts about the thoroughness of his study of the *Critique* Brouwer wrote:

> You told me you could not be entirely sure whether I had studied Kant enough to express an opinion. Of course, I cannot give you that certainty, but I can tell you that I have read the whole of the *Kritik der reinen Vernunft* and have repeatedly and seriously studied many sections, including those which concern the subject of my dissertation. (Brouwer to Korteweg, 05.11.06, [DJK20]).

At the same time Brouwer defends himself against an alleged accusation by Korteweg that he should not refer to Kant in a mathematical thesis:

> As to your comment that the name Kant is out of place in a mathematical dissertation, you will see that Russell's *Foundations* refers continually to Kant, and that Couturat's *Les principes de Mathématiques* is supplemented with an appendix of more than 100 pages on Kant... Poincaré calls the present battle about the foundations of mathematics a continuation of the old mathematical-philosophical battle between Kant and Leibniz. (ibid.).

There is, however, in Brouwer's own notes ([BMS3A]) little evidence of a thorough and detailed critical study of the *Critique* to which for example he subjected Russel's *Essay*. In another letter to Korteweg he claims to be less concerned with the writings of pure philosophers:

> 'The main reason why I have so exclusively considered Russell's book is that he is the only one who writes about the philosophical foundations of mathematics and, at least most of the time, uses an exact language with which one can battle on mathematical grounds. This is not done by Hegel, Schopenhauer, Lotze and Fechner, to name but a few. If I had to discuss them I would have to move on a pure philosophical plane; this you do not want and I do not wish to do either. (Brouwer to Korteweg, 07.11.06, [DJK23]).

It is worth noting Brouwer's rather negative approach to references; he seems to imply that only those with whom he 'can do battle' are worth investigating. Rarely does he quote mathematicians or philosophers who support his views (except in his mystical writings) and there are few or no references indicating an external source of inspiration.

Interesting also is his preference for Russell over pure philosophers on account of his 'exact language'. In *Life, Art and Mysticism* he castigates the language of philosphers, their endless distinctions especially in the epistemological debate. Yet the fundamental questions to which he sought an answer were the very problems, though couched in the professional jargon, that occupied the minds of contemporary philosophers. However original and revolutionary Brouwer has been, his interests and even his prejudices belong to his age. In spite of his independence and insistence on solitary thinking, his peculiar philosophy and his philosophy of mathematics followed in the wake of a broad movement in European thinking.

In his Inaugural Address ([1912A]) Brouwer describes the controversy concerning he foundations of mathematics as between 'the French Intuitionist School' and 'the predominantly German Formalist School' (p.7). In spite of many disagreements on detail, especially the role of language and logic, he sides firmly with the intuitionist revival, to which he refers as '*Neo*-Intuitionism'. When later he had established himself as the father of mathematical intuitionism he refers to these neo-intuitionists as 'the *pre*-intuitionsist school mainly led by Poincaré, Borel and Lebesgue' (e.g. [1952B], p.140). Using the term 'neo-intuitionism', Brouwer clearly showed his awareness of a long tradition of 'intuitionism', a strand of philosophical thinking, particularly strong in France, based on the dualistic principle of separation of mind and matter and on acceptance of intuition as a faculty complementing intelligence. Ryle describes the separation of mind and matter as the 'doctrine of the ghost-in-the-machine' and 'Descartes' myth' (RYLE [1949], Ch.I). In the Cartesian universe 'matter' is the physical world, extending in time and space, 'mind' includes the human mind as well as God. Knowledge of the physical world is never certain, neither is knowledge resulting from reasoning, which only takes us from premises to conclusions. Certainty can only be found in consciousness of the Self, in the 'clear and distinctive ideas', in Intuition, described by Descartes as 'lumen naturale'.

Kant rejected the possibility of direct knowledge, of knowing 'noumena', things-in-themselves, and considered the Cartesian Intuition a prerogative of God.

The Cartesian view, however, remained alive in France. Right through the nineteenh century there are eminent philosophers such as Maine de Biran, Ravaisson, Lachelier, Boutroux and Bergson, who in various ways uphold the principle of an intuiting mind. Maine the Biran maintained that knowledge of the Self exceeds the bounds of the pure phenomenon and reaches reality in itself, a reality which Kant had declared inaccessible to human thought. Ravaisson made a clear distinction between the abstracting and reasoning intelligence and intuition, 'which grasps the nature of being' (RAVAISSON [1867]); he introduced a parallel distinction between two philosophical methods: the first uses analysis, proceeding to decompose by successive dissection and abstraction, trying to explain the higher through the lower elements, the second proceeds by synthesis, reaching for the unity that binds these elements together and for life itself.

Life, and in particular human life, which was outside the scope of positive science and escaped the mechanistic and scientific methodology of Kantian philosophy, became the central problem of the intuitionist movement towards the end of the nineteenth century, and especially for its main proponent, Bergson.

*Henri Louis Bergson*, was born in Paris on 18th October 1859. Like Brouwer, he was a brilliant student, excelling in the classics, the arts as well as the sciences. His masters recognized his mathematical genius and with regret accepted his decision to study philosophy ('You might have been a mathematician and you will only be a philosopher').

The chief concern of his philosophical masters Boutroux and Ravaisson was the dichotomy in the physical universe between determinism and contingency: on the one hand the necessity of the laws of nature as interpreted by reason, and on the other hand 'freedom' and contingent action, which Reason fails to grasp but consciousness recognizes as a fact of life. They sought an explanation of contingent phenomena and action by modifying the notion of scientific necessity and resort to a Divine, free Being.

During the seven years of isolation that followed his graduation Bergson struggled with the problem of physical reality and its mechanistic solution by nineteenth-century scientism, and developed his 'new philosophy'. Time, continuity and movement emerged as the central issues. These essential components in the make-up of reality had been considered incapable of being grasped by Reason and therefore been relegated to the realm of the unknowable. Bergson's philosophy starts from the conviction that we can know reality in its fullness. He accepted the dual aspect of reality, but recognized a corresponding duality in the human faculty of knowing: on the one hand materiality or matter, which is static and known through sense-perception and Reason, and on the other hand spirituality, vitality or duration, which is known directly by the faculty of Intuition, 'by means of that which is spiritual in ourselves we come into sympathy with life and spirit'. Reason and Intuition are complementary aspects of the faculty of human knowledge; their object and scope mark the boundaries of the realm resp. of science and philosophy. Science is confined to the mechanical, the inert, the static; its instrument of knowledge is Reason, the analytical mind which gathers information from sense perception and reasoning, acquires knowledge which is discursive, compound, relative and expressible in symbols. Living reality, however, the 'fluent', remains outside the limited scope of Reason and is the domain of philosophy and metaphysics. It is knowable, not by dialectic or logical processes, 'imprisoned in categories', but directly by consciousness through immediate intuition. Intuition transcends all the imperfections of analysis; it is a dynamic union of knower and known; through the veils of phenomena it sees the living reality directly. It yields knowledge of the Self 'through the immersion in the indivisible flow of consciousness'. Very much under the influence of Darwin's theory of evolution, Bergson sees intuition as derived from and akin to instinct, which is 'natural knowledge', not acquired through experience. In man instinct is raised to intuition. Like instinct, intuition is direct insight, a sympathetic attitude to reality by which one places oneself inside an object; it is knowledge which is simple, absolute and languageless.

Philosophy embraces, as does human knowledge, the two complementary aspects of the one living reality. It does not dispense with reasoning, but it subordinates analysis and discursive thought to intuition.

This was the message of Bergson's doctoral thesis *Essai sur les données immediates de la conscience*, published in 1888. In subsequent publications, *Matière et Mémoire* 1896 and *Introduction à la Métaphysique* (in *Revue de Métaphyisique et de Morale*, 1903) the notions of Intuition, Time, Mind and Free Will were further developed. *L'Évolution Creatrice*, published in 1907, introduces Bergson's theory of 'becoming', the dynamic continuum of life, movement and time. As in cinematography, the optical illusion of movement may be given through a succession of stills, but movement itself can never be analysed. Movement is not produced by the juxta-position of pause and pause, but is a continuous process of 'becoming', only grasped by intuition.

By the turn of the century Henri Bergson, now Professor of Philosophy at the College de France, had become the inspired leader of the intuitionist trend in philosophy. The role of Intuition, which his predecessors had taken for granted and vaguely described as 'the Heart that knows' (Pascal) or 'the guiding Light' (Descartes), had been made more precise and was established as the centre of a philosophical theory. Against Kant's critical method, which limited the knower and the knowable, Bergson had placed 'the intuitive method', which recognized the dual faculty of Reason and Intuition probing a universal reality which includes life, change and time; a method which was to bring about:

> a reconciliation between metaphysics and science, and supporting the one by the other without sacrificing either, after having sharply distinguished between them. (BERGSON [1903]).

### 3.7 French Mathematical Intuitionism

The revival of mathematical intuitionism, or neo-intuitionism as Brouwer would have it, must be seen against the background and as part of a general growing reaction against the mechanistic, analytic and intellectualist tendencies of nineteenth-century philosophy and science; a new trust in and respect for man's intuitive ability to grasp truth directly. In the philosophy of mathematics the 'analytical' tendency was represented by the Logicist School. In the vacuum following the deposition of Euclidean space as the a-priori intuitive foundation of mathematics, and with the advances in Set Theory and Logic, Leibniz's logicist thesis seemed to present a plausible alternative. There were renewed attempts to seek the ultimate foundation of mathematics in logic. The new logicists maintained that the whole of mathematics is analytic, can be derived from logical a-priori principles by deductive reasoning.

Neo-intuitionism was primarily a reaction against the claims of logicists which reduced mathematics to a formal science, a sterile manipulation of logical symbols. Brouwer's historical surveys all name Poincaré, Borel and Lebesgue as the leaders of the 'Neo-intuitionist' or 'Pre-intuitionist' School.* They are united in their conviction that mathematics has more

---

* Addressing his Berlin audience in the controversial year 1927, Brouwer added 'Kronecker'

than formal significance, that it has what Heyting would call 'Inhaltliche Bedeuting'(HEYTING [1934], p.3) and relates to some mathematical reality; they further maintain that linguistic expression remains an essential element in the definition of mathematics, 'definitions in disguise' or 'conventions' providing the tenuous link between reality and its description. Brouwer referred to this as the 'modified observational standpoint' (e.g. [1952B], p.140). Equally common to them all is a resort to 'Intuition' in some form, as is indeed the vagueness as to what constitutes this Intuition and this reality.

*Poincaré* is almost as ambiguous on Intuition as Kant (see further 4.1.3). Apart from Intuition in the sense of Kant's 'Empirical Intuition' – in the senses or the imagination – (POINCARÉ [1905]), or as a simple mental generalization through some kind of induction (e.g. by extending the visual image of a simple geometric figure in the imagination to a more complex one), he speaks usually of Intuition in the sense of a purely mental grasping of fundamental principles or of reasoning as a whole. More than Russell's or Couturat's Intuition, needed for the initial principles and axioms of logic but dispensed-with at a later stage, Poincaré insists, as did Descartes, on the need for Intuition as a guide at all stages, in selecting axioms as well as the sequences of a logical argument:

> Which of all these possible routes is the best way to our final object? Who will tell us which to choose? We need a faculty that will show us from afar the final goal, and that faculty is Intuition. It is necessary to the discoverer in helping him to choose the way.(POINCARÉ[1905], p.26).

Intuition sometimes becomes identified with synthetic a-priori judgments basic to mathematics, fundamental truths which are immediately clear and cannot be proved analytically, i.e. reduced to logic. Such an a-priori synthetic judgment e.g. is the principle of complete induction, but also the axiom of choice (POINCARÉ[1902] and [1906], p.313). Intuition is also identified with mathematical reality, 'the deeper geometry'.

*Borel.* There are many references in Borel's work to Intuition, but none that explain the exact nature of Intuition. He stresses, as did Poincaré, the necessity of Intuition in mathematics and seems to identify Intuition with the 'deeper mathematical reality':

---

([BMS32]); Kronecker's name was again left off the list in the *The Cambridge Lectures* [BMS51].

> The negative 'freedom from contradiction' is not sufficient for mathematics, which must be founded on the deeper mathematical reality... (BOREL [1898], p.222);

and

> ... purely logical arguments, such as Hilbert's, have only verbal value; the constructed edifice does for the moment not bear any relation to reality...(op.cit., p.176).
>
> These logical arguments are symbols which do not correspond to any Intuition. (op.cit., p.181).

In his later work the demand for reality becomes the condition that the mathematical object can be thought in the concrete human mind (BOREL [1947], p.765) and can be expressed in a finite number of words (BOREL [1946]) (cf. also Lebesgue's 'nommable'). Sometimes Intuition is linked to so-called 'notions claires', which again are not further defined.

Of the Neo-intuitionists Poincaré has undoubtedly been the greatest influence in Brouwer. We have already mentioned Brouwer's letter to Korteweg in which he speaks of the deep impression Poincaré made on him, 'the highest ideal for any mathematician'. During the early years Poincaré's ideas were frequently the topic of conversation of Brouwer and his dialectical partner Mannoury. There is ample evidence in Brouwer's *Notes* [BMS3A] and *Foundations* [1907] of his study of Poincaré's work to-date. As most of Brouwer's references, they are critical and negative, and reveal fundamental differences in their appreciation of language. More radically different, however, were their attitudes towards science. Poincaré's universal interest embraced science as well as mathematics; he relished the mathematical challenge which the problems of science inspired. He aimed at a philosophy which would safeguard science from 'geometric empiricism' and both science and mathematics from axiomatic mechanization. Brouwer did not share Poincaré's interest in science; his sole and full-time concern was mathematics, which he saw in danger of trivialization at the hands of logicists and of subservience to science. He felt attracted by the Neo-intuitionists' appeal to Intuition as a higher authority. Their appeal to Intuition, however, was occasional; they felt safe in their reliance on Intuition by the assurances of their pure-philosopher friends. Brouwer probed and speculated on the precise nature of Intuition and its role in mathematics and, as Bergson had done for philosophy, made it the key and basis for a systematic philosophy of mathematics.

Brouwer's general philosophy, in which his 'Intuition' is grounded, has a strong Cartesian and Bergsonian flavour; yet in important respects his interpretation of notions such as 'mind', 'consciousness' and 'the exterior world' is special and warrants further consideration.

## 3.8 The Self, Consciousness and Mind

In Brouwer's philosophical analysis of the human make-up and man's interaction with the world outside distinctions are not mere philosophical and psychological conventions, they reflect a definite difference and separation in reality. There is first the almost polar opposition between spirit and body, and a chasm between the real world of the individual Self and the world outside. As to the internal structure of 'Mind', the various states or 'phases' which result from certain crucial happenings, such as the first sensitive experience, are disjoint stages in a stepwise evolution. The changes or 'transitions' from one phase to another are transformations as fundamental as the metamorphoses of the butterfly. Equally distinct are the human activities of mathematics, science and language, each pegged-to and characteristic of one of these states. This evolutionary process and the transitions are not wholly historical, taking place in succession of time, the order of events is ontological, follows the nature of things and actions.

The only continuous element, the unitary constant in this dynamic process, and the subject of these mental states is the Soul. In the most extreme Platonic and Cartesian form of spiritualism and in nearly all Western and Eastern religions the Soul is a spiritual entity, temporarily inhabiting a body but capable of leading a disembodied existence. According to the Pythagorean doctrine of metempsychosis, transmigration of souls, and the Eastern doctrine of re-incarnation and Karma, the Soul exists before its re-incarnation; its present state and destiny being phases of moral retribution continuing through eternity. Christianity believes in the creation of the Soul at the moment of human conception or birth. The Christian Soul is immortal, after death it continues to exist in its pure form, is united with God in pure contemplation or condemned to eternal seperation from Him. The nature and 'life' of the disembodied Soul are the subject of considerable speculation of scholastic theologians; as pure spirit it lacks the faculties of perception, its knowledge is direct union, providing mysteriously a criterion of self-identity and a source of happiness through 'Anschauung' of God.

Brouwer professes his firm belief in his own immortality:

> These my convictions lead to and include my immortality. ([BMS1A], 1).

Frequent references to 'Karma' in [1905] are evidence of his sympathies with the Hindu belief in re-incarnation, as is his active participation in spiritist seances during the years 1914-1916 *. In Brouwer's philosophical writings these mystical views receive only marginal attention and much of it can be considered to be private view, irrelevant to his philosophy of mathematics. Some aspects of Soul, however, do bear on his notion of Intuition.

His analysis of 'life in the mind' (e.g. [1948C]) accepts a pre-perceptional stage in the internal being of the Soul, a state of 'stillness' yet unaffected by sensations, an existence outside the realm of time and space, where the only knowledge is awareness of the Self (Du. 'zelf' or 'ikheid', i.e. 'I-ness'). For this self-awareness Brouwer does not need the intermediary 'cogito' of Descartes; in the singular universe of the Soul, 'the deepest home', the Soul and the I or Self are identified, it is the strongest form of knowledge: the direct union of the Soul and the Self, and, unlike any other, active knowledge, it is stillness.

The realm of the Soul is not bound by time or space:

> For time and space belong to my percepual images, and my I-ness is wholly independent of them. ([BMS1A], 14).

In *Life, Art and Mysticism* Brouwer confesses that no more can be said about the nature of the Soul or the Self:

> What this Self is we cannot further say; we cannot even think about it, for we know full well that all thinking and speaking is at a great distance from the Self. Neither can we get any nearer to the Self by means of thoughts or words, but only by 'turning into the Self' as it is given. (p. 13).

In Brouwer's metaphysical analysis the historical event of the Soul's incarnation – conception or birth – does not play any role. The first momentous happening is the event of perceptional experience. Absolute stillness has gone; the Self has left its pure solitary being and through this metamorphosis become 'Consciousness'.

---

* Testimony of Brouwer's stepdaughter Miss A.L.E. Peijpers; cf. also van Eeden's 'Diary' 15.05.16.

This and subsequent transformations are described as steps, transitions in a deteriorative process of moving-away from the inner being of the Soul towards the exterior world, an 'exodus from its deepest home'. Some of the transitions are so radical and fundamental that return is impossible, they are 'irreversible'. The first sensation has effected such a profound transformation, it is the Self's entry into time and conscious life. In identifying the Self and consciousness Brouwer does not suggest some Lockean identification of the knower and the content of consciousness, or personal identity and memory. Brouwer's 'Consciousness' is the name of the Self in its new state, which was caused by and started with the first perceptional experience.

While little could be said about the nature of the Soul or Self in its pre-perceptional state, the subsequent transformations of Consciousness, the genesis and development of mathematical, scientific and social activity were explored and described in detail in : *Mathematik Wisschenschaft und Sprache* [1929], *Willen, Weten and Spreken* [1933] and especially *Consciousness, Philosophy and Mathematics* ([1948C] and [BMS54]); and briefly in *Discours Final* [1950C]. The main ideas can be traced to *The Profession of Faith* [BMS1A] and *The Foundations of Mathematics* [1907] , especially in its first version [BMS3B].

Brouwer distinguishes between three main phases of consciousness:
1. the 'naive phase', opened with the creation of the world of sensations;
2. the 'isolated causal phase' of scientific activity;
3. the 'social phase'of social acting and language.

All three phases are moves away from the inner Self, 'consciousness in its deepest home'. At the *naive phase*, however, innocence is not yet lost. The only object of active attention is the Self, in stillness. Intermittently it receives its images through sensations, passively and spontaneously. The events of images are not yet linked and the spontaneous attention given to sensations is reversible. There is stillness between sensations; any reaction to sensations is direct and spontaneous, as yet there is no active, outward movement away from the Self, no e-motion, no activity of will:

> Consciousness in its deepest home seems to oscillate slowly, will-lessly and reversibly between stillness and sensation. ([1948C], p.1235).

[1950C] speaks of 'un monde-reve', a dream-world. It is a state when the transcendental values of goodness, truth and beauty still prevail:

> In the first phase of the Exodus there is beauty in the joyful miracle of the self-revelation of consciousness, appearing in the reversible formation of

egoic elements of the objects found in forms and forces of nature, in human figures and features. ([BMS54], p.5. This is a literal translation of the Dutch version of *Consciousness, Philosophy and Mathematics*. In the English, published version 'in the reversible formation of' is left out and the Dutch 'gelaatstrekken', i.e. facial features and expressions, probably in error translated by 'human destinies'. The word 'egoic' here means 'directed towards the Self'. Ed.)

It is immediately after this stage that Brouwer places the 'primordial happening'. So far the only *action* of consciousness has been self-awareness, the objective world does as yet not exist; we are still at the pre-reasoning stage, the a-priori stage of consciousness; the dormant faculties of memory, reason and will have as yet not been activated.

The 'primordial happening' is the first act of consiousness, the initial phenomenon. What sets it off is not clear; for Brouwer, who does not accept causation and necessity, it simply is a phenomenon:

> It seems that only the status of sensation *allows* the initial phenomenon of this transition. ([1948C], p.1235).

The primordial happening is a becoming aware of discreteness in time, an insight, an act of consciousness, which by its retaining and linking faculties creates its first object, the number two. It is a momentous, irreversible event which transforms consciousness from being active only in its self-revelation and passive in its undergoing sensations into an active, creative agent. The primordial happening is like a new birth, the conception of something new, to which Brouwer gives the name 'Mind'. He uses the English word 'mind' even in the Dutch original version ([BMS54]) of *Consciousness, Philosophy and Mathematics*, which describes the primordial event and the birth of Mind in Brouwer's usual pregnant way:

> By a move of time one sensation present in consciousness gives way to another sensation in such a way that consciousness retains the former as a past sensation, and moreover through this distinction between present and past, recedes from both and from stillness and becomes *mind*. (p.1235).

Mind is consciousness acting mathematically, creating the natural numbers and constructing further mathematical objects:

> As mind it takes the function of a subject experiencing the present as well as the past sensation as object. And by re-iteration of this two-ity phenomenon, the object can grow to a world of motley plurality. (ibid.).

At first only the pure mathematical objects exist in the microcosm of Mind, and no moral stigma is attached to these objects:

> There is the fullest constructional beauty, the introspective beauty of mathematics... where the basic intuition is left to free unfolding. (p.1239).

Yet the same mathematical ability of Mind plays an essential part in the creation of 'the exterior world of the subject', the onset of the second, 'the Causal Phase' of Exodus. Mind begins to interpret sensations mathematically: it links different sensations into sequences, and again links such sequences:

> It performs an identification of different sensations and different complexes of sensations, and in this way creates *iterative complexes of sensations*. (p.1235).

Special cases of repeated sequences are *things* and *causal sequences*. A 'thing' is no more than a sequence of sequences of sensations in which the elements, the sensations, do not necessarily occur in the same order in time: 'whose elements are permutable in time'. Other human beings, 'individuals', and even the body the Self inhabits, 'the home body of the Subject' are also 'things'; and the 'whole of things is called *the exterior world of the subject*'.

At first sight Brouwer's interpretation of the exterior world seems to resemble Hume's sceptical view of things as bundles of sense-perceptions, or at least Kant's phenomenalist doctrine, which rejects knowledge of things-in-themselves. There are, however, important differences. First of all, whereas Kant's exterior world is a public universe, based on a universal objectifying structure of the human understanding, Brouwer's exterior world is a wholly private one: it is the creation of the Mind, 'the exterior world of the Subject'. His argument against 'the plurality of mind' and 'the hypothesis of a collective super-subject experiencing an objective world' is based on the privacy of the exterior world, created by the individual Mind or Subject.

He distinguishes between Mind at the stage when it creates its exterior world and Mind at the next phase, when reasoning, causal acting and later when cooperative acting have set in. A public, objective exterior world of things, the same to all individuals, can only be the result of cooperation with other individuals, which takes place after each individual has created his own exterior world of objects. There is no guarantee that the exterior world of the subject, 'the contribution of the subject-individual relative to things', is the same as that of other individuals:

> In all cooperation which involves acts of the Subject, it appears to the causal attention within the system of cooperative causal acting, that the contribution of the Subject-individual in his relation to things is of the same nature as that of the object-individuals involved in the cooperation. ([BMS54], p.6).

Brouwer does not deny that other individuals exist nor that these individuals have some form of inner capacity, a 'capacity for action', which distinguishes them from other 'things'. His contention is that to the Subject these individuals are 'object-individuals', and that their 'inner capacity' and the Mind of the Subject are of an ontologically different order:

> The languages of the civilized nations, which are mainly 'languages of cooperation', have blurred (Du.'verdoezelen') the essential difference between the acts of the Subject and acts of object-individuals, by designating a word for the special capacity for action which distinguishes these individuals from other things. It would not be unreasonable to use the word 'reason' for this, but definitely unreasonable to use the word 'mind', because this implies that the Subject in his scientific thinking in fact places in each individual a mind dependent on that individual and with a free will, and thereby elevates itself to a mind of second order which would experience other, totally separate consciousnesses (which therefore are in fact disciousnesses) as sensations. Quod non est.' (op.cit., pp.6-7).

The notion of the Subject's own exterior world seems paradoxical, implying as it does on the one hand a separation from the Subject, the Self, and on the other hand a belonging-to the Subject. As so often, Brouwer has given the word 'exterior' his own special meaning. The metamorphosis from naive consciousness to Mind takes place as the objects are born. At that point it seems there are in the Mind two elements: the Self and the object, and there are two distinct directions of the Mind's attention, one towards the Self and referred-to as 'egoic', the other away from the Self and described as 'receding'. It is characteristic of the naive phase that directions are constantly reversed between Self and sensations: 'Consciousness in its deepest home seems to oscillate between stillness and sensations'. When, however, objects are created the attention can become irreversibly concentrated on such objects, the object becomes 'estranged' from the Self, there is a 'loss of egoicity'. *Will, Knowledge and Speech* [1933] seems to imply that there is a reversibility even of the major transitions, including the Primordial Happening:

> Every human being has the power to dream away time-awareness and

separation between the I and the perceptional world, or vice versa, re-call into being the perceptional world, the condensation of things. (p.46).

But he admits that:

> Regression from the second and third phase to the naive phase is difficult to realize, perhaps only a temporary flowing back to the deepest home, by-passing the naive phase... ([1948C], p.1238).

The exterior world of things therefore belongs to the Mind, yet it is exterior in the sense of its irreversible directedness away from the Self.

In the original manuscript of *Consciousness, Philosophy and Mathematics* Brouwer added in the margin a passage which was included in the later published version. It refers to the focus of 'egoicity' as 'Soul', and seems to point to the possibility of recognizing Soul in a Subject-individual by signs of egoic movement:

> The whole of egoic sensations indissolubly connected with the individual is called the Soul of the corresponding human being. The Soul connected with the subject-individual is rather latent but manifest in sensations of vocation and of inspiration. (p. 1235).

### 3.9 Causal Attention, Will and Cunning Activity

The movement away from the Soul towards 'objects' as means-to-an-end is an action of Mind, identified with Will. The transition from Mind to Will is intitiated by a phenomenon referred-to by Brouwer as 'causal attention': Mind viewing the objects in his world of sensations mathematically, interpreting them as 'causal sequences'.

A causal sequence is a sequence of sequences of sensations, in which the sensations, which form the elements of each sequence, follow each other 'necessarily' in constant time-order:

> A repeated complex of sensations of constant order in which if one of the elements occurs all following elements also occur, and this in the right order and without interference of the subject, is called a *causal sequence*. ([1948C], p. 1235).

In [1905] and [BMS3B] the seeing of causal sequences is referred to as an 'act of the Intellect' or of 'Reason'; [1907] speaks of 'viewing the world intellectually, i.e. with a mathematical, causal eye' ([BMS3B], p.1). In his later, more detailed analysis of human thinking Brouwer seldom uses the word 'reason' and prefers the term 'mathematical viewing', 'causal attention', 'causal thinking' or 'mathematical thinking', as to emphasize its mathematical origin and its distinctness from 'logical reasoning', which is a by-product of language, occurring at the third, the 'social phase'.

Causal attention, although described as 'intellectual observation' (e.g. [BMS3B], 2.10), is more than an act of perception. The regularity observed is not part of the reality of 'Nature', not even of the reality of the 'things' of the Exterior World of the Subject. It is a creative act which generates a causal coherence in the exterior world of the Subject, and is vaguely described as a 'phenomenon' or 'an expression of man's will':

> This 'seeing' is a human act of externalization... the seeing originates in man, is an expression of man's will alone ... ([BMS3B], p.35).

Causal attention is the 'phenomenon' at the threshold to the second phase in the Exodus of Consciousness, the phase of *causal activity* or *Mathematical Activity*; it lies in 'the dawning atmosphere of fore-thought', the Subject sees the possibility of predicting the future and acting purposefully.

In the paradisical state of naivety the Subject's reactions to sensations had been spontaneous, 'instinctive'. Causal attention enables man to see objects as desirable not for their own sake, but as a means to an end. It is distinct from the actual purposeful pursuance of objects, 'causal', 'mathematical' or 'cunning' acting, which is wholly identified with 'free will'. Causal attention may lead to, but does not necessarily evoke causal acting:

> Causal attention *allows* the conative activity of the Subject to develop from spontaneous effort to fore-thinking enterprise by means of the free-will phenomenon of the mathematical act.
> The mathematical act consists in this, that in a causal sequence a later element, which cannot be achieved by spontaneous conative action but which is nevertheless desired (the aim) is realized forcedly and indirectly by shrewd calculation. This is done by effecting an earlier element of the sequence (the means), which in itself probably is not desirable but which can be achieved, in order to procure ultimately the realization of the desired element as a consequence. ([BMS54], p.2; cf. [1948C], p.1236).

Mind is changed fundamentally and morally by acting upon knowledge gained from causal attention, the activation of the free will, or the 'phenomenon of *cunningness*, which creates *the acting man*' ('le phénomène d'astuce créant l'homme agissant', [1950C]). The 'cunning act' or 'mathematical act'* is the original sin in Brouwer's vision of cosmic events:

paradise is lost and naivety gone. Activity of the subject at this stage, mathematical or causal activity, is 'cunning', 'shrewd' and 'calculated', 'not beautiful', i.e. morally evil. The word 'morality' is avoided in this context, as it smacks too much of the bourgeois code of accepted manners and decency, but Brouwer leaves no doubt as to his conviction of the moral evil inherent in 'causal activity'; it is the main message of *Life, Art and Mysticism* and a constantly recurring theme in *The Foundations* and in his post-1927 publications. The pessimism which is characteristic of these periods in Brouwer's life is a fatalistic acceptance of the free will as part of human nature and of causal activity as unavoidable. Such pessimism, based on the role of the free will, is also prominent in the work of Schopenhauer, with whom Brouwer has much in common. Both consider the free will as the source of all evil, as do indeed many moral philosophers. Traditional ethics considers the will to be a faculty of free choice; whether the chosen act is good or evil is determined by individual conscience, often consulting the collective judgment of society; the will is capable of freely choosing a good object. Schopenhauer's Will is the creator of his 'real world', which is inherently evil. Brouwer identifies 'Will' with the mind pursuing an object as a means to an end and condemns all such acting as evil, causal acting and evil are identical; the creation of his 'exterior world of the Subject' precedes activity of the Will, is an activity of Mind not yet corrupted by causal acting, by Will.

The impact of causal attention and causal acting on Mind is such that 'regression to the naive phase' is hard to realize:

> Mind, having once entered the phase of causal attention, remains in a lasting state of causal thinking, i.e. searching for causal sequences and for possibilities of creating and securing causal sequences, and of causal acting, i.e. activity resulting from causal thinking. ([BMS54], p.3).

Redemption, a return to the naive phase of 'playful' and 'beautiful' acting, even with perceptional images, is possible but only if one succeeds in abandoning the purposeful or causal approach:

> ...if elements of causal activity founded on the primary forces of nature are stripped of their utilitarian character, i.e. their conative content, and of the resulting restrictions of their combinations so that they are given the possibility of free unfolding and producing things of perceptional beauty. ([BMS54], p.6).

---

\* In the original Dutch version of [1948C] Brouwer simply speaks of 'de wiskundige handeling' or 'the mathematical act' (p.2); in the corresponding passage of the published English version this is changed to 'the cunning act' ([1948C], p.1236).

## 3.10 The Phase of Social Acting and Language

The characteristic feature of the third phase of Consciousness, the 'Social Phase', is described in [1929] and [1933] as 'the enforcement of will (on others) by means of sounds' ('die Willensauferlegung durch Laute'). Again, the Social Phase does not follow the preceding stages necessarily, it is 'an anthropological phenomenon'.

The fundamental happening which effects the transition to this phase is not the creation of language as some condensation of human thinking, it is a change in the direction of Will.

'Will' itself already implies a move away from the Ego, a movement towards an object as a means to an end. At the simple causal stage the move orginates in the Mind, it is cunning activity to serve the immediate purposes of the Subject. There is a further move away from the Ego if cunning activity serves the purposes of other, 'non-egoic objects'. Cunning activity in the service of others, 'the sub-servience of mathematical attention and mathematical activity to non-egoic objects, is called labour' ([1933], p.49).* The Subject is forced to 'act mathematically' by the will of other individuals, 'the collective will of a group of people, or of the whole of mankind'. The transformation is from 'cunning Mind' to 'social Mind', i.e. Mind acting under pressure of others, though still acting freely, i.e. cunningly. Brouwer speaks of:

> preparedness for labour... evoked in the Subject either directly by suggestion, i.e. by striking terror or fright, by seduction, brute force or by arousing the imagination, or indirectly by so-called training [Du. 'dressuur' exclusively used for training animals to perform tricks. Ed]. (ibid.)

The essence and function of 'language ' is no more than a 'means of will-transmission'. The role of (written) language as 'an aid to memory' is mentioned as an after-thought, but its origin and purpose is the enforcement of will:

> At the most primitive stages of civilization and in the most primitive relations of man to man this transmission of will from one individual to another is brought-about by simple gestures, in which the primitive natural sounds of the human voice play a predominant role... ([BMS54], p.4).

---

* This distinction provides an interesting key to understanding Brouwer's interpretation of the respective role and the 'liberation' of man and woman in *Life, Art and Mysticism*.

A more sophisticated language is created with the development of human society:

> ... when in a more radical organization of such groups the acts to be imposed become too differentiated and complicated and can no longer be indicated by a simple call. In order to be able to continue to conduct business in these circumstances by means of auditive or visual signals of request or command, the whole of rules, regulations, things and theories is now subjected to causal attention... (ibid.; cf. also [1948C], p.1237 and [1933], pp.49-50).

Brouwer's views on the nature of language and logic will be further discussed in the following chapters, as will his views on Intuition, Mathematics and Science. In the development of Mind language does not play an essential role; its only function is that of instrument, a defective instrument in human inter-action. Mind remains a private independent domain closed to others, a 'Cita del Vaticano'. Direct communication with other minds, if these exist, is impossible; language 'only touches the outside of an automaton'.

Brouwer was well aware that his development of Mind through all its stages does not represent the historical course of events and that none of these phases describes the present reality as it affects the individual or society. It is his metaphysical analysis of the human mind in its essential elements and activities, the basis of a philosophy in which science, mathematics and logic can be allocated their proper place and role, a philosophy which was inspired by and remained true to his romantic vision of a good natural world, spoiled by the cunningness of man, a nostalgia for Paradise. If there is any change in Brouwer's attitude from the early days of *Life, Art and Mysticism* to his later years, it is perhaps a growing confidence or hope that some of this Paradise can still be regained:

> Even at the second and third stage there is beauty in remembering the wonders of past naivety, a memory evoked either by direct reflection through a mist of wistfulness and nostalgia or by works of visual or aural art or certain forms of scholarship and science; the kind of science which, with great reverence, reveals to the yet unarmed senses the primary forces, forms and laws of Nature, and that without tools or with modest tools that do not violate Nature, or which shows them in their full glory by precise registration; a science which, because of its deep respect, rejects any expansion of man's domination of Nature. ([BMS54], p.5; cf. [1948C], p.1238).

# CHAPTER IV
# BROUWER'S PHILOSOPHY OF MATHEMATICS

## 4.1 The Primordial Intuition of Time

*4.1.1 Introduction*

Brouwer's search for the 'Genesis of Mathematics' started as a critical examination of current philosophies of mathematics, the various forms of 'formalism' as well neo-intuitionism, identifying and rejecting elements which according to his philosophy originated in the base or 'sinful' tendencies of human nature, i.e. elements of science as causal acting and linguistic elements as part of social acting. His interpretation of mathematics as constructive thinking on the basis of the Primordial Intuition alone emerged by a process of elimination of these elements and by pure philosophical reflection on the nature of reasoning and the problem of time.

He shared the gut reaction of the French neo-intuitionists against current logicist and formalist trends, their conviction that mathematics needed to be anchored directly in the human mind, but he was more radical in his search for the precise nature of that contribution and in his acceptance of the consequences of his foundational convictions. The neo-intuitionist interpretation of 'intuition' had remained vague and of little consequence to mathematical practice, a kind of synthesis of ideas, owing its significance and applicability to to these concepts being subconsciously associated with and rooted in experience. Language still played a positive role in adding to individual experience, communicating a collective experience or cultural heritage, and more directly, as a medium in which ideas are isolated and capable of being manipulated, or even used as a condition for mathematical existence.

Brouwer began his search for the origin and nature of mathematics at the start of his mathematical career. He did not draw on a long experience as a mathematician; he could afford to be radical, he did not have a paradise to lose. His commitment to mathematics as mind-activity was total, as was his rejection of language as contributing in any way to the creation of mathematics. Moreover, he was not prepared to accept a contribution from man's exploration of the physical world. His aversion from science and applied mathematics made him more clearly dissociate mathematics from experience.

### 4.1.2 *Genesis of the Concept and Definitions*

There are clear indications in *The Foundations* that Brouwer's peculiar notion of the Intuition of Time arose from a critical examination of the physical sciences, inspired by a conviction of their immorality and a will to expose their true nature.

In his pure-mathematical survey, Chapter I *The Construction of Mathematics,* the Primordial Intuition is only mentioned in the context of the continuum; the sequence of ordinal numbers is accepted as 'known' and 'intuitively clear'. The first account of the *Primordial Intuition of Time* appears in Chapter II, originally headed 'The Genesis of Mathematics related to experience'. In its original, unexpurgated version Chapter II is mainly a critical exposure of science, the words 'mathematics' and 'mathematical' are hardly used. In the opening statements the origin of science is placed in 'man's will to destroy and rule' and science is analysed as 'seeing in the world causal sequences in time'. These statements are immediately followed by the account of the Primordial Intuition, the first and only one in the whole of *The Foundations*, apart from a further mention in the final summary. It only occupies six lines, but it does contain all the essential features of what would become Brouwer's standard definition:

> The primordial phenomenon is no more than the intuition of time, in which repetition of 'thing-in-time and again thing' is possible, but in which (and this is a phenomenon outside mathematics) a sensation can fall apart in component qualities, so that a single moment can be lived-through as a sequence of qualitatively different things. ([BMS3B], p.2).

In the published version the clear 'moral' introduction is left out – against Brouwer's will – and the Intuition of Time is introduced as the fundamental phenomenon of causal attention or 'mathematical viewing':

> Man has a faculty which accompanies all his interactions with Nature, a faculty of viewing his life mathematically, seeing in the world repetitions of sequences, causal systems in time. The fundamental phenomenon here is no more than the intuition of time, which makes repetition of 'thing-in-time-and-again-thing' possible and on the basis of which moments of life fall apart as sequences of qualitatively different things; these are subsequently

concentrated in the intellect as mathematical sequences, not sensed but perceived. ([1907], p.81).*

The word 'here' is significant, it links man's causal interpretation of nature directly to the fundamental phenomenon. Significant also is the reference to 'things', which in both versions of *The Foundations* are elements of the Primordial Intuition. In Brouwer's later interprepations 'things' as well as causality are themselves sequences of sequences, part of the exterior world of the Subject and upstaged to the second phase of consciousness. (Cf. 3.9).

Brouwer's later versions of the Primordial Intuition add little new to the cryptic description given in *The Foundations*. There is a change in the prominence given to the Primordial Intuition and its role in pure and applied mathematics. The account of the primordial happening takes the central place in *Intuitionism and Formalism* [1912A] and omits all reference to application; the Primordial Intuition of Time is the 'fundamental phenomenon' of pure mathematics:

> ...the intuition of two-oneness, the primordial intuition of mathematics which immediately creates not only the numbers one and two but all finite ordinal numbers... (p.12).

Post-1927 publications revert to the original pattern: causality again seems foremost in Brouwer's mind, the account of the Primordial Intuition as the source of pure mathematics emerges after a lengthy moral introduction on causality, a pattern which may well reflect the route by which Brouwer arrived at his notion of Primordial Intuition.

*The Foundations* suggests an independent development of causal viewing and of pure mathematics, both on the basis of the Primordial Intuition:

> The process of objectifying the world through the primordial intuition of 'repetition in time' and 'following in time' gains in generality by the construction of mathematics from the *same* primordial intuition without reference to direct applicability. ([BMS3B], p.3).

This distinction and the almost incidental reference to 'mathematics' and 'mathematical' reflect a stage in Brouwer's thinking when he had solved

---

* The translation of the *Collected Works* omits the word 'and' (Du. 'en op grond waarvan...'), giving the wrong interpretation that 'moments of life break up' as a consequence of 'repetition in the form of thing and again thing'; in the original version this clause depends directly on 'the intuition of time'.

the problem that had pre-occupied him from the days of the *Profession of Faith*, the problem of scientific necessity, and began to realize that this solution was also the key to and basis of a new philosophy of pure mathematics. Scientific necessity, causality is the main topic of Chapter II. Causality is analysed and found to be no more than human observation and interpretation of events following each other in time, the necessity of science ultimately reduced to the time-relation of before-after, 'the only a-priori element in science is time'. *Life, Art and Mysticism* had blamed the loss of the paradisical state of spontaneous acting on the emergence of the Intellect 'which breaks down the independence of fleeting phantasies by linking them ... providing a develish service by linking two phantasies as means and end '(p.19). The *Foundations* exposes the precise nature of this link:

> The primordial phenomenon is the becoming aware of 'thing-in-time and again thing'... the Intuition of Time. ([BMS3B], p.2).

The possibility of constructing the sequence of ordinals is implied but not spelled out; in the course of the argument Brouwer even observed that the intuition of many-oneness is sufficient for the whole of mathematics. The emphatic final conclusion, however, refers to science and to time as its sole a-priori:

> Since scientific experience originates in application of intuitive mathematics, and there is no science except experimental science, using only properties of intuitive mathematics, we can only call a-priori that which is on the one hand common to all intuitive mathematics and on the other hand sufficient for the construction of the whole of mathematics, the intuition of many-oneness, the primordial intuition of mathematics. And since this intuition is nothing but the becoming-aware of time as no more than change we can say: *the only a-priori element in science is time*. ([1907], p.99; italics are Brouwer's).

### 4.1.3 Brouwer's Primordial Intuition of Mathematics

Although bound-up with and arising from the analysis of science and causality, Brouwer's Primordial Intuition of Time as the only a-priori of mathematics is perhaps the most momentous discovery of the *Foundations*; the discovery not only that from awareness of time numbers and the whole of mathematics can be generated, but that discreteness and awareness of discreteness can only exist in time as a dynamic continuum and is based on its fundamental order-relation of before-after.

In spite of Brouwer's own assurance that all that Intuitionism did or had to do was 'to strengthen Kant's a-priori Intuition of Time', his interpretation differs in almost every respect from Kant's a-priori Form of Intuition. It also differs from Bergson's philosophy of time and his conception of intuition.

Kant's a-priori Intuition of Time is a 'pure form of intuition'; although described as a contribution of the mind to knowledge, it is a rather passive characteristic of human thinking, the essential temporality of consciousness,

> ...a pure intuition, which even without any actual object of the senses or of sensation, exists in the mind a-priori as a mere form of sensibility. (KANT[1787], B.35).

Brouwer's a-priori intuition of time is invariably described as a 'phenomenon', or 'a happening' (Du. 'oergebeuren'). Unlike Kant's pure form, which is a permanent element of the human mode of thinking, the primordial Intuition is an event in the genetic order of things. Consciousness in the pre-mind stage had responded spontaneously but automatically and determinately to sensations; after the primordial event the initiative for action passes to the mind, consciousness as mind acts freely. The event of the Primordial Intuition is the simultaneous presence in consciousness of two sensations, one recognized as past, the other as present. The seeing of these, linked and yet separate, different in as much that one is recognized as 'before' the other, is almost a self-generating act of mind. Circularity of self-generation is avoided by careful use of the word 'phenomenon'. Yet the Primordial Intuition is a human act, an intuition, a direct grasp of the time-relation before-after; it is primordial or primitive in the sense that its constituent elements are essential but non-isolable aspects of the single, most fundamental act of mind, the instantaneous grasp of the before-after, past-present in the one-dimensional single-sense vector space of time.

Brouwer side-steps the issue of what initiates the 'phenomenon'. His ontological analysis, concluding the sole a-priori of time, leads to a genetic hierarchy of 'events' in the 'life of the mind of the Subject', the 'first' being the becoming-aware of time. Science, all reasoning and the whole of mathematics can ultimately be reduced to no more than the ordered sequence of numbers; since the only necessary and sufficient ingredient of number and of discreteness is the relation 'before-after', the becoming-aware of this aspect of time must genetically be the first, the primordial happening.

Bergson's intuition of time is the immediate grasp of time as continuum, as pure fluidity, 'real time'. His solution of the dichotomy of time as continuum and time as a succession of isolated points, indivisible moments in a homogeneous medium, was to attribute all aspects of discontinuity, discreteness and measurability not to real time but to mechanical, mathematical, spatialized time. Whereas real time is pure continuity, fluidity, and is grasped by 'pure spirit' in intuition, mathematical time is a homogeneous spatial frame superimposed on time by the analytic intellect, 'spatialization of time'. Quantitative measure of time and discreteness presuppose the static, spatial mathematical continuum and number; number is based on the juxtaposition of units in homogeneous space:

> In space alone is a juxtaposition of this kind possible, because space alone, being homogeneous, permits units to be differentiated and identified at the same time, that is to be added together. (BERGSON[1889], p.77).

Brouwer recognized the distinction between intuitive time and measurable time, to which he referred as 'scientific time':

> Scientific time certainly is a-posteriori, appears only in experience, and as a one-dimensional coordinate with a one-parameter group structure can be introduced for the cataloguing of phenomena. ([1907], p.99 footnote).

To him, however, measure is not as elemental as number; he isolates number and measure. His intuitive time is a non-metric topological linear continuum, and the invariant relation before-after sufficient for the genesis of number. Moreover, Brouwer did not accept the a-priori of space w.r.t. number, nor Bergson's mathematical, static continuum; he realized that human awareness of juxtaposition of units in homogeneous space requires 'the move of time' and his mathematical continuum shares the dynamic character of the 'time continuum'.

Both Bergson and Brouwer describe the Intuition of Time as a direct grasp of 'change'. In Bergson's interpretation change is purely qualitative, 'from this to something else'. Brouwer seems to agree with Bergson when he states that 'the Primordial Intuition is the becoming-aware of time as nothing but change' ([1907], p.99), and that this change is qualitative:

> ...the falling apart of moments of life into qualitatively different parts to be re-united while remaining separated by time. ([1912A], p.12).
> ... the becoming aware of time as the falling apart of life's moment into two qualitatively different things of which the one makes room for the other but in spite of this is retained by memory. ([1933], p.45).

In his definitions, however, there is a greater emphasis on the attendant difference in time as a succession of heterogeneous moments: the simultaneous presence in the mind of images of 'this then' and 'that now'.

Brouwer leaves aside the problems of simultaneous presence in the mind of an image in memory and a present sensation, the link between them, and their distinctness, problems which were central to Bergson's philosophy. He states their distinctness, 'apartness', and concentrates on its all-important time-element: 'consciousness retains the former as a past sensation'.

The essential step to pure-mathematical abstraction is described in [1912A] as 'abstraction from every emotional content' and in [1929] and [1933] as 'abstraction from all content' leading to 'temporal two-ity...temporal three-ity and the temporal sequence of arbitrary multiplicity'. Little is said about the abstracting move from temporal to mathematical discreteness. [1948C] simply states that 'divesting of all quality' results in 'the empty form of the common substratum of all two-ities' (p.1237).

The mathematical abstraction of the Primordial Intuition preserves a 'natural' order-relation, a generalization of the 'before-after', in [1908A] referred-to directly as 'first and 'second':

> In the Primordial Intuition of two-oneness 'first' and 'second' are being held together. ([1908A], p.569).

The essential 'before-after', 'first-second' relation makes the Primordial Intuition more than a constructive principle of sets, more than a 'collecting into a whole' of juxtaposed units: it is the first step in the construction of a sequence, the principle of induction.

Succession in time is also implied in yet another facet of the Primordial Intuition, the 'again', which, given the freedom of the human mind, is extended to 'and-so-forth', leading in 'unlimited unfolding to the creation of new mathematical entities in the shape of predeterminately or more or less freely proceeding infinite sequences.' ([1948C, p.1237).

### 4.1.4 The Discrete and the Continuous in the Primordial Intuition

From his discussions with Brouwer and from reading the *Foundations* Mannoury had formed the impression that Brouwer's philosophical foundation of mathematics was 'the certain, universally human intuition of continuity, identified by the author with the intuition of time...' (MAN-

NOURY[1907B], p.175). In spite of some references in the *Foundations* linking the continuous and the discrete, the emphasis in the account of the Primordial Intuition of time is exclusively on the discrete. Identification of the time intuition and continuity seems more in line with Bergson's view, well-known to Mannoury.*

While Bergson sought to isolate the continuous and the discrete, Brouwer tried to encompass both within Intuition, 'a unity of the discrete and the continuous'. When he seems to echo Bergson's view on the impossibility of constructing a continuum out of indivisible points, he rules the premises of the argument out of order: neither the discrete nor the continuous can be thought of as wholly isolated, independent of the other:

> Since in the Primordial Intuition the continuous and the discrete appear as inseparable complements, each with equal rights and equally clear, it is impossible to avoid one as a primitive entity and construct it from the other, posited as the independent primitive. ([1907], p.8).

The complementary aspect of the continuous and the discrete is not to be taken as simple negation: an essential requirement of discreteness, apartness of moments in time is the existence of a 'between', on the other hand awareness of change, 'the move of time', is only possible by recognition of a 'before and now', a past and a present moment. Particularly significant is Brouwer's claim that the discrete cannot be wholly isolated as the primitive, 'the original entity'. His Primordial Intuition is not just a further, philosophically based justification of the natural numbers, accepted by other constructivists as their starting point, nor is it the intuition of a set of units, not even a set of discontinuities. Numbers, ordinal numbers, are conceived in Intuition as discrete points in a linear time continuum, 'the discrete in the continuous'. This descendancy enabled Brouwer later to clarify the structure of the 'dynamic continuum'.

The 'primordial' nature of the time intuition does not allow it to be further decomposed into more primitive, self-contained and wholly isolable elements; any aspect is bound-up with others in the one act and is itself apparent in various facets. This applies in particular to the continuous aspect of the Primordial Intuition. It represents the fluid of the continuum and 'constancy in change' ([1907], p.179), consciousness as the constant bearer of change as well as its active role of linking a past

---

\* There is in the remains of Mannoury's estate evidence of his study and analysis of Bergson's philosophy and of a meeting of Bergson and Mannoury in December 1906.

sensation to a present one, 'the thinking-together of more units' ([1907], p.8).

The continuous as the synthetic, linking aspect of the Primordial Intuition is the fundamental binary construction, complementary to discreteness as breaking-up or 'falling-apart':

> In the Primordial Intuition of two-oneness the intuitions of continuous and discrete meet; 'first' and 'second' are held together, and in this *holding-together* consists the intuition of the continuous (continere = hold together).([1908A], p.569).

The breaking-up of a present moment into a present sensation and the memory of a past sensation is as essential as the joining of the two. Even the further process of constructing 'three' and other numbers and entities requires a breaking down of one of the two constituent elements:

> In the temporal two-ity emerging from time-awareness one of the elements can again and in the same way fall apart, leading to temporal 'three-ity', or three-element time sequence is born. Proceeding this process, a self-unfolding of the primordial happening of the intellect, creates the temporal sequence of arbitrary multiplicity. ([1933], p.46).

Brouwer is at pains to point out that not the unit, individuality, not even the singularity of the present moment is the basis of number. The new entity resulting from linking past and present constituents, the fundamental number is 'two-ity':

> In the first constructive act there are two though-together discrete things (as indeed Cantor says in his Vortrag auf der Naturforscherversammlung in Kassel 1903). F. Meyer says that *one* thing is sufficient since I can add as second thing the fact that I think it. This is incorrect since precisely the *adding to it* (i.e. positing while retaining what was thought before) *presupposes the intuition of two* and this mathematical structure is only afterwards applied to the original thing as thought and to the I who thinks it. ([1907], p.179).

Whereas the continuous as the unifying, linking aspect in the creation of two-ity is primarily a function of consciousness, the continuous as 'the between' reflects the underlying continuity of time on which the mathematical continuum is modelled. Brouwer clearly distinguishes between these two aspects of the continuous:

> We recognize that the continuum intuition, the 'fluid', is as elemental as the thinking-together of more things into one'. ([1907], p.8).

Both aspects play a role in the creation of the mathematical continuum; but whereas the linking aspect is most prominent in 'constructing a scale on the continuum' and in 'making it measurable', the fluidity of time, time as a linear continuum as grasped by consciousness in the Primordial Intuition, is the immediate source of the mathematical continuum. The statement of *The Foundations* that 'the Continuum is intuitively given' and Brouwer's continued identification of the Continuum and 'the intuition of the between' do not conflict with his constructive demands nor with his later development of the continuum as 'medium of free becoming'.

If the accounts of the Primordial Intuition are rather cryptic and vague about the process of creating two-ity from time-awareness (cf. 4.1.3), they are even less specific about the process of creating the mathematical continuum from the intuition of time. In the Primordial Intuition time as 'fluidity' is grasped directly together with the discrete 'then' and 'now'; the distinction of moments in time is only possible because of the 'between' and yet the 'between' presupposes the existence of endpoints. No specific mention is made of the process of abstracting the mathematical continuum from the time-interval, the temporal 'between'. The mathematical continuum of [1907] is immediately identified with the 'between', a mathematical 'between' which has retained some characteristics of the temporal 'between' but has gained its own distinct identity with a new balance and new emphasis. The mathematical abstraction in particular has preserved a 'natural' order, the generalization of the 'before-after', the dynamic order of the irreversible move of time. The emphasis of the mathematical 'between', however, is on its infinite potential for insertion: 'the between that is never exhausted by the insertion of new units' ([1907], p.8). The dynamic, growing aspect of the mathematical continuum is the constructive action of the Subject, generating sequences of insertions which do not necessarily follow the 'natural' order.

It is this aspect of the mathematical continuum, consistently referred-to throughout Brouwer's works as 'the intuitive between', which became the basis of his new interpretation of real number in 1917 (see further Chapter VI).

## 4.2 Intuitive Thinking and Mathematics

The use of the words 'intuitive' and 'instinctive', especially in Brouwer's earlier writings, has led some commentators to conclude that his intuition carries the more general connotations of 'instinctive knowledge' or 'recognition of self-evident truths'.

Even if Brouwer's 'intuition' and 'intuitive' extend and apply to mind-activity beyond the Primordial Intuition, it can in no way be identified with instinct in the sense of knowledge inherent in the human race or with individual knowledge which dispenses with mental effort, a kind of hunch, 'inspiration', a knowledge off-hand. The 'instinct' and 'instinctive' refer to the direct, determinate response to sensation characteristic of the 'naive' stage which precedes the free stage of Mind.

Neither can Brouwer's 'intuitive' be identified with 'self-evident' as is suggested in KÖRNER[1960]. Self-evidence implies complete elimination of effort on the part of the Subject, the evidence being supplied by and intrinsic to the known object. Apart from the fact that Brouwer never uses the word 'self-evident', he did not recognize 'truths-in-themselves' (e.g. [1948C], p.1243 and [1955], p.113). Any analytic statement which could claim such self-evidence is a dismissed as a non-statement or 'meaningless'.

Brouwer does, however, use 'intuition' in the sense of the human faculty of knowing directly, independent of reasoning. Indeed, the Primordial Intuition is nothing but this faculty in action; it lies at the root of all thinking and is the ultimate foundation of 'true' reasoning.

In his genetic conception of life, Mind from its birth in the primordial happening is active along two distinct lines:

1. It uses the content of sensations and the mathematical intuition to create 'things', 'the exterior world of the Subject'; it operates combinatorially on the objects of this world and projects its findings and expectations on the physical reality, a process referred-to as 'externalization'.
2. Restricting itself exclusively to the elements given in the Primordial Intuition and the tools provided by it, Mind constructs new entities in 'unlimited unfolding'. The full constructive power of the reasoning intellect is employed, but since it operates only on and with objects which originate ultimately and exclusively in Intuition, this mind-activity can rightly be called *intuitive thinking*.

Pure mathematics is identified with intuitive thinking; it embraces the acts of Intuition as well as the intellectual constructions with the a-priori data of the Primordial Intuition. Taken in this sense, intuition circumscribes and delimits the domain of mathematics, defines its restrictions and freedom. As intuitive activity it is 'introspective', inward-looking, completely independent of the content of sensations; it is restricted and debarred from using 'extraneous' elements, including ideas which are 'iterative complexes of sensations', the exterior world of the Subject, including also such extraneous elements as language and logic. On the other hand there is complete freedom to act within the domain of the a-priori consciousness. Mathematical, intuitive thinking is free from the restrictions imposed by extraneous elements, free also fro the arbitrary restrictions self-imposed by formalization. Both this freedom and the limitation of acting within the a-priori data and with the tools of intuition are the ultimate criteria for intuitive mathematics:

> The only possible foundation of mathematics is to be found in this intuitive construction, or in the obligation to observe carefully what intuition does and does not allow. ([1907], p.77).

Intuition, therefore does not only play a role in the initial stages of the conception of number, it is a constant guide and determines the structure of mathematics at every stage. The use of the word 'primordial' (Du. 'oer') and the emphasis on the 'primordial happening' might lead one to interpret Brouwer's intuition as some pre-historic, once-and-for-all event of which only the effects have survived in the form of some basic constructs such as number. The latter could then equally well and more conveniently be adopted as primitive concepts, leaving the name 'intuitionist' as no more than a nominal or historic epithet. Such interpretation, however, ignores the fundamental unifying and justifying principle which underlies Brouwer's philosophy of mathematics, and misrepresents his Intuition. As was stressed above, 'primordial' is to be taken in the sense of a metaphysical a-priori; it recognizes the inadequacy of combinatorial reasoning and the ultimate need for some principles which are established as true directly and independently of sensation and of reasoning. Intuition is primordial in the sense that it underlies all reasoning and every mathematical structure. The mathematician, at any stage of his construction in need of basic constructs and tools, can consult Intuition, not by relying on some vague feeling, but by 'introspection', re-living and reflecting on the primordial happening, 'a regression which is hard to realize' and requires 'detachment concentration' ([1948C], p.1248). The final summary of *The Foundations* lists some of 'the constructional elements which can be read from the Primordial or Continuum Intuition, notions such as *Continuous, unity, again, and-so-forth...*' ([1907], p.180).

Intuitive thinking or intuitive mathematics demands utmost rigour and discipline, to be constantly on guard against the intrusion of 'extraneous elements'. Much of Brouwer's Intuitionist campaign was negative in the sense that it had to eradicate non-intuitive elements which had crept into mathematical practice. But on the positive side Brouwer could claim Intuitionism to be a coherent philosophy of mathematics providing a foundation on which the whole of mathematics can be constructed:

> From the Primordial Intuition all mathematical systems including spaces and their geometries can be constructed independently of experience. ([1909A], p.5); and:
> Mathematics is a free creation of the mind independent of experience; it develops from a single a-priori intuition. ([1907], p.179).

## 4.3 The Nature of Mathematics

> Intuitionist mathematics is an essentially language-less mental structure which comes into being by the self-unfolding of the abstraction of two-ity as the Primordial Intuition. ([1947], p. 339).

The problem of the relation between knowledge actually in the individual mind and the collective body of knowledge to-date, recorded and systematized, is not peculiar nor wholly confined to mathematics. This double aspect of any knowledge is generally understood and does no enter into the definition of the discipline concerned.

Brouwer's identification of mathematics with mathematical thinking rests partly on the unique place he claimed for mathematics among human activities, partly on his views on language and cooperation. These led him to distinguish sharply between:
1. knowledge as it is gained first-hand and held by the individual mind;
2. its recording in a symbolic, physical medium 'to aid the memory';
3. the communication of such recording between individuals and the recording and cataloguing of so-called collective knowledge.

He insisted that any philosophical consideration of a particular form of knowledge, its nature and distinctive characteristics, should concentrate exclusively on the first phase, the manner of its conception and the distinguishing features of its operation in the individual mind. He stressed that subsequent, arbitrary recording in an alien medium does not change, add-to or clarify the essential characteristics of this knowledge. As to collective knowledge, it relies in its formation on recording and communication and lacks a basis of reality.

While this distinction holds for all branches of human knowledge, it is particularly relevant in the case of mathematics. At the first level, as knowledge originating and residing in the individual mind, pure mathematics can claim truth and reality as defined by Brouwer:

> Truth is only in reality, i.e. in the present and past experiences of consiousness. ([1948C], p.1243).

Such truth and reality is also attributed to naive experience, i.e. simple sensations as such, but not to scientific or causal thinking, the creation of the exterior world in the mind of the individual. Even at the first, individual level scientific knowledge lacks the transcendental values of truth and of beauty, i.e. moral goodness; recording and communication further compound errors and immorality. Mathematical thinking, originating exclusively in the Primordial Intuition and wholly confined in the individual mind, has these transcendental qualities in full measure. They are the characteristic attributes of mathematical, intuitive thinking, but not necessarily of its subsequent recording. In particular the special a-priori truth, traditionally claimed for mathematics, is the exlusive prerogative of 'language-less activity of the mind exclusively based on the Primordial Intuition'.

Brouwer's world of pure mathematics is the individual mind acting on the Primordial Intuition, self-contained and in total isolation from the physical world, 'the thought-world of solitude' ([1933], p.52), 'exclusively a matter of individual consciousness of the Subject' ([BMS47], p.2; app.8). In analogy with Brouwer's 'exterior world of the Subject' one could refer to it as *the interior world of the Subject* (although Brouwer has not used this expression himself). This conception of mathematics determines all the distinctive features of intuitionist mathematics: the constructive nature of its operation and of mathematical existence, as well as its cautious attitude towards contributions introduced by 'application' and logic.

Within the world of mathematics, the interior world of the Subject, there is unlimited scope for further construction with the elements of Intuition:

> ...when the primordial Intuition is left to free unfolding; this free unfolding

is not bound to the exterior world and thereby to finiteness.([1948C], p.1239).

However, since mind creates its exterior as well as its interior world there is the possibility and danger of straying outside the boundaries of the pure-mathematical domain. Already *the Foundations* shows Brouwer's awareness of this danger, of the need to stay within the system and the precise boundaries of what he called 'first-order mathematics'.

Pure, first-order mathematics is rigorously restricted to the actual constructive activity with the elements of Intuition, 'the pure construction of intuitive mathematical systems' (p.173). Constructive activity with non-intuitive elements is obviously excluded from the pure-mathematical domain. More surprising is Brouwer's distinction, already in 1907 between first-order mathematics and the meta-mathematical consideration of its properties. Discussing a-priori synthetic judgments on the basis of the Primordial Intuition (and listing some, such as 'the possibility of mathematical synthesis, the possibility of intercalation and the possibility of proceeding ad infinitum') he states:

> They are the result of viewing the Primordial Intuition mathematically, and therefore presuppose the Primordial Intuition in what is being viewed as well as in its viewing; they belong to what in the next chapter we shall call mathematics of the second order. ([1907], p.119).

This demarcation between 'pure-mathematical activity' and meta-mathematical consideration of its properties applies a-fortiori to any consideration of mathematical activity as expressed in symbols. In *The Foundations* Brouwer analysed Hilbert's formalization and referred to such meta-mathematical consideration as 'mathematics of the third order':

> ...a mathematical system of the third order, which [supposedly] justifies the construction of the system and shows its non-contradictority. (p.175).

Brouwer was justified in claiming in [1928A] his 'priority' over Hilbert for authorship of the notion of meta-mathematics. More importantly, he exposed the non-viability of Hilbert's Programme long before Gödel's Incompleteness Proof. His arguments were based on the inadequacy and irrelevance of formalization as a foundation of mathematics and on his conception of mathematics as a closed system and the need for its justification from outside the system; a justification not to be found in meta-mathematical consideration of mathematical activity, but in the authority of the intuiting mind grasping some aspects of the reality of time.

Brouwer was well aware of the implications of his philosophy of mathematics. The essentially subjective nature of intuitive mathematics was his deliberate choice and was not considered to undermine the necessity and certainty of mathematics, the importance of 'objectivity' or the lack of it depending on one's valuation of cooperation and collective agreement. (see further 4.6). A more serious problem, appreciated by him more in later years, is human weakness of mind and memory. The highly acclaimed freedom of mathematics left to infinite unfolding becomes inevitably restricted by the limitations of human memory:

> The language-less constructions originating in the self-unfolding of the Primordial Intuition by virtue of their being present in memory alone are exact and true. Unfortunately, the human faculty of memory, which has to see and hold these constructions, is by its very nature limited and fallible. ([1933], p. 58).

On the positive side, the conception of mathematics as 'intuitive construction' provided a clear answer to the question of mathematical existence and truth, and a secure foundation of a programme of re-constructing mathematics: the domain of mathematics is well defined, and as constructive, intuitive thinking it is given its distinctive identity different from any of the sciences and from deductive logical thought.

## 4.4 Reflection on the Nature of Mathematics

One of the characteristic features of Brouwer's thinking is his remarkable gift for recognizing essential characteristics and seeing clear lines of demarcation. In some cases such distinctions were partly inspired by pet aversions or bias against particular forms of thinking or activity such as logic, language or the application of mathematics, leading to campaigns for radical separation, exclusive pre-occupation with one and avoidance of the other. More often than not, however, they were the result of sharp analysis, and brought rigour and discipline into his topological as well as his foundational work. His demand for rigour particularly impressed Carathéodory, who wrote in an aside to Hilbert:

> The length of my manuscript is due to my trying to be absolutely rigorous; you know how often mathematicians sin in this respect, with the exception of course of Brouwer... (05.05.12).*

---

* Letter in Nieders. Staatsbibl. Göttingen.

Brouwer recognized that in his investigation into the foundations of mathematics he was acting outside the proper domain of 'first-order' mathematics.

Since in the 'mathematical viewing' of pure mathematics the mathematical objects become 'things' or 'causal sequences', he considered such meta-mathematical study to be applied mathematics:

> Strictly speaking, the construction of intuitive mathematics by itself is an act, not a science; it becomes a science, i.e. a survey of causal sequences repeatable in time, as mathematics of the second order, which is mathematical viewing of mathematics proper or of the expression of mathematics in language;... here, as in theoretical logic, we are dealing with an application of mathematics. ([1907], p.98).

From his vantage point outside the domain of mathematics Brouwer considered the nature of mathematics with more than a 'mathematical' eye and his investigation also included other areas of human activity and their relation to mathematics. Probing beyond mathematics and seeking a meta-rational justification of Reason in direct knowledge, Brouwer was clearly and consciously moving in the domain of philosophy. When Korteweg commented on Chapter II of *The Foundations* and remarked that some parts 'could at most be decribed as philosophy', Brouwer accepted the criticism and insisted that the search for the roots of mathematics should concentrate on human nature:

> You agreed to my choice of subject 'provided there would be enough mathematics in it', probably suspecting that it would move me strongly in the direction of philosophy, which indeed it has done...
> What I bring you now concerns itself exclusively with *how mathematics is rooted in life and what therefore should be the starting point of the theory*; all other parts of my dissertation derive their significance from this main thesis.
> (Brouwer to Korteweg, 05.11.06 DJK20; italics are Brouwer's).

He distinguished between his 'work on the foundations of mathematics' and his 'contributions free of philosophy' ([1919D], p.798). Significantly, some of his major contributions such as [1908C], [1933] and [1948C] appeared in pure-philosophical journals.

As to certainty and truth, this reflective thinking may lack the specific truth of mathematical intuitive thinking but it shares a common source in direct insight, the ultimate basis of certainty.

Intuitive mathematics does not derive its truth and certainty from its meta-mathematical or philosophical consideration, which is descriptive, non-regulatory and in no way to be seen as an attempt to prove the consistency of mathematics; nor does reflection on the nature of mathematics intrude into the pure-mathematical domain and threaten its autonomy. No one was more anxious than Brouwer to safeguard the independence of mathematics against intrusion of alien principles of phenomenal sciences such as logic. Yet nowhere in his writings is there a trace of fear that the autonomy of mathematics is threatened by philosophical considerations except those which surrender its independence and allow exterior elements to interfere in the internal conduct of mathematical activity.

To some of Brouwer's friends and followers his philosophy was an embarrassment and an expendable by-product of his Intuitionism. To Brouwer the foundation of mathematics was only to be found in 'human life', by introspection; his philosophy was a source of inspiration and a guiding light – a role which Descartes and Poincaré identified with intuition itself –, reminding the mathematician constantly of the principles, the freedom and the limits of his discipline.

## 4.5 Constructiveness and Existence

### 4.5.1 Brouwer Construction

Brouwer's favourite and almost exclusive metaphor in characterizing intuitive mathematics was 'building'. In his earlier Dutch papers he almost invariably uses the words 'bouwen' (to build),'opbouwen' (to build up) and 'gebouw' (a building), words which in line with mathematical tradition are usually translated into English as 'construct' and 'structure'. When referring to intuitive mathematics Brouwer avoids the Dutch words 'construeren', 'constructie' and 'structuur', which are associated with a certain complexity, artificiality and which lack the 'down-to-earth' and dynamic notion of progressively building upwards with bricks on a solid foundation. An example of such an artificial construction is given in [1907], p.140, where the words 'construeren' and 'samentimmeren' (hammer together) are used to describe the formation of non-Archimedean and non-Pascalian geometries.

With the present proliferation of notions of constructivity, Brouwer's naive conception of construction, i.e. building, cannot be over-emphasized. Mathematical thinking is building upwards step by step on the foundation of Intuition, using only the materials and tools found in Intuition and those constructed from them 'previously', i.e. at an earlier stage of the construction.

The notion of mathematics as constructive activity suggests a natural distinction between:
1. the Subject, the creating Mind;
2. the act of mathematical creation, in its most elementary form consisting of a 'splitting of one' and 'a linking of two';
3. the concepts, elementary ones created in the Primordial Intuition as well as those resulting from complex constructions.

In Brouwer's conception, however, neither the constructive act nor the concepts can be entirely isolated from the creating Subject, nor can a mathematical concept be cut loose from the act which brought it into existence. The Subject is not just a builder nor a parent who has given birth to an independent being; the mathematical object remains alive only as a thought, a part of consciousness. Moreover, the 'creative Subject' is totally involved in the act of construction: as Consciousness acting through Will he remains the free agent within the domain of mathematics, the vital element of any construction, in particular of those which are usually described in mathematics as 'operations', 'relations' and 'functions'.

The existence of a mathematical object, therefore, is its presence as a thought in consciousness; mathematically existent is: having been created in the Primordial Intuition or being or having been built on and with the elements of Intuition. The distinction between concepts in the process of being created by mind and those held in memory, mind in its sustaining function, is no more than a distinction between present and past.

The ambivalent words 'construction' and 'building', denoting a constructive act as well as its result, are most appropriate names for the Brouwerian mathematical concept, which not only includes the process of its construction but is wholly characterized by it.

### 4.5.2 *Classification of Constructions*

The 'subjective' element in Brouwer's building, the 'free arbitrariness in the self-unfolding of mathematics' ([1907], p.119) and the Subject's continued recourse to intuition make a simple and complete classification of pure-mathematical concepts virtually impossible. Indeed, nowhere in Brouwer's writings is there a complete classification of concepts or constructions. There are some broad distinctions and instances of a detailed hierarchy of concepts, as e.g. in the construction of species (see further 6.3.7).

A hierarchical classification of concepts, identified with their constructive processes, can only be based on their genetic order. Such a classification emerges from *The Foundations* and distinguishes between:
1. elements of construction;
2. elementary constructions;
3. a hierarchy of more complex constructions.

In his Summary Brouwer refers to primitive elements as 'building elements which can be read off from the Primordial Intuition'. In parallel with primitiveness and non-derivability in the medium of language and logic he points to an atomicity and unconstructibility of certain elements of the Primordial Intuition:

> There are elements of mathematical building which in the systems of definitions must remain non-derivable, therefore in communication must find expression in a single word, sound or symbol. They are the building elements which can be read off from the Primordial Intuition; concepts such as continuous, unity, again, and-so-forth are not derivable. ([1907], p.180).

Although recognized as constituents, 'building elements', the primitive elements do not genetically precede, they come into being in and with the primordial constructive act. Brouwer carefully uses the phrase 'can be read-off from. (Compare also his comment on the genetic priority of 'two-ity' over 'unit' as quoted above 4.1.4.)

[The primitive elements of construction are distinct from what Brouwer calls 'the synthetic a-priori judgments of constructibility which can be read-off from the Primordial Intuition, such as the possibility of mathematical synthesis, the possibility of insertion, the possibility of proceeding ad infinitum...they are the result of viewing the Primordial Intuition mathematically and belong to mathematics of the second order.' ([1907], p.119).]

In the genetic hierarchy of constructions the Primordial Intuition of two-ity is clearly the first construction, the first constructive act. Brouwer describes it as 'the first act of building' ([1907], p.12). It is an elementary 'intuitive' construction in the sense that it is directly and wholly rooted in Intuition and represents a single constructive act to be used in further constructions. The basic successor construction of two-ity is used to create the number three and each natural number and the concept of sequence (see 6.2.2); in the more general form of the 'again' it is the constructive act of repetition.

There are other elementary constructions, constructive acts requiring direct recourse to Intuition and based on distinct elements of Intuition, such as:

The free act of 'insertion', based on the intuitive 'between' and used by the Subject to generate rationals, intervals and the order-type $\eta$ (see 6.2.4.3) and, in Brouwer's later treatment of the continuum, used to generate 'real elements of the continuum (see 6.3.11);

The operation of 'fitting-in one construction into another', which is based on the 'linking aspect' of Intuition and forms an essential element in the construction of 'property', 'relation' and of mathematical negation or 'absurdity-construction'. After 1917 Brouwer became increasingly more aware of the fundamental role of property, then re-named 'species', 'at every stage of the development of mathematics' ([BMS53], p.2 e.a.), and claimed it to be a 'modality of the Primordial Intuition'. (see further 5.8.2 and 6.3.5-7).

Mathematical constructions, although all based on the Primordial Intuition and referred-to as 'intuitive' constructions, are not necessarily simple:

> Some constructions that we build are so complex that we cannot grasp them at once. ([1907], p.126).

The lack of simplicity and uniformity in the domain of mathematical constructions is mainly due to the freedom of the Subject to create ever more and more complex constructions, using Intuition and previously constructed entities and tools in 'a free unfolding'. The condition 'previously constructed' is not just a restriction, it is a sanction of their mathematical birth-right to be exploited to the full. For example, [1912A] points out that in the construction of the infinite ordinal $\omega$ 'of course, every previously constructed set of every previously performed constructive operation may serve as the unit.' (p.13).

In this free-unfolding, however, the 'previously acquired' can only be a general principle for a genetic hierarchy of complex constructions, and was used as such e.g. in Brouwer's hierarchy of species (see 6.3.7). A comprehensive hierarchical classification of all mathematical constructions must remain illusory.

The identification of mathematical entities with their processes of construction and the active role of the free Subject do not allow a neat 'horizontal' classification of constructions based on philosophical categories and reflected in grammatical distinctions between nouns, verbs, adjectives etc. Brouwer claims such distinctions to be purely linguistic and their use in mathematics to be a historical accident; they refer to a phenomenal world of things with properties and actions. By detaching the mathematical from the phenomenal world and from language, confining it wholly to the inner world of mind, Brouwer had been able to avoid the problem of knowledge of phenomenal 'things-in-themselves'. In his interpretation of mathematics intuition of mathematical noumena is possible, i.e. direct contact of mind and its own pure-thought creations. In the pure world of thought-construction the phenomenal distinctions between substance, property, relations and action do not apply. All mathematical entities are 'things of the interior world of the Subject', concepts or thought-constructions, each with its own individuality and identified with the whole of its construction; and yet their common, most appropriate and comprehensive characterization is as 'creative acts of the Subject'.

Brouwer makes distinctions and uses existing language referring to numbers, sets, species, functions etc. Further analysis, however, reveals that these notions are given wholly new interpretations and that his distinctions are less absolute than their classical counterparts. For example, Brouwer-sets or spreads, determinate sequences as well as functions are defined essentially as constructive procedures or algorithms, sometimes referred-to as 'laws', used by the Subject to generate new mathematical entities. Neither is the distinction between procedures and entities wholly exclusive since entities themselves are identified with the processes that brought them into existence.

Because of the central role of the acting, free Subject the important distinctions in Brouwer's Intuitionist mathematics are based on factors such as: whether or not the generating processes have been or can be completed, the measure of freedom the Subject has left himself when constructing his procedures, confining the domain of his 'choice'; the former, for example, leads to distinctions between finite and infinite constructions, between truth and absurdity, the latter to the distinction between determinate sequences and choice-sequences.

### 4.5.3 Distinctive Features of Mathematical Constructive Thinking

The definition of mathematics as 'constructive thinking of the Subject exclusively based on the Primordial Intuition' determines its special 'idealized' character and sets it apart from other human activities and forms of thought. Unfortunately, Brouwer remains rather vague about the ideal reality of pure-mathematical construction. In his Intuitionist campaigns the emphasis was mainly negative: mathematics is not to be confused with its application, mathematics is essentially language-less, mathematical constructive thinking is not to be identified with 'logical thinking', nor with the historical , psychological processes of thought of the flesh-and-blood mathematician.

Chapter V will treat in detail Brouwer's philosophy of language and logic and his campaign to free mathematics from its confusion with symbolic expression and logic. Suffice it here to state his major arguments:

1. Any recording of a pure intuitive construction in the physical medium of language or symbols does not convey its true nature nor carry its characteristic certainty;
2. The formalist conception of mathematics severs the vital link between the symbol and its conceptual antecedent, and identifies mathematics with structure of symbolic systems. *The Foundations* frequently refers to the activity of logicists and formalists as 'construction' or 'building', but a construction in the domain of language resulting in 'purely linguistic structures';
3. Traditional logic distilled from existing mathematical recording a pattern which was mistaken for a universal structure of mathematical thinking. It imposed and perpetuated a habit of thought which was not characteristic of intuitive thinking nor even of the thought-processes of the flesh-and-blood mathematician.

In a letter just before the publication of his thesis he attempts to persuade Korteweg that 'mathematical reasoning is completely different from logical reasoning' and illustrates this by comparing the logical proof of the acuteness of the isosceles triangle with the reasoning of the mathematician:

> The theorem 'if a triangle is isosceles it is an acute triangle' is expressed as a logical theorem: the predicate 'isosceles' in the case of triangles is considered to imply the predicate 'acute'...
> but the thoughts of the mathematician, who because of the poverty of his language formulates his theorem as a logical theorem, proceed in a way quite different from this interpretation. He imagines that he is going to construct an isosceles triangle, and then finds that either at the end of the construction all angles appear to be acute or that on the postulation of a right or obtuse angle the construction cannot be executed. In other words

he thinks the construction mathematically, not in its logical interpretation. (Brouwer to Kortweg, 23.01.07 [DJK35]; see app.13).

In this passage Brouwer constrasts 'logical' thinking with the general thought-processes of the mathematician, the idealized pure-mathematical construction of the Subject as well as the 'historical' thought-sequence in the life of the flesh-and-blood mathematican.

The historical course of events in the development of mathematical concepts and mathematics, individually and collectively, is affected by factors outside the interior world of the Subject and by human weakness; it includes inspiration by physical phenomena, periods of gestation, guesses, hypotheses, and above all, learning through human communication. In the reality of pure-mathematical construction there is no room for human weakness; it is time-based but free of all 'external' and 'social' elements.

The distinction between the pure-mathematical construction and the historical, psychological thought-process was clearly recognized by Brouwer even at the time of writing his *Foundations*. Contrast, for example, the highly abstract Primordial Happening and his account of the historical event of the child learning numbers, as given in the opening lines of Chapter One:

> One, two, three,... we all have learned by heart the sequence of these sounds... ([1907], p.3).

Gilbert Ryle makes a distinction between 'theory building' or 'the construction', the route by which the originator arrives at his theory, and 'the didactic account of the theory', the verbal, logical account intended to convince others. (RYLE[1949], p.271). Brouwer was well aware that the neat and final account of a mathematical theory does not necessarily follow the path along which the flesh-and-blood mathematician has travelled (as is manifestly shown in any one of his manuscripts with their countless alterations and systematic crossings-out, in the Brouwer household known as 'gatemaking'; see for example illustration p.482). But more important to him was the disparity between the 'verbal account' and the pure thought-construction within the limits of Intuition, which alone carries the ultimate certainty of mathematics.

## 4.6 The Subject and the Intersubjectivity of Mathematics

*4.6.1 Brouwer's Idealism*

Brouwer's philosophy of the physical world can with some reservations be described as idealism. It shares many features with the anti-materialist tradition in its various forms from Platonism, Berkleian solipsism to the idealism of Kant, Schelling and Schopenhauer. Yet for an understanding of the ideal or idealized nature of Brouwer's mathematical construction and his 'Subject' the idealistic connection could be misleading and confusing. For example, idealists like Hume place an emphasis on the 'subjective' nature of human knowledge, including mathematics. Their subjectivism, however, is just another aspect of their empirical scepticism, which is extended to mathematics as geometrical abstraction and generalization. At the other extreme, in the Platonic version of idealism or immaterialism, mathematics is conceived as an 'idealized' reality, free from the limitations of a physical world and the limitations of sense perception, free also and independent from a knowing subject.

The process of 'idealization' is an attempt at perfection, a reaching for the perfect reality which underlies our human knowledge. Platonic idealization is a total exclusion of every aspect of the physical, contingent reality including time, action and individuality. The mathematical domain, as the whole of the Platonic universe, is timeless, static and universal, existing independently of human knowledge.

Brouwer upheld the a-priori of mathematical certainty and the 'spirituality' of the mathematical reality, transcending the limitations of a material world, but a reality whose exclusive seat is to be found in the individual human mind. Unlike Plato he held that perfection can be achieved without abandoning time, action and human individuality. Indeed, to Brouwer these are the essential components of mathematical reality. His idealization is the process of exluding from human thinking all elements except those 'found in the Primordial Intuition', resulting in pure intuitive thinking. It is more than the Aristotelian or Platonic abstraction, which is an epistemological process enabling man to distil from observation of a physical reality his universal concepts reflecting an objective metaphysical reality. For Plato this reality was the spiritual universe of Forms or Ideas; for Aristotle the metaphysical reality remained embedded in the physical universe: form and matter are the essential components of all beings, by abstraction the mind isolates form from matter, it does not create new entities. In Brouwer's genesis of mathematics the only contact with the physical universe takes place at the threshold of mathematical thinking. Sensations are a necessary condition for the initial phenomenon to take place, but the abstraction from the content of sensations is total, the mere fact of having sensations being sufficient for time-awareness. In the Primordial Intuition a new entity is created, which only exists in the mind, the sole mathematical universe. Mathematics then is the activity of Mind, not with Aristotelian abstracts but with newly created elements of Intuition alone. At no stage of the mathematical construction can elements be introduced which are the result of abstraction from a physical universe and lack this pure intuitive pedigree.

## 4.6.2 *The Idealized Subject*

The Brouwer domain of mathematics is well defined by the process of idealization, i.e. the process of excluding from human thinking all extraneous elements except those found in the Primordial Intuition. Such ontological analysis, the separation of various aspects and exclusive consideration of some, applied to human thinking is feasible and accepted philosophical practice. Idealization of the individual Subject, however, meets with fundamental problems and is by many considered to be impossible and contradictory. Personal identity is generally seen to be inextractably bound-up with personal history; in the Humeian monistic tradition individual identity is no more than the totality of experiences, past and present. In Brouwer's philosophy these experiences belong to the 'exterior world of the Subject' and are excluded from the 'inner life of the Subject'.

Brouwer's mystic belief in the Soul and the possibility of its disembodied existence was to him a satisfactory solution of the philosophical problem of personal identity (see 3.8). His analysis of mathematical activity, however, remained free from such speculation. His theory of the Idealized Mathematician or the Creating Subject is founded on three main convictions:
1. the existence of the 'I', the 'Self';
2. the essentially individual nature of mathematical activity;
3. the possibility of ontological separation of those elements of Mind which are sufficient for mathematical activity.

Without following the Cartesian 'ergo' Brouwer accepts the existence of the Ego and self-awareness, even at the naive phase, which is 'The joyful miracle of the self-revelation of consciousness' ([1948C], p.1238). Leaving aside all questions of origin and nature, the Ego at the onset of mathematical activity is simply given; introspection is its natural form of knowing and inner conviction the primary and ultimate criterion of truth. The Ego, the originator of mathematical activity is not the disembodied Soul, but Consciousness transformed to Mind or Mind born in the primordial happening and existing in time.

The Primordial Intuition, on which and with which the whole edifice of mathematics is built, is direct insight, introspection by and in the individual mind, the source of the unique nature of mathematics and the unique certainty of mathematical knowledge. Other spheres of knowing originate in a sensual perception of the physical world, their concepts are the result of abstraction and analogy, or to use Brouwer's words, of 'causal viewing' and 'induction'. It is this difference in the human faculty of knowing which brings about the separation between the two worlds of mind: on the one hand introspection, intuition, leading to awareness of the Self and of time and to the construction of mathematics, on the other hand a combination of sensual perception and 'mathematical viewing', creating the exterior world of the Subject in which are included the identities of other beings, other minds as well as second-level mathematical concepts originating in other minds and communicated through symbols.

Both the inner and the exterior world of the Subject are the private domain of the essentially individual Subject. The privacy of the exterior world of the Subject is one of the main arguments in Brouwer's sceptic interpretation of science. 'Solitude' ([1933], p.52) is even more characteristic of the idealized inner world of the Subject, from which 'things' and other minds are banned. The notion of a Platonic, independently existing prototype, 'a hypothetical omniscient being' is rejected out of hand ([1955], p.114). But so is any form of generalization of mind:

> ...the Subject would be elevating itself to a mind of second order, experiencing incognizable alien consciousnesses as sensations. Quod non est. ([1948C], p. 1239).

A 'collective mind' lacks any base in reality; it is condemned as 'a collective hypothesis of a collective supersubject experiencing some objective world' (ibid.), based as it is on the assumption of a plurality of mind and the possibility of real communication between minds. In the inner world of the Subject, the universe of first-level mathematics, there are no other minds and there is no universal concept of mind.

The idealized Subject is the individual human being, stripped of the physical and emotional faculties which create his exterior world, but maintaining his individual personality and his creating and retaining faculties sufficient for mathematical generation. Brouwer tends to reserve the word 'will' for the human faculty of action away from the inner Self towards the exterior world. In the context of pure mathematics he prefers the word 'freedom', which expresses not only the absence of external restrictions but more positively the dynamic aspect of the Subject in his continued unfolding of the Primordial Intuition.

The idealization of the Subject is the ontological isolation of the pure-mathematical Self from the flesh-and-blood mathematician. In founding mathematics on intuition Brouwer had made the individual mind the creator and sole arbiter of mathematical truth, and so provided the flesh-and-blood mathematician with a source and authority within himself for true mathematical construction. But the flesh-and-blood mathematician in his struggle towards mathematical perfection has to discipline himself, set aside the whole of his exterior world, suspend part of his thinking and himself, or rather transcend every aspect of his physical reality, and the physical human condition imposes its limits. Reaching middle-age, Brouwer became more and more aware of the limitations of the flesh-and-blood mathematician: the distractions of a physical environment, the limitations of memory and the inadequacy of any physical symbol in representing a mathematical concept and aiding the memory:

> The language-less constructions originating in the self-unfolding of the Primordial Intuition, on the basis of their presence in the memory alone, are exact and correct. Unfortunately, the power of human memory, which holds and oversees these constructions – even when it resorts to using linguistic symbols – is by its very nature limited and fallible. In a human mind equipped with an unlimited power of memory pure mathematics, practised in solitude and without the use of symbols, would be exact. ([1933], p.58).

### 4.6.3 Mathematics, the Life of the Idealized Free Subject

The question remains whether this philosophical analysis of the Idealized Subject is more than sterile speculation, has any relevance for the practice of mathematics or reveals a significant characteristic of mathematics not found in other disciplines. Similar investigations into the nature of the sciences or forms of art will also reveal certain 'idealized' and 'subjective' aspects in man exploring and describing the physical universe or creating art. Indeed, the 'ideal' aspect of science was highlighted by Brouwer and used as an argument against the 'exactness' of science; scientific theories were denounced as merely representing the views of their makers of a physical reality distorted by their 'causal eye'.

Further analysis, however, shows that the concept of the Idealized Subject lies at the heart of Brouwer's philosophy of mathematics. Both the 'idealized' and the 'subjective' are implied in his conception of Intuition; they are also the most fundamental tenets of his philosophy and characterize his Intuitionist Mathematics completely.

Whereas idealization is one of the many aspects of scientific and artistic endeavour, it encompasses the whole of intuitive mathematics. The complete isolation of intuitive mathematics from the physical environment and from the exterior world of the Subject, and its exclusive foundation on introspection, the Intuition of Time, determine both its 'ideal' and its individually-human characteristics. The Idealized Subject in turn delimits the domain of mathematics and characterizes its elements: mathematics is the intuitive constructive life of the Subject, 'life in the Mind', proceeding in time. Mathematical constructs, born from the Intuition of Time, are just parts of, episodes of that life; their temporal nature is that of the live Subject, who exists in time, i.e. who in his movement along the time continuum has reached a certain point, the present, but a present which is forever shifting. What precedes the present is past, the completed actions of the Subject and as such held by memory. The future does not exist except as an open possibility for further free action:

> ...truth is only in reality, i.e. in the present and past experiences of consciousness... expected experiences are only true as anticipations and hypotheses. ([1948C], p.1243).

Since the Subject has a beginning and his constructive acts are discrete events, the past of the Subject as a totality of sequential completed activity is finite.The order of the constructive events in the life of the subject, the successor relation of before-after, is the priciple of coherence and validity of the mathematical construction (constructive events in the individual life of the Subject, not the collective 'state of human knowledge at some given time... on the basis of empirical observations about the history of mathematics' à la Posy. (POSY[1974]).

Existence in time allows the Subject to survey only past constructive events; he is imprisoned in time, 'a slave of time', he cannot leave his time-dimension, does not have the divine power of overseeing the whole of time from the outside. In 6.2.2 we shall discuss the fundamental changes this entails for the classical concept of sequence: on the one hand a reduction of its claims to actual infinity, on the other a widening of its general scope.

Post-Brouwer formalization of the 'creative Subject' has rightly concentrated on the sequential aspects of constructive events. What, however, any formalization must fail to grasp – apart from any other essential characteristic of the creative agent – is the dynamic aspect of the present and the special indeterminacy of the future. Any recording of mathematical construction concerns itself with past, completed constructions: it freezes the present. As to the future, formalization no more than states possibilities, it cannot predict the singular sequence of constructive acts to be determined freely by the Subject, nor grasp the range of his free actions except in terms of restrictions made in the past.

In its time-aspect mathematics may be compared to music: as harmonious sound it only exists in time and proceeds with time; it is distinct from the musical score and even from its frozen recording on disc or tape. Unlike mathematics, music can exist independent of its creators, the composer and the performing artists, e.g. when it is mechanically reproduced. But even in the recorded sound process there is a present, a past and a future, a future wholly pre-determined by the recording. Continuing this analogy one might liken the Creating Mathematical Subject to improvising jazz-musicians: the part of the composition played so far exists only in their memories, the sounds to-come, though building on preceding sounds and harmonies, are yet to be created.

Mathematics as constructive pure mind-activity can only be alive, i.e. have a 'present', in the live Subject. Brouwer is not concerned with the general philosophical problem of time; his time, as the whole of mathematical activity and the Creative Subject, is idealized, i.e. isolated to its role in mathematics at the primordial stage as well as its further construction. The present is the only real moment, the present existence of the Self holding alive in memory the whole of constructions as successive acts, the past of the Subject. The reality of the past, its existence and truth is 'having been constructed'. The future is no more than the openness of the present, the definite possibility of continuing the constructive activity. It is not surprising that the terms of constructed mathematics fail to capture the future: as 'not having been constructed' the future implicitly lacks reality and existence. The only 'real' aspects of the future are past constructions and restrictions affecting succeeding activity. The Subject's awareness of the present allows continuation of past constructive activity even beyond the finite past, 'thinking repetition as being continued indefinitely' ([1912A], p.12). Since mathematical reality is 'being thought' or 'having been thought', sequences generated in this way have a mathematical reality, but unlike the past finite constructs the denumerably infinite sequences are incomplete. In Brouwer's ideal time the past is an interval consisting of a finite sequence of constructive acts and open at the end-point, the present; its dimension is the denumerably infinite.

The present of the live Subject, however, is more than the possibility of past constructions being repeated or implemented; what distinguishes the live Subject from an automaton or computer is his creative freedom. The negative term 'indeterminacy', referring to an absence of restrictions, does not adequately characterize free future mathematical activity. Freedom presumes absence of constructed restrictions, of exterior influence and even ulterior motives, i.e. causal thinking. More positively, it expresses the exclusive and sufficient power of the Subject to create mathematics at will. The present is the dividing point between determinacy and freedom, between past and future. Whereas the past in ideal time consists of singular acts, decided and completed, the free future is wide open. From the perspective of the present the live Subject oversees the past, the totality of constructed mathematics, a world of linearly ordered, sequential constructs. At the threshold of the free future he faces a world which lacks every characteristic of the constructed past: there are no definite discrete acts. The free future lacks the simple structure of the sequentially ordered past constructs, its 'dimension' transcends the denumerable.

In his search for the foundations of a-priori mathematics Brouwer had identified human awareness of past and present moments as the necessary and sufficient foundation of discrete number and combinatorial constructions. This first insight, 'the first Act of Intuitionism', was the result of his analysis of mathematical thinking, isolating it from other aspects of human thinking, leaving and needing no more than certain aspects of individual humanity and of time.

The 'Second Act of Intuitionism' was his recognition of the implications of the live Subject's freedom for the 'non-separable' parts of mathematics, his recognition that the topological structure of the 'free future' differs essentially from that of the determinate past (as well as from that of the future of pre-determined sequential activity), and characterizes the continuum.

Philosophers have found it impossible to describe freedom in terms of phenomenal reality, the empirical person and time as past experience. Kant resorted to a 'homo noumenon'*. For Bergson the problem was 'free causation in real time'. Brouwer's freedom is the present ability of the Subject to continue building on and with the elements of Intuition and constructions 'previously acquired', the freedom of the ideal Subject in ideal time where the future is wholly and exclusively subject to the decision or choice of the Subject. The Second Act of Intuitionism accepts the future as a legitimate domain of mathematical activity, in particular it recognizes infinite sequences split by the present into a completed part and an infinite 'tail' of elements yet-to-be-decided. (see further 6.3.10 and 6.3.11).

Brouwer's first use of the expression 'the Creating Subject' in [1948A], in which he made a relatively minor modification to free-choice sequences, has made some commentators conclude that herewith 'Brouwer had developed a new theory of the Creating Subject' (KLEENE[1965], KREISEL[1967]). Others associate the Creating Subject exclusively with Brouwer's development of free-choice sequences (TROELSTRA[1965], POSY [1974]).

The element of 'choice' in Brouwer's new characterization of the continuum undoubtedly highlights the need for a free, acting Subject. Yet the Creating Subject or Idealized Mathematician, is as indispensable in Brouwer's philosophy of separable mathematics as it is for his concept of free-choice sequence and the continuum. In past, completed constructions the free role of the Subject is less conspicuous; although created freely at the time of construction, as past constructions they are finished and determinate, including the algorithms which generate denumerably-infinite sequences. With reservations as to complete accuracy, such past constructions can be recorded in language, and recorded without reference to Creating Subject and so assume a quasi-independent existence. But there is no doubt as to Brouwer's conviction, right from the start, of the essential free role of the Subject in the generation of all mathematics and the 'idealized' aspects of individual man creating mathematics and keeping it in existence. Mathematics as constructive mind-activity requires the individual Subject as a home base; its a-priori, the independence from the phenomenal world as well as from internal influences of motivation, spe-

---

* 'The freedom of the intelligible character... signifies the spontaneous, original choice, accomplished outside time, of the homo noumenon.' (KANT[1785], p.245).

cifies the nature of the Subject: 'idealized' in the sense of acting exclusively in the domain of intuition, and free at the moment of construction.

Awareness of these essential characteristics of the Subject is evident in Brouwer's earlier writings. The 'idealized' and 'subjective' are clearly implied in his definitions of Intuition and mathematics as intuitive construction. 'Free' is a favourite epithet of the Subject and of mathematics already in [1907] (the 'free unfolding' or 'free arbitrariness' of mathematics, e.g. p.119). Freedom is also exercised in particular constructions, for example [1907] refers to the freedom of the Subject to construct a dense scale on the continuum in such a way that no point of the scale coincides with a given point of the continuum:

> We can make sure that the approach to the point by an infinite binary fraction takes place according to a law of proceeding, freely chosen and conceivable*. ([1907], p.9; cf. also pp. 62-67).

[1912A] even considers the possibility of free-choice sequences:

> If the Intuitionist on the basis of the linear continuum admits a fundamental sequence of free choices as an element of construction he shows easily that any such constructed non-denumerable set has no other power than that of the continuum. ([1912A], p.23).

At this stage, however, free choice as an element of construction is still hypothetical; in the context of the traditional notion of set as a completed totality Brouwer's constructive demand is uncompromising ('the intuitionist can only construct mathematical sets that are denumerable...' ibid.). Only when in 1917 he made a complete break with traditional set theory did free choice become accepted as a constructive act in the generation of point of the continuum. (See further 6.3.4).

The most explicit account of the Creating Subject is given in the 'philosophical' papers *Will, Knowledge, Speech* [1933] and *Consciousness, Philosphy and Mathematics* [1948C]. These two papers re-state and develop Brouwer's earlier views on mathematics as mind-activity and place the individual-human and ideal aspects of the creator of all mathematics firmly in the centre of his philosophy.

---

\* The translation of the *Collected Works* 'arbitrarily given' interprets 'willekeurige' as an adverb qualifying 'denkbare'. Such interpretation would in Dutch have required 'willekeurig'. The Dutch 'willekeurig', more than the English 'arbitrary' reflects 'the choice of will' ('wil' and 'keuze'). Given Brouwer's precise somewhat puritanical use of language, 'willekeurige' should be interpreted in its original meaning: 'resulting from choice of wil'. See further 6.3.10.

## 4.6.4 Objectivity and Inter-subjectivity

To those familiar with Brouwer's almost obsessive concern with truth and certainty the emphasis on the role of the individual Subject may seem paradoxical. Intuitive mathematics as conceived by Brouwer is wholly subjective and lacks what is usually identified with truth: objectivity and intersubjectivity. Indeed, intuitive mathematics is in every respect non-objective, whether one takes 'objective' in the Brouwer sense of belonging to the exterior world of the Subject or in the generally accepted sense of 'belonging not to the Consciousness or the perceiving or thinking Subject but to what is presented to this, or the non-ego, external to the mind' (Oxford Dictionary).

No objectivity is claimed for mathematics by Brouwer. The objectivity he is prepared to ascribe to the physical sciences is only 'an invariance in our image of nature relative to an important group of phenomena' ([1907], p.95). Objectivity of knowledge presupposes the existence of objects external to the mind as well as a human faculty of apprehending them, sufficiently accurate to ensure agreement between individuals. The a-priori as antecedent to experience cannot be objective.

Like Kant, Brouwer founded a-priori mathematics on some ability of the individual mind to grasp reality directly. To safeguard the inter-subjectivity of mathematics Kant assumed the 'Forms of Intuition' to be inherent and the same in the perception of all human beings. To Brouwer the assumption of the existence of other minds introduced an a-posterior element for which there is no place in the a-priori domain of pure intuitive mathematics. In [1948C] he exposes the fallacy of 'knowing other minds through sensations' and introducing 'object individuals' in intuitive mathematics, which is tantamount to:

> ... the mind elevating itself to a mind of second order experiencing alien consciousnesses as sensations. Quod non est. And which moreover would have the consequence that the mind of second order would causally think about the pluralified mind of first order, then cooperatively study the science of the pluralified mind, and in consequence of this study assign a mind of second order with sensation of alien consciousnesses to other inviduals. And so on. Usque ad infinitum. And this nonsense would still go further...
> ([1948C], p.1239).

Körner maintains that 'Intuitionists regard mathematical constructions as inter-subjective experience' (KÖRNER[1960], p.136). This Brouwer has never stated. Subjectivity, the closedness of the ideal inner world of the

Subject, rules out inter-subjectivity as a characteristic of intuitive mathematics. There is, however, a place for other minds in the causal world of the flesh-and-blood mathematician and in the world of cooperative acting; they have an existence peculiar to the exterior world of the Subject. Intersubjective 'mathematical' activity is acceptable to Brouwer, but with all the limitations attached to activity at the third phase of consciousness. At the second, 'the isolated causal phase', the solitary flesh-and-blood mathematician is already hampered by the delusion of an exterior world and by the limitations of memory. The third phase, of cooperation between individuals, of inter-subjective activity, introduces further pitfalls; it presupposes a sameness of intuitive thinking in different minds and relies on communication between individuals in an alien medium:

> A very essential hypothesis in the mathematical viewing of fellow-men is the supposition of the presence in each of them of a mathematical-scientific mechanism of viewing, acting and reflecting similar to one's own. ([1933], p.48).

Symbolic recording, even in some private language for the benefit of the flesh-and-blood mathematician alone, 'as an aid to memory', is inadequate a representation of the Subject's pure mind construction. Intersubjective mathematics, moreover, is based on a common language, presumes not only the sameness of intuitive thinking of different minds but a common agreed linguistic system which represents it:

> ... words which they would use only as invariant symbols for definite elements and relations between elements of pure-mathematical systems created by them. (op.cit., p.58).

The creation of such a 'pure language' was the chief aim of Brouwer's Signific campaign, his temporary involvement in the Signific movement. His first and life-long pre-occupation was with the growing confusion of mathematics and its linguistic representation; his first major campaign was 'the disengagement of mathematics from language and logic'. His search for the nature of pure mathematics led to a clear distinction between a-priori mathematics, isolated in the inner world of the Subject, and all other activity of mathematical man in his physical and social environment, including the cunning acts of the individual, the struggle of the flesh-and-blood mathematician towards mathematical perfection as well as his attempts to communicate with others in a physical medium.

Having established the proper role of communication, language and intersubjective cooperation, Brouwer showed his appreciation of their value and made his significant contribution. The manuscripts of his publications, many times re-worded and revised, are witness of his constant search for precise description of his thoughts. His active participation in the international mathematical life, his own major contribution to the collective treasure of mathematics and even his intuitionist campaign are evidence of the value he placed on the inter-subjective activity of mathematicians. To the flesh-and-blood mathematician the achievement of others communicated to him are a source of inspiration, they do not relieve him of the need to re-construct them in his own mind. In the ideal world of the Subject there is no 'authority', except the authority of his own Intuition.

### 4.7 The Application of Mathematics

> The construction itself is an art, its application to the world an evil parasite.([BMS3A], *So-called Philosophical Foundations*, IV.26).

In the Intuitionist philosophy the applicability of mathematics may be of secondary importance, it was the central issue at the most critical stage of Brouwer's development, a source of personal conflict which focused his attention on the foundations of mathematics.

The conflict started as a romantic, environmentalist rebellion against an intellectualist and utilitarian establishment, a reaction against the 'applied' emphasis of his department in its early stages of growing from a service section into an autonomous mathematics department. To the radical and uncompromising student Brouwer the willingness of his professor and 'promotor' to accept an ancillary role was moral surrender. Korteweg was still part of the Aristotelian-Euclidean tradition which considers all mathematics to be an abstraction from the physical world; application of a generalized mathematical system to a physical reality was a natural return to the real world from which it all sprang. Also he had his roots in school-teaching; as a pedagogue he was imbued with the need of the adolescent to anchor abstract thinking in a concrete reality. His inaugural address, *Mathematics as an ancillary science* (KORTEWEG[1881]), was a statement and a programme. He recognizes the inspiration of problems prompted by science, and also defends the right of the pure mathematician to concern himself with problems not immediately or obviously capable of application: 'Man builds not only house to live in, but also monuments' (p.5). Mathematics originates in 'man's desire to know and understand nature; the added power to dominate nature to the benefit of mankind is accidental, not its chief objective, ... but welcome and to be considered as a fitting and well-deserved bonus' (p.3).

Brouwer's response was hostile and extreme: the power to dominate nature, rather than a benefit to mankind, is 'its curse'. *Life, Art and Mysticism* declares 'this power of the Intellect to be the source of all evil' and evil itself. Confrontation with the intellectual establishment turned to inner conflict when his further analysis of reasoning exposed a close link between man's 'cunningness' and his mathematical ability; a conflict between his moral prejudice and his commitment to his chosen subject, which focused his attention on the relation between mathematics, science and the physical reality. This conflict and the development in Brouwer's thinking are reflected in the changing emphasis in his doctoral investigation. Originally planned and entitled 'The Value of Mathematics' (Brouwer to C.S. Adema van Scheltema, 04.07.04) it became 'The Foundations of Mathematics'. At first 'the role of experience' was to be considered only in 'The Genesis of Mathematics', one of the six planned chapters (cf.2.4). In the original draft and in the published text 'Mathematics and Experience' is the second of three chapters and was considered by Brouwer to be the most momentous part of the work. It contained his first attempt to define mathematics as intuitive thinking, but only as part of an investigation into the relation between mathematics, science and the physical world.

Much of the criticism in *The Foundations* and elsewhere is directed at scientific practice and at the philosophy of practising scientists or mathematicians as he saw it or imagined it to be: a naive realism which accepts the evidence of the senses and scientific instruments as the ultimate criterion of man's knowledge of the physical reality; in its crudest form, a philosophy which considers the application of mathematics to be an isomorphic correspondence relating mathematical entities to physical facts, and mathematical rules and operations to physical interactions.

'Moral' considerations continue to play an important part in Brouwer's analysis of science; his identification of evil with purposeful or 'cunning' acting leads to a moral bias against science, apparent also in his later work.

Yet from his general criticism of applied mathematics and his comment on particular applications such as space and logic emerges a coherent alternative philosophy of empirical science, its nature, principles and truth, and its relation with pure mathematics.

### 4.7.1 Brouwer's Critique of Scientific Practice

In his criticism of past and current practice and thought Brouwer points at a number of flaws in the scientific perception and interpretation of the physical reality:

The so-called scientific laws attributed to nature are due to the uniformity of the scientist's measuring instruments (cf. Eddington's later reference to 'pointer-readings'):

> The physicist concerns himself with the projections of the phenomena on his measuring instruments, all constructed by a similar process from rather similar solid bodies. It is therefore not surprising that the phenomena are forced to record in this similar medium either similar 'laws' or no laws. For example, the laws of astronomy are no more than the laws of our measuring instruments when used to follow the course of heavenly bodies. ([BMS3B], p.10; cf. also Brouwer's letter to Korteweg 13.11.06 [DJK26], published here as app. 13).

The scientist is further accused of wilfully distorting his image of the physical reality. In his observation he is guided by his will, which in various ways interferes with his perception; he is selective in his use of sensations, ignores differences and extends his perceptional world beyond the actually observed by imagination or induction:

> The strategy of objectifying the world forces man to eliminate 'deviating' influences and thereby to abnormalize his environment. ([BMS3B], p.4).

> The observed sequences do not exclusively consist of phenomena observed independently of the human will: they are complemented by what man himself has evoked. ([1907], p.83).

The influence of will is even more apparent when causal attention offers the prospect of achieving 'a desired end' by effecting the 'means', i.e. of causal or mathematical acting (3.9).

Brouwer's moral disapproval and condemnation of 'cunning activity', however, does not just concern certain scientific practices. All science is reprehensible, because by its very nature it is causal or mathematical acting. There is even resentment that, although not wholly reliable, it usually 'works':

In general these tactics of observing sequences and then jumping from the end to the means prove effective and are a source of human power...In this way one succeeds in getting on Nature's weak side. ([BMS3B], p.2).

However, Brouwer's most fundamental criticism of the 'scientific image of the world' is its ideal nature: the objects of science and its laws are created by the Intellect; they are a product of man's mathematical power, a structure existing only in the mind of man. The observed regularity is not part of or inherent in Nature itself, but forcibly superimposed on it. He refers to this process as 'anthropomorphization of Nature', also as 'externalization', giving a physical form to an ideal structure:

> ... man impregnating Nature with the human self and repressing other one-sided developments. This externalization by man, making his environment subservient to the full development of his humanity, appears to us as a process whereby nature itself becomes linear and regular and all other life repressed or adapted to mankind. ([BMS3B], p.1).

Brouwer's bias and his obsession with the moral issue should not obscure his major contribution to the debate on the nature of science nor do they invalidate his arguments. His doctrine of the Exterior World of the Subject is a radical departure from the naive-realist tradition. Well in advance of Eddington, Whitehead and Weyl it recognized the ideal and constructive aspects of perceptual science, a radical departure too from the Kantian tradition, which maintained the a-priori of categories for all empirical knowledge.

### 4.7.2 The Ideal Nature of Science

In Brouwer's global view of man's universe we distinguished between three different worlds:
1. the physical reality, 'Nature', existing independently of man and perceived directly by the senses;
2. the domain of pure mathematics, 'the Interior world of the Subject';
3. the world of concepts which the human intellect has derived from sense-perception by means of his mathematical Intuition, 'the Exterior World of the Subject'. (cf. 3.8).

The fundamental thesis of Brouwer's philosophy of science is the complete identification of science with the exterior world of the Subject; the question of existence and nature of 'scientific truth' is a matter of relationships between these three worlds.

The objects of science are the 'things' of the Exterior World of the Subject; the observed regularity, the laws of science, are causal sequences. Both are the scientist's mathematical interpretation of the physical reality: sequences of sequences of sensations. In *The Foundations* the emphasis is on 'causality' and on the a-priori of mathematics w.r.t. science: the mathematical interpretation of the physical world is 'causal thinking' and the objects are referred-to as special 'causal sequences':

> Objectifying the world is seeing in the world causal systems in time. ([BMS3B], p.4);
> We create the mathematical systems in the exterior world ... Since 'object' is nothing but a causal sequence constant with respect to other variable causal sequences, we can say that we create the objective world in all freedom. (Op.cit., p.11).

Scientific necessity and the 'causal coherence of Nature' are exposed as figments of imagination:

> It follows naturally that a causal sequence does not exist except as a phantasy accompanying the tendency of the human will towards mathematical acting. Therefore, there cannot be any question of there being a causal coherence in the world independent of man. ([1933], p. 47).

Causal coherence is the characteristic structure of the exterior world of the Subject. The essence of causal thinking is to be found in the common constant time-order of sequences; experience is interpreted as sequences of discrete elements ordered on the basis of a time-coordinate and linked to other sequences of similar elements and the same time-order. In the sense that the pure-mathematical sequence underlies this ordering and linking, all causal thinking is an application of mathematics, although historically causal thinking may well precede pure-mathematical thinking (Brouwer even hints at a direct descendency of causal thinking from the Primordial Intuition, cf. 4.1.2).

[In *The Foundations* application of mathematics is altogether identified with causal thinking. Moreover, the domain of applied mathematics is not necessarily restricted to the world of sensory perceptions and the domain of logic; the Subject may consider the constructs of his 'Inner World', his pure-mathematical experience, with a causal eye and attribute to them a causal necessity. Such systematization is 'second-order' mathematics or applied mathematics:

> Mathematics becomes a science, i.e. a linking-together of sequences repeat-

able in time, causal sequences, in the domain of second-order mathematics, which is *mathematical viewing of mathematics* and of *the language of mathematics*, causal coherence in the first instance exists in the way mathematical systems succeed one another, in the second instance in the way mathematical symbols, words or notions succeed each other. But this, like theoretical logic, is an application of mathematics, an experimental science. ([1907], p.98)]

The identification of mathematical application and 'causality' in *The Foundations* should be seen in the context of Brouwer's argument for the Primordial Intuition of Time as the only a-priori condition for all intellectual activity against Russell's demand for some 'form of externality', 'a higher-dimensional world of objects or things'. The notion 'object' in this context is an 'invariant other than time':

> ... Russell tries to show that time alone is not sufficient on the ground that otherwise *objects* could not be perceived. We reply: in reality only time as a one-dimensional intuititive continuum is the mathematical substratum of the simplest causal sequences perceived by man. It does not matter that no objects i.e. invariants other than time itself play a role here. Objects occur only in parts of experience of a more complex mathematical substratum, in the same way that we find invariants only in mathematical systems of some complexity. ([1907], pp.104-105).

To avoid confusion with Russell's notion of 'object or thing' and to emphasize the sole a-priori of time, Brouwer speaks of 'simple causal sequences', sequences of sense-impressions repeatable in time. The need for a 'substantive' element in the nature of object, which in Russell's quoted argument discreteness of time could not provide, is met by the synhesizing role of the Subject, who as in the Primordial Intuition is the source of 'permanence in change'.

In his later work ([1929], [1933] and [1948C]) a clear distinction is made between 'things' or objects and causal sequences (and mathematical constructs under meta-mathematical consideration). Both objects and causal sequences have sense-impressions as their basic elements; objects are 'things...interative complexes of sensations', distinct from causal sequences, their elements are 'permutable in time' ([1948C], p.1235).

Most important, however, is the ideal nature of objects and causal sequences; as sequences they are constructed by the Subject and constitute his Exterior World:

> The Subject creates an iterative complex of sensations... ([1948C], p.1235)

The ideal, constructive nature of the Exterior World of the Subject is Brouwer's main argument against 'scientific truth'. His scepticism is not based on doubts about the accuracy of sensations. 'External experience', the apprehension of Nature through simple sensations, is part of the independent universe of Nature and the naive phase of consciousness; such direct knowledge is accepted as good and real:

> For my Ego only these images exist and as such they are real... ([BMS1A], p.1).

The characteristic structures of the Exterior World of the Subject, however, are wholly man-made, conceived by Mind. The chasm between this ideal world and Nature rules out any notion of truth as an isomorphic correspondence, the scientific image accurately reflecting Nature, following and predicting its course. The mathematical model-relation cannot be attributed to two such alien and isolated worlds, their elements and their characteristic modes of operation.

### 4.7.3 The Process of Application and its Mathematical Truth

In Brouwer's universe Nature and Mathematics are two wholly self-contained worlds; the exterior World of the Subject parasitically draws its life blood from its two independent partners (and from man's cunning will). Much of Brouwer' energies were given to proving on the one hand the independence of Mathematics from the Exterior World of the Subject in its various aspects, in particular logic, and on the other hand the dependence of science and logic on Mathematics. The argument for the latter in *The Foundations* revolves round the a-priori of the Primordial Intuition of Time. With respect to mathematics the a-priori of Time is its necessity and sufficiency for the construction of the whole of mathematics. The a-priori of the Intuition of Time as a 'necessary condition for' is extended to percepual science, since 'it originates in the application of intuitive mathematics' (p.98); indeed, it is 'the only a-priori element of science' (p.99).

The process of applying mathematics takes place in the limbo of the Exterior World of the Subject or 'the mathematical receptaculum of experience'; a mathematical system is superimposed on sensory perceptions:

> ... a projection of some mathematical system on experience... a free act like the construction of mathematics. ([1907], p.179).

Unlike the pure mathematical construction, this projection uses sense-perceptions as basic elements; as activity of the Subject it consists in a choice of one mathematical system 'from the stock of 'unreal' causal systems waiting to be projected on reality' (op.cit., p.83) and matching it to simple experience.

'Space' is after logic the most thoroughly argued case of mathematical application in *The Foundations*. Its treatment there and in *The Nature of Geometry* [1909A] is a good illustration of Brouwer's theory of application.
In line with his general vision of three separate universes he distinguishes between 1. geometries, 2. Nature as naively observed and 3. 'empirical space'.
All geometries are synthetic, i.e. pure-mathematical systems constructed on and with the Primordial Intuition alone and independent of experience, independent also from empirical space:

> If a-priori is taken in the sense of independent-of-experience... it follows from its intuitive construction that the whole of mathematics is a-priori, non-Euclidean as well as Euclidean geometry, metric geometry as well as projective geometry. ([1907], pp.97-98); and:
> From the Intuition of Time, independent of experience, all mathematical systems are built-up, including spaces and their geometries. ([1909A], p.5).

In particular, Euclidean geometry is the Cartesian coordinate system.
As to Nature itself and our naive sensation of Nature, it lacks the mathematical structure or 'perception' of space:

> The space of animals and of trees is not a Euclidean space, has no transformation-group...([BMS3A], *So-called Philosophical Foundations*, III.9); and:
> We receive our experiences without any mathematics and therefore without any perception of space. ([1907], p.115).

Indeed, our perception of space, 'empirical space', is an application of a mathematical structure, a structure freely chosen by the Subject and in an act of creation of the Intellect super-imposed on experience:

> Mathematical classification of groups of experiences, and therefore also the creation of our perception of space, are free acts of the Intellect. We can choose to relate our experiences to this cataloguing or undergo them unmathematically. (ibid.).

No particular spatial structure and its underlying geometry can claim to be the unique substratum of the phenomena of Nature:

> There is no definite empirical space: we can catalogue all phenomena in any space with as many dimensions as we like, curved as bizarrely as we wish... Empirical science is tied to mathematics, but experience can never force us to adopt a particular mathematical system. ([1907], p.117); also:
> The choice as to *which* mathematical system is not determined a-priori, but a matter of convenience, taste and habit. ([1909A], p.13).

Moreover, the chosen spatial structure can never be a wholly representative and reliable model of Nature, not even of man's sensation of Nature. In the ideal model discrete and finite experiences can be extended by induction and supplemented into continuous and infinite systems, but there is no guarantee of the existentence of 'real' experiences corresponding to those produced in the model.

In the context of geometry and space Brouwer poses the fundamental question of mathematical application: the relation between scientific theories and the mathematical structures applied. His denial of a 'model' relationship is again based on the complete isolation of mathematics constructed on the Primordial Intuition alone, which rules out isomorphism in the full sense. Scientific theories, even when supplemented into complete systems by the creation of 'possible experiences', are hybrids, they rely in their creation on experience. Mathematical truth, as having been constructed exclusively on and with the Primordial Intuition, cannot be attributed to them. Neither can results in the perceptual theory guarantee the existence of corresponding truths in the pure-mathematical structure. Brouwer's idealization, the complete independence of mathematics, includes the denial of such a contributory role in the construction of mathematics to any part of the perceptional world. His defense of the a-priori of mathematics w.r.t. the physical sciences and conceptual space in *The Foundations* may seem somewhat overdue; it was partly a reaction to practical attitudes and to Russell's attempt to preserve some a-priori role for space. These applications were no longer a foundational issue. They did, however, clarify Brouwer's general case for the purity of mathematics, keeping out 'extraneous' elements.

A real threat was posed by the continued and new claims on behalf of logic, considered by Brouwer to be no more than just another application of mathematics. Already in 1908 Brouwer posed the fundamental question of application in the context of logic, the legitimacy of moving from one domain, or 'medium' as he called it, to another. Here the movement starts in mathematics, is proceeded in the medium of logic and returns to a conclusion in the domain of mathematics:

> There still remains the more specific question: can one in pure-mathematical constructions and transformations temporarily neglect the constructed mathematical system and move within the accompanying language structure... confident that whenever one returns to the mathematical construction, every part of the reasoned argument will be confirmed? ([1908C], p.4).

Almost 50 years later he posed the same question, but with the emphasis on the existence of a mathematical construction parallel to the continued logical argument:

> Suppose that an intuitionist mathematical construction has been carefully described by means of words and then, the introspective character of mathematical construction being ignored for a moment, its linguistic description is considered by itself and submitted to a linguistic application of a principle of classical logic. Is it then always possible to perform a language-less mathematical construction, finding its expression in the logico-linguistic figure in question? ([1952B], p.141).

The argument for Brouwer's negative reply to these questions is based on the unique nature of the Inner World of the Subject: pure-mathematics, its objects and its own modes of operation, which are in no way parallelled in a physical universe nor in the Subject's Exterior World, and are not subject to the rules or laws derived from its application.

Brouwer's philosophy of logic (see further Chapter 5) forms part of his philosophy of mathematical application. One might speak of an Intuitionist Applied Mathematics as one does of Intuitionist Logic and reconstruct science, applying only the legitimate structures of pure mathematics and accepting Brouwer's reservations as to 'truth'. Little is known about Brouwer's own practice and there has as yet not been any serious investigation into Brouwer's teaching of applied mathematics and possible implementation of his Intuitionist principles. Yet for five years of a most formative period of his life he held the chair in 'Mechanics and Projective Geometry'. Reports suggest that his teaching was traditional. His only published contributions *Aviation and Photogrammetry* [1919R] and [1920F] were considered not important enough to be included in the *Collected Works*.

In his later years Brouwer showed greater appreciation of 'scientific thinking' and even gave it his moral approval. Scientific thinking is acceptable, provided it is not done 'purposefully' and not applied to 'reality'. It is even 'beautiful', i.e. morally good' if done 'playfully, i.e. without inducement of either desire or apprehension or vocation or inspiration or compulsion' ([1948C], p.1236) and:

> Such science, evoking beauty, reveals or playfully mathematizes naively perceptible forms and laws of nature, after having approached them with attentive reverence and with a minimum of tools. Furthermore, there is *constructional beauty*, which sometimes appears when the activity of constructing things is exerted playfully. ([1948C], p.1238. Cf. also his reference to 'logical play' in [1950C], discussed in 5.7.2).

# CHAPTER V
# LANGUAGE AND LOGIC

## 5.1 Society and Communication

In the world of mathematics as intuitive constructional activity, the 'Inner World of the Subject', language does not play a part. The emphatic mention of the negative characteristic of 'languagelessness' in all Brouwer's definitions of mathematics was no doubt justified at a time when the dominant trend was towards a closer and essential link between mathematics and linguistic formalism. Yet his dismissal of language can hardly be described as a corollary to his main thesis on the intuitive nature of mathematical activity, an afterthought of a disinterested philosopher. His views on language, as those on the application of mathematics, are rooted in prejudice, and they in turn predetermined his intuitionist conception of mathematics.

The moral bias against application may well be traced to student sympathies with fashionable romanticism, the low regard for language has its deeper source in Brouwer's character and temperament.

Analysing a man's character and personality is an even more daunting and impossible task than comprehending and interpreting his ideas; even a biography can do no more than give a view, a portrait which bears some approximate resemblance. On the other hand, views so passionately and consistently held as were Brouwer's on language can hardly be understood when isolated from their author.

Brouwer was a man of strong convictions and emotions, none more so than those which concerned the relationships between himself and his fellow-men. The brash comment of the *Profession of Faith:* 'To me other human beings are the ugliest part of my world of images ... for most of them I don't care two-pence ...' ([BMS1A], p.5) leaves no doubt about the feelings of the adolescent Brouwer. *Life, Art and Mysticism*, written some 7 years later, reveals a persistent lack of appreciation of human company: the presence of others, rather than a source of joy and enrichment, is an intrusion and evokes irritation and resentment. The misogeny of *Life, Art and Mysticism*, undoubtedly due to some personal inadequacy, is part of a universal contempt for all humans. A student article *On Morality* seems to describe their immorality as essential to their status of being 'other people': 'In their very existence lies their necessary immorality.' (Lau van der Zee (pseudonym), On Morality, *Propria Cures* 16, no 10, p.110; [BMS2]). The reader is urged 'to turn into himself' and 'seek the solitude of the desert or a deserted island'. Solitude was actively sought during these early years, Brouwer's most productive period: *The Foundations* and practically all his topological papers were written in The Hut, his lonely cottage in the forests of Laren.

His enthusiastic entry into public life in 1912, however, puts a question mark over the seriousness of his quest for solitude and the sincerity of his views on human society. He becomes frantically involved in academic and national politics and seems to thrive in the cut and thrust of the power game. There is a complete U-turn in his attitude to social reform: condemned in [1905] as 'messing about with (Dutch: 'knoeien met') society to be left to idiots-with-ambition' (p.95), it now becomes the main objective of *The Academy of Practical Philosophy and Sociology*. The optimism of what we described as 'the Signific Interlude' is the new conviction that social reform can be achieved by linguistic engineering; an optimism which seems to extend to his views on language itself.

The hope of a new society and a new language were abandoned by the end of the twenties. There is disenchantment and a return to his early pessimism; the silence of the thirties is only interrupted by the publication of *Will, Knowledge and Speech*, which more emphatically than ever spells out the evils and deficiencies of society and of language.

One might dismiss these changes of mind as purely emotional and of little consequence to his intuitionist vision of mathematics. They were, however, radical changes and account for some apparent inconsistencies in his statements on language and even mathematics. Underlying the changes there is in Brouwer's character a constant factor which determined these changing attitudes in changing situations. There is a fierce sense of independence, a conviction of self-sufficiency and superiority, which slanted his views on society and language in line with his own position in the social hierarchy. During the periods when he was on the receiving end of authority or when he felt deprived of power and influence, the emphasis of his writing is on the freedom of the individual and his separate identity; society is the oppressive presence of 'others', a threat to individual freedom. The 'others' are: an authoritarian schoolmaster-father, the 'power-thirsty church' and later ' the hypertrophied world of cooperation... enforcing the serfdom of the individual'. There is no recognition of any benefits of society, of a collective wisdom or a cultural and scientific heritage passed on to the individual. The exclusive function of society is power, the enforcement of its collective will, making the individual conform to its straitjacket of behaviour and thinking; and language is the instrument of this 'imposition of will'.

When Brouwer had secured a position of authority for himself the same pride and sense of superiority became over-confidence and contempt of-others. The naive and arrogant claims of the *Signific Manifesto* are based on the express conviction of the members of their own superiority, the superiority of 'pure-feeling, independent thinkers' over 'the un-thinking masses'. At that time he seems to accept the collective wisdom of the elite Circle, ' a group of pure-feeling independent thinkers in concord ', yet he still refuses to accept them as an authority, and safeguards his own rights by expressly 'excluding from the outset any out-voting of a minority view by a majority' ([1919C], p.5). Language still is the instrument of power, now to be wielded by some superior minds for the good of the masses, a human artefact deteriorated and worn, to be clinically re-created. The implied contempt for language was noticed by Buber; declining the invitation to join the Circle he defends the mystery of language evolving through history and carrying the wealth of man's experience and knowledge:

> The making and creation of the word is to me one of the most mysterious processes in the world of thought; indeed, in my view there is no essential difference between what I call the creation of the Word and what has been referred to as the appearance of the Logos. The evolvement of the word is a mystery which takes place in the excited and blossoming soul of man, in his discovery of the world and in his thoughts and poetic vision of the world. Only the word that in this way is born in the mind can flourish in man. Therefore, in my opinion it cannot be the function of some committee to *make* the word. (Buber to the Secretary of the Academy 17.03.17, published in the *Communications of the International Institute for Philosophy in Amsterdam* I, p.29).

Brouwer's reply points at 'the Western languages which lack words of exclusively spiritual value, words that can direct human society in line with the breathing of the 'Weltgeist to the tranquility of Tao'. He declares his trust in the ability of some men to create such words 'provided they are men of clear perception and sharp mind, acting together in mutual understanding, though remaining materially distant from each other'. (Brouwer to Buber, 04.02.18, published in the same issue of the *Communications*).

The optimism of the Signific Interlude is Brouwer's belief in the possibility of creating a new, better language and in some form of cooperation of superior minds. That optimism was shattered by what he experienced as loss of power and recognition. The two publications of his Silent Years return to the unreliability and immorality of language and human cooperation. Confidence in human cooperation never recovered, but in later years there are signs of a renewed trust in the ability of the individual to manufacture a new and better language. In his history of the Signific Movement ([1946A]) he writes of his convictions, turning Buber's argument somewhat on its head:

> Brouwer still maintained the standpoint which considered the creation of a new stock of words bringing verbal intercourse and in consequence social organization within reach of the spiritual tendencies of life as the primary task of Significs. But he expressed his gradually increasing doubts as to the effectiveness of cooperation to attain this aim. He had come to the conclusion more and more that Buber was right in his denial of the creative power of collective work in this domain. (p.206)

### 5.2 The Purpose and Origin of Language

In his pre-occupation with language Brouwer followed the trend of his time. The originality of his views is due to his refusal to join any particular school and to his mathematical habit of work, reducing a problem to its most fundamental and primitive form. While contemporary philosophies analysed language in its many and complex aspects, he searched for the primordial origin of the language phenomenon. In his vision of the ontological order of things the genesis of language is placed at the start of the social phase, the extension of the phase of causal, cunning i.e. purposeful actvity.

At the social phase cunning activity is extended to man's fellow-creatures; they and their activity become the means to his ends, subordinated to his purpose. Social life, social activity is the complex of cunning interactions in a group of individuals, ' a power-grid of will-transmission' ([1933], p.50). Reduced to its primordial and simplest form each social act is an 'imposition of will', inducing an act of a fellow individual in pursuance of the end desired by the Subject. The responding activity of the fellow-creature, 'the object individual', is causal or mathematical activity: its immediate aim is relief of pressure imposed by the authority, its ultimate purpose, however, is 'non-egoic', lies outside the scope of the acting person. Brouwer refers to such mathematical viewing and acting in the service of others as 'labour'. The process of inducement to labour is described in detail in his paper on the origin of language:

> Preparedness to labour is evoked in man either directly by suggestion, i.e. by giving fright or striking terror, by temptation, by subjugation as one does of animals or by arousing phantasies, or indirectly by training, i.e. by influencing the experience of the person to be reduced to servitude in such a way as to lead him to a mathematical view which makes labour acceptable as a means to pleasure or avoidance of pain. ([1933], p.49)

The primordial origin of language is to be found in this process of inducement to labour, its essential purpose is 'transmission of will'. Language in its widest sense is an instrument of will-transmission. As any instrument it is the product of man's cunningness. It is surprising that Brouwer's descriptions of 'cunning activity' fail to mention man's creation of tools. Tools and instruments in their creation and use are essentially cunning: their own physical being is wholly subordinated to the purpose of their maker and user, no more than a means to an end. They are the concrete distillation of the purpose of cunning man, his 'mathematical view' and will. Cunningness is compounded by the essential purpose of language', i.e. the inducement to labour.

The physical appearance of language, as of any instrument, is only determined by its specific purpose. As long as it is effective in achieving this purpose, any physical entity or act can serve:

> At the most primitive stages of civilization and in the most primitive man-to-man relationships the transmission of will to evoke labour and servitude is brought about by simple gestures of all kinds especially and predominantly the emotive natural sounds of the human voice. ([1933], p.51).

In [1905] language is reduced to 'crude sounds whose purpose is to keep the will of separate human beings on one path' (p.38).

Transmission of will remains the 'essential purpose of language' in all Brouwer's philosophical writings; it is one of the main themes of [1933] and [1948C]. Even at the height of his Signific optimism he wrote:

> Words are no more than commando-signals,... means of training and controlling (Du. 'dressuurmiddelen') [BMS22B]; and
> All verbal utterances are more-or-less developed verbal imperatives, i.e. speaking can always be reduced to commands or threats, and understanding to obeying. ([1916C], p.1). cf. also:

> In dialogue there is on two sides 'the will to dominate each other's will, like the wild war cry of Red Indians which accompanies the will to break the will of others. ([1905], p.40).

In his analysis of the origin of language Brouwer frequently uses words associated with the training of domestic and circus animals, thereby stressing the singular purpose of language as an instrument of will and power, provoking action rather than thought. The difference seems to be only a matter of response: the animal response to human command is irrational, automatic reaction, the human response is calculated and 'cunning'.

> [the result of] mathematical viewing which makes labour acceptable as a means to securing pleasure or avoidance of pain ....even though such mathematical considerations may often disappear from consciousness once it has produced servility, which then continues as automatic habit. ([1933], p.49).

The growing complexity of human society demands a more complex and subtle code, the essential purpose of language, however, remains transmission of will:

> In a more radical organization of human society labour to be imposed becomes too differentiated and can no longer be started and kept going by a simple cry... ([1933], p.51).

One might well argue that by placing the origin of language in the human command, restricting its primordial use to the communication of will, Brouwer has left the fundamental philosophical problem of language unresolved, the nature of the sign as an artificial carrier of meaning. However, Brouwer is not a philosopher of language, the problems of philosophical/linguistic analysis were not his first concern. By tracing the human activities of thinking and speaking back to their primordial origin he hoped to reveal their true and separate identities. In the ontological development of human thought there is a gradual move away from the 'pure consciousness in its deepest home' through mathematical viewing to enforcement of will in social cooperation. The primordial origin of language was found at that very extreme, as remote as possible from thinking in its purest form.

## 5.3 Language and Pure Thought

The separation of thought and language is the 'first recognition of Intuitionism' and its most far-reaching thesis. It is based on Brouwer's solipsist principle of the absolute privacy of mind, the closedness of the world of individual consciousness and its monopoly of thought. Thinking is the exclusive act and function of individual man; the world of thought is the world of the Subject, the Inner as well as the Exterior World. Speaking by its very nature is an activity of social man, a physical instrument of physical man for provoking action by others,

> ...touching only the outer wall of an automaton ... there is no exchange of thought ... Thoughts are inseparably bound up with the Subject. So-called communication of thoughts to someone else means influencing his actions. ([1948C],p.1240).

The complexity of the processes of perception, the creation of the Exterior World of the Subject confuses the issue; it obscures the ideal and private nature of thought behind an apparent common foundation in an 'objective world'. When thought is taken in its purest form, 'consciousness in its deepest home ', the wholly spiritual activity at the pre-rational stage, it demonstrates most clearly its essential subjectivity and the impossibility of communication. The soul lacks organs of communication and between souls there is no direct link. [1905] starts the chapter on 'Language':

> From life in the mind follows the impossibility to communicate directly with others, instinctively through and beyond gestures and looks, or even more immaterially through the separation of distance. People then try and drill (Dutch: 'dresseren') their off–spring in some form of communication through signs, crude sounds, painfully and helplessly; for never has anyone been able to communicate with others, soul-to-soul. (p.37).

The pure inner life of the Subject cannot be communicated, the purer the thought, i.e. the further remote from cunningness and will, the less appropriate the use of language. Words cannot capture the aspects of the inner life nor even intimate emotions. Moreover, they are useless when there is no harmony of will and superfluous if harmony of will between people already exists:

> Ridiculous is the use of language of so-called philosophers or metaphysici discussing morality, God, consciousness, immortality and the free will, people who don't even love each other...

Ridiculous is language when there is an argument...
Ridiculous is the language of conversation ...
Comical is the language of boys and girls, when there is already a harmony of will and no need for the help of language. ([1905], pp.38-39).

The real about-turn of the 'Signific Interlude' was Brouwer's acceptance of the power of language in changing society and the possibility of at least some communication between men, an optimistic belief that words could express the pure content of consciousness, express 'spiritual values'. The 1918 *Manifesto* remains vague about the precise nature of these 'spiritual values'. The reference to the failure of the Western languages to create a store of words of deep spiritual values (Du. 'bezonnen', i.e. the result of reflection) and Brouwer's correspondence with Buber, Gutkind and H.Borel point to differences in Western and Eastern thinking and their languages. The Eastern appreciation of the mystical and 'philosophical' is reflected in the language, its structure and words which are incapable of direct translation into the languages of the Western world. These in turn are the product of a materialistic mind obsessed with work and productivity. Somewhat naively Brouwer presumed that such Eastern words could be transplanted into Western languages, or equivalent words be created. When Gutkind objected that 'words like the Chinese Tao or the Hebrew Tora, so strikingly analogue in meaning and sound, can never be grown on a European foundation and presuppose a radical reform of society, require a higher society', Brouwer replied: 'The world cannot wait for 'higher society' while the 'higher society waits for the word.' (Correspondence Brouwer-Gutkind published in the *Communications of the International Institute for Philosophy in Amsterdam*, 2nd issue).
The aim of what he termed 'Intuitive Significs' was:

1. To create words of spiritual value for the languages of the western nations and thus make those spiritual values enter into their mutual understanding;
2. to point out and brand those words of the principal languages which falsely suggest spiritual values for ideas ultimately originating in the desire for material safety and comfort, and in so doing to purify and correct the aims of democracy towards a universal commonwealth. ([1918A], p.4).

Neither Brouwer's Signific vision nor his enthusiasm for language-reconstruction were shared by the new academic members of the Circle, Mannoury and the 'linguist' van Ginneken. In the 2nd issue of the *Communi-*

*cations* 'spiritual values' are no longer mentioned. There is a pragmatic acceptance of language as a fact of life; Significs is now 'an investigation of language, partly descriptive , partly compilation'. The Signific aim remains 'better mutual understanding by improving, purifying language'. As a working basis a division of language in 'stepped levels' (Du. 'taaltrappen') is agreed, representing a historical/psychological development of language from a primitive to a high level of sophistication. Even if Brouwer at this stage is prepared to compromise and agrees that 'this division is perhaps provisionally acceptable', he still feels compelled to warn that such gradation to 'higher' levels remains a downward movement on the spiritual scale: '...the intellectually higher levels of language are spiritually lower'.

Brouwer's most mature views on language are to be found in his publication of the 'silent years', *Will, Knowledge and Speech* [1933]. Political and Signific ambitions have been abandoned and there is no further speculation on language capturing pure thought, 'spiritual values'. The emphasis again is on will-transmission as the primordial and essential function of language; its necessary deficiencies are analysed and traced to its limitation as instrument of will. The use of language for other purposes is conceded, even in the thought-processes of solitary man. The dominant use of language in social life has imposed a habit on his thinking, already corrupted by cunningness. Moreover, given the limitations of flesh-and-blood man, language has its usefulness as 'an aid to memory':

> Even if language in its origin and in the first place is a function of the activity of social man it plays a significant role also in the reflective thinking and mnemotechnical processes of the solitude of singular man. This is due to the habit-automatism of using language as well as to the role which science and social organization continue to play even in the thought-world of solitude. (p.52).

In the world of pure thought language remains an alien, no more than an unnatural 'accompaniment', a satellite, following the movements of the inner life of the Subject at a distance and in an alien medium.

## 5.4 The Programme of Reform

In his search for the primordial origin of the human faculty of thought and language Brouwer had moved into the realm of ideological and metaphysical speculation. Yet in spite of his mystical sympathies and his romantic world-weariness he can hardly be described as a detached metaphysicist probing the ultimate nature of things for its own sake. He was a reformer and a revolutionary, convinced of the erroneous ways of the world around him and convinced of his own mission to put them right. But, rather than accept the dogma of an established doctrine or philosophy as the basis for his reform, he insisted on establishing for himself a clear conception of the fundamental parameters. In his questioning of fundamental principles and using the results of his own reflective analysis as a basis for reform he turned his back on the positivist, scientific tradition and returned to the 'method' of individualist metaphysici like Descartes.

In Brouwer's approach to reform we can distinguish 3 definite stages:
1. the metaphysical exploration of man's universe, revealing the nature of its basic aspects in their purest form: thought as the exclusive activity of individual consciousness, mathematics as 'intuitive' constructive activity, language as a physical instrument of will-transmission etc.;
2. a critical and diagnostic examination of current theories and practice, recognizing fallacies and tracing the deeper source of their aberration;
3. a positive programme of re-construction.

In both the critical and the positive parts of Brouwer's programme the fundamental concepts and principles of his metaphysic remain a constant guide, an 'idealized' norm.

The limitations of the human condition are accepted, in particular the needs of the flesh-and-blood mathematician to use language in his construction of mathematics. But they are recognized as limitations, not inherent in the pure, idealized activity of thought or mathematics, due to the weakness of his mental powers and to the inefficiency of language, an instrument essentially designed for a different purpose. The need for the use of language in the earth-bound practice of mathematics is conceded; *Will, Knowledge and Speech* [1933] even touches on some fundamental questions concerning the 'language of mathematics'. Brouwer's treatment of language and its role in mathematics remains predominantly negative: he highlights imperfections and pitfalls, and the attention is soon diverted to logic, ontologically analysed as no more than a mathematical application in the physical domain of language.

It is a pity that the momentum of Signific enthusiasm did not move Brouwer to a systematic philosophical analysis of 'mathematical language' as the basis for his positive re-construction of mathematics. Fundamental questions such as the relations between language, the idealized mathematical construction and the thought-process of the flesh-and-blood mathematician remained unanswered or described in vague metaphors such as 'accompaniment'. His investigation into the problems of language and semantics was mainly conducted as part of his critical program of reform and remained selective: topics such as 'meaning' chosen for their importance in counter arguments.

## 5.5 Brouwerian Semantic Theory

### 5.5.1 The Privacy of Language

The philosophical basis of Brouwer's stand on language is his doctrine of the Subject, the individual mind's absolute monopoly of thought. His fundamental thesis on signification and the nature of language is a simple one: in language there can only be two elements: 1. the thought of the Subject and 2. the word, the arbitrary name attached to it by the Subject. In its exclusion of any other elements or 'realities' this thesis is essentially negative. Since such other 'realities' are fundamental elements in almost all traditional semantic theories, it is not surprising that Brouwer's comment on language is predominantly negative and critical. Moreover, his arguments are not concerned with subtle details of linguistic analysis, they expose the fundamental reliance of the classical tradition and current philosophies on the existence of a universal objective world. The simple denial at this stage of the argument of any universal objectivity of 'concepts', 'content' and 'meaning' may seem dogmatic, but it follows directly from Brouwer's philosophy of 'reality' and 'objects'. There is no universal objective world to which language would directly refer. Words are names given by man to 'things' as he knows them, the products of his mathematical viewing, and to the pure constructs of his inner world. Brouwer's objects are subjective concepts, inhabitants of the thought-world of the Subject.

> In complete misunderstanding of the nature of language the view has grown and established itself that words are names indicating entities of fetish-like character which lead a permanent and immutable existence, independent of language and independent of man's causal attention, so-called concepts... [1933], p.56); and
> There reigned a conviction that a mathematical assertion is either false or true, whether we know it or not, and that after the extinction of humanity mathematical truths, just as laws of nature, will survive. ([BMS59], p.1; cf. also [BMS 3B], p.11 and [1912A], p.87).

Dismissed here is the existence of a Platonic world of name-souls and of any objective conceptual world, including the idealist world of common concepts held by a plurality of minds. The Aristotelian and Kantian traditions had recognized the active role of man's mind in perception and the formation of concepts, but in their quest for certainty they searched for a source of universal objectivity either in the physical world, the common domain of abstraction, or in the process of cognition for an element common to all men, a guarantee of the universality of concepts.

In Brouwer's subjective thought-world the Subject's Intuition of Time is a sufficient a-priori basis of his mathematical activity and guarantee of its truth. Furthermore, it is the primordial basis of his mathematical view of the physical world, the 'Exterior World of the Subject', the source of its existence, its truth. The thoughts of the Subject, his own past and present experience are the only reality to which language can refer. Knowledge of the experience of others lacks this real basis and is on a par with the Subjects own non-confirmed guesses, no more than hypothesis:

> From this and especially from the rejection of the hypothesis of the plurality of mind it follows that *truth* is only in *reality*, i.e. the present and past experiences of consciousness...expected experiences and experiences attributed to others are true only as anticipations and hypotheses; in their contents there is no truth. ([1948C], p.1243; the original Dutch version simply says: 'Expected experiences as such and the reputed experiences of others as such are not truths... There are no non-experienced truths.' (p.11)).

Brouwer's rejection of the plurality of mind and of any form of communal thought carries with it a total denial of the existence of 'meaning' in the usual sense of 'an objective content directly associated with the word capable of being the common property of several thinkers'. There are few references in Brouwer's work to 'meaning'. In his Signific writings he prefers the vaguer 'value' (Du. 'waarde'), and when he does speak of 'meaning' he qualifies it by the word 'so-called': 'the so-called meaning of words' ([1919C], p.6). If we wish to give 'meaning' a place in Brouwerian philosophical semantics it can only be that of the individual thought of the Subject as assigned by him to an arbitrary physical object. Brouwerian meaning is private and unique and language essentially 'private language'.

### 5.5.2 *The Instability of Language*

The purely ontological analysis of the nature of language and its origin, the 'genesis of language', leaves aside the historical and psychological facts. As in Brouwer's analysis of the genesis of mathematics the sequence of historical events in the life of flesh-and-blood man are irrelevant in his study of the essential characteristics of language, the 'genesis' of language. Again as in the case of mathematics, Brouwer was well aware of the historical and psychological processes of learning a language. Indeed, the division into 'language levels' of his *Signific Investigation of Language* ([1919C]) is an attempt to analyse and trace the development of language from the primitive 'basic language (Du. 'grondtaal') or child-language' to the sophisticated levels of 'scientific and symbolic language'.

Adoption of a convential sound-sequence or notation does not invalidate the concept of private language. In the process of signification the choice of symbol is arbitrary. It is irrelevant whether the Subject makes up his own notation or uses a sound-combination already available. Historically a meaningless sequence of sounds or visual marks may well precede, it becomes a word only when the Subject has created his 'object' and linked it to a material sign. In the ontological order the Subject's thought is a-priori to language and meaning is private.

In his obsession with 'cunningness' Brouwer made the Subject's ulterior motives the primary purpose and function of language, thereby ignoring the more immediate purpose, inherent in language and an essential part of the process of signification: to re-call a thought from memory through some inverse mapping from symbol to thought. This element, as well as the process of signification itself and the privacy of language, are more apparent in the simpler context of what Brouwer calls the 'secondary'role and function of language, which is two-fold:
1. an aid to the memory of the flesh-and-blood Subject;
2. a medium of communication with others.

These two distinct functions of language, and their different claims to reliability, are mentioned in practically all Brouwer's writings on the foundations of mathematics. E.g. [1907] speaks of the language of mathematics as:

> ...only a defective means of communicating mathematics, and an aid to memory (p. 169);

in its first version [BMS3B] as:

> a means of accompanying our own memory of a construction, or the communication of it to others ([BMS3B], pp.34,35);

and in [1947]:

The role of mathematical language can only be that of an aid to help remember mathematical constructions or construction methods... or to suggest them to others, sufficient for most practical purposes but never completely safeguarding against error. (p.339; cf. also [1929], p.157, [1952B], p.141, etc ).

*Will, Knowledge and Speech* [1933], discusses the separate aspects of language in its 'social function' and in its 'mnemotechnical function in the solitude of individual man'. It, naturally, concentrates on language recording mathematical constructions, but the distinctions made are equally valid for all language.

In language as 'a prop-to-memory' the problems of communication do not arise, it is private language in its simplest form. The need for a prop-to-memory stems from the limitations of flesh-and-blood man, his inability to have clearly present in his mind all the details of past constructions and 'objects'. Whereas to the idealized mathematician past and present thought form one single though ordered reality, past activity in the mental life of the flesh-and-blood mathematician is not all prominently present. His power of present awareness has a limited range and his power of recall from his memory-store is limited too; and even when a past thought has been re-called there is no guarantee that the thought is 're-minded' in its original form. Brouwer simply states: ' the human power of memory, which has to oversee these constructions, is by its very nature limited and fallible' ([1933], p.58). Associating his thought with a name, 'an aural or visual thing', the Subject hopes that future perception of this name will help him recall the thought or thought-construction from the obscurity of memory into the limelight of present-awareness. The association of the concept with its code is brought about by an act of will, it is an arbitrary link between two heterogeneous universes with no other correspondence beyond the willed association. The link that really matters in this process of signification is between the present and past thought of the flesh-and-blood Subject, the thought originally assigned to a symbol and the thought evoked in him by the perception of that symbol. [BMS49] specifically demands 'unaltered retention by memory' (p.4). For the idealized Subject – who does not need the aid of language – these two thoughts would be the same in every respect except for their distinct identities as past and present, pure repetition and the basis of the intuition of time. In [1933], analysing logical principles, Brouwer speculates on 'accurate language' in an ideal situtation based on the hypothesis of unlimited mental powers of the Subject:

> A hypothetical human being with unlimited power of memory who would use words only as invariant symbols for fixed elements and relations between elements of pure-mathematical systems constructed by him. (p.58).

In the life of the flesh-and-blood Subject such sameness is never wholly achieved. Brouwer uses the word 'instability' for the lack of this sameness or congruence. In [1919C] language seems to be blamed for this instability: 'the changeability and relativity of meanings belong to the essence of language' (p.6). [1933], however, points to 'memory' as the deeper source of the essentially unstable nature of language:

> ...the human memory, which has to oversee these constructions is by its very nature limited and fallible, even when it calls in the help of linguistic signs. (p.58).

Instability takes on a different dimension when language is used as a medium of communication between two different individuals. They use a 'common' language, i.e. in each of their private languages they employ the same symbol or word, which unstably recalls their original thoughts assigned by them to that word. There is, however, no basis for the assumption of a sameness of their respective original thoughts nor of their unstable replicas evoked by the same word. The instability of private language used as a-prop-to memory is a gradual property approaching stability, based on the continuity of the Self; the instability of common language is more fundamental, it is based on the absolute individuality of thought, 'thought which is inseparably bound-up with the Subject' ([1948C], p.1240). The creator of the word, 'the speaker', and the interpreter, 'the hearer', are two different individuals who each have created their private thought-world. The use of 'common language' is an attempt by the speaker to evoke a thought in the mind of the hearer similar to the one in his own mind:

> By means of sounds or symbols they try to evoke in others copies of their own reasoning and mathematical constructions. ([1907], p.128).

Because of the difference of mathematical thought-construction in different individuals the Subject needs to verify, make 'true', all constructions communicated to him and:

> This verification may in different persons lead to different results, because they will match the words of the conclusion to different mathematical systems in their minds which correspond to these words. ([1907], p.136).

The instability of 'common' language, the lack of congruence between the thought of speaker and hearer using the same word, was already diagnosed in [1905]:

> ...when using the same words even in the most restricted sciences and referring to the fundamental concepts from whch they have been constructed, two people will never think the same. (p.37).

[1933] is less categorical, it does not presume difference of thought but insists that concepts are subjective, the product of the Subject's reasoning i.e. mathematical viewing, and that sameness is a hypothesis:

> ... so far we have discussed rational consideration, i.e. mathematical viewing of the perceptional world...It is an essential hypothesis of all mutual understanding between people that such rational consideration has the same uniform structure for all individuals. (p.52; cf. also p.48).

Even between two 'idealized Subjects' stable communication is impossible, exactness is lost:

> In a human mind empowered with unlimited memory pure mathematics, practised in solitude and without the use of linguistic symbols would be exact. However, this exactness would again be lost in mathematical communication between individuals, even between those empowered with unlimited memory since they have to rely on language as a means of communication. ([1933], p.58).

It may seem surprising and ironic that Brouwer concedes that common language is least unstable when used to describe mathematics. In his early writings this is hardly a compliment: mathematics is still identified with 'mathematical viewing' and cunningness, the common source of its immorality and that of language.

> Mathematics is part of the technique of culture made commercial... prize-competition and a sack-race of the village-fair... common and worthless. Mathematics is that substratum of externalization in which mutual understanding works well. That's why originally language as a means of communication is mathematical' ([BMS3A], Mathematics and Society I.35).
> Mutual understanding between two people is a matter of degree, but in mathematics their mutual understanding is either total or it does not exist, like being-asleep, one either is or is not. ([BMS 1B], p.4).

Brouwer's discovery at the time of writing his thesis, 'his primordial recognition', of an honourable status of pure mathematics distinct from its applications led to his campaign of separating mathematics from language. In the first version of his thesis the three issues: pure mathematics, mathematical application and language are still somewhat confused, but the subjectivity of the pure-mathematical system is recognized and the agreement between systems in different individuals only a matter of similarity:

> The creation of language is itself an example of a human uninstinctive act based on mathematical knowledge. *Common* mathematical systems are not desirable for their own sake (mathematical systems certainly lack a psychological basis for accordance), only as a means to an end. Again: the emotional representations evoked by the same word in different individuals may well be different. This, however, does not affect the general efficiency of language. Because of the similarity of the relations in the mathematical systems the speaker can still force the activity of the hearer with sufficient accuracy in the direction he desires, even if the representations associated with this activity are totally different in different individuals. ([BMS 3B], p.5).

Acceptance of the *practical* efficiency of language in every-day life isalso found in later writings, in spite of its instability it usually 'works', like other applications of mathematics:

> Stability and exactness of language are not needed in every-day practice since the collective will and training-automatism have made people good understanders who only need half a word. (a literal translation of the Dutch expression equivalent to the English 'to a blind horse a nod is as good as a wink') ([1933], p.52).

Pure intuitive mathematics, the inner life of the Subject is part of personal experience, in the above quotation referred to as 'emotional representations different in different individuals'. The wider sense of the term 'emotional' is clear from Brouwer's description in [1919C] of the 'second language level', '*Emotional language*' [Du. 'stemmingstaal']

> ...the language of modern western society, constructed in subject-predicate form, but only in as far it restricts itself to personal experience and emotions, popular speech, poetic language, eastern picture-writing and the language of non-pasigraphical, non-applied mathematics. (p.8).

The division into language-levels is partly based on whether the thought, 'the meaning', is directly linked to the word ('the word speaks directly to the imagination') or indirectly via other words and their meanings, 'word-combination' [Du. 'woordkoppeling'] which affect the mind of the hearer through the memory of other words' (p.8). At the higher language-levels, 'the language of Society', 'the language of science' and 'symbolic language', reliance on word-combination increases; the link with thought becomes weaker, and almost vanishes at the highest level of symbolic language:

> there is hardly any question of 'meaning' at this level (p.7).

Since word-combination can be regulated and organized at an inter-personal level it may seem that there is stability at the higher language-levels, but:

> [their instability] what we described as the changeability and relativity of meaning, will be re-discovered by self-reflection and by observation of language. (ibid.).

The essential privacy of language and its instability were fully recognized in [1948C]. In the opening lines Brouwer expresses his resignation that his lecture 'will remain a soliloqui' and the central theme is the Subject's monopoly of the reality of thought, the rejection of a universal objectivity and of a plurality of mind. 'Meaning' even of common language is analysed exclusively in relation to the Subject:

> Truths are only in reality, i.e. the present and past experiences of consciousness... they may be linked to audial or visual things, words; linked in so far that such a word is assigned a *meaning*, i.e. the mind sees itself subject to a rule that whenever a word occurs he calls into present-awareness a certain truth, and that he notices that he follows this rule and in social cooperation his fellow-men act in conformity with this rule. ([1948C], original Dutch version [BMS54], p.11 ).

There are in other words three aspects in the notion of 'meaning in common language':
1. a thought or truth recalled in the mind of the Subject by the word;
2. the Subject's awareness of the link between thought and word and of his ensuing activity;
3. the Subject's observation of similar activity in other individuals effected by the same word.

In every one of these aspects the Subject is the sole agent; there is no mention of other minds, only of a similarity of the external actions of other individuals observed by the Subject.

### 5.5.3 Words, Sentences and Truth

We described Brouwer's investigation of language as a metaphysial inquiry in which linguistic problems are given little and only critical attention. In this global approach to 'the phenomenon of language' some major issues of semantic theory are bracketed with those of linguistic science, treated as part of 'linguistic structure' and dismissed as historical accident. The lack of clear grammatical and syntactical distinctions in his more positive speculation seems almost deliberate. The *Word* is the fundamental representative and unit of language. There is no reference to sentences or propositions, only a vague mention of 'word-groups' or 'word-complexes'.

This exclusive concern with 'the word' is partly due to a habit of reducing problems to their simplest, 'primordial' form. But there is a more fundamental reason for Brouwer's reluctance to give special status to the linguistic structure of the sentence or accept the classification of words as characteristic of the thoughts and thought-processes they name. The active, constructive nature of all reasoning does not allow the notion of an atomic substance, nor a simple division into categories such as action, substance or property. Even the 'elementary concepts' represented by 'elementary words' ([1933], p.52) or 'basic words' (Du. 'grondwoorden', [1919C], p.7) are active mathematical constructions of the Subject, 'sequences of sequences'; the only atomic elements in the constitution of concepts are single sense-impressions and the elements of the Primordial Intuition. The great variety of constructions in the exterior as well as the interior world of the Subject, 'developed in free unfolding', defy simple classification (cf. 4.5.2).

The single word as an arbitrary label for a mental construction represents signification in its simplest and essential form. The sentence or any group of words is no more than a complex name for an often but not necessarily more complex construction. The particular form of joining words into sentences, as the creation of single words, is a matter of history. Investigating the historical development of language one will recognize the subject-predicate structure to be the product of 'mathematical viewing', a simple mathematical structure which has become part of the standard modes of language. Brouwer did not analyse this structure, he simply states that 'modern Western civilization has constructed its languages in subject-predicate form' ([1919C], p.8). His criticism of linguistic structure, both of the more elaborate logical combination of sentences and of the subject-predicate sentence, is that it does not necessarily model the thought-process it names. Sentence-structure is accepted as part of the language and its characteristic deficiency; it is part of

> ...emotional language... constructed in subject-predicate form, but only in as far it expresses personal experiences and emotions (popular language, poetic language... and the language of non-applied mathematics). ([1919C], p.8).

The crucial distinction here, and in Brouwer's notion of truth, is between the use of language to report, record or simply name a thought or thought-process and a 'meaningless' combining of words, a playing with words usually inspired by a conscious use of 'linguistic' structure. The existence of thought or thought-construction in its relation to the linguistic expression could be described as 'meaningfulness', taking meaning in the restricted Brouwer-sense. Brouwer identified it with Truth, in the words of the oft-quoted passage of [1948C]:

> Truth is in reality, i.e. in the present and past experiences of consciousness. (p. 1243).

Brouwer-truth therefore is not a property of linguistic expressions themselves, based on their internal structure or on a verifiable correspondence of its components with those of an objective reality (not even the correspondence of its components with the elements of the Subject's thought-construction, to which Brouwer refers as 'exactness'). It is the simple presence of 'meaning', the thought or thought-process associated by the Subject with the verbal expression, its characteristics are those of 'meaning': Truth is not the prerogative of sentences; a word or any combination of words is true as long as it expresses the thought of the Subject. Truth is essentially subjective, defined entirely in relation to the Subject as speaker or hearer; a statement is verifiable i.e. made true only by introspection, the Subject's own construction or its re-call. In his student-notebook Brouwer bluntly states:

> The riff-raff does not know that truth of language i.e. in human communication does not exist. ([BMS1B], p.4).

There are no universally true statements, so-called true assertions; assertions become true only when they have been 'experienced' by the Subject.

> The 'truth' of an assertion has no other meaning than that its content has in fact appeared in the consciousness of the Subject. ([BMS47]; p.1; app.8).

As far as the Subject is concerned all words and word-complexes which do not express or report his own thought-construction are meaningless; they include statements made by others and not or not yet re-constructed by him, as well as his own guesses, hypotheses and purely linguistic constructions.

One of the most far-reaching consequences of Brouwer's notion of truth is the distinction between simple absence of truth and falsehood which became the basis of his new definition of negation and his rejection of the Principle of the Excluded Middle. Once he had established a constructive notion of negation, falsehood was no longer the alternative truth-value in the domain of all statements, it became the experience of a constructive impossibility, essentially within the category of truth.

### 5.6 Brouwer's Analysis of Mathematical and Linguistic Activity

We have argued that Brouwer's views on language and on the application of mathematics, although justified by argument and developed into a coherent philosophy, were inspired by prejudice. The driving force behind his campaigns on both counts were passionately held convictions of good and evil, based on personal likes and dislikes. His views on logic, even if they found a place in this philosophy consistent with his conception of language and mathematics, are more detached and can boast an academically more respectable origin. His, more effective, campaign against what he saw as abuses of logic may by many be considered passionate and extreme, it is not a moral crusade. Indeed, no moral stigma is directly attached to logic neither at the beginning nor towards the end of his career when he even described logic as ' a thing of exceptional beauty and harmony ' ([1955], p.116). In Brouwer's student days logic was the topic of 'dialectical' discussion with his friend Mannoury, who shared his social and moral concern but took an independent and positive line on logic.

It seems that only during the later stages of writing his thesis Brouwer started his 'genetic' analysis of logic and formalization. In the early planning of the thesis a treatment of language and logic is notably absent. Of the six planned chapters (see 2.4) two deal with the construction and genesis of mathematics, one with its philosophical significance and the other with its moral value. Only one chapter is planned to treat 'The foundations of mathematics on axioms'. From Brouwer's own Notes [BMS 3A] under the heading 'Axiomatic Foundations' it appears that the main attack was to be directed against the axiomatic treatment of mathematics; the emphasis is on its artificiality and on the restrictions it imposes on the free construction of mathematics:

> ...the axioms are a straitjacket in the construction. (I.27);
> axiomatic mathematics is concerned not with is architecture, it is experimental pottering i.e. it treats first-order mathematics like chemistry treats matter. (III.18);
> ...a monster and threat to free construction. (IV.15).

There is no reference to Hilberts *Grundlagen*; criticism is still aimed at the 'concrete' axiomatic method:

> ... axioms were not primary in mathematics, they were found afterwards; once we captured nature by chance we started searching for axioms in the constructed system. (VIII.56).

Clearly evident is Brouwer's conviction of the artificiality of the axiomatic structure and the triviality of the logical modes of thinking, quite uncharacteristic of the free constructive thought-processes of the mathematician.

Current trends towards an even more essential role of logic in mathematics prompted Brouwer to analyse in detail logic and the processes of formalizing mathematics. The first and most detailed analysis is found in the published version of *The Foundations*. It traces the 'genesis' of logic and formalization, the steps taken in moving from pure-mathematical activity to levels of constructive activity of increasingly 'higher order'. Each higher level is based on the lower one and generated by three steps in a process of linguistic engineering:
1. the recording in language of constructive activity at the lower level;
2. viewing this recording mathematically, observing its (linguistic) structure;
3. isolating the symbols and the linguistic structure from their corresponding elements in the constructive activity of the lower level, and using both symbols and structure in new constructive activity at the higher level.

The analysis of [1907] starts from the level of pure intuitive mathematics and proceeds to 'mathematics of the third order', and Brouwer adds:

> one could of course go on, but the mathematical systems of even higher order would roughly be copies of each other; there is therefore no point in continuing the process any further. (p.175).

His object here is to reveal the difference between various levels and the inconsistent movement between levels by logicists and to some extent by Hilbert.

The distinction between higher-order levels is not found in later writings (except for a brief reference to 'second-order mathematics' in [1928A] in which Brouwer claims priority of the notion of meta-mathematics). The all-important distinction is between the first-level of pure-mathematical activity and the other levels which all are products of linguistic engineering. This was already recognized in [1907]:

> ...other phases have no mathematical significance. Mathematics belongs only to the first phase. (p.175).

The first steps in the 'genetic' analysis are the basis of Brouwer's critical campaign and the source of his most spectular claims relating to logic. They analyse the relation between pure mathematics and logic in genetic order:
1. the recording of mathematical constructions (the language of mathematics);
2. observation of structure in the linguistic recording and conscious use of this structure (classical logic);
3. isolating symbols and structure from their mathematical content and using them in pure-formal constructions (formal logic).

## 5.6.1 The Language of Mathematics

In the genetic order the pure-mathematical activity of the Subject stands independent and in splendid isolation. It precedes and affects all activity of linguistic recording and construction but as pure-mathematical construction it is not in any way affected by them.

The whole of linguistic activity is part of the imperfect world of the flesh-and-blood mathematician. Genetically first, and nearest to the pure-mathematical construction comes its recording by the mathematician in his language; in the words of the analysis in [1907] 'the linguistic parallel of mathematics: mathematical speaking or writing' (p.173). Brouwer often refers to it as 'the language of mathemematics' (e.g. [1907], p.169, [1919C], p.8, [1933], p.54 etc.) or 'mathematical language' (e.g. [1907], p.129), [1933], p.54, [BMS51], p.4 etc.). Since it describes 'first-order mathematics', it is a 'first-order' language (cf. [1933], p.53).

It is clear from the context that Brouwer refers not to language as a system with its vocabulary and rules, but to the recording of mathematics in language. Other references spell this out:

'the language accompanying mathematics' ([1908C], p.3),
'language accompanying the mathematical mental activity' ([1923B], p.3), or
'the linguistic description of mathematics' ([1952B], p.141).

In Brouwer's descriptions of the relationship between pure mathematics and its recording in language the emphasis is on the negative aspects, i.e. the absence of any affinity and especially of relations accepted by others. The only positive reference is to 'accompaniment' (Du. 'begeleiding'), a metaphor which Brouwer used persistently and without further explanation. One could speculate on the Brouwerian meaning of 'accompaniment' and exclude e.g. any suggestion of 'guidance' as implied in the modern usage of the Dutch 'begeleiding', or the musical notion of playing a different but complementary part in harmony. But even the most general notion of a parallel movement meets with problems, suggesting as it does some homogeneity of partners and medium. It seems that Brouwer chose the metaphor 'accompaniment' as representing the loosest and vaguest relation between movements or activities, thereby leaving the only emphasis on the dynamical aspects of both mathematics and language, i.e. the 'record' in the process of being made or being read rather than the static printed word itself. The description of the 'first step' of Brouwer's analysis only refers to the former i.e. the process of expressing the mathematical construction in words; mathematical language is: 'the linguistic parallel of mathematics: mathematical speaking or writing' ([1907], p.173). However, in this passage mathematical language is considered in its relation to the further processes of logical interpretation and formalization. Elsewhere [1907] is quite specific about the essential purposes of the language of mathematics as 'a medium of communication and an aid to memory' (p.169), an essential component of which is the act of hearing or reading. Mathematical language, or 'mathematical language of the first order', therefore is the whole of the active processes of writing/speaking and reading/hearing relating to particular mathematical constructions; as the mathematical activity itself they proceed in time. The 'frozen', printed text by itself, like a musical score, record or tape, or Karl Popper's deserted wasp's nest, reverts to its pure-physical status of ink-on-paper.

[In section 4.6.3 we made the comparison between mathematics and music. One could extend the comparison and liken the relation mathematics and mathematical language to that between the act of composing music and its written score. In VAN STIGT[1971] I expressed surprise that Brouwer never made this obvious comparison. In 1976, however, I discovered the manuscript of a report on meetings of the Signific Circle in 1924 which was published later as *Signific Dialogues* [1937]. There Brouwer writes: '...language which accompanies mathematics as musical notation accompanies a symphony by Bach(!) or an oratorio by Handel' ([BMS23B]). In the printed version this was replaced by 'the weather map describing the atmospheric conditions'. It was pointed out in 4.6.3 that the comparison with music holds for some aspects: the time-existence of music, the difference between music as live sound and its recording, in particular the relation beteen the activity of composing and the musical score. Brouwer's change of mind may be due to concern that such comparison could lead to misunderstanding and confusion: it could be argued that the reality of music exists in it being performed rather than its existence in the mind of the composer; the music score can be used to generate and re-vive such live music.]

Mathematical language suffers from the inadequacy inherent in all language. In particular it suffers from instability and inexactness, admittedly to a lesser degree (5.5.2). Brouwer's final verdict is that:

> ...also for pure mathematics there cannot be an infallible language i.e. language which in communication excludes misunderstanding and in its mnemotechnical function offers a guarantee against errors i.e. confusion of different mathematical entities. ([1933], p.53).

While Brouwer was prepared to accept the even higher degree of instability of language in other spheres of life, the instability and inexactness of 'mathematical language' remained a life-long obsession and became the key-issue in his campaign for the separation of mathematics and mathematical language.

The Dutch word for mathematics, 'wiskunde' (certain knowledge), reflects a traditional acceptance of certainty as an essential attribute of mathematics. According to Brouwer's historical interpretations this special claim of a-priori certainty became extended to the written recording of mathematics. His new conception of mathematics had re-established the a-priority of mathematics by founding it on the human thought-processes alone, eliminating all external elements. The independence of mathematics was the very basis of its a-priori and 'languagelessness' one aspect of this independence. In Brouwer's metaphysical arguments the absence of any direct link between mathematics and language, and the characteristic instability of language remain the fundamental reasons why the recording of mathematics in language lacks such guarantee of certainty or exactness. In his criticism of current practice, however, the emphasis is often, especially during the early years, on the inadequacies of existing language as the system of words and linguistic structure available to the flesh-and-blood mathematician; the distinction between 'mathematical language' as the first-order expression in language of a mathematical construction and language as a system, an available tool, becomes somewhat blurred. Vocabulary and linguistic structure are the legacy of a logical tradition, 'the language of logical reasoning', which restricts the mathematician in expressing his pure constructions and even lead him into error. In his correspondence with Korteweg Brouwer speaks of 'the poverty of present languages' and '...the use of these languages which has led mathematics on false trails' (23.01.07, [DJK35]). He seems to hold out the possibility of 'a pure language'; his version of the Signific ideal was the 'creation of a pure language' – a lexicon – also for mathematics. When hopes for such an improved, pure language faded, distinctions became clearer and the instability of language more clearly recognized as the fundamental reason for the impossibility of an exact recording of a mathematical construction i.e. mathematical language. [1929] and [1933] point out that instability, i.e. failure to guarantee congruence between the mathematical construction originally ascribed to the word and the one evoked by that word would remain even if a 'pure language' were used:

> Even when an exact mathematical language were to become a reality, the possibility of misunderstanding, i.e. confusing the mathematical constructions expressed in this exact language, would in no way be eliminated. ([1933], p.54).

Indeed, the ultimate reason for the imperfections of existing language and the impossibility of creating a pure language is to be sought in the instability of the language process and in the essentially 'posterior' character of language, the prior existence of the mathematical construction. Brouwer mentions vaguely the 'inertia' of existing language in perpetuating 'logical reasoning' ([DJK35]), he fails to point out that existing language in its vocabulary and structures can only reflect older constructs and construction-processes and will always run behind the new demands of mathematics in its 'free unfolding'.

At the practical level Brouwer resigns himself to using existing languages carefully, constantly improving the texts of his manuscripts, but always reminding himself and the reader of the 'essential inadequacy of language. In a cryptic note in the margin of his 'species construction' he writes:

The recording in language, which is essentially inadequate, often remains somewhat vague; as long as it is unambiguous in its instructions for constructions [Du. 'handelingsvoorschriften'] in intuitionist mathematics. ([BMS32], p.11; note dated 5.3.47, probably for use in one of his Cambridge lectures.)

The absolute separation of mathematics on the one hand and 'mathematical language' and all structures built on it on the other, is perhaps the most distinctive feature of Brouwer's Intuitionism ('The first act of Intuitionism which completely separates mathematics from mathematical language' ([1952B], p. 140)). It puts him in opposition to logicists and formalists on every front, it also sets him apart from his intuitionist predecessors, the 'pre-intuitionist school'.

The pre-intuitionists Poincaré, Borel and Lebesgue, although rejecting the logicist and formalist thesis of the sufficiency of symbols and logic, still considered expression in language or symbols as an essential part of mathematics. Poincaré distinguished between mathematical invention, the mental process of search, inspiration and construction, and expression in words and logical proof. Invention plays a fundamental role in the creation of mathematics, but 'expression in a finite number of words' is Poincaré's criterion for mathematical existence and logical proof, the only guarantee of truth of a theorem:

> Logic, which alone can give certainty, is the instrument of proof; intuition is the instrument of invention (POINCARÉ[1905], p.29).

Poincaré was particularly critical of a publication by Leroy which expressed views very similar to those of Brouwer:

> Science contains two elements of very different character, an *act* by which it is inspired and which is its origin, and a *representation* which expresses the acquired results as well as is possible. Progress is due to the former, the latter provides a language more or less well-adapted, but never a true source of verification. ( LEROY[1905], p.203).

This radical view on the split between mathematics and mathematical language was re-iterated by Brouwer, who in turn reproached Poincaré:

> He does not go to the heart of the problem, which lies much deeper, i.e. the confusion of the act of building mathematics and the language of mathematics... How little Poincaré is concerned in his criticism with the intuitive construction of mathematics as its only foundation is evident from his words: 'Les mathématiques sont indépendantes de l'existence des objects materiels; en mathématiques le mot exister ne peut avoir qu'un sens, il

signifie exempt de contradiction'. It almost reminds us of his opponent Russell... mathematical existence means: having been constructed, and whether the accompanying language is free of contradiction is not only unimportant, it does not constitute a criterion of mathematical existence. ([1907], p.177).

Poincaré's demand that a concept is only truly mathematical if it can be defined in a finite number of words (Poincaré, Scientia XXIV, p.6, quoted by Brouwer in [1912A], p.13), was reinforced by Borel and Lebesgue. They do not wholly identify mathematical clear thinking with 'verbal thinking' in the manner of linguistic phenomenologists, but presume that if a mathematical notion is sufficiently sharp and clear it can be expressed in language, and that therefore expressibility in language is a test of mathematical legitimacy. Lebesgue demands of mathematical objects that they are 'nommables'; Borel restricts 'real' mathematical objects to 'those we know so well that we can speak about them without danger of error or misunderstanding'. Brouwer's conception of constructive mathematics demands a precision and exactness well beyond that of Borel's notion of 'clearness', but neither his constructs nor his construction methods need the sanction of verbal expression. The question whether precise concepts can always find a verbal or symbolic expression is quite irrelevant in view of the instability and inexactness of language. Poincaré's insistence on verbal expression and proof reduced the mental constructive processes, 'invention', to mathematical preliminaries, in Brouwer's view not merely an undervaluation but a total denial of the constructive nature of mathematics.

### 5.6.2 *The Language of Logical Reasoning*

Brouwer's metaphysical inquiry into the genesis of language and its relation to thought centred on 'the word' as its fundamental representative and unit; his historical and critical accounts recognized the complex reality of 'existing language', rooted in Western culture, with its vocabulary, idiom, the syntactical and grammatical structure of the sentence, as well as the stylistic and logical conventions governing the composition of sentences. In particular the classical conventions of logic are identified as part of the linguistic inheritance, the tool available to the flesh-and-blood mathematician. Brouwer refers to it as *the language of logical reasoning*, the syllogistic combination of sentences which has become the habitual framework for the individual in expressing his thoughts. Analysing its structure he exposes the syllogism as 'the simple mathematical structure of whole-and-part'. Referring to the the classical example of Socrates' mortality:

> In such a mathematical system it is a mathematical tautology that if all elements with the predicate 'human' are part of those with the predicate 'mortal', that the element 'Socrates' of the first group is also part of the second. We are dealing here with one of the simplest forms of mathematical reasoning, i.e. by means of a tautology passing from one relation to another. ([1907], p.131).

It should be noted, however, that although originating in 'mathematical viewing' – as indeed do all other aspects of language – the language of logical reasoning is an 'unconscious' use of the mathematical structure. In the same way that the native speaker does not need to be aware of the underlying grammatical structure of the language he uses, the classically and mathematically educated individual can express his mathematical constructions in the framework of logical syllogism unaware of this structure, he is not guided, only restricted by it; it is the language of 'intuitive logical reasoning' ([1907], p.127).

The language of logical reasoning is often quite unrepresentative of the construction the mathematician is trying to express:

> That even in parts of mathematics where the relation of whole and part does not enter into the mental construction, one often, for the sake of communication in words, transforms the relations that were in the mind into relations of whole and part – so that the usual language of general mathematics is full of the modes of expression of logical reasoning – , can only be attributed to a centuries' old tradition of logical terminology in language. ([1907], p.128).

The language of logical reasoning is a kind of literary form, a cultural convention rather than something universally and essentially human. Brouwer leaves open the possibility of a culture with different logical conventions:

> ...that with the same organization of the human intellect, and therefore with the same mathematics, a different language of understanding had emerged in which there was no room for our language of logical reasoning. Probably there are people living outside our cultural sphere where this is the case. Neither can we preclude the possibility that at a further stage of development, logical reasoning will lose its position in the language of our culture. ([1907], p.129).

This passage of *The Foundations* led to a heated exchange between Brouwer and Korteweg, who remarked that 'it must be a rather strange race that does not think logically'. We quote Brouwer's reply at length because of its relevance to his views on logical reasoning and the fundamental distinction he wishes to make between the constructive thought-processes and the logical structure imposed by conventional language:

> At the beginning of the chapter I show that mathematical reasoning is *not* logical reasoning, that only because of the poverty of language it makes use of the connectives of logical reasoning, and perhaps thereby will keep alive the linguistic accompaniments of logical reasoning even long after the human intellect has outgrown logical reasoning. Far from it being 'a strange race that does not think logically', I believe that it is only a sign of the inertia of language that the relevant words still exist in modern languages. One seldom meets a pure use of these words; their impure use in every-day life has led to all kinds of misunderstanding, and in mathematics to the false notions of set theory. These notions have come about not from a lack of mathematical insight but because mathematics for lack of a pure language had to make do with *the language of logical reasoning*, whereas its thoughts proceed not as logical reasoning but as mathematical reasoning, something quite different.
>
> The theorem: 'if a triangle is isosceles it is an acute triangle' is expressed as a logical theorem: the predicate 'isosceles' in the case of triangles is considered to imply the predicate 'acute', i.e. one imagines all the triangles of a given plane represented by the points of an $R_3$, and one then sees that the domain of $R_3$ representing isosceles triangles is contained in the domain representing all acute triangles. This is in fact true, and logical formulation and logical language can therefore safely be applied. But the thoughts of the mathematician, who because of the poverty of his language formulates this theorem as a logical theorem, proceed in a way quite different from the above interpretation. He imagines that he is going to construct an isosceles triangle, and then finds that either at the end of the construction all angles appear to be acute or that on the postulation of a right or obtuse angle the construction cannot be executed. In other words he thinks the construction mathematically, not in its logical interpretation.
>
> It is precisely the main theme of Chapter 3 to show that careless use of a logical language instead of a mathematical language has led mathematics in some of its parts on a false trail.
>
> May I just explain why I believe that logical language is out of date; in fact we already discussed this the day before yesterday. The mathematical systems which are applied to the outside world, and therefore alone should be considered for representation into language, should through its mathe-

matical theory be able to teach us something practical. But the mathematics of whole-and-part does not teach us anything new for every-day life. Once this system is applied to a part of the perceptional world, even a very mediocre mind can read off directly all the consequences; he does not need any intermediary logical reasoning. One knows very well nowadays that if something about the exterior world is deduced by logical reasoning, something that was not so directly a-priori evident, it is then for this very reason wholly unreliable. (Brouwer to Korteweg, 23.01.07 [DJK 35]).

The use of the language of logical reasoning and its attending axiomatic structure seems to effect an intermediary stage in the thought-processes of the mathematician: a conversion process of the completed mathematical construction into the framework of logical reasoning, a logical afterthought adapting the construction to the peculiarities of conventional language. Brouwer seems to accept such adaptation and the language of logical reasoning as a legitimate representation of the mathematical construction, of course, with reservations as to appropriateness, exactness etc. due to the inevitable 'poverty' of language. The linguistic recording, in spite of its logical 'literary form', is acceptable as an accompaniment of first-order, pure mathematics provided it is an attempt to describe a preceding pure-mathematical construction, completed in the mind of the mathematician. Euclidean Geometry for example is considered justified, provided it is such an adaptation of a pure construction in a logical form:

> ...if Euclid considered his mathematical construction as completed (i.e. as a Cartesian space with a transformation-group), and if his reasonings merely served as an accompaniment of some transition from clearly-seen relations (i.e. subordinate constructions) through a series of tautologies to new, not directly seen relations, in other words as an accompaniment of an exploration of a construction he had already built himself. In that case his work would be purely mathematical, the fact that he did not introduce or use coordinates is only a methodical imperfection. ([1907], p.135).

The crucial distinction is between a description of a completed mathematical construction, albeit in the form of logical reasoning, and the logical construction which starts from pure symbols-without-meaning and proceeds by means of logical principles, the formal-axiomatic structure. The latter is no more than 'a linguistic structure' (Du. 'taalgebouw'), creating the 'pathological geometries' of which Hilbert and others, and possibly also Euclid, are guilty. Continuing his analysis of Euclidean Geometry Brouwer speculates:

> It is of course also possible that Euclid did not see it like that, and that he fell into the trap like so many others who thought they could reason logically about things and mathematical systems other than those they had themselves constructed; who overlooked, for example, that when logic uses the word *all* or *every*, these words, in order to have any meaning, include the silent restriction *in so far they belong to the mathematical system that one has constructed before*. ([1907], p.135).

In the universe of the Idealized Subject the boundary between mathematics and any linguistic activity is absolute: mathematics is languageless. In the working life of the flesh-and-blood mathematician the recording in conventional language – the language of mathematics and the language of logical reasoning – is an inevitable accompaniment of his mathematical construction; as such it can be considered to be part of his 'first-level' mathematical activity. In this context the boundaries run between the mathematical construction with its embodiment in language and what we have termed 'linguistic engineering', the mathematical viewing of dead symbols and the construction of systems at 'higher levels'. In the move to activity at the second level (cf. 5.6.1) mathematical language is divorced from the pure-mathematical reality and its outer shell becomes the object of scientific investigation. The resulting activities are the autonomous sciences of logic and formalization. Brouwer did not question their autonomy, their legitimacy nor their mathematical origin. His criticism was directed at the confusion of their separate identities and that of mathematics, and the inconsistent movement between these different disciplines.

## 5.7 Brouwer's Analysis of Logic

### 5.7.1 Logic, the Science of Reasoning

Brouwer's [1907] analysis of mathematics and 'linguistic' activities of various kinds is the more remarkable if one takes into account the time and circumstances of writing. He lacked the obvious advantage of the observer/historian, who with hindsight can draw clear lines of demarcation and present a balanced and comprehensive view, based on established and recognized facts and achievements. *The Foundations* was conceived during the turbulent years of the *Foundational Crisis* when the protagonists in the great debate were still staking their positions, at the start of a period of major change and expansion of both mathematics and logic. In the case of logic these changes were so fundamental that without exaggeration one can speak of a change of identity; they affected Brouwer's response and the development of his ideas, and are reflected in a changing emphasis and terminology. It is greatly to his credit that in the midst of the turmoil he was able to analyse the change, to identify distinct categories of human activity and their ensuing products, and make his unique critical contribution.

From the [1907] Analysis emerges a division into three main areas of logical activity, each with its distinctive features and its special relationship with mathematics, and each the subject of his special criticism:
1. pure logic as science and abstraction;
2. the application of logic (including the errors of traditional logic);
3. formal logic.

Developments and Brouwer's own 'new insights' caused shifts of focus and emphasis particularly in his criticism and his suggested remedies. The conviction of the separate identities of these interpretations of logic remained and, if anything, grew stronger as time went by.

## 5.7.2 Pure Logic as Science and Abstraction

The Brouwerian interpretation of science distinguishes between 1. simple mathematical viewing, the observation of a mathematical structure in a group of phenomena, and 2. causal acting, the 'cunning' use of this knowledge which includes the wilful manipulation and extension of the phenomena and practical applications beyond the range of the phenomena observed (cf. 3.9 and 4.7). This distinction is more than a moral criterion by which some aspects of science are judged to be good or 'beautiful' and others evil; more also than academic discrimination between a respectable practice of pure science for-its-own-sake, 'done playfully', and the mercenary pursuit of applied science, 'a means-to-an-end'. It is a criterion of legitimacy, determining the boundaries of valid practice. For all science these are the boundaries of the 'Exterior World of the Subject', its subject-matter is the perceptual world, the mathematical creations from phenomena as observed by the Subject. For each branch of science these boundaries are the limits of the system constructed solely on the basis of a group of phenomena actually observed. Excluded in particular are: supplementing the system with phenomena not observed and superimposing the structure of one system onto a different group of phenomena.

In spite of Brouwer's generally negative attitude to logical practices there is in his philosophy a place for a legitimate science of logic, to which one could refer as 'pure logic'.

Brouwerian 'pure logic' is a science, a mathematical viewing of a range of physical phenomena as observed by the Subject; its special subject-matter is the verbal account of the Subject's reasoning:

> People, who want to view everything mathematically have done this also with the language of mathematics... the resulting science is theoretical logic... an empirical science and an application of mathematics... to be classed under ethnography rather than psychology. ([1907], p.129);
> ...the scientific consideration of the symbols occurring in purified mathematical language, and the rules of manipulation of these symbols.([BMS49], p.2; app.7).

In his [1907] analysis, which examines the relation between pure mathematics and logic, Brouwer restricts himself to 'the pure construction of intuitive mathematical systems' and 'its linguistic parallel'; logic is:

> ...the mathematical viewing of this language... observing a (logical) structure in the verbal edifice... ([1907], p.173).

Since it reflects the 'Idealized' domain of pure-mathematics we may refer to it as 'the idealized linguistic record' which holds all and only the completed constructions of the Subject to-date, his well-constructed mathematical systems.

In later papers on language and logic the subject-matter of logical investigation is the verbal account of any reasoning: [1955] includes 'common-sensical thought' and [1933] refers specifically to constructions in the External World of the Subject: 'the mathematical viewing of finite systems of objects of the objective space-time world' (p.54).

It cannot be over-emphasized that language in this context must be understood in the restricted Brouwer-sense of 'a report on a completed thought-construction'. The subject-matter of Brouwerian pure logic is not the totality of all possible symbols and symbol-combinations, whether relating to an objective world of concepts, 'a chimerical all of ideal truths', or divorced from any meaning; nor even conventional language as the tool available to the Subject. It is the 'verbal edifice' resulting from the attempt by the Subject to express his reasoning in words. Even if logic searches for 'linguistic structure', the only legitimate domain of its search are the verbal accounts of the Subject's own reasoning:

> ... logic is observing structure in this edifice...but the elements of these verbal

edifices are linguistic accompaniments of mathematical constructions and relations. ([1907], p.173);
We must emphasize that the syllogism and other logical principles can be considered to be valid in the language of logical reasoning which treats finite or denumerably infinite groups of elements or domains in the continuum, but in any case: mathematical systems that have been constructed.' ([1907], p.132).

There are few direct references to the methods of 'pure logic'. As in his general treatment of science Brouwer is more concerned with the fundamental abuse of not confining oneself to the domain of investigation, applying the logical structure found in the description of one system to that of another or to the domain of all possible symbols and symbol-combinations divorced from their meaning. From his 'critique' one can, however, derive the notion of pure logic and its legitimate practice. [1933] describes the process of logical investigation in a primordial setting:

... human beings with unlimited memory, who have recorded their constructions in shortened form in a suitable language, surveying the strings of their affirmations in this language and then able to see the occurrence of the linguistic figures of logical principles in all their mathematical modifications. (p.59).

The elements of the verbal account are here referred to as 'affirmations'; elsewhere in [1933] as 'correct affirmations i.e. effectively indicating actual mathematical viewing' (p.54), in [1948C] and [1955] as 'assertions' or 'true assertions' thereby emphasizing, in Brouwer's terminology, the 'real' or 'experienced' constructions which precede and underlie the verbal expression.

The logical regularity observed is the uniform occurrence of a sequence of affirmations, 'the linguistic figure (Du. 'taalbeeld') of a logical law', for finite systems defined as

... certain transitions from correct affirmations to other correct affirmations [i.e. also 'effectively indicating actual mathematical viewing' Ed.] which one calls the laws of identity, contradiction, tertium exclusum and syllogism. ([1933], p.54).

In this context the word 'principle' is a misnomer, the pure-logical principle is not a guide nor a generative source, it is no more than a generalized pattern, observed in existing accounts of reasoning and due to the reasoning represented. *The Foundations* for example refers to the absence of contradiction in mathematical accounts as obvious, resulting naturally from the mathematical construction described:

> Classical logic studies the linguistic accompaniment of logical reasoning... for mathematical, constructed systems, and we know from the fact that we can see these mathematical systems, that *there* the sequence of sentences which follow one another logically will never produce contradictions since they accompany mathematical acts of construction... (the logicists, however, turn things round and start from these principles...). ([1907], p.159).

Equally obvious to Brouwer, or at least not surprising, is the occurrence of contradictions when logical principles are used to generate symbol-combinations:

> Logical principles are only valid for words with mathematical meaning. As to Russell's logic, exactly because it is only a system of words without a pre-constructed mathematical system to which it refers, there is no reason why contradictions should not occur. ([1907], p.163).

The frequent occurrence of the syllogistic structure observed by the pure-logician especially in mathematical accounts – though uncharacteristic of the mathematical construction described – is due to the Western tradition of using the 'language of logical reasoning'. Its validity is based on the tautology implicit in the relation 'whole-and-part'. (cf. 5.6.3)

The Principle of the Excluded Middle (PEM) will be discussed in detail in sections 5.8. Suffice it here to say that in the strict context of 'pure-logical' investigation the PEM presumes that all mathematical problems have been solved.

In his attempts to refute the PEM Brouwer had to move outside the domain of pure, constructed mathematics and that of pure-logical investigation of past mathematical accounts. Like other principles of classical logic the PEM is a statement about assertions, symbol-combinations representing completed constructions as well as mere hypotheses of possible future constructions.

The fundamental weakness of Brouwerian Pure Logic is inherent in his notion of pure science. 'Mathematical viewing', observing regularity, is deriving laws, which by their very nature extend to the not-yet-observed. Brouwer's solution was to reject the universality of scientific laws and their useful application.(cf. 4.7). They remained hypotheses, in constant need of verification. When Brouwer developed 'a theory which with some right may be called Intuitionist Logic' he stressed the need to verify any application of his alternative principles of judgeability, testability and reciprocity of complementarity:

In Intuitionism, of course, all three of these principles, being assertions about assertions, are only then 'realized', i.e. only then convey truths when these truths have been experienced. ([1954A], p.3).

As a theory of assertions of hypotheses, the possibility or impossibility of future constructions, the new Brouwer Logic operated outside the domain of pure-logic i.e. the verbal accounts of completed mathematical reasoning.

'Pure Logic' in this restricted sense may be of little general interest, its significance in Brouwer's philosophy lies in its relation to both puremathematics and further logical activities: distinct from both, dependent on mathematics and an essential phase in the genesis of applied logic and formalization. In the ontological order the observation of a structure in a linguistic record is a distinct activity which presumes the existence of that record and the reasoning recorded, and precedes the application of the observed structure. The post-factum nature of logic w.r.t. mathematical reasoning and its scientific character of 'mathematical viewing' were the basis for Brouwer's claim that: 'whereas mathematics is independent of logic, logic is dependent on mathematics' ([1907], p.127). This claim for independence extended well beyond the primordial genesis of mathematics and included the denial of any contributory role of logic and its principles in the practice of pure mathematics.

The [1907] analysis of logical activity points to an intermediate phase between the pure scientific observation of regularity in the mathematical record and the application of logical principles, the creation of what one might call 'abstract logic'. The scientific observation of regularity in the physical domain of the mathematical record is followed by a deliberate act of 'thinking away' the meaning of words and symbols. A new logical system is constructed, modelled on the architecture observed in the mathematical record by applying the observed regularity to symbols and words isolated from their meaning in pure, constructive mathematics, a mathematical system of 'second order':

> [Phase3] ...The mathematical viewing of written or spoken mathematics: one observes logical linguistic structures... but the elements of those linguistic structures are still linguistic accompaniments of mathematical constructions.
> [Phase 4] Thinking no longer of the meaning of these elements of the above logical structures and with them constructing a new mathematical system of 2-nd order imitating these structures. ([1907], p.173).

In his later years Brouwer refers sympathetically to this 'abstract logic' as 'logical play', 'done for its own sake, not arising from fear, compulsion, desire or from professional considerations'. ([1950A]; cf. also [1955]). It is distinct from the scientific phase of observing regularity in the mathematical record, it is:

> ...logical play, replacing the objects of perception by fictitious and purely indicative objects. (ibid.);

distinct also from pure constructive mathematics which

> ...abstracts completely from objects... and which is obviously richer because of the total freedom that mathematics enjoys. (ibid.)

Both the 'scientific' and the 'abstract' phases will play an important role in Brouwer's contribution to an 'Intuitionist Logic' (see further 5.10). His first concern was the classical confusion of these distinct activities and the application of logical principles in mathematics, the result of 'the phenomenon of cunning which creates the acting man'. (ibid.)

### 5.7.3 *The Application of Logical Principles*

Brouwer's Intuitionism is perhaps best remembered for his sensational refutation of the Principle of the Excluded Middle (PEM), his alternative philosophy of mathematics as thought-construction almost forgotten. It is ironic that his campaign against the PEM was prompted by a desire to reduce the emphasis on logic in mathematics and that his arguments developed from an attempt to expose the triviality of the logical principles.

There was a considerable development and change of emphasis in Brouwer's tactics and his attitude towards logic, in particular the use of the PEM. There is also a remarkable consistency in the grounds of his opposition, his ontological argument, the a-priori of thought-construction with respect to its subsequent recording and the nature of logic as a post-factum investigation of this record. Formulae produced by mechanical i.e. 'thoughtless' use of logical principles by definition lack mathematical existence.

Brouwer's initial opposition to the use of logical principles stems from a general dislike of the 'cunning' use of reasoning; the basis for its rejection is the same as that of all applied science: the structure observed in a physical reality is the result of mathematical viewing, cunning activity distorts this reality further by moving outside the domain of investigation.

His first challenge of the use of logical principles is a general attack on the Aristotelian notion of logic as 'organon', a universal tool in the discovery of new truths. As in the case of language, the use of logical principles is condemned as particularly inappropriate and unreliable when investigating 'Nature' or reflecting on the 'transcendental' values of life. The main theme of *Life, Art and Mysticism* is the excellence of direct insight, mystic knowledge, which even in its description lacks all logical structure:

> Nowhere in mysticism is there a thread or appropriate sequence; every sentence stands by itself and does not need another to precede or follow it. ([1905], p.76).

The lack of logical structure in the language of 'wisdom' is stressed again in the now famous paper *The Unreliability of Logical Principles* [1908C], and the absence of mathematical viewing identified as the source of this lack:

> In *wisdom* there is no mathematical viewing,...the language of introspective wisdom appears disorderly, illogical, because it can never move along the systems of suppositions which have been imprinted on life, it can only accompany the breaking-down of those systems and perhaps in this way reveal wisdom, which is the source of this breakdown. (p.3).

Indeed, [1908C] continues the mood and the arguments of *Life, Art and Mysticism*, which is also clear from Brouwer's own repeated references. [In the Dutch original publication four of the six footnotes in the first few pages refer to *Life, Art and Mysticism*; it is unfortunate that they have been omitted in the English translation of *The Collected Works*.]

*The Unreliability of Logical Principles* [1908C] is divided into three sections, investigating the reliability of logical principles resp. in wisdom, science and mathematics. It is interesting to note that in Brouwer's argument the mathematical connection determines the measure of reliability. Least reliable is the use of logical principles in wisdom, where 'there is no mathematical viewing'. In science, the investigation of Nature, the classical trust in the use of logic is misplaced. The logical principles are part of the mathematical structure imposed on man's perception of Nature, and:

> ...a mathematical system of suppositions when divorced from the perceptions it formed, and extended indefinitely, can hardly remain reliable in guiding along these perceptions. (p.2)

Moreover, [1908C] considers the mathematical structure of classical logical reasoning to be that observed in Euclidean geometry, subsequently transferred and used to generate new truths about Nature by manipulation of words:

> The approach of the classics, who in their experience-based geometry saw that their reasoning from accepted premises and according to logical principles led only to unassailable conclusions, was to extend by induction logical reasoning into a method for the construction of science, and the logical principles to a human tool in the construction of science. But the geometrical reasonings are only valid for a mathematical system, constructed in the Intellect, independent of any experience. (p.2).

The same argument against the reliability of logic as a universal tool in Science and the reference to the scientist's awareness of the need for verification can be found in [1907] and in Brouwer's letters to Korteweg:

> ...the overvaluation of logic, an Aristotelian and scholastic idea, still strong in Spinoza and to a lesser degree in Kant, and only in the 19th century abandoned by philosophers, that by means of logic one could discover secrets of Nature which were not a-priori evident. In fact the conclusions which one reaches in this way are valid, not for Nature itself but only for the mathematical systems which are arbitrarily projected on it. (Furthermore, only part of it covers what has been directly experienced while the rest is an extension by induction.) That these conclusions *correctly* apply to Nature must in each case be verified. ([1907], p.136).

The mathematical origin of the principles of logic is the basis of Brouwer's rejection of their use in transcendental philosophy -'wisdom'- and in science. His challenge of the use of logical principles in mathematics is only modest by comparison. [1908C] concludes:

> Summarizing:
> In Wisdom there is no logic.
> In science logic is often but not durably effective.
> In mathematics, it is not certain whether all logic is permissible and it is not certain if it is possible to settle the question whether all logic is permissible. (p.7)

Yet in spite of its relatively modest claims, [1908C] is revolutionary and marks an important change in Brouwer's views on the application of logic in mathematics. [1907] starts from a conviction that logic cannot contribute to mathematics, a conviction supported by the ontological argument of the a-priori of mathematics w.r.t. language and logic. The contribution of logic to classical mathematics is exposed as restricting and trivial:

'logical reasoning' is merely an ill-fitting strait-jacket in the formulation of mathematics, the target of attack is the axiomatic and syllogistic structure; the principles of logic, including the PEM, are no more than tautologies:

> The theorem 'a function is either differentiable or not differentiable' does not say anything; the same is expressed by 'if a function is not differentiable it is not differentiable'. But the logician who looks at the words of this sentence and discovers a regular behaviour in the succession of those words and other such sentences, projects also here a mathematical system and calls such a sentence an application of the Principle of the Excluded Middle. (p.131).
> [Withdrawn in [1917A]: 'On pages 131-132 the application of the PEM in mathematics are called tautologies giving no information at all. This is an error.' (p.442).]

To interpret the instance of regularity in a genuine mathematical account as an application of logic or logical principles is 'like considering the human body to be an application of anatomy.' (p.130).
The application of logical principles as constructive operators is dismissed in [1907] as resulting in mere 'verbal edifices' which have nothing to do with mathematics:

> Linguistic edifices , sequences of sentences which follow one another according to the laws of logic... Even if it appears that these edifices can never show up the linguistic figure of a contradiction, they are only mathematics as linguistic constructions and have nothing to do with mathematics, which is outside this edifice, e.g. with ordinary arithmetic or geometry. (p.132).

Examples of such verbal edifices are:

> ...the pathological geometries of Hilbert. (p.140),
> 'the logical constructions certainly of Bolyai and possibly also Lobatcheffsky' (p.136),
> Cantor's transfinite numbers and Dedekind cuts;... they are the result of considering language - which is only a defective means of communicating mathematics and has nothing to do with mathematics itself except as an accompaniment - to be an essential part of it, and the laws which govern the succession of sentences, the logical laws, as the real guide in the activity of mathematical construction. (p.141).

A mere verbal edifice, which starts from meaningless symbols and is constructed by applying logical principles, obviously lacks mathematical existence in Brouwer's interpretation of mathematics as 'thought-construction on the basis of the Primordial Intuition alone'. The strict requirement of thought-construction would also invalidate thoughtless linguistic construction even on a real mathematical foundation. [1907], which assumes the logical principles to be tautologies, could do no more than insist on the sanction of a completed mathematical construction:

> The question now is: suppose we have somehow, without thinking of a mathematical interpretation, proved that a logically constructed system on the basis of some linguistic axioms is non-contradictory i.e. at no stage of development of the system do we meet two contradictory theorems; if moreover we find a mathematical interpretation of the axioms ... does it then follow that such a mathematical construction *exists*? But nothing like this has ever been proved by the axiomatici. (p.141).

In [1908C] the argument shifts from questioning the mathematical existence of the product of 'thoughtless' logical construction to doubts about the validity of the classical instruments of logical construction, the tautological nature of the principles of logic. In the question as posed in [1908C] it is assumed that the pure-mathematical construction has been completed, and in an experiment of logical building on part of the mathematical construction the logical results are checked against those of the underlying mathematical system:

> Can one in the case of pure-mathematical constructions and transformations temporarily ignore the thought-reality of the constructed mathematical system and move in the accompanying linguistic construction guided by the principles of syllogism, contradiction and tertium exclusum, and be confident that, every time one recalls the thought-reality of the mathematical construction on which one has been reasoning, each part of the reasoned argument will be vindicated ?

Brouwer's answer is that:

> It will be shown that this trust is well-founded as far as the first two principles are concerned but not in the case of the latter principle. (p.4).

The PEM is no longer dismissed as trivial; its apparent tautological character exposed as based on a confused notion of negation and a confusion of mathematical systems (see further 5.9). It should be noted, however, that in conceding the 'reliability' with restrictions even of the PEM, Brouwer did not abandon his absolute demand for 'thought-construction' nor admit a limited form of logical construction as a safe substitute. Even during the euphoric years of Signific Reform and 'the Brouwer Logic' the result of logical construction requires:

> ... empirical corroboration practically or theoretically [i.e. in the physical world or in the pure-mathematical construction] ([1923B], p.2).

In [1933] logical principles are: regularity observed in genuine mathematical accounts, on reflection recognized by the philosopher as 'to be expected':

> Suppose these hypothetical human beings with unlimited power of memory, who would use words only as invariant signs for definite elements and for definite relations between elements of already constructed pure-mathematical systems...and who recorded their constructions in shortened form in some suitable language, surveyed the strings of their affirmations in this language and then would be able to see the occurrence of the linguistic figures of logical principles. Careful rational reflection would then show that as far as the principles of identity, contradiction and syllogism are concerned such an occurence could be expected. (p.59).

Brouwer's inquiry into the application of logical principles in mathematics remains mainly a critical investigation of past and current practice. To prove his fundamental claim that 'logical construction' cannot generate new mathematical truths he searches for flaws in existing practice, the use of the traditional principles in classical mathematics. The abuse uncovered is that of a principle derived from one system applied in a different domain: the PEM observed in the account of a finite mathematical system, used as a logical operator in an infinite system. In spite of his enlightened view of classical logic as an arbitrary, cultural convention, the question of applying logical principles as posed by Brouwer in [1908C] (quoted above) and later in [1933], [1952B] and [BMS51] remains confined to the principles of classical logic.

Using his interpretation of logical principles one might raise the more general question of application: 'Can the logical principles derived from the verbal account of a mathematical construction be used mechanically, as logical operators to produce statements which express new results in the same mathematical system?' As to Brouwer's unwillingness to accept any logical construction as a means of generating mathematical truths there can be no doubt. The general case of logical construction, however, is not open to refutation by counter-example and could only be disproved by the metaphysical argument of the a-priori of mathematics and the fallibility of language.

In Brouwer's argument of the absolute independence of mathematics with respect to language, 'accompaniment' is a one-way relation. Even an ideal language can do no more than faithfully reflect a completed mathematical construction and the 'linguistic figure' of a logical principle no more than follow a preceding step in the thought-construction. A constructive move in the linguistic medium can only relate to a parallel constructive move in the thought-world of mathematics if such a construct exists, i.e. has been thought or 'experienced'. Mechanically produced linguistic results such as hypotheses, guesses and the accounts of 'reputed experience of others', become meaningful only after the Subject's parallel thought-construction, the sole norm of truth in the Idealized world of Mathematics. One might argue that the close parallel between the mathematical constructive steps and the 'logical figures' based on them should at least provide a present guarantee that the Subject's continued future thought-construction would confirm these linguistic results. Brouwer did in fact speculate along these lines. [BMS51] and [1952B] explore the continued logical construction on a part-completed mathematical construction and question the possibility of performing a mathematical construction confirming its linguistic parallel. 'Possibility' is to be taken here not as a vague might-be, but a definite present guarantee that such a construction can be performed.

> Suppose that an intuitionist mathematical construction has been carefully described by means of words, and then, the introspective character of the mathematical construction being ignored for a moment, its linguistic description is considered by itself and submitted to a linguistic application of classical logic. Is it then always possible to perform a languageless mathematical construction finding its expression in the logico-linguistic figure in question? (p.141).

The earlier [BMS51] version comes dangerously close to admitting a Platonic objective real world of mathematics:

> Suppose that in mathematical language, trying to deal with an intuitionist mathematical operation, the figure of an application of one of the principles of logic is, for once, blindly formulated. Does this figure of language then accompany an actual language-less mathematical procedure in the actual mathematical system concerned? (p.5).

The affirmative answer – identically the same for both questions – is qualified: apart from the universal qualification of the 'inadequacy of language' there is a reservation as to the PEM and the answer, as the questions, is confined to the traditional principles of classical logic:

A careful examination reveals that, briefly expressed, the answer is in the affirmative as far as the principles of contradiction and syllogism are concerned, if one allows for the inevitable inadequacy of language as a mode of expression and communication. But with regard to the principle of the excluded third, except in special cases, the answer is in the negative. ([BMS51] p.5, [1952B] p.141).

The nature of this qualified reliability of some principles – 'accompaniment', or a definite possibility of a future mathematical construction confirming the logico-linguistic figure – and its implications are not further explored. The relation of accompaniment here is introduced to serve the argument refuting the PEM, showing the impossibility of 'each application of the Principle of the Excluded Middle accompanying some actual mathematical procedure' ([BMS51] p.5). In any case, the question of logical construction by means of valid principles remains a hypothetical one in view of the the over-riding qualification of 'the inevitable inadequacy of language' and ultimately the indispensable requirement of thought-construction.

The search for reliable logical principles is part of a search for a reliable mathematical language, described in [1933] as 'a programme of linguistic reform', 'a programme of reviewing and renewing the architecture of the verbal edifice corresponding to the mathematical thought-constructions' (p.57). The constructive use of logical principles, 'the use of logical principles to deduce new linguistic affirmations without thinking of the mathematical content indicated by these derivations..' is: 'pure linguistic application' (p.54). The programme of reform will never be wholly succesful nor will logical construction ever be reliable since 'for pure mathematics an infallible language cannot exist' (p.53).

The modest claim of Brouwer's Intuitionist linguistic and logical reform was to improve the language of mathematics as medium of communication and aid to memory. Improvement was achieved by removing the excesses of the PEM, by a more precise definition of negation and even by 'pure-logical' analysis, the 'Brouwer Logic' and an 'Intuitionist Logic'. The fundamental function and weakness of language, however, were never lost sight of; improvement remained a cosmetic exercise and the improved language and logic reliable only for the practical purposes of the flesh-and-blood mathematician:

> It follows that the necessarily inadequate, limited and in-its-use-always-uncertain language of communication between human beings, whose power

of memory is limited, even if it has been fashioned with all possible refinements and precision, will only then be able to play its mnemotechnical, thought-economical and communicative role with *practically* acceptable reliability...(p.59).

Acceptance of a limited reliability of logical principles never included a constructive role in the generation of new truths. Brouwer's last major statements are unequivocal in their insistence on thought-construction as the exclusive source of mathematical truth:

> Logic is not a reliable instrument to discover new truths...Intuitionist mathematics deduces theorems exclusively by means of introspective construction. ([1948C], p.1243).

[1954A] speaks with pride of the intuitionist improvement of logic with its own principles, 'a theory which with some right may be called Intuitionist Mathematical Logic', but it adds a warning note:

> In Intuitionism, of course, all these principles ...are only then 'realized' i.e. convey truths, when these truths have been experienced. (p.3).

In the final paragraph of [1948C] Brouwer summarizes his contribution to logic and his views on the use of logical principles in his characteristic cryptic way:

> I should like to terminate here. I hope I have made clear that Intuitionism on the one hand subtelizes logic, on the other hand denounces logic as a source of truth. (p.1249).

## 5.8 Brouwer Negation

In Brouwer's criticism of traditional logic the notion and use of negation are the most central issue. His pre-occupation with negation goes back to the early days of *The Foundations*, even before his challenge of the PEM. The dominant use of negation in traditional logic, particularly in the 'productive' principles of contradiction and excluded middle, and the increasing emphasis on non-contradictority in contemporary foundation theory aroused his suspicion. Moreover, the vague classical notion of negation could not serve as a constructive element in mathematics as he conceived it.

Brouwer showed great courage in daring to challenge a concept so fundamental in logical and mathematical practice and never questioned, not even by the French Intuitionist School. Yet he never dispensed with negation altogether. His initial probe into the nature of mathematics led to the conviction of the need for negation as an essential element of construction and to an interpretation that could be accommodated within his conception of mathematics as thought-construction. The implications of his new, strict interpretation of negation for the PEM were almost immediately clear and provided welcome ammunition in his campaign against the use of logic and logical principles in mathematics.

The implications for logic itself, the opportunity of developing an alternative logic were not seized upon until 1923. 'Brouwer negation' remained the corner-stone, but the absolute distinction between mathematics and logic became somewhat blurred at a time when 'Signific' considerations allowed language and logic a more positive role and when the task of re-constructing mathematics demanded a practical code of formulation. Negation, 'absurdity', became an operator in the domain of pure mathematics and that of logic, generating a logical calculus of absurdities and giving rise to a new, 'Brouwer Logic'.

### 5.8.1 *Brouwer's Constructive Notion of Negation*

The vagueness of traditional negation reflects a lack of precision in the classical 'All', an amorphous Universe of concepts, words and word-combinations, in Brouwer's terminology a 'chimerical all', embracing the 'real' as well as the 'unreal', i.e. the thought as well as the yet-unthought and the inconceivable; a universe in which every entity, property and word provides an automatic partitioning into two equally well-defined complementary domains.

In the Brouwerian universe of mathematical thought-constructions, distinct from the logical universe of word-construction, distinct also from 'the language of mathematics' recording mathematical constructions, negation can only be a constructive act or its resulting construct, ultimately based on the Primordial Intuition. But while elements of negation are apparent in constructions at all levels, attempts to trace it to its most fundamental form are easily trapped in circular argument.

'Affirmative' constructivists, who claim to avoid negation by relying entirely on relations such as ' $<$ ' and ' $>$ ', seem unaware that the exclusion of equality remains an essential and most-used element of such 'difference'-relations or *in*equalities. Brouwer moreover refuted the affirmative thesis by producing 'two real numbers which are different while neither can be shown to be greater or smaller than the other.' ([1948A], p.963).

Dummett signals circularity in the definition of negation in terms of contradiction, $p \wedge \neg p$ (DUMMETT[1977], p.13); but some circularity is also present in the standard Intuitionist definition of negation based on the difference between 0 and 1 and in Brouwer's definition of 'difference': 'Two mathematical entities will be called different if their equality proves to be absurd; the notation for difference will be $\neq$' ([1954A], p.5 ).

The problem of the 'primordial origin' of negation is not solved by a constructive interpretation of complementarity in terms of a universal set restricted to all the past constructions of the Subject. Complementarity would still require an act of discernment, setting one construct apart from each and every other construct, an act which presupposes the power of discrimination between constructs that are equivalent and those that are not. One might argue that the individual identity of each construct, or the Primordial Intuition as awareness of discreteness is the ultimate and sufficient source of negation. But individual identity and discreteness can be claimed e.g. for two constructions of two-ity of the Subject at different moments of time and recognized by him to be equivalent (cf. Brouwer's use of the word 'discrete' in the context of species: 'a species is called discrete if any two elements can be proved either to be equal or to be different' ([1954A], p.5)).

Negation is not listed among Brouwer's primitive notions, the 'elements of mathematical construction which have to remain underivable...to be read off immediately from the Primordial Intuition' ([1907], p.180). Yet the power of discrimination, of linking and splitting elements, recognizing equivalence and distinctness, is presupposed in the primordial phenomenon. Equivalence is needed in the final stage of the creation of 'two', 'the common substratum of all two-ties'; as relative sameness of two distinct elements it presupposes the even more primordial 'distinctness' which is implied in the recognition of 'before-after', and 'the falling-apart of a moment of life as a sequence of different things'.

### 5.8.2 Absurdity as Impossibility of 'Fitting-in'

There is no evidence in Brouwer's writings of speculation on the nature of negation in terms of elements of the Primordial Intuition. His search for the genesis of negation starts from the reality of mathematical constructions present in the mind or memory of the Subject at a given time, and concentrates on the process of constructing relations between them.

In the strict Brouwerian interpretation these 'real' mathematical entities or properties are all completed, 'experienced' or 'previously acquired', constructions and the relations, when established, are new constructions. The pre-constructional stage is described in [1908C] and [1923B] as 'hypothesis' and the (flesh-and-blood) mathematician attempting 'to fit a system *b* in a certain way into a system *a*' (Du. 'inpassen'); the systems *a* and *b* are both completed constructions, in Brouwer's terminology 'mathematical entities previously acquired'. The successful attempt by the Subject is: 'constructing the completion of the conjectured fitting-in' ([1908C], p.5). The vague 'fitting-in' should be interpreted as establishing a correspondence between elements and operations of the two constructions, [1923B] speaks of:

> ...properties of systems i.e. the possibility of fitting systems into other systems according to prescribed *incidence of elements* [Du. 'elementenincidenties'). (p.1).

In [1907] the process is described in more detail:

> Within a constructed system it is often possible simply to fit-in new constructions quite independent from the original form of its construction. As their elements one uses the elements – or systems – of the old construction in a new arrangement, bearing in mind, however, the arrangement of the old system. The so-called 'properties' of a given system are no more than the possibility of such building of new systems in certain agreement with an existing system. It is precisely this *fitting* of new systems in a given system which plays an important role in the construction of mathematics. (pp.77-78).
> As example Brouwer then quotes the group-structure of certain given systems.

The successfully completed 'fitting-in' of one system into another clearly constitutes a construction in the strictest sense. Brouwer's problem of negation stems from the apparent conflict in a mathematical reality which is constructed and yet growing and which, moreover, is a-priori true. In the constructed reality of the Subject the only alternatives are the successfully completed 'fitting-in' and negation by definite completed construction. Yet the Subject, aware of the dynamic, 'becoming' nature of his constructive world, never-finished at any time of his life, recognizes the open question as an alternative outside his domain of constructed mathematics. The full implications of the 'becoming' nature, not only of the universe of mathematics but of some mathematical constructions, were not clear to Brouwer until later. His initial concern was with negation in the context of constructed mathematics.

The synthetic a-priori nature of mathematics requires relations, although freely created by man, to be truths of necessity; the mathematical 'is' and 'is not' are: 'must be' and 'cannot be'. The succesful fitting in of a construction $b$ in a construction $a$ establishes the relation '$a$ is $b$ or $a$ has property $b$ and $a$ must have property $b$ ; a-fortiori it shows that $b$ can be fitted into $a$, that the structures $a$ and $b$, as created by the Subject, are inherently compatible. The alternative '$a$ does not have property $b$', i.e. $a$ cannot have property $b$, similarly requires a construction. The 'cannot' is not an admission of weakness on the part of flesh-and-blood mathematician, it refers to an incompatibility of the two constructs which makes a successful fitting-in impossible. One might compare the attempts of the Subject to fit structure $b$ into structure $a$ to a home-made game of shoveboard: failure to slide disc $b$ through the slot $a$ may be due to lack of skill on the part of the player, it could also be due to the relative size of disc and slot in which case the impossibility could conclusively be shown by measurement of both.

### 5.8.3 Constructed Impossibility

Brouwer remains vague about the nature of the constructive 'impossibility' and its ultimate source. He uses strong words such as 'absurdity' (Du. 'ongerijmdheid') and 'hitting upon impossibility' and *examples* of obvious and simple absurdities such as $1=2$. The first description of 'Brouwer negation' simply says:

> I just observe that the construction does not go further [Du. 'gaat niet', i.e. it does not work], that in the main edifice there is no room to be found for the posited construction. ([1907], p.127).

From the context it is clear that failure is due to the constructions themselves, some incompatibility. The word 'incompatibility' is used in [1954A] instead of the usual 'absurdity':

> ... an incompatibility e.g. the identity of the empty two-ity with an empty unity. (p.3).

Brouwer seems reluctant to recognize even a potential incompatibility inherent in constructions for fear that such analysis might detract from the active, constructive role of the Subject, even though these constructs and their limitations are the Subject's own free creations. Incompatibility is not observed but constructed by the Subject, negation is

> a *construction* of an incompatibility ([1954A], p.3),

and in [1908C] constructive negation is described (literally) as:

> ...the construction of the hitting-upon impossibility of the fitting-in. (p.5).

Such a construction can be interpreted as a sequence of constructive steps leading to a position where any further move meets with an obstacle, a kind of check-mate. Apart from 'stuiten op' (hitting upon, hitting a snag) Brouwer uses expressions like 'vastlopen' (running aground, get stuck [1923B]), translated in the German version as 'Hemmung' (being arrested), a term also used in the context of choice sequences to express the predetermined termination of the finite sequence. These expressions and the use of the neutral passive voice obscure the identity of the source of the failure. Unlike the chess-game in which a sequence of bad moves causes the failure, the construction leading to absurdity was doomed from the start, there is no winning strategy; failure has its roots in the two constructs themselves.

Yet in tracing the source of incompatibility to the constructs themselves one does not necessarily negate the active role of the Subject: all aspects of these constructions are the Subject's own creation, including those that determine the succes or otherwise of an attempt to relate them. Moreover, the individual make-up and existence of each construct are not sufficient, the notion of 'incompatibility' requires a 'linking' of the two, a simultaneous presence of both in the mind of the Subject and his attempt to relate them in a definite way.

If fitting one system into another consists in constructing a certain correspondence between their elements, 'incompatibility' of two systems is established by successive elimination of every possible pairing-arrangement. The number of possible arrangements is determined by the nature of the constructs and, as Brouwer points out, in the case of finite systems complete elimination is possible:

> Within the limits of a definite finite main system one can always test, that is prove or reduce to absurdity the properties of the system, i.e. test whether a system can be fitted into another according to prescribed incidence of elements since the fitting-in as determined by the property can in every case be executed in only a finite number of ways, which each in turn can be undertaken and pursued either until it is successfully completed or until it gets stuck. ([1923B], p.1).

Whereas in the successful completion of the attempt of fitting-in the unsuccessful trials can be discarded as irrelevant, they play an essential role in the construction of 'incompatibility' of the two systems: the sequence of frustrated attempts must be complete. All possibilities must have been exhausted and every avenue closed. By successive elimination the construction of incompatibility of systems $a$ and $b$ is reduced to simpler forms, until finally the Subject is left with e.g. the attempt to equate 1 and 2.

In the final analysis, however, the absurdity of $1 = 2$ and each elimination in the construction of incompatibility is a recognition by the Subject that a particular constructive move is blocked, more in the nature of an insight than a construction. The Subject's constructive activity in the process of negation is confined to the steps leading to 'absurdity': '... constructing the hitting upon impossibility'. Absurdity ultimately remains a primordial-primitive relation. Brouwer does not dwell upon the nature of 'absurdity' nor on the range of absurdities that are intuitively clear. Underlying his notion of negation is the assumption that for a single constructive act in a simple finite system the Subject can immediately grasp 'absurdity', but that in the case of more complex systems he can only conclude 'absurdity' after the attempt of fitting-in has been broken down into single constructive moves which each are found to be absurd.

This demand for completeness refers to 'the ways system $b$ can be fitted into system $a$' and does not necessarily require the systems themselves to be finite. In his concern with the 'trustworthiness' of the logical principles Brouwer stressed the finiteness of the systems which guarantees the finiteness of the number of possible ways of fitting-in. Speaking of *possible* future constructions – distinct from the real, those already completed – he remarks:

> The case that $\alpha$ has neither been proved to be true nor to be absurd, but a finite algorithm is known leading to a decision either that $\alpha$ is true or that $\alpha$ is absurd obviously is reducible to the first and second cases. This applies in particular to assertions of possibility of some construction of bounded finite character in some finite mathematical system because such a construction can be attempted only in a finite number of particular ways and each attempt proves successful or abortive in a finite number of steps. ([BMS59B], p.1; app.9).

### 5.8.4 Mathematical Absurdity and Other Notions of Negation

Extending his investigation to the logic of possible future constructions and pure-linguistic assertions Brouwer moved outside the domain of mathematics proper, the universe of completed constructions; each extension generating a negation by simple complementarity. In the quoted passage for example one can distinguish between various kinds of negation: in the domain of assertions all those expressing *un*proven hypotheses, 'the case that $\alpha$ has neither been proved to be true nor to be absurd', and within this domain of unproven hypotheses a distinction can be made between those which have as yet not been completed but which can be completed because of the finiteness of the systems and those which lack this guarantee. In his development of 'becoming' mathematics Brouwer moreover introduced *incomplete* constructs, partly completed entities which in as far as they have been completed belong to the domain of 'pure mathematics'. Most of the counter-examples refuting the PEM are based on such incomplete constructions. In that context and elsewhere Brouwer uses the terminology of negation ('not defined' etc.) and a notion of negation appropriate to the domain of inquiry. Within the domain of 'pure', constructed mathematics, however, negation is mathematical absurdity; as a relation between constructs it is confined to the domain of property-constructions. In this completed universe there are only 'affirmative' and 'negative' constructions i.e. the successful completion of fitting $b$ into $a$ and the absurdity of fitting $d$ into $c$.

'Completion' of the absurdity-construction, the claim that all constructive possibilities have been eliminated, is primarily a reference to the past and present, the only reality of Brouwerian 'pure mathematics'. Its finality, however, implies a 'negative' claim on the future which is more clearly expressed in the word 'impossible'. Impossibility, as usually understood, precludes action now and at any time in the future, it is 'cannot now and ever'. Impossibility in this sense is equivalent with absurdity, indeed, 'cannot now and ever' is an expression used in Brouwer's later work instead of 'absurdity' (e.g. [BMS47]; app.8). The specific mention of 'now' and 'ever' is more than an emphatic hyperbole, it distinguishes absurdity from a 'weaker' form of impossibility, the 'cannot-now'. Impossibility-now includes what at the present stage the Subject cannot do for lack of knowledge or because of the 'growing' nature of some mathematical constructions ('A mathematical entity in its free growth may at some time acquire a property it did not have before.' ([BMS59B], p.1; app.9)).

In Brouwer's attempts to refute the PEM and to reform logic these different notions of impossibility became somewhat confused (see further 5.8.3), but that confusion did not extend to his views on mathematical absurdity. The strict requirement of completion of the property-construc-

tion applies even in the case of a growing mathematical entity, provided the absurdity relates to its completed state, in Brouwerian parlance: 'the initial segment'.
The admission of 'freely-growing entities' may well raise questions as to their precise mathematical status, it does not affect the notion of mathematical absurdity.

One might speculate on 'principles', constructive procedures available to the Subject in the pure-mathematical domain, in particular those derived from absurdity. For example, it follows from the nature of a completed construction that constructions both of affirmative and negative property-relations are conclusive and exclusive, i.e. the Subject is aware that having constructed the relation $R(a, b)$ it is pointless and impossible to prove that the relation $R(a, b)$ is absurd and vice versa (the mathematical equivalent of the logical principle of contradiction). Such awareness, however, is an essential aspect of the *completed* construction; considered in isolation and referring to a hypothetical alternative, it could be described as a meta-mathematical 'Principle of Conclusiveness', leading in the non-mathematical domain of logic to the Principle of Contradiction. In the domain of mathematics it is no more than an emphatic statement of the completion of the construction, not a newly emerging truth.

Neither can pure-mathematical absurdity be used as a generalized operator on single constructs, generating mathematical entities. Each property relation, affirmative as well as negative, is a binary construction, requiring two existing constructs, wholly dependent on the nature of the two constructs and the particular way of fitting-in.

Brouwer's general critism of symbolic notation, that it does not model the mathematical constructive reality, applies in particular to the conventional notation of negation. E.g. a notation such as $\sim R(a, b)$ suggests negation to be an operator applied to the existing mathematical construct $R(a, b)$. But in the pure-mathematical domain only $a$ and $b$ can claim mathematical existence, $R(a, b)$ has only the status of attempt at fitting-in. If the notation $\sim R(a, b)$ were to be adopted it could only be as an abbreviation of 'the Subject has constructed the absurdity of fitting b in a certain way into $a$'.

Double negation fails to have any meaning in this domain and in the terms of the notation (further construction of negation again requires two completed constructs). The interpretation 'Impossibility of the construction of absurdity', non-contradictority, lies outside the scope of Brouwerian 'first-order' mathematics, belongs to the wider domain of the Subject's thoughts which includes incompleted attempts, hypotheses and metamathematical proof-theory. The Subject's awareness of the impossibility of constructing absurdity might be due to his successful attempt of fitting $b$ into $a$, it could also be the simple realization that complete elimination of all fitting-in arrangements is impossible because of the infinity of the systems concerned. In the latter case the completion of the construction $\sim R(a,b)$ is ruled out while the Subject has as yet not succeeded in his attempt of fitting $b$ in the required way into a.

## 5.9 The Refutation of the PEM

The logician's Principle of the Excluded Middle, i.e. that every categorical statement is either true or false, is rooted in a philosophy so fundamentally different from Brouwer's that a meaningful debate on its validity is well-nigh impossible. Indeed, even if the PEM became the prominent target of Brouwer's attack, the real issues were the metaphysical posits concerning objectivity and truth. Against the almost universal tradition of accepting an objective reality as the collective source of human knowledge and the norm of its truth, Brouwer proclaimed the total autonomy of the individual, creating knowledge and truth. In his attempt to win over his contemporaries to his views and convince them of the error of their ways, he searched for evident fallacies resulting from their misconception of truth and the role of logic. In [1908C], the PEM was identified as such a fallacy, illustrating his general thesis that logical principles are unreliable. As the full implications of his own definition of mathematical truth and falsehood became clearer, the PEM was no longer a meaningless tautology but an obvious misrepresentation of fact: every arbitrary statement does not necessarily represent a proven truth or a proven falsehood, in Brouwerian terminology a construction of an affirmative or negative property-relation. The simple existence of unproven hypotheses, unsolved mathematical problems, is a refutation of the PEM. A few examples of such unsolved problems are given in [1908C]:

> Is there in the decimal expansion of $\pi$ one digit which continues to occur more frequently than the others?
> Do infinitely many pairs of consecutive equal digits occur in the decimal expansion of $\pi$? etc. (p. 7).

In [1955] Fermat's Theorem and each yet-unsolved mathematical problem are quoted as "yielding a refutation of the principle of the excluded third." (p. 114). Such arguments, however, convince only the already converted, those who accept Brouwer's restricted notion of truth and falsehood as the constructive experience of the Subject to-date. Apart from [1908C], it is used by Brouwer only in his later years, when the PEM is defined as:

> [the claim] that every mathematical assertion (i.e. assignment of a mathematical property to a mathematical entity) is either a truth or cannot be a truth" ([1948C], p. 1243).

At that stage he could also draw his conclusion as to the time-bound nature of truth: a mere hypothesis becomes a truth or falsehood only from the moment truth or falsehood is constructed:

> An assertion which is in the fourth case may at some time pass into one of other cases because further thinking may generate a construction accomplishing this passage. ([1955], p. 114)

The freedom of the Subject to continue constructing, 'unfolding', mathematics, is a quarantee that at any time of his life there will be unsolved mathematical problems.

To prove the unreliability of the PEM to a mathematical public convinced of the objectivity and the timeless nature of mathematical truth, the young graduate Brouwer had to shift the argument to the domain of possible truths and the logicans' domain of assertions. His attempts to refute the PEM during the coming years proceeded on two fronts:
1. by devising counter-examples of essentially unsolvable mathematical problems;
2. by revealing contradictions arising from the application of the PEM in mathemical construction.

### 5.9.1 The Unreliability of the PEM, 1908

The question of the PEM as raised in [1908C] is not whether every mathematical assertion is either true or false in the Brouwer sense, it concerns the much wider ranging assumption of decidability, the assumption that every mathematical problem *can* be solved, i.e. be proved to be true or proved to be false:

> As to the Principle of the Excluded Middle, this requires that for every hypothetical fitting-in of systems into each other in a certain way one can always construct either its successful completion or the hitting-upon-impossibility. (p. 5).

The domain of investigation here is not that of 'first-level', i.e. constructed, mathematics, not even the linguistic record of completed mathematical constructions, but the domain of classical logic: assertions, the linguistic linking of words or symbols representing completed mathematical constructions into categorical statements which may as yet represent only the hypothesis of a property construction. Within the domain of hypothetical property-constructions Brouwer distinguishes between those which because of the finite nature of the systems concerned can one day be solved and those which do not carry that guarantee. Pointing to the present lack of such guarantee in the case of infinite systems [1908C] makes the modest claim that the solvability of every mathematical problem has not yet been proved and that:

> therefore for infinite systems the Principle of the Excluded Middle is for the time being not a reliable principle. (p.6).

At this stage Brouwer had to concede that reductio-ad-absurdum was not a weapon available to him in his refutation of the PEM:

> And yet its unjustified application will never lead to contradiction; one will therefore never discover that such reasoning is groundless by the occurrence of a contradiction. Contradiction would require the completion and the contradictority of fitting-in both to be contradictory at the same time, which the principle of contradiction does not allow. (ibid.).

After 1918 the domain of non-contradictoricity of the PEM became more restricted. The above passage was quoted in [1923B] but with an added qualification:

> The unjustified application of the PEM to *properties of well-constructed mathematical systems* can never lead to contradiction. (p.3 footnote *).

Apart from the restriction to 'well-constructed mathematical systems' there is a new stricter interpretation of 'property-construction' as proof of 'element-hood of species' or its absurdity for a single 'entity' (cf. 6.3.6). If a statement $\tau$ expresses the successful completion of such a property-construction and the statement $\sim\tau$ of its constructed aburdity, no contradiction can arise from $(\tau \vee \sim\tau.)$ The absurdity of the PEM in this case, the absurdity of $(\tau \vee \sim\tau)$, i.e. neither $\tau$ nor the absurdity of $\tau$, would result in the 'contradiction $\tau$ is absurd and the absurdity of $\tau$ is absurd'.

The case of contradictoricity of the PEM as claimed by Brouwer after 1928 rests on a wider interpretation of the property construction (see further 5.9.4 and 6.3.6). Non-contradictoricity of the PEM for 'a single assignment $\tau$ of a property to a mathematical entity', now referred-to as *the Simple Principle of the Excluded Third*, is re-affirmed and the argument repeated literally:

> If it were contradictory the absurdity of $\tau$ would be correct and absurd at the same time, which is impossible. ([1948B], p.1239; cf. [1948C], p.1244; also [1928A]).

The non-contradictority of the PEM in these instances remained a source of irritation to Brouwer; e.g. in [1933] he writes:

> That these [examples] have not simply put an end to the classical practice of mathematics is due to the fact in-its-favour that, even though the principle of the excluded middle is incorrect, it does not lead to contradiction if one restricts oneself to finite groups of properties. Intuitionism in its battle against the errors of classical mathematics, therefore, finds itself deprived of the most fashionable means of repressing errors of thought: the reductio ad absurdum, and has to rely exclusively on exhortation to reflect and reason. (p.63).

The freedom from contradiction in such applications of the PEM deprived Brouwer of the standard weapon of proving its incorrectness, it also showed up the weakness of non-contradictority: even incorrect theories can be non-contradictory. The conclusion of [1933], 'the Intuitionist has to rely exclusively on exhortation to reflect and reason', is an appeal to a higher authority than the rule of logic, which like the legal system of justice is arbitrary and self-defined:

> An incorrect theory does not become less incorrect because it cannot be refuted by contradiction, in the same way that a criminal policy is not less criminal because it cannot be stopped by a court of justice. ([1923B], p.3).

### 5.9.2 *Mathematics Without the PEM, 1917*

The PEM remained somewhat in the background during the years of 'frenzied topological activity'. When in 1912 Brouwer turned his attention again to foundational matters, both the PEM and the incomplete, the 'growing' or 'becoming' aspects of mathematics became the key issues in his programme of re-construction. His involvement with Schoenflies's work on set-theory had convinced him of the need for re-constructing set theory and analysis on a sound philosophical basis. In a stroke of genius he now brought together his earlier doubts about the classical assumption of determinacy underlying the PEM, his conviction of the active, free role of the Subject constructing mathematics in time, and his deep understanding of the topology of the continuum. In 1917 Brouwer launched his new Set Theory, based on the 'admission' of free choice of the Subject as a legitimate mathematical procedure and a construction characteristic of the continuum. The fundamental 'revolution' is his acceptance of free growth, indeterminacy, in mathematics; it negates the classical postulate that all mathematical entities and properties are necessarily determinate, a postulate in Brouwer's mind embodied in and identified with the PEM. This identification is reflected in the title of his first major publication on Intuitionist Set Theory, [1918B] and [1919A], it is *The Foundations of Set Theory independent of the logical Principle of the Excluded Middle*'. In spite of this title [1918B] is not a polemical treatise: the PEM is not refuted, the classical postulated PEM and determinacy are simply abandoned and the resulting greater freedom is exploited in the development of a new analysis.

Identification of 'dependence on the PEM' and 'determinacy' presupposes that no other parts of classical logic had been affected by the assumption of determinacy. Indeed, there is around this time a marked shift from Brouwer's original 'total rejection of all logic'; opposition for the time being concentrates exclusively on the PEM, and logic without the PEM becomes an acceptable alternative. Moreover, 'the use of the PEM' becomes the litmus-test in the re-construction of analysis and function-theory, identifying where classical mathematics had gone astray. *The role of the principle of the excluded middle in mathematics, especially in function-theory*, [1923B], equates 'incorrect logic' with 'the use of the PEM' and lists some of the classical theorems which are untenable because of their use of the PEM:

> Of fundamental importance in this incorrect 'logical' (i.e. using the PEM) mathematics of infinity, especially in the theory of real functions (mainly developed by the Paris school) are the fundamental properties which follow from the PEM:
> 1. that the points of the continuum form an ordered point-species,
> 2. that every mathematical species is either finite or infinite. (p.3).
> Useless and untenable are:
> The Paris-school notion of integral, the so-called L-integral...
> The extended disjunction principle and... the Bolzano-Weierstrass theorem...

The Heine-Borel covering theorem... but also older, more consolidated theories in the mathematics of infinity, e.g. that of the convergence of infinite series...etc. (pp.3-5).

### 5.9.3 Brouwer's Counterexamples

A simpler, constructive analysis based on the new Brouwer-continuum would justify the postulated abandonment of the PEM and counter Hilbert's pragmatic argument of the handicap this would be to the mathematician, '... the boxer denied the use of his fists'. However, to silence his immediate critics in the early twenties Brouwer needed a proof of the invalidity of the PEM acceptable even to those who recognize as legitimate only mathematical entities determined by law, a counterexample which would refute the assumption that every mathematical problem is solvable. He searched for a mathematical entity which is incomplete, growing in time, yet one whose constituent elements are determined by orthodox mathematical facts, and such that neither the affirmative nor the negative construction of a given property can be completed, in the terminology of [1923B] a property that cannot be 'tested'.

In [1923B] the first attempt was made at constructing such a counterexample:

> Let $d_v$ be the $v$-th digit to the right of the decimal point in the decimal expansion of $\pi$ and let $m = k_n$ if, as one proceeds with the expansion of $\pi$, it happens for the $n$-th time at $d_m$ that the segment $d_m, d_{m+1}, ..., d_{m+9}, ...$ of the decimal expansion forms the sequence 0123456789.
> Further, let $c_v = (-\frac{1}{2})^{k_1}$ if $v \geq k_1$, otherwise let $c_v = (-\frac{1}{2})^v$. Then the sequence $c_1, c_2, c_3, ...$ defines a real number $r$ for which none of the relations $r = 0, r > 0, r < 0$ holds. (p.3).

We should note that in the conclusion the alternatives are not those of Brouwer's strict interpretation of the PEM: affirmation or absurdity of one property; further, that the three alternatives are negations, though not negations in the strict Brouwer-sense of mathematical absurdity. Taken in the strict Brouwer-sense e.g. $r \neq 0$ would require a proof that the occurrence of the sequence 0123456789 in the decimal expansion of $\pi$ is absurd. Negations here are simply statements of the absence of proof.

In Brouwer's next attempt that year, [1923C], the same real number $r$ is used but the question is confined to one property, the rationality of $r$:

> If we call a real number $g$ rational if two integers $p$ and $q$ can be calculated

whose quotient equals $g$, then $r$ cannot be called rational, but on the other hand the rationality of $r$ cannot possibly be absurd, for in that case $k_1$ could not possibly exist, which would imply that $r = 0$ and therefore is rational. (p.877).

The irrationality of $r$ is 'shown' to be absurd, but again the alternative, the rationality of $r$, is negated on the basis of absence of proof. The use of the word 'cannot' is ambiguous, it suggests absurdity but in this case relates to our present inability to calculate $p$ and $q$ because of the yet-unknown value of $r$ (which in each of the alternative cases is rational). 'Cannot' in the the strict interpretation of mathematical absurdity requires a present proof of absurdity of the rationality of $r$, a mathematical solution of the problem and a denial of its unsolvability.

It is significant that this counterexample forms part of Brouwer's paper [1923C], which introduces his 'Calculus of of Absurdities' and 'the Brouwer Logic'. We shall show in section 5.10 that the 1923 absurdities include a new notion of negation to which we will refer as 'logical absurdity', distinct from mathematical absurdity and seemingly stronger than the mere absence of proof.

In the 1923 counterexamples and the pre-1940 versions of their generalization, negation of 'testability' is invariably expressed by 'cannot'. A generalization of the hypothetical occurrence of 0123456789 in the decimal expansion of $\pi$ was already hinted at in [1923B]:

> Of course, we could define $r$ also in terms of another arbitrary property $x$, such that one cannot indicate a particular finite number that has property $x$ and neither can one prove the impossibility of property $x$ for every finite number. (p.3).

This property was coined 'eine fliehende Eigenschaft', a 'fleeing property', for the first time in [1929]; it appears as an undated addition in the margin of the Dutch manuscript of the *Berliner Gastvorlesungen* ([BMS32], p.5)] and is defined as:

> A fleeing property is a property [of natural numbers] for which in the case of any arbitrary natural number one can derive either its validity or its absurdity, whereas one cannot indicate a natural number which has the property, nor can one prove its absurdity for all natural numbers. ([BMS32], p.5; also [1929], p.161 and [1933], p.60).

'Cannot' suggests an impossibility of *ever* completing the construction, and this indeed was the object of Brouwer's search: an essentially unsolvable mathematical problem. It could be argued that because of the inclusion of impossibility of completing the affirmative construction in the definition ('one cannot indicate a natural number which has the property'), the Fleeing Property represents an unsolvable problem at least by definition. Such logical impossibility lacks the mathematical-constructive basis of proof of absurdity; moreover, in the constructive reality the Fleeing Property and its attendant impossibility could be temporary, as was admitted later by Brouwer ('a property may lose its fleeing nature').

The search for an unsolvable mathematical problem was effectively abandoned by Brouwer at some stage during the 'Silent Years'. [1933] still quotes the Fleeing Property in its 'cannot'-form; but the PEM is rejected on the grounds of 'reasoning and reflecting' on the nature of mathematics and language. The main theme of the paper is the essential difference between mathematical constructions and assertions. Interesting is the mention of 'the confusion of absurdity and practical impossibility'.

In his post-war papers Brouwer returns to the clear orthodoxy of the early years, the absolute distinction between the realm of linguistic assertions and the reality of completed thought-construction. There is now a recognition that the 'possible' can only be given proper mathematical status when defined in terms of present knowledge of the Subject. 'Can' no longer represents a vague possibility of future construction but the Subject's present knowledge of 'a finite algorithm leading to the construction either of truth or absurdity'. ([BMS59B], p.1; also [1955], p.114). Its negation is recognized to be lack of knowledge of such algorithm: 'we do not know an algorithm'. The 'cannot' of the Fleeing Property is replaced by 'no way is known' and 'is not known'. In [BMS47] the difference between absence-of-an-algorithm and absurdity is 'the difference between *cannot-now* and *cannot-ever*.' (p.1). A particular case of 'cannot-now' is 'the mathematical assertion which cannot be tested, i.e. no method is known to derive its absurdity or the absurdity of its absurdity' and a-fortiori its truth. Indeed, 'cannot' in this case interpreted as absurdity would make the notion of non-testability a contradiction.

The strict interpretation of the mathematically 'real' and 'possible' is the basis for a division of assertions, 'statements assigning a mathematical property to a mathematical entity', into 1. those that have been proved, 2. those that are provable in the strict sense, i.e. an algorithm is known, and 3. the unproven and not-yet-provable:

In mathematics no truths can be recognized which have not been experienced and for a mathematical assertion α the two cases formerly exclusively admitted [true and false] have been replaced by the following four:
1. α has been *proved to be true*;
2. α has been *proved to be false i.e. absurd*;
3. α has neither been proved to be true nor to be absurd, but an algorithm is known leading to a decision either that α is true or that α is absurd;
4. α has neither been proved to be true nor to be absurd, *nor do we know an algorithm leading to the statement either that* α *is true or that* α *is absurd*. ([1955], p.114).

Notably absent from this division is the category of unprovable assertions. Indeed, in his post-war refutation of the PEM Brouwer abandons his search for essentially unsolvable problems. His refutation of the PEM is now based on:
1. the strict interpretation of mathematical truth and the guaranteed existence of unsolved mathematical problems;
2. contradictions arising from the PEM when applied to the indeterminate but legitimate objects in his intuitionist construction of the continuum.

Given the Brouwer interpretation of true and false as 'having been proved', the PEM is an obvious misrepresentation of fact. For the present, Fermat's Theorem and 'any mathematical assertion in the fourth case yields a refutation of the principle of the excluded third'. ([1955], p.114). The freedom of the Subject and of mathematics guarantees the supply of unsolved problems at any time in the future. The Fleeing Property has survived but is now a typical example of an unsolved problem; the solution of a particular Fleeing Property may one day well be found but there will always be others:

> Obviously, the fleeing nature of $f$ is not necessarily permanent, for a natural number possessing $f$ might at some time be found or the absurdity of the existence of such a natural number might at some time be proved.
> ... Nevertheless there will always be Fleeing Properties...
> ... and as was remarked above of Fleeing properties not having lost their fleeing nature there will always be an abundance. ([BMS59B], p.2; cf. also [BMS51], p.6).

## 5.9.4 *The Contradictority of the PEM*

A proof of the *contradictority* of the PEM is first given in [1928A]. It is based on Brouwer's interpretation the real number as a 'spread' and on what he describes as the 'Fundamental Theorem of Finite Sets (see further 6.3.13). According to this theorem 'any splitting of the unit-continuum into two results in one part-species which is identical with the unit-continuum and another which is empty.' ([1928A], p.378). A division of the unit-continuum generated by applying the PEM to the assertion of a property such as 'rationality' would then lead to the alternatives that all the points of the unit-continuum have or cannot have the asserted property, viz. all points of the continuum are rational or all points of the continuum are irrational. The same counterexample is quoted in [1948C]:

> The simultaneous supposition for all real numbers $c$ that the assertion '$c$ is rational' has been proved to be either true or contradictory does lead to a contradiction. (p.1244).

In that same year, however, Brouwer had already developed a new counterexample which, he claimed, provided an even stronger refutation of the Principle of Reciprocity of Complementarity and the PEM. It was again based on his interpretation of the real numbers and the relations between them of 'order' and 'virtual order'.

[1948B] and [1951] list a number of relations between real numbers together with their 'non-contradictorities':

| I   | 1. $a = a$      | I   | 2. either $a \leqslant 0$ or $a \geqslant 0$ |
|-----|-----------------|-----|----------------------------------------------|
| II  | 1. $a \neq 0$   | II  | 2. either $a < 0$ or $a > 0$                 |
| III | 1. $a \geqslant 0$ | III | 2. either $a = 0$ or $a \mathbin{\diamondsuit} 0$ |
| IV  | 1. $a > 0$      | IV  | 2. $a \mathbin{\diamondsuit} 0$.             |

In each pair the second member is non-negative, and its non-contradictority is represented by the first member. ([1951], p.358).

In the special Brouwer interpretation of order-relations each of the above non-negative assertions and the operation of 'absurdity-of-absurdity' generates its non-contradictority, but the non-negative assertion and its non-contradictority are non-equivalent, i.e. their equivalence is absurd. In case IV e.g. the non-contradictority of the relation '$\diamondsuit\ 0$' is given by '$> 0$', but the equivalence of '$\diamondsuit\ 0$' and '$> 0$' is absurd (this absurdity is in [1949B] the basis for the assumption that 'the intersection theorem of plane Euclidean geometry is contradictory').

The Principle of Reciprocity of Complementarity which asserts the equivalence of a statement and its non-contradictority is therefore false, and Brouwer claims:

> Each of the above non-negative assertions demonstrates the theorem of *contradictority of the principle of reciprocity of complementarity* (a fortiori

that of *contradictority of the equivalence of truth and non-contradictority in mathematics*) whose force considerably exceeds that of the theorem of contradictority of the complete principle of the excluded third (a fortiori that of contradictority of the solubility of all mathematical problems) which was deduced previously. ([1951], p.358).

## 5.10 Logical Negation and 'the Brouwer Logic'

### 5.10.1 The Calculus of Absurdities

*Intuitionist splitting of the fundamental notions of mathematics* [1923C] marks an important shift in Brouwer's involvement with logic. His investigations and campaign so far had centred on the separate identities of mathematics and logic and on the role of logic in traditional mathematics. In particular the logical Principle of the Excluded Middle was identified as unreliable. A beginning was made with the reform of mathematics, a re-construction of mathematics without the use of the PEM.

[1923C] is Brouwer's first attempt at positive reform of logic itself. It arose from the need for a new formulation in his programme of re-constructing mathematics, a need which was reinforced by the claims and ideals of the Signific Movement; apart from the conviction of the general corrupting effect of a logical tradition on current language there was a more immediate concern about the corruptions resulting from confused negation and the PEM. The formulation of Intuitionist Mathematics and in particular Intuitionist Analysis required new words which adequately reflected the new relations generated by mathematical absurdity, replacing those which embodied the fundamental assumptions of the PEM.

[1923C] was intended as an exercise in generating such new words. The underlying logical theory, which sparked off the international debate on 'the Brouwer Logic' and initiated an Intuitionist logic, occupied only a few lines in the introductory paragraph. They do not represent, nor were they intended to represent a complete alternative logical theory or even an axiomatic basis for such an alternative system; they were the corrections needed in classical logic – as in traditional mathematics – following the discovery of a fundamental error. Moreover, in deriving his alternative logical principles Brouwer follows the route of the genesis of logic as he saw it, moving at the different levels of mathematical construction and 'pure-logical' observation and abstraction. The proposed corrections are an attempt to establish at the logical level the implications of the strict mathematical negation. What was not fully appreciated, at least not spelled out in detail in [1923C], was that the rigorous distinction between mathematical construction and logical theory leads to forms of necessity, truth and absurdity which are characteristic of the logical level and different from those of the domain of pure-mathematical construction.

The proposed corrections of [1923C] are chiefly negative, refuting the main corollaries of classical negation and excepting those that can be preserved. As in all his pronouncements on logic or non-mathematical matters, Brouwer does not use symbolic notation as-if to emphasize their non-mathematical nature, and possibly because of the preferred ambiguity of ordinary language. 'Absurdity' and 'correctness' are spoken-of in the abstract and are used as general operators in a Calculus of Absurdities. It is clear, however, from the use of the word 'predicate' and from the context that they are applied to 'assertions of property' (the terms used in the almost identical proofs of [1948C] and [BMS51]).
I. The opening lines express the most categorical rejection of the Principle of the Excluded Middle and of what is described as 'a special case of the PEM', the complementarity of truth and absurdity of the property-relation:

> The Intuitionist conception of mathematics not only rejects the Principle of the Excluded Middle altogether, but also the special case contained in the Principle of Reciprocity of Complementary Species. (p.877).

Total rejection of complementarity obviously invalidates the classical equivalence of double negation and truth. The surprise move of [1923C] is that it gives some recognition to double negation, 'non-contradictority', hitherto dismissed as meaningless. It is introduced without further clarification as 'absurdity-of-absurdity' and is allowed a place in the amended logic.
II.i. Rejected is:

> The classical assumption that for every property the two exclusive alternatives are correctness and absurdity, i.e. that correctness and absurdity-of-absurdity are equivalent.

II.ii. Equally unacceptable is the complementarity of 'absurdity and absurdity-of-absurdity.'

III. But conceded is that:

> 'In the Intuitionist conception absurdity-of-absurdity follows from correctness.

IV. Perhaps most surprising, especially in the light of II, is the acceptance of the mechanical logical 'cancelling' of double negations in sequences of three or more absurdities:

> Such striking out of absurdity-predicates is allowed. It is based on the following theorem:
> Absurdity-of-absurdity-of-absurdity is equivalent to absurdity.

Whereas I, II and III are given as simply following from the rejection of the Principle of Complementarity, for theorem IV a proof is given, and a proof on traditional logical lines:

> Proof a. If the property $y$ follows from the property $x$, then from the absurdity of $y$ follows the absurdity of $x$.
> Therefore, since absurdity-of-absurdity follows from correctness, absurdity must follow from absurdity-of-absurdity-of-absurdity.
> b. Since from the correctness of a property follows the absurdity-of-absurdity of that property, then, as a special case, from the correctness of absurdity must follow the absurdity-of-absurdity-of-absurdity of the property. (p.878).

V. Brouwer concludes:

> The Intuitionist may, on the basis of this theorem reduce a finite sequence of absurdity-predicates to either absurdity-of-absurdity or to absurdity. (p.878)

The reference to a legitimate mechanical operation of 'striking-out' symbols representing absurdity is a clear indication that Brouwer had entered the domain of logic proper. [1923C] is a practical exercise in 'pure' and 'abstract' logic, seemingly grafted on the 'correct' parts of the classical tradition. It also adopts a modified form of logical necessity, 'correctness' and absurdity; unfortunately, [1923C] remains silent about their precise nature and indeed the distinctions between the mathematical and logical are somewhat blurred.

My interpretation of the Brouwerian notions of logical truth and absurdity in the following sections is based on the fundamental thesis of

Brouwer's philosophy, the complete separation of mathematics and logic, in particular his interpretation of the genesis of logic as given in [1907] and re-affirmed in his post-1929 papers.

## 5.10.2 Logical Necessity and Truth

The Brouwerian notion of logical necessity follows directly from the nature of pure logic as a science, the observation of regularity in the physical domain of the Subject's record of mathematics. It is the necessity characteristic of 'mathematical viewing', the scientist's interpretation of observed repetition of sequences of phenomena, extended by induction into a law, his conviction that the phenomena will and must continue the observed pattern. The necessity of logical laws, derived only from observation of patterns in the symbolism of the mathematical record, as of all scientific laws, is a necessity a-posteriori.

A distinct feature of the 'science of pure logic', however, is that the domain of investigation is not the mere physical appearance of symbols.

Brouwer insists that at the first phase of pure-logical investigation symbols are not divorced from their conceptual content. The Subject-logician reads his mathematical record with understanding; to him the symbols are 'invariant signs for certain elements and certain relations between elements of pure-mathematical systems constructed by him', ([1933], p.58). They include the symbols which represent the successful completion of a mathematical property construction or its absurdity and the a-priori necessity of the mathematical constructive steps. Through the imperfection of language he recognizes what is presented as true, what 'must-be' on the basis of the recorded mathematical construction. For these a-priori mathematical truths appearance in the Subject's complete mathematical record to-date is the norm of truth to the reading Subject-logician.

This conviction of certainty or truth is distinct from what Brouwer describes as the pleasurable feeling of certainty resulting from the 'causal' interpretation of phenomena, the certainty or necessity of physical laws. In particular the necessity of pure-logical laws is the logician's causal interpretation of the appearance or non-appearance of certain symbol-combinations following other symbol-combinations earlier in the record. By means of these laws the logician extends the domain of truth to the not-yet observed in the mathematical record:

> ... passing from correct affirmations (i.e. effectively expressing actual mathematical viewing) to other correct affirmations. ([1933], p.54).

To the truths reported in the record he adds those that 'should' appear at some time in the future or that 'can never' appear in the record, and this on the basis of logical laws such as the 'Principle of Contradiction' and the 'Principle of Syllogism'. The Principle of Contradiction is derived from 'pure-logical' observation of the completed record and the Subject's awareness of the 'conclusiveness' of the completed construction: he knows that having successfully completed the property construction $(a \supset b)$ the mathematical absurdity $\sim(a \supset b)$ cannot occur later in the record.

The necessity of the 'if-then' of the classical Modus Ponens according to [1907] is based on tautology and 'the simple mathematical structure of whole-and-part'; in the Brouwerian new species theory it is based on the concept of property as substructure or 'species' construction itself and the relation subspecies. The completed species-construction of 'fitting-in' a construction $y$ into a construction $x$ represents a mathematical i.e. necessary truth; if applied to two species such as e.g. divisibility by 6 and divisibility by 3, the proven fitting-in is the constructive basis of the implication $x \supset y$.

A similar constructive interpretation can be given to the implication $(\sim y \supset \sim x)$; however, it is hard to see how $(\sim y \supset \sim x)$ could be 'fitted into' $(x \supset y)$ and the implication $(x \supset y) \supset (\sim y \supset \sim x)$ be given this constructive interpretation. The same can be said for $(x \supset \sim \sim x)$ and $(\sim \sim \sim x \supset \sim x)$. These implications can only be based on an extensional interpretation of species: the 'element-species' or 'Brouwer-class', and be identified with the relation 'subspecies' (see further 6.3.6 and 6.3.10).

In [1923C] Brouwer did not challenge the classical notion of implication nor attempt to give a constructive interpretation of the general notion of logical implication. One might argue that 'necessarily following' in its most general form expresses a sequential structure common to both mathematical and logical construction, which with some reservation would form a sufficient basis for such generalization. To be meaningful, however, this general interpretation would need to reflect a correspondence between the modes of construction of the mathematical and logical implication and between the restrictions on their components. Such a correspondence is already ruled out by the Brouwerian thesis of the complete separation of mathematics and logic, the different nature and constructive coherence of mathematical and logical reasoning. In the case of 'implication' these differences are particularly obvious and rule out any meaningful correspondence. Whereas the mathematical '$a \supset b$' is a *construction* of $b$ on and with $a$, the classical-logical '$a \supset b$' concerns the hypothetical occurence of symbols in the written record and requires the relation between $a$ and $b$ to be no more than the impossibility of the occurrence of $\sim b$ following an antecedent occurrence of $a$, and vice versa.

The ultimate reason why Brouwer in 1923 did not challenge classical implication or attempt a comprehensive reform of logic is his continued conviction of the separate identities of mathematics and logic. His involvement in logical reform, as indeed his other campaigns during the early twenties for the reform of language, university structure and politics, remained a temporary distraction, the intervention of a 'wise outsider' in a non-mathematical domain. In spite of a more sympathetic attitude towards logic and some active participation, the gap between mathematics and logic was as wide as ever: two separate and autonomous disciplines, each with its own constructive coherence and modes of operation which do not allow adequate translation or interpretation either way. Brouwer's Signific colleague de Haen was criticized for 'trying to find a logical coherence of concepts in accordance with the logical composition of words' ([1916C], p.2). On the other hand a comprehensive reform of logic, reflecting accurately the restrictions of mathematical structure, must have seemed a pointless exercise in view of the 'fundamental inadequacy of logic'. Brouwer seemed resigned to accept logic for what it was, or what he considered it to be. His 'intervention' was restricted to the classical interpretation of negation, which had led to fundamental errors in logic and mathematics.

### 5.10.3 Logical Absurdity

In Brouwerian philosophy the origin and nature of negation or absurdity in logic, as in all science, is the necessity resulting from 'mathematical viewing', the conviction that the phenomena follow one another in constant sequence, referred to as a 'law'. Science considers the occurrence of a phenomenon or its absence not in isolation but in relation to other phenomena; logic in particular does not concern itself with the simple, contingent occurrence of certain symbols or their contingent absence but with the occurrence or absence of certain symbols or symbol-combinations as forced by their predecessors in the written mathematical record. The logical negation or absurdity $\sim x$ is a statement that the symbol-combination $x$, representing a mathematical hypothesis, is necessarily absent from, cannot appear further in the mathematical record, and this on the grounds of 'laws' derived by the subject logician in his pure-scientific investigation of the mathematical record. Logical absurdity as an aspect of scientific necessity lacks the a-priori of the pure-mathematical construction, it also differs from mathematical absurdity in its genesis and composition.

While the mathematical absurdity $\sim(a \supset b)$ is a constructed incompatibility, ultimately based on the introspected content of $a$ and $b$, the logical absurdity is generated from preceding statements and ultimately requires some axiomatic propositions, i.e. statements not generated by application of logical laws. In Brouwerian philosophy these are not arbitrary axioms but reports of 'experienced truths', i.e. completed mathematical constructions; nor does the logician need to restrict himself to a minimal number of such axioms: any mathematically constructed truth or absurdity can form the basis of further logical generation. In particular, the logician, 'reading the mathematical record with understanding' and applying what we have called 'the Principle of Conclusiveness', can interpret the report of a completed absurdity construction such as $\sim(a \supset b)$ as the logical absurdity 'the statement $(a \supset b)$ cannot occur in the mathematical record now and ever'. Indeed, in the restricted domain of his investigation this can be the only interpretation. Moreover, the rejection of the classical Principle of the Excluded Middle and in particular of what Brouwer calls 'its corollary, the Principle of Reciprocity of Complementary Species', ensures that no new assertion of mathematical truth or absurdity can be generated by mere application of logical laws. In spite of the [1923C] semantic confusion of 'cannot now' and 'cannot ever', the laws spelled-out in the Calculus of Absurdities do not allow the generation of a single negative property assertion unless the mathematical construction has been given explicitly or implicitly. This also applies to the case IV, $(\sim \sim \sim x \supset \sim x)$, where a single absurdity is generated from an odd, apparently weaker string of absurdities. The proof is based, not on some allowable cancellation of double absurdity, but on the constructive implication of $\sim (\sim \sim x)$: if $x$ is the property-assertion $(P, a)$, i.e. $a$ has property $P$, then any proof of the absurdity of $(\sim \sim x)$ includes elimination of all possible ways of fitting $P$ into $a$.

The Brouwerian logician, therefore, can treat recorded mathematical absurdity as equivalent with logical absurdity either presumed as an axiom or derived by logical law.

On the other hand, logical absurdity is not subject to the restrictions of mathematical absurdity as a binary construction; the 'cannot-appear-in-the-record' is a simple qualifier which can be applied to formulae of any kind; in particular, absurdities can be compounded, generating a new logical 'truth-value', absurdity-of-absurdity or non-contradictority.

The obviously circular definition of absurdity by contradiction stems from an attempt to set aside all semantic considerations and define absurdity purely formally. Whereas the mathematician can trace the source of contradiction ultimately to an incompatibility of two existing mathematical concepts (cf. Brouwer's identification of contradiction and mathematical absurdity:'.. where they pronounce a contradiction I simply observe that the construction does not go further, that in the main system there is no room for the construction posited.' ([1907], p. 127)), for the logician, starting from hypothetical property assertions as atomic elements, incompatibility in its simplest form is a relation between assertions, and the principle of contradiction – the incompatibility of $x$ and $\sim x$ – the most primitive formal incompatibility. However, the expression $\sim(x \wedge \sim x)$ clearly shows its dependence on the notion of absurdity, which must remain as a primitive logical notion, i.e. undefinable in the terms and language of logic. The semantic interpretation of absurdity as 'cannot appear in the record' is based on the nature of logic as a science, on the presumption and claim of science that certain phenomena follow one another necessarily, and on the dichotemy inherent in the concept of necessity and 'law': any claim that phenomena follow one another in one constant sequence ipso facto excludes deviance, declares others to be impossible.

Moreover, scientific laws are descriptive of man's interpretation of a necessary physical reality. In particular the logical law or principle of contradiction is a description of one of the necessities in the domain of formulae. It could be argued that as a generalization the law of contradiction is based on 'absurdity' alone. Such a generalization, however, is a further abstraction, eliminates the notion of logical absurdity, and is a further step away from the constructive reality of mathematics.

In his 'analysis' of [1907] Brouwer questions the status and generative power of the formulae expressing logical principles. He distinguishes on the one hand between these formulae and their 'intuitive' content, and between these and other, proper formulae of the logical construction. Principles may be expressed in symbols but the resulting formulae are not part of the logical construction itself. Moreover, in their application it is not the formulation, i.e. the logical 'form' which generates further logical formulae, but the intuitive content:

the [logical] system is to be split into the actual construction and the principles by which the construction develops itself. Even if these principles can also be formulated in symbols, such formulations must be considered as heterogeneous with respect to further formulae; these are generated by application of the principles not as formulations but as intuitive acts of which the added formulations are only linguistic accompaniments. ([1907], p.174).

The 'intuitive act' underlying the principles of logic is the recognition and acceptance of necessity characteristic of mathematical viewing, the constancy of sequences to the exclusion of others, and the 'conclusiveness' of the mathematical construction.

### 5.10.4 Non-contradictority and Absurdity-of-absurdity

Brouwer's change of views on logic in 1923 brought about some considerable changes in his interpretation of 'non-contradictority'. In one fundamental respect, however, his views remained wholly consistent: the non-mathematical status of non-contradictority was never in doubt. Even when in the post-1923 period some recognition was given to absurdity-of-absurdity, the domain of application remained restricted to logic.

In the early campaign the single, simple issue was the completely separate identity of mathematics on the one hand and that of language and logic on the other. 'Contradiction', as the word implies, is a purely linguistic phenomenon, a 'linguistic figure' (Dutch: taalbeeld), and is dismissed as a symptom which may signal an error in the work of the flesh-and-blood mathematician but which is usually due to the weakness inherent in language and 'logical building'.

Any non-contradictority, or 'freedom from contradiction', attributed to a faithful record of mathematical construction is only due to the mathematical construction. The completed mathematical construction itself, its 'Conclusiveness', is a sufficient guarantee that no future construction could undo it and produce a contradiction in the mathematical record.

> ... in the linguistic accompaniment of mathematical constructed systems... the sequence of sentences which follow one another logically will never produce contradictions since they accompany mathematical acts of construction. ([1907], p.159).

Brouwer's argument against logicists and formalists is that they have 'turned things round', have made non-contradictority the norm of mathematical existence and truth, and tried to reform mathematics by creating a linguistic framework in which contradiction cannot occur:

> The final aim of these investigations, still far out of reach, in which the mathematical thought-world is disregarded, consists in the construction of a linguistic mechanism, which, apart from a few not too drastic amputations, is capable of providing a verbal image of the whole of mathematics known so far, however, a linguistic mechanism that excludes the occurrence of the linguistic figure of the contradiction. ([1933], p.58).

In his later work he reports that the Formalist hope of crowning their work one day 'with a proof of non-contradictority was never fulfilled, and nowadays, in view of the results of certain investigations of the last few decades, has, I think, been relinquished.' [BMS49], p.2).

At the end of his *Foundations* Brouwer dissociates himself from the French Intuitionists, who still maintain an essential role of language in the construction of mathematics. In particular Poincaré, although much admired and even worshipped, is reproached for his insistence on non-contradictority as the criterion of mathematical existence and truth:

> How little Poincaré considers the intuitive construction of mathematics as the only foundation is clear from his words: 'Les mathématiques sont indépendables de l'existence des objets materiels; en mathématiques le mot exister ne peut avoir qu'un sens, il signifie *exempt de contradiction.*' (POINCARÉ[1905], p.819).
> It almost sounds like his opponent Russell. Mathematics, certainly, is wholly independent of the physical world, but *existence* in mathematics means *having been constructed*. Whether an accompanying language is free from contradiction is in itself of no importance, it certainly does not constitute a criterion of mathematical existence. ([1907], p.177).

Non-contradictority here is to be understood in the usual sense of consistency, the guaranteed freedom from contradiction of a set of axioms and the whole logical system constructed from it. Hilbert's programme of logical proof of the consistency of mathematics is challenged by Brouwer in [1907] on the grounds that consistency proofs ultimately are based on the non-contradictority of intuitive mathematics; moreover, he questions the mathematical existence of systems generated by pure logical-linguistic models:

> 'The following question arises: suppose we have in some way proved, without thinking of any mathematical interpretation, that a logical system constructed from some linguistic axioms is non-contradictory, i.e. that at no stage of development of the system two contradictory theorems can occur.

If we then also find a mathematical interpretation of these axioms, does it then follow from the non-contradictoriness of the logical system that such a mathematical system *exists*? But that has never been proved by the axiomatici; ...e.g. it has nowhere been proved that, if a finite number satifies a set of conditions which can be shown to be non-contradictory, this number indeed exists. ([1907], p.141).

The linguistic nature of the axiom system and its non-contradictoriness is the central argument in Brouwer's challenge of what he referred-to as 'Formalism' in his Inaugural Address, [1912A].

*'Non-contradictoriness' as a logical 'truth-value'* is not seriously considered until 1923, although there is evidence that as early as 1908 Brouwer was aware of the implications of his discovery that there are assertions that have not been proved to be true and yet cannot be proved to be absurd. In a footnote to [1908C] he states:

Among the theorems that in mathematics are usually considered to have been proved one should distinguish between those that are correct and those that are non-contradictory; to the latter e.g. belongs the theorem that a closed point-set can be split into a perfect and a denumerable set. ([1908C], p.7).

As long as the 'separation of mathematics and logic' was the single, all-important issue, 'non-contradictoriness', exposed as a purely logical attribute, was simply dismissed as mathematically irrelevant, as indeed were logical truth and absurdity. Only when Brouwer started to take logic seriously and investigate the implications of his rejection of the PEM as a valid principle in logic, did non-contradictoriness emerge as a special notion with its own interpretation and rules, distinct not only from mathematical truth and existence, but also from logical truth.

The logical significance of non-contradictoriness is recognized in [1923C] where it is introduced for the first time as 'absurdity-of-absurdity'under the heading of par.1: *'Truth-predicates'* (Du. 'juistheidspredicaten'). We commented above on the logical nature of the 'truth- and absurdity-predicates' and on the lack of clear distinctions in [1923C]. As to Brouwer's conviction of the logical, non-mathematical nature of 'absurdity-of-absurdity', his later comments leave no doubt. In his classification of assertions absurdity-of-absurdity constitutes a special case of unsolved mathematical problems, assertions which belong to:

...case 3b, those that have neither been proved to be true (1) nor have been

proved to be impossible now-and-ever (2)...and without the existence of a method which must lead to (1) or (2).
In some cases it happens that one can show the impossibility-of-the-impossibility although one cannot prove its truth. ([BMS47], p.1).

The non-quivalence of absurdity-of-absurdity and truth, i.e. logical truth, is the fundamental thesis of Brouwer's Calculus of Absurdities. In formal-logical terms it is a primitive proposition, laying down a rule of logical manipulation, in this case the abolition of the classical rule $\sim\sim a \supset a$ for all $a$. In the Brouwerian conception of pure-logic such fundamental rules are derived from 'scientific' observation of the mathematical record and need justification. The classical rule of equivalence of truth and non-contradictority, 'the thoughtless passage from non-contradictority to truth' ([BMS59], p.2), based as it was on the complementarity of truth and absurdity, could no longer be justified when the PEM was found to be unreliable. [1923C] speaks of 'reciprocity of complementarity of truth and absurdity' as 'a special case of the PEM'; later publications refer to it as 'a corollary of the PEM'. Indeed, although Brouwer's rejection of the PEM was based on his rejection of an 'objective' mathematical reality, the counterexamples refuting the PEM are a direct challenge of the Principle of Reciprocity of Complementarity. The favourite example is the rationality of the real number $b_f$ generated by the fleeing property $f$: 'the assertion '$b_f$ is rational' is non-contradictory without being true' ([BMS59], p.2). Later all affirmative, 'non-negative', assertions are claimed 'to demonstrate the contradictority of the Principle of Reciprocity of Complementarity', and the contradictorityof this principle becomes the norm and an essential part of 'non-negative assertions':

> We define a non-negative assertion as an assertion for which with regard to the elements of that species the principle of reciprocity of complementarity is contradictory. ([1951],p.358).

The rejection of the equivalence of logical truth and non-contradictority was a definite devaluation of non-contradictority, and Brouwer recognized that the 'loss of this opportunistic axiom' meant 'the loss of the use of indirect proof' ([BMS47], p.2). On the other hand, in the hierarchy of logical truth-values non-contradictory assertions rank 'higher' than those which Brouwer in his later work called 'not-tested', 'assertions which are neither true nor false nor contradictory' ([BMS66], p.1). Moreover, non-contradictority generated useful logical relationships for the new language, needed in Brouwer's reconstruction of analysis. It is unfortunate that Brouwer's pre-occupation during the twenties with the generation of new logical relations and distinctions detracted his attention from the question of the mathematical relevance of non-contradictority.

[Although in a letter to Heyting (26-06-24) he stressed his self-restraint in generating new relations (' I imposed restrictions on myself... ways of extending my tables [of [1923C] ] are legion.'), the 'splitting of notions' during the following years proliferated and contributed to the seizing-up of Brouwer's re-construction of mathemtics. The clear 'intuitive' vision, so characteristic of his topological work and his philosophical inquiry, seems to desert him when he leaves his mathematical home-ground and tries to beat logicians and lawyers at their own game.]

The interpretation of 'non-contradictority' follows directly from its composition and the interpretation of logical absurdity, 'the operation of declaring the absurdity of' (Brouwer to Heyting, 26.06.24). If $x$ is a formula expressing the hypothesis of a property-construction, $\sim \sim x$ is to be interpreted as: 'the absurdity construction $\sim x$ is necessarily absent from the record'. The logician reaches this conclusion on the strength of logical laws and his 'meta-mathematical insights'. As maintained above, the assertion of absurdity-of-absurdity does not express a proper construction in the idealized reality of mathematics; it does,however, reflect reasoning at the pre- and post-mathematical stage. At the intuitive pre-mathematical stage e.g. of attempting to fit $b$ into $a$, it may be the recognition by the flesh-and-blood mathematician that $(a \supset b)$ cannot be constructed, that the indeterminate nature of the constructs does not allow 'complete elimination of all ways of fitting $b$ into $a$. The conviction of the absurdity-of-the-absurdity of $(a \supset b)$ may also at the post-mathematical stage result from the successfully completed fitting of $b$ into $a$, an application of 'the Principle of Conclusiveness'. In either case the mathematical significance of non-contradictority is questionable. Concluding the (weaker) non-contradictority of an assertion from its truth, although a valid logical move used also by Brouwer in his proof of [1923C] IV, is trivial whatever interpretation is given to mathematical truth and non-contradictority. In the more usual sense of 'not proven mathematically, i.e. to be true or to be absurd, but the absurdity of its absurdity has been shown', non-contradictority may seem at least to eliminate one of the two lines of inquiry or construction. But, although distinct in its outcome, the 'constructive' activity of the flesh-and-blood mathematician consists in the same attempt of 'fitting one construct into another', whether it leads to successful completion or 'complete elimination'. Recognition that complete elimination is impossible is no more than the certainty that his trials can only be halted by succesful completion but may well continue ad infinitum.

'Untested' hypotheses, i.e. those that have not been proved to be true or absurd nor shown to be non-contradictory, as yet lack that guarantee and leave the option of absurdity still open.

The 'intuitive' interpretation of non-contradictority is less clear and distinctions become somewhat blurred in the complexities of relations generated by the logical 'splitting' of notions and relationships between relations, and when Brouwer recognizes a limited 'field of validity' of the PEM and the Principle of Reciprocity of Complementarity. As a result, non-contradictority in some cases is found to be equivalent to truth:

> Some non-contradictorities of constructive properties $\zeta$ may within a certain species of mathematical entities either be given a constructive form (possibly but not necessarily because the Principle of Reciprocity of Complementarity applies in the case of $\zeta$) or the form of absurdity of a constructive property...
> But for other non-contradictorities there seems to be little hope that such equivalent will ever be found. ([1948B], p.387; cf. [1948C], p.1249).

The argument is supported by examples; no norm is given, either in terms of logical form or based on an intuitive ancestry, for differentiating between various non-contradictorities, as indeed between formally negative and genuine ('essentially') negative properties. Examples in both cases are intuitive mathematical constructions; attempts to define distinctions logically lead to somewhat circular definitions, such as quoted above of the non-negative property.

## 5.11 The Crisis in Brouwer's Intuitionism

One of the attractions of Brouwer's philosophy of mathematics as 'languageless thought-construction' is its apparent simplicity: it avoids the pitfalls of epistemology and semantics and determines clear criteria of mathematical truth and existence. And yet in Brouwer's programme of re-constructing mathematics it is the simple 'total separation of mathematics and language', 'the first act of Intuitionism', which proved most difficult if not impossible to implement. Indeed, the very attempt to present and publish an 'Intuitionist Mathematics' compromises this separation; it also implies some recognition of a collective domain of mathematical truth. By expressly renouncing any such claim Brouwer could uphold his theory of separation. However, in his re-construction of mathematics he did not succeed in keeping the mathematical pure-thought construction wholly isolated from its verbal accompaniment and in following the 'genetic' course marked out in his *1907 analysis*, which starts from the idealized pure-thought construction. On the other hand, the practical confusion of language and mathematics helped to conceal the fundamental conflict between the Brouwer philosophy of mathematics as completed thought-construction and the element of indeterminacy introduced in the 'Second Act of Intuitionism'.

Brouwer's failure to resolve these conflicts in his theory and practice was, in my opinion, the main cause of the demise of his programne. It led to confusion and to the 'unbearable awkwardness of his Intuitionist Mathematics', to his loss of self-confidence and his 'silence', and allowed the emergence of a new Intuitionism, which abandoned his philosophy and in particular his doctrine of absolute separation of mathematics and language.

In the following sections we consider some of these conflicts and developments.

*5.11.1 The Use of Language and Logic in Brouwer's Reconstruction of Mathematics*

Much of Brouwer's elaborate logical armoury was set up to confute the traditional logical assumptions, in particular the PEM. Unfortunately, the battle against the PEM not only diverted much of Brouwer's time and energy from his task of re-constructing mathematics, it also deflected the programme from the clear constructive course he had set himself, a course wholly determined by the intuitive constructs, 'by their way of development from the basic Intuition' ([1948C], p.1244), to be followed by a search for a suitable linguistic-logical expression. Reading Brouwer's post-1923 papers one cannot avoid the impression that the spawning relationships and distinctions find their source in a logical splitting of notions and are subsequently given a mathematical meaning and justification.

In Brouwer's ontological analysis there is a clear and absolute distinction between mathematical and logical activity, and between their respective necessity, truth and absurdity. These fundamental differences naturally result in different modes of construction. Indeed, it was the gut-feeling of a disparity in 'mathematical and logical thinking' that sparked off Brouwer's investigation into the nature of mathematics and logic (cf. Brouwer's letter to Korteweg [DJK35], app.13). Evidence of constructive methods which are 'natural', characteristic of mathematics, is clearly found in his treatment of arithmetic and topology. Weyl singles out his treatment of the continuum as particularly 'natural':

> Mathematics with Brouwer gains its highest intuitive clarity. He succeeds in developing the beginnings of analysis in a natural manner, all the time preserving the contact with intuition much more closely than had been done before. (WEYL[1949], p. 54).

The failure of Brouwer's programme of re-constructing mathematics is blamed by Weyl on 'an almost unbearable awkwardness resulting from the inapplicability of the simple laws of classical logic' (ibid.). It seems, however, that failure was partly due to a loss of 'contact with intuition', in particular in his analysis where logic plays a more active role than it was accorded in Brouwerian philosophy. Indeed, there is a failure to provide an adequate philosophical basis for the element of indeterminacy, introduced as the key to the continuum, a failure to adapt or extend the Brouwerian concept of mathematics to accommodate the 'Second Act of Intuitionism'.

Indefinitely Proceeding Sequences, determined by free choice of the Subject alone, are an eminent characterization of the dynamic, growing and free aspects of mathematics and its time-existence; the active, creative role of the Subject is more predominant here than in the construction of finite and law-like entities. And yet their indeterminacy, their incompleteness, which is claimed to give a precise characterization of the point-of-the-continuum, disqualifies them as legitimate mathematical constructs in the strict Brouwer sense. The freely-proceeding sequence, as the life of the Subject, is divided by the 'present moment' into two distinct parts or 'segments': the first consisting of the past and present acts of the Subject, the second of the potential future acts included in the 'open present' (cf. 4.6.3). In the case of law-like infinite sequences each and all elements are generated by a law or algorithm, a result of a past constructive act of the Subject. The identity of each future element of the free-choice sequence – possibly within a self-restricted domain of choice – is wholly dependent on a future act of the Subject (cf. e.g. [BMS59], p3; app. 9). In the Brouwerian interpretation of mathematical existence as 'having been constructed' only the first part, the completed segment can claim mathematical existence and is therefore the only part that can be used as a legitimate construct, 'previously acquired', on which further constructions (including property and absurdity constructions) can be built.

Freely-proceeding sequences, like complex numbers, straddle two different realities: the mathematical real past and present and the possible future. Yet in the Second Act of Intuitionism they are introduced as single genuine 'mathematical entities', representing specific points of the continuum and serving as a legitimate domain for further functional construction. While the completed initial segment provides stability and strict-mathematical legitimacy, the most essential role of securing 'continuity' and the measurability of the continuum is played by the indeterminate tail; indeterminacy, incompleteness, is the dominant characteristic of the Indefinitely Proceeding sequence.

It is ironic that the use of language and logic enbled Brouwer to develop his theory of choice sequences. Words are timeless, they do not discriminate between past and future existence of the concepts they refer-to, and 'logical truth' spans both past and future. In particular, words can be used to 'define' hypothetical, future acts, make the indeterminate more manageable and treat the ambivalent Indefinitely Proceeding Sequence as an existing single entity. During the twenties Brouwer constructed a verbal and logical framework which accommodates the new hybrid, but which also conceals a conceptual deficiency and a conflict with his philosophy of mathematics as completed thought-construction. His general statements during this period and after continue to insist on the absolute separation of mathematics on the one hand and language and logic on the other, and on the need for mathematical as distinct from logical construction. The role even of the reformed, 'intuitionist' logic is still no more than that of a useful predictor of possible results. All discussions of Intuitionist Logic end with a reminder that such results 'are only then realized, i.e. only then convey mathematical truths when they have been experienced' (e.g. [1948C], p.1245, [1954A], p.3), i.e. have been verified by a completed mathematical thought-construction. That the verifying thought-construction is more than a purist formality or a subjectivist requirement but follows a different constructive route, is clear from his warning that even legitimate logical truths 'cannot always be experienced' ([1948C], p.1243). Unfortunately, Brouwer remains silent about these constructive processes, so different from those expressed in the wording of the logical construction, and in particular about the constructive reality underlying indeterminacy. Neither did he modify his philosophy to give the Indefinitely Proceeding Sequence full mathematical status.

The ambivalence of the Indefinitely Proceeding Sequence led to some remarkable Intuitionist theorems (see further 6.3.13), but also contributed to the 'unbearable awkwardness' of Brouwer's analysis and possibly to his abandoning the programme of re-construction. It is significant that the unfinished manuscript [BMS37] of one of his last major reconstruction of analysis ends abruptly in Part Two, 'Fundamental notions of real functions of one variable' after an introductory section on the nature of functions.

### 5.11.2 The Pragmatism of Post-Brouwer Intuitionism

The problem of the ambivalence of the Indefinitely Proceeding Sequence, and more generally the conflict between the 'conceptual' and the 'logical' in Intuitionist Mathematics, was solved by Brouwer's 'followers' in different ways. They are united in dismissing his philosophy, his 'metaphysics', as irrelevant and in selecting parts of the Brouwer doctrine and practice based on that philosophy.

Radical constructivists like Bishop welcome Brouwer's 'disengagement of mathematics from logic' and adopt some of his constructive principles and methods. They do not share his views on the conceptual nature of mathematical objects nor his pre-occupation with the dimensional difference between the discrete and the continuous, and reject his ambivalent solution:

> Brouwer became involved in metaphysical speculation by his desire to improve the continuum. A bugaboo of both Brouwer and the logicians has been compulsive speculation about the nature of the continuum. In Brouwer's case there seems to have been a nagging suspicion that unless he personally intervened to prevent it, the continuum would turn out to be discrete. (BISHOP[1967], p.6).

Bishop has shown how much can be achieved in re-constructing analysis on Brouwerian lines without resorting to Indefinitely Proceeding Sequences. Pragmatically, he avoids the fundamental problem of the continuum and accepts the role of the 'hypothetical' in mathematics as long as it has a computational interpretation.

The Intuitionist-Logical tradition, led by Heyting, takes the whole of Brouwer's Intuitionist Mathematics and his reformed logic as its starting-point. They do not subscribe to Brouwer's general subjectivist philosophy, and effectively abandon the most fundamental parts of his philosophy of mathematics: the ultimate authority of individual Intuition, the pure 'mental' nature of mathematical objects, the absolute separation of mathematics and language, as well as Brouwer's claim of the monopoly of truth for his philosophy and the mathematics based on it. The philosophical problem of the genesis of mathematics is dismissed as not directly relevant, and the role of language and logic is raised to that of an indispensable component of mathematical practice. Their Intuitionism is not so much a philosophy and foundation of mathematics, more a form of practice, 'a cultural phenomenon'.

This interpretation of Intuitionism is wholly due to Heyting; it is the product of his personality and development. It could succeed and become the established form of Intuitionism because he was the right man at the right time and place. He had the qualities that made Brouwer promote him in 1930 as the spokesman for his Intuitionism, when he himself wearily abandoned his programme: tactfulness, modesty and a loyal and lasting interest in the Intuitionist cause.

Heyting's famous paper on the formalization of Intuitionist logic (HEYTING[1930A and B]) was not the rebellious coup it is sometimes made out to be. Indeed, it was supported by Brouwer (see further 5.12.4), but it came at a time when Brouwer withdrew from the public debate, allowing Heyting to keep the Intuitionist issue alive and gradually develop his own version of Intuitionism. The change of direction and emphasis again reflects Heyting's personality: cool and matter-of-fact, he kept clear of Brouwer's metaphysical speculation; his regard for the contribution of others and human cooperation moved him towards a more positive stand on language, and to practical collaboration. The 'new spirit of peaceful cooperation which gained victory over that of ruthless contest' (HEYTING[1960]) was very much due to him. He kept Intuitionism alive by his participation in the international debate and by 'making school', promoting an interest and encouraging his best students to research into various aspects of Intuitionist Mathematics (for details see *Bibliography* under Troelstra, van Dalen, van Rootselaar, Dijkman, Ashvinikumar and Gibson).

Heytings neo-intuitionism diverges from the Brouwer orthodoxy on three fundamental issues: the philosophical foundation of mathematics, the nature of mathematics as thought-construction, and the role of language and logic in mathematics.

Heyting consistently maintained that one could practise Intuitionist mathematics without subscribing to Brouwer's general solipsistic philosophy and that philosophical search for the genesis of mathematics is fruitless psychologism and irrelevant. Already in his dissertation – under Brouwer's supervision – he takes the independent line that 'all attempts to found mathematics on metaphysics must come to grief' (HEYTING[1925], p.1).

He adheres to the Brouwer doctrine that mathematics is constructive thought-activity (cf. e.g. the opening line of his [1930A]: 'Die intuitionistische Mathematik ist eine Denktätigeit...'), and accepts a notion of a mathematical thought-reality distinct from and prior to logic, but 'without attempting to seek a philosophical justification' (HEYTING[1930C], p.957).

Notably absent from Heyting's writings is the Brouwerian account of the genesis of mathematics, the primordial happening of the individual Intuition, and the consequent emphasis on the time-bound nature of mathematics. HEYTING[1956] even states clearly that ' Mathematics begins after the concepts of natural numbers and of equality between the natural numbers have been formed' (p.15). The formation of these purely mental concepts is reduced to the stage of 'pre-mathematics' and not further analysed:

> As far as I know, psychology has not discovered mental atoms. Every notion may be analysed, none is comprehensible by itself; any notion depends for its explanation upon its relation to other notions. The notion of the natural numbers is no exception to this rule. (ibid.).

Less concerned with the problems of epistemology, Heyting could move away from Brouwer's extreme views on the subjective and time-bound nature of mathematical thought and the absolute separation of mathematics and language, and adopt a more positive stand on mathematics as collective thought-activity and the role of language. In his inaugural address he declares:

> Intuitionist mathematics therefore emerges as mind-activity which is shared by a number of people bound-together and supported by ordinary language as well as by the special language of mathematics and the language of symbols. (HEYTING[1949], p.17).

Heyting's statements on the role of language in mathematics show a gradual but definite break-away from the Brouwer orthodoxy. His early writings still echo Brouwer's total separation:

> Intuitionist mathematics is thought-activity and all language is only a means of communication... (HEYTING[1930A], p.42).

Indeed the motto of [1930A] in its original version (Heyting's prize-winning entry of the 1928 competition of the Dutch Mathematical Society) is the biblical reference 'Stones instead of Bread', stressing the utter barrenness of pure formalization. But HEYTING[1949] raises doubts about the pure-mental, languageless aspect of Brouwerian mathematics:

> It seems to me that the view that Intuitionist mathematics in its pure form would exist only in the thoughts of an idealized mathematician and that language would represent only a very loose connection with these otherwise independent mental constructions, conflicts with reality. (p.14).

This doubt is finally resolved in HEYTING[1956] and language recognized as an indispensable component and representation of mathematical reality; the word, not the memory of the Subject, guarantees the continued existence and identity of the mathematical construction:

> I am even forced to revise somewhat our notion of a natural number. If a natural number were nothing but the result of a mental construction, it would not subsist after the act of its construction and it would be impossible to compare it with another natural number constructed at another time and place. It is clear that we cannot solve this problem if we cling to the idea that mathematics is purely mental. In reality we fix a natural number, $x$ say, by means of a material representation. (p.15).

Instead of total independence and heterogeneity of mathematics and language-and-logic, Heyting proposes a phenomenological partnership: words and logical operators are the embodiment of thought-constructions. The conceptual 'content'of a mathematical object is as essential as the mind is in the human make-up, but it is equally inseparable from its physical representation. Indeed, in the collective enterprise of mathematics its physical appearance is the immediate and common reality.

Its pre-verbal conception in some individual mind is acknowledged – Brouwer's first phase of pure-mathematics – but left as a necessary but mysterious preliminary, a creative invention à la Poincaré. Real mathematics concerns thought only after it has been expressed in words or symbols and become common property. 'Thought' remains the vital component of mathematical objects and determines the behaviour of their physical appearance and their relative structure.

The 'individual' origin and nature of thought is recognized:

> ...my mathematical thoughts belong to my individual life and are confined to my personal mind... (HEYTING[1956], p.8),

but there is sufficient 'analogy' between the thoughts of individuals to justify a notion of 'common thought':

> ...other people have thoughts analogous to our own and they can understand us when we express our thoughts in words... (ibid.).

In this sense, 'thought-construction' can be upheld as the characterization of mathematics as collective activity, and is therefore an essential aspect of mathematical and logical investigation. The Intuitionist not only examines and systematizes mathematical symbolisms, he is concerned with its 'constructive interpretation', the common thought-content corresponding to these symbols and their logical structure.

['Constructive interpretations' of the logical operators were attempted by Heyting in [1930A]; of the Intuitionist 'proposition' by Heyting and Freudenthal (cf. 'the Heyting-Freudenthal debate in TROELSTRA[1983B]); of the 'Idealized Mathematician' by Kreisel in [1968] and Troelstra in [1968]; and of 'Choice Sequences' mainly by Troelstra in [1968], [1969B], [1977] and [1982], by van Dalen and Troelstra in [1970], by Kreisel and Troelstra in [1970] and by Posy in [1975].]

Yet in spite of the declared aim of linking symbols to mathematics, which is 'thought-activity' (HEYTING[1930A], p.1) and 'construction by Reason' (HEYTING[1930C], p.958), these constructive interpretations almost exclusively concern functional behaviour in the domain of logical symbolism. Concepts and the processes of reasoning are referred-to but not further analysed. For example, the constructive interpretations of logical implication and absurdity are described by Heyting as:

> A proof of $a \to b$ is a construction which transforms any proof of $a$ into a proof of $b$...
> A proof of $\neg a$ is a construction which transforms any possible proof of $a$ into some fixed false statement such as $0 = 1$. ([1934]).

Both interpretations take the notion of 'construction' as understood and describe the construction underlying the logical operators as 'a transformation', which seems more characteristic of the process of logical derivation than of the Brouwerian constructions of property and absurdity. (Cf. also our comment in 5.8.1 on the circularity of defining absurdity as a procedure reducing a hypothesis to a known absurdity.)

Indeed, the search for constructive interpretation in the sense of 'thought-construction' is bound to be frustrated and remain trapped in linguistic, circular definition by the Heyting compromise of essential partnership of thought and word, his refusal to consider the 'pure mental'.

In Brouwer's philosophy of mathematics circularity was avoided by placing the reality of mathematics and its ultimate authority outside the domain of language. The reliability of the 'linguistic record' of mathematics and its collective practice were inevitably sacrificed; but the a-priority and transcendence of mathematical truth were safeguarded, grounded as it was in the concept-creating faculty of man.

Insisting on 'material representation' as an essential element of mathematical existence, Heyting and his successors abandon the principal feature of Brouwer's characterization of mathematics, and with it his distinctive philosophical foundation of mathematics. They return to the pragmatism of the French pre-intuitionists, who recognized the need for some vague intuitive content but whose practical and exclusive concern was symbols and their logical structure. The existence of an underlying and preceding thought-construction is presupposed; in so far post-Brouwer Intuitionism distinguishes itself from 'Formalism', which starts from meaningless symbols and an arbitrary set of axioms. Its axiomatic basis is the mathematical creative thought of its Founder expressed in words or symbols. (Cf. Myhill's definition of Intuitionist Mathematics: 'roughly the practice of Brouwer and Heyting'.)

## 5.12 The Formalization of Mathematics

### 5.12.1 The 'Formalist' Schools and Formal Axiomatic Theory

Brouwer's early writings present the foundational debate as polarized between 'Intuitionism' and 'Formalism' on the two related issues of the nature of mathematics and the role of language in mathematics. 'Formalists' are all those who identify mathematics with its linguistic or logical form:

> To the consistent Formalist mathematics is nothing but a series of relations without meaning which have mathematical existence only when they have been expressed in spoken or written words, together with the mathematical laws which govern their development. ([1912A], p.9).

Exclusive concern with linguistic form, the rejection of any conceptual content characterizes the 'logicist' interpretation of mathematics as well as the 'formal axiomatic' approach of Hilbert's *Grundlagen*. After 1928 Brouwer refers to them all as the 'Old Formalist School':

> ...the *Old Formalist School* (Dedekind, Cantor, Peano, Russell, Couturat, Hilbert, Zermelo), who maintain that a rigorous treatment of mathematics dispenses with all elements outside language and logic. ([BMS32], p.2).

In later versions:

> ... the *Old Formalist School* (Dedekind, Cantor, Peano, Russell, Couturat, Hilbert, Zermelo) for the purpose of a rigorous treatment of mathematics *and* logic rejected every extra-lingual element. ([BMS41], p.2; cf. also [BMS49], p.2; [BMS51], p.2; [1952B], p.139).

The distinguishing and redeeming feature of the 'New Formalist School' is its acceptance of 'the benefactions of Intuitionism' ([1928], p.337), recognizing the need for Intuition at least in 'meta-mathematics', or 'mathematics of the second order':

> ...the New Formalist School (Hilbert, Bernays, v. Neumann), which unlike Old Formalism accepts in confesso the Intuition of the natural numbers and induction... ([BMS32], p.3).
> Hilbert founded the *New Formalist School*, which postulated existence and exactness independent of logic, it's true not for mathematics proper, but for meta-mathematics or mathematics of the second order, i.e.the scientific consideration of the symbols occurring in the purified mathematical language and the rules of manipulation of these symbols. This New Formalism, in contrast with Old Formalism, consciously and *in confesso* made use of the intuition of the natural numbers and complete induction. ([BMS49], p.3).

By conceding the need for Intuition in the construction of a formal system the New Formalists effectively abandoned the claim that 'formal structure' is the ultimate foundation of mathematics. Brouwer's main, ontological argument against Formalism, however, is the non-pure-mathematical nature of the whole process of logical investigation and formalization. The 'pure-logical' investigation of past mathematical records and the formal systematization of the language of mathematics are 'scientific', i.e. mathematical viewing applied to the physical domain of language, no more than attempts at improving the instrument of communication. The fundamental error of Formalism is its confusion of mathematics and language:

> ... but no attention is paid by New Formalism to the circumstance that

between the perfection of mathematical language and the perfection of mathematics proper no clear connection can be seen. (ibid.).

The emphasis of Brouwer's criticism of Formalism and formalization varies over the years in line with his changing mood and attitude to language and logic. At the time of disillusionment with Signific and Intuitionist reform the emphasis is on the inevitable failure of the formalist programme because of the essential inadequacy of language. During the early years of rebellion and in the post-war period 'axiomatic method' is one of the main targets of attack.

Although directed at the 'Formalist', his criticisms concern more widely shared practices and objectives; they are particularly relevant as to Brouwer's views on the nature of formalization, its legitimate role, its value and limitations.

In [1933] formalization is described as:

> ...an attempt to make the language of mathematics purely mechanical (p.54), and
> ...the construction of a linguistic mechanism which excludes the occurrence of the linguistic figure of contradiction (p.58).

Its legitimate, scientific role is restricted to the description of mathematics already constructed:

> ... a linguistic mechanism capable of providing a verbal image of the whole of mathematics known so far... (ibid.).

The immediate purpose of formalization is perfection of the language of mathematics, the ultimate formalist aim the perfection of mathematics itself. Such 'linguistic reform of mathematics', however, is ruled out not only on ontological grounds, even the immediate objective, the perfection of the language of mathematics, can never be obtained: language is essentially inadequate:

> There is still a wide-spread belief in the possibility of ensuring the total exactness of pure-mathematics by making the language of mathematics purely mechanical. Such belief, as we have shown, can only be due to failure to appreciate the real nature of language. (p.54).

The essential weakness of language is the tenuous, unstable link between thought and physical symbol, the instrument of language is incapable of perfection:

> One cannot make provision against this inadequacy of language, as has been tried by the Formalist School, by subjecting the mathematical language, i.e. the system of symbols intended by the speaker or writer to evoke pure-mathematical constructions in others, in turn to mathematical viewing, and through complete overhaul providing it with an exactness and stability as one finds in physical instruments or the phenomena of exact science, thereby making use of a meta-language... (p.53).

There is disagreement between the Brouwerian philosophy of mathematics and every aspect of the formalist characterization of mathematics as 'formal axiomatic theory' and its claims of consistency, completeness and decidability.

Axiomatization in its classical and 'modern' form comes under heavy fire in the *Foundations*. Modern axiomatici and in particular Bolyai, Lobatcheffski and Hilbert are criticized for making the arbitrary selection of axioms the sole basis of a mathematical system and for their obsession with consistency, independence and minimality:

> They, and in particular Hilbert, have been busying themselves with constructing linguistic edifices of pathological geometries, trying to show which properties are preserved and which are not when one drops some of the axioms. They set themselves the aim to restrict the necessary axioms to a minimum for each of these logical edifices. ([1907], p.133).
>
> (cf. also [BMS3A] VII.20: 'One can never show that axioms do not contain anything superfluous.')

Brouwer's objection to axiomatic structure is part of his general argument against 'logical building of verbal edifices', 'forced building' (Du. 'dwangbouw'): it does not reflect or represent mathematical construction:

> It does not follow from the consistency of the axioms that the supposed corresponding mathematical system exists. Neither does it follow from the existence of such a system of mathematical reasoning that the linguistic system is *alive*, i.e. that it accompanies a chain of thought, and even less that this chain of thought is a *mathematical* construction. ([1907], p.138).

In particular, the axioms of current systems do not represent the mathematical primitive building bricks, the Subject's 'primordial recognitions of Intuition'. Classical mathematics had erred by including among the axioms ('a limited stock of evidently true assertions') those based on spatial observation:

For some familiar regularities of outer and inner experience, which with any attainable degree of approximation seemed invariable, absolute and sure invariability was postulated. These regularities were called axioms and were put into language. ( [BMS49], p.1).

Moreover, classical mathematics and especially modern axiomatici introduced arbitrary, 'opportunistic' axioms, based neither on inner nor outer experience:

> To introduce a species of real numbers which can represent the continuum...classical mathematics had to resort to some logical process starting from anything-but-evident axioms. The axioms currently most widely used for this purpose are those defined by Hilbert. ([BMS47], p.2).

But the most compelling reason for Brouwer's rejection of 'axiomatics' is the artificial restriction which any axiom system imposes. The setting of arbitrary and final restrictions in advance of mathematical construction is a complete denial of the freedom of the Subject to draw constantly on Intuition, the 'freedom of mathematics' which 'can never be exhausted in anyone system and develops in self-unfolding guided by free arbitrariness.' ([1907], p.119).

The restrictiveness of axiom systems is the dominant theme of two chapters of the preliminary *Notes* to Brouwer's dissertation (headed 'Axiomatic Foundations' and 'Examples of Unicity-proofs of axioms' [BMS3A]). His strength of feeling can be gauged from his outspoken comment:

> ...in the construction [of mathematics] axioms are a straitjacket (stating terms before building has started, as one does in the case of houses)...(I.27) Axiomatic Foundations probably originated because of carelessness in this forced building... a good check for erring minds, but a disaster for the free spirit... I refuse to take part in such mathematical logic... (IV.15) etc.

Axiomatics is further criticized for being unaesthetic, often futile and trivial:

> ...axiomatic construction is unaesthetic... (VIII.71);
> Veronese says that the first axiom of geometry is the possibility of distinguishing between different points. What futility! (V.24);
> ...puzzle research of trivialities... (III.18);
> ...'Two things, equal to a third, are equal to each other'. One can only utter such trivialities if one considers the linguistic form by itself. (IX.16);
> ...The fools who do not see anything but a system of relations ...an axiomatic system is nothing but combinatorics... (VII.13).

These blunt comments, made in the privacy of personal notes, show the young Brouwer's dislike of axiomatic method; they focus on what he considered to be errors and abuses in current practice, mainly the overrating of axiomatic structure and its role in mathematics. They also reveal his views on the true nature and the legitimate function of formalization: a 'scientific' process of analysing a mathematical record, a 'post-mortem' examination distinct from live mathematics:

> It is equally stupid and simple to consider mathematics to be just an axiom system as it is to see a tree as nothing but a quantity of planks. ([BMS3A] I.28)

This remained Brouwer's consistent view of formalization, even if during the following years there are changes in the immediate target of his criticism and the severity of his attack. The criticism of 'axiomatics' in [1907] pp.133-142 concerns almost exclusively the axiomatization of geometry; in [1912A] it is the axiomatization of set-theory, and in particular Zermelo's axioms of 'selection' and 'inclusion', which bear the main brunt. With his involvement in set-theory during the years 1912-1917, and with his changing fortunes, his attitude towards axiomatics seems to change. In the new wave of Signific optimism there is at least a dispassionate acceptance of language and its various aspects, including its axiomatic form, as a fact of life, and even a hope of improving it. Most surprising perhaps is Brouwer's choice when in 1917 he was in a position to reorganize his department, creating for himself the chair in 'Set-theory, function-theory and axiomatics'. These anomalies, however, do not signal fundamental changes or a U-turn in Brouwer's view on axiomatics and formalization. The conflict in this case is between theory and practice, between Brouwer's critical philosophical judgment and the traditions of his chosen profession. A wholly re-constructed mathematics in line with his philosophy was not an available alternative. [There is also, as with so many anomalies in Brouwer's life, the conflict between his philosophical and moral convictions on the one hand, and what he expediently chose or resigned himself to do on the other. Compare e.g. his moral crusade in [1905] against health baths, theosophy, mathematics, foundational study of mathematics, etc. while indulging in all these activities himself; also, his philosophical rejection of the scientific application of mathematics and his acceptance in 1913 of the chair on Mechanics at Amsterdam University. His cynical justification at that time is: the practical need for 'cunningness', for 'meedoen' (i.e. 'joining-in' with the connotation of 'running with the hounds'), in this world where paradise is lost. (cf. *Student Notebook* [BMS1B], app.2).]

As to axiomatics, the inconsistency between critical philosophical judgment and mathematical praxis was already evident in Brouwer's early work. Shortly after condemning non-Euclidean geometries as 'pathological' and 'pseudo-geometries', he published an orthodox account of *Non-Euclidean Geometry* [1909D], and made his considerable contribution to the further development of the axiomatic theory of non-Euclidean space (e.g. in *The characterization of the Euclidean and non-Euclidean motion groups in $R_n$* [1909E]).

The testimony of the students of Brouwer we have been able to interview (Heyting, Euwe and Struik), moreover, confirms that in his undergraduate courses Brouwer followed a strictly traditional line. His regular course ('Cycle') in 'Axiomatics' was recorded in detail by Heyting ([*Dictaat van Brouwer's Colleges in Axiomatiek*] in the Heyting Academic Estate, Mathematisch Instituut, Amsterdam University); it is a rigorous but orthodox treatment of axiomatic geometry without reference to foundational issues and without any attempt at 'intuitionist re-construction'.

An explanation of the continued use of 'old methods' in his original mathematical publications is given in 1921 when Brouwer claims:

> ... in my philosophy-free mathematical papers published in this period I have made regular use of the old methods, but trying as much as possible to derive only those results which could be expected to find a place, after the systematic construction of an intuitionist set-theory, in the new system, retaining their main significance may-be with some modification. ([1919D2], p.798).

### 5.12.2 The Formalization of Intuitionist Mathematics

Brouwer's views on a legitimate form and practice of formalization are clear from his attitude to Heyting's work: on the one hand showing an interest in, even encouraging and supporting the development of an Intuitionist axiomatization and formalization, and on the other hand staying aloof and dismissing it as a 'sterile', mathematically unproductive, exercise. In a recorded interview with van Dalen and myself Heyting gave an account of his emerging interest in Intuitionism, the circumstances surrounding his choice of research-topic, Brouwer's role and his reaction to the results.

In spite of Brouwer's presence, the Mathematics Department of Amsterdam University during the early twenties was not an academic centre promoting Intuitionism. There was no post-graduate programme of seminars and research in foundational studies, nor did Brouwer directly encourage his students to undertake research in intuitionist mathematics.*  When Heyting completed his undergraduate studies in 1922 and registered for a doctorate, the initiative of doing research in Intuitionist Mathematics came from him. The work for the dissertation was carried out in complete isolation (while Heyting was teaching full-time at a secondary school in Enschede, far away from Amsterdam) and without further help from Brouwer. But significantly, in view of our comment on Brouwer's attitude to axiomatics, the subject itself, *Intuitionist Axiomatics of Projective Geometry*, was suggested by Brouwer. Indeed, Brouwer seemed so keen on the subject that he made the same suggestion to someone else; Heyting recalls:

> ... I had read Brouwer's publications and began to take an interest in Intuitionism.
> First I went to see de Vries – no student dared to approach Brouwer – but de Vries referred me to Brouwer. Brouwer was rather surprised, especially because I had already done something on intuitionist lines. He then suggested that I try to 'make the axiomatics of projective geometry intuitionistic'.
> Typical of Brouwer: I had done a considerable amount of work all by myself, when one day I heard that Brouwer had given the same subject to someone else. I quickly wrote to him. He apparently thought that I had not done anything about it and had given up ... (Interview Heyting 24 September 1976).

In the Introduction of *Intuitionist Axiomatics of Projective Geometry* ([1925]) Heyting outlines his interpretation of 'the Intuitionist conception of mathematics' and an admissible form of axiomatics, the guiding principle in his work of 'bringing projective geometry into concord with the Intuitionist conception of Mathematics'. Setting aside other, metaphysical considerations he starts from the Brouwerian principle that 'one can only reason about a mathematical system after it has been thought, i.e. after one has constructed the system in one's mind.' (p.1). As to axiomatics, he takes the agreed Intuitionist view that it cannot 'serve as the foundation

---

* The research of the international following which Brouwer had gathered around himself, almost exclusively concerned topology; only few of his own students completed their doctoral studies with him (Heyting, Belinfante, Haalmeyer, Geldoff and de Loor).

of mathematics'; but he dissociates himself from Weyl's radical position of [1921] which rejects 'all general set-theory and general statements about functions' (p.70) and therefore denies axiomatics any meaningful place in mathematics. Indeed, the possibility of constructive generalization, Brouwer's principle of constructing species of higher order (cf. 6.3.6), is Heyting's justification of axiomatization and is its guiding principle:

> We take the view that Professor Brouwer's principle of *Begründung der Mengenlehre* I, p.4, by which from species of order $n$ a species A of order $n + 1$ is formed, has a clear meaning. Axiomatics is an application of this principle: the property defining the new species A is the existence of the relations between the elements of its elements as specified by the axioms. (HEYTING[1925], pp. 2-3).

In his disagreement with Weyl, Heyting concurs with Brouwer. To the passage quoted by Heyting (WEYL[1921], p.70) and the passage p.71 ('Arithmetic and Analysis contain only general statements about numbers and free-becoming sequences, no general theory of functions and sets of independent content.') Brouwer reacts in a private note:

> This restriction of mathematics to mathematical entities and species of the lowest order is totally unjustified. This is obviously a reference to p.70 where he [Weyl] dismisses my theory of species as meaningless, and it shows that in the end Weyl only half understands what Intuitionism is about.

The constructive principle of species was not used by Brouwer for the purpose of axiomatization nor indeed did he himself ever attempt to axiomatize Intuitionist Mathematics. HEYTING[1925] recognized the essential role of the species-construction in any axiomatization in line with Brouwer's philosophy of mathematics and language; its treatment is a model of such axiomatization. In the axiomatic 're-organization of the mathematical record' certain general statements may well become initial statements or axioms and others be derived from them, but as true mathematical statements they must all reflect preceding constructions, in the case of a general statement the construction of the 'common property' of 'elements previously acquired'.

Heyting's axiomatization in [1925] accordingly starts with 'the construction of the linear continuum', 'the coordinate projective plane' and 'coordinate projective space' ( I, §§1-4) before determining the axioms, or rather examining and adapting Pieri's axioms as given by Whitehead ([1913]):

> §5. Investigation of Pieri's axioms.

We shall now examine in how far the projective geometry as we have constructed above satisfies Pieri's axioms. (p.21)

Chapters II-IV are an axiomatic development of Intuitionist projective geometry based on Brouwer's characterization of continuity and order.

In the choice of axioms and the whole treatment the conceptual, the mathematical 'content', is the determining factor and the ultimate justification; the axiomatic method of HEYTING[1925] is 'inhaltlich' rather than 'formel'.

Brouwer approved; commenting on the first draft he writes: 'it is very much to my satisfaction.' (Brouwer to Heyting 26.06.24). He registered his approval of the final version by awarding Heyting the doctorate 'cum laude', and in the degree ceremony declaring the work to be 'a valuable contribution to Intuitionist mathematics, the mathematics of the future'.

### 5.12.3 Heyting's Formalization of Intuitionist Logic

Perhaps the main significance of Heyting's dissertation is its effect on Heyting himself, generating an intimate knowledge of intuitionist principles and issues and a life-long interest and commitment.

The first major step in Heyting's gradual move towards an independent interpretation of Intuitionism is his formalization of Intuitionist logic, *Die formalen Regeln der Intuitionistischen Logik* (HEYTING[1930A]). Fraenkel claims it to be 'the most decisive step since the establishment of neo-intuitionism in 1907' (FRAENKEL[1958], p.228). It is not within the scope of this book to evaluate Heyting's formalization or to analyse and survey post-Brouwer developments of Intuitionism and its formalization. [We have listed some relevant literature at the end of this section; for an up-to-date survey of formal Intuitionist mathematics we refer the reader to TROELSTRA AND VAN DALEN [1988].] Our concern is with the impact Heyting's formalization made on Brouwer and with his reaction. We therefore confine ourselves to aspects of the paper and to the circumstances which are relevant in this context.

Heyting's attempt at formalizing Intuitionist logic follows in the wake of the debate sparked off by the publication in 1923 of Brouwer's 'Calculus of Absurdities' (5.10.1). The 'debate' was conducted in the *Revue de Métaphysique et de Morale* and the *Bulletin de l'Academie Royale de Belgique* with contributions by Wavre ([1924], [1926A], [1926B]), Levy ([1926A], [1926B]), Barzin et Errera ([1927]), E. Borel ([1927]), Avsitidysky

([1927]), Khintchine ([1928] and Glivenko ([1928], [1929]). The most significant contributions were those of Glivenko, who made the first attempt at formalizing Intuitionist logic, stating and listing the axioms acceptable to Brouwer and proving some of Brouwer's results. In addition he proved: 1. the universal non-contradictority of the PEM ('... 'the proposition $p \vee \neg p$ is absurd' is absurd in the Brouwer sense'); 2. that absurdity provable in the classical sense, is also provable in the Brouwer sense.

[It is most unlikely that Brouwer or Heyting were aware of Kolmogorov's paper, published in Russian in 1925 ('On the Principle of the Excluded Middle') and its important contribution to the formalization of Intuitionist logic.]

Brouwer stayed aloof from the debate but took an obvious interest, flattered being the centre of controversy and being heralded by some as the revolutionary innovator of the century in the field of logic. (He asked the editors for special off-prints, and years later listed the contributions of Levy, Wavre and Borel among the five recommended papers as 'useful ... although rather ancient' preliminary reading for his paper at the Canadian Mathematical Congress. [Brouwer to Williams 23.07.53].)

The original version of Heyting's formalization was his prize-winning entry, *'Stones instead of Bread'*, in a competition of the Dutch Mathematical Society, organized by Mannoury. With his usual modesty Heyting recalls the events and describes his method of work:

> It was in response to Mannoury's competition.
> I find it really rather annoying that people keep going back to that paper. All I did was to work out an intuitionist formalization of logic, and I happened to be the first to do it.
> For this I made use of Russell's system, the only system available at the time... I had as yet no experience in that field. [Interview 24.09.76].

Brouwer's immediate reaction was pleasant surprise, whole-hearted approval and an offer of publication in the Mathematische Annalen:

> Dear Heyting,
> Your manuscript has been of the greatest interest to me. I am sorry you want me to return it so quickly...
> I have already such a high regard for your work that I invite you to prepare a German version (expanded rather than shortened). Perhaps you could then also make a sharp distinction between *original* symbols and those introduced by means of definition (as abbreviations of other symbolic

> expressions)... Perhaps in connection with §13 the notion 'law' could also be formalized. But these are minor points... [Brouwer to Heyting 17.07.28].

The 'expanded' and improved version was published in 1930 through Brouwer's good services, because of the 'Brouwer-Hilbert affaire' not in the *Mathematische Analen* but in the *Sitzungsberichte der preuszischen Akademie von Wissenschaften*.

Brouwer's approval of HEYTING[1930A and B] concerns the work as a whole, with its support and endorsement of Brouwerian philosophy and its expressed reservations as to the function and value of formalization. On the main issue, the role of formalization in creative mathematics, Brouwer remained consistent and uncompromising. Questioned on various occasions about Brouwer's reaction to his formalization Heyting invariably remembers it as 'negative' and dismissive:

> He always maintained that formalization is unproductive, a sterile exercise. He never changed his mind about that... [Interviews 1967 and 1976].

Heyting's own views on the function and value of formalization gradually moved from full endorsement of Brouwer's stand to a more positive appreciation of its advantages and even acceptance of its necessity:

> As to your question about Brouwer's attitude towards formalization, I would like to add that he always maintained that formalizing mathematics is unproductive since mathematics is constructing-in-the-mind, of which language, and therefore also a formal system, can only give inadequate representation. I have become more and more convinced that at least in the communication of mathematics formalization has its great advantages. From recent research into the notion of choice sequences it has become apparent that for any sufficiently clear representation formalization is even necessary. I refer here to the survey given by A.S. Troelstra in his Principles of Intuitionism. [Heyting to Van Stigt 29.10.69].
> (cf. also: 'Formalization is in fact no more than scientific text-research (Du. 'wetenschappelijk tekstonderzoek'), but it can and has already produced important results.' [Heyting to Van Stigt 14.10.71]

HEYTING [1930A] and [1930B], however, unequivocally conform to the Brouwer view that formalization is post-factum analysis and that no formal system can be an adequate tool in the creation and description of new mathematics:

> Intuitionist mathematics is thought-activity and all language, including

formal language, is only a means of communication. It is in principle impossible to construct a system of formulae that would be equivalent to Intuitionist mathematics, because the possibilities in human thinking cannot be reduced to a finite number of rules to be determined in advance. (Opening lines HEYTING[1930A]).

As to the formal system proposed in HEYTING[1930A], Brouwer accepted it as an adequate representation of his own reformed logic. His letter to Heyting of 17 July 1928, quoted above, suggests improvements on only a 'few minor points'. Reminded of his promise to round off the debate on the 'Brouwer Logic' in an article ('setting out my views on the publications about intuitionist logic', [Brouwer to de Donder 13.06.30]), Brouwer refers the Editor for an authorative representation to Heyting's papers:

> While preparing a paper on Intuitionism for the Bulletin of the Académie Royale de Belgique I was pleasantly surprised to find that one has already appeared from the hand of my pupil Heyting [HEYTING[1930C]] which clarifies in a masterly way all the points which I had wanted to clear up. I believe that after Heyting's paper there remains nothing more to be said on the questions under discussion. [Brouwer to de Donder 09.10.30].

We have already commented on the significance of Heyting's contributions in changing the direction of Intuitionism (5.11.2). The more immediate effect of the publication of his Formalization was better and wider understanding of the claims of Intuitionism; it made Intuitionism accessible to a wider mathematical audience, and this also by the simple use of symbolic notation as distinct from words and ordinary language. Although a revolutionary departure from Brouwer's practice, the use of symbolic notation in expressing mathematical and logical constructions is not a matter on which Brouwer and Heyting disagree in principle. Both agree that between ordinary language and symbolic notation there is no essential difference: In 1907 Brouwer writes ' Like all mathematical language, this language [of classical logic] can also be condensed to symbols without any difficulty.' ([1907], p.159); and Heyting in the introduction of [1930A] quoted above attributes to 'formal language' the same limitations as to 'ordinary language'. But whereas Brouwer remains reluctant to adopt symbolic notation, Heyting recognizes its advantages:

> The attempt to express the most important parts of mathematics in formal language is therefore only justified by its greater conciseness and precision when compared to ordinary language, properties which make it a particularly suitable medium in helping to understand the intuitionist notions and their use . ([1930A], p.42)

Heyting's trust in the greater precision of symbolic notation was not shared by Brouwer, not even during the years of Signific belief in the possibility of improving the instrument of language. Much of the alleged incomprehensibility of Brouwer's writings is due to his refusal to conform to the common practice of using logical and set-theoretical symbols. His use of symbolic notation remains confined to traditional mathematical operations and relations (the elaborate system of notation introduced in his new analysis is restricted to relations of order and denumerability).

Brouwer explained his opposition to symbolic notation, according to Heyting, by stressing that expression of a mental construction would be further complicated if first an explanation had to be given of the agreed symbol; moreover, that this explanation had to be given in ordinary language. (Heyting Interview 1970).

No doubt, association of symbolic notation with Logicism and Formalism has also been a contributory factor to Brouwer's dislike; symbols more than ordinary language can be used as primitive constructs without any prior meaning. But perhaps the most fundamental reason for Brouwer's reluctance to use symbolic notation is an a-priori unwillingness to accept the possibility of precise formulation of mental constructions and the possibility of producing mathematical truths mechanically by means of a logical calculus. In contrast with living language, in which words have various connotations and shades of meaning, formal symbols are presumed to be precise. Paradoxically, because of the very vagueness of living language, Brouwer claims in *Life Art and Mysticism*, truth can be found through language and particularly through poetry and other works of art. The power of living language to convey truth transcends the immediate precise meaning of words or symbols.

In [1933] the emphasis is on the dangers of 'linguistic mechanisms' and the fallacy of 'the belief in the possibility of ensuring the total exactness of pure mathematics by making the language of mathematics purely mechanical.' ([1933], p.54).

In the mildness of his declining years Brouwer could speak of the 'sumptuous symbolic logic ' as 'a thing of exceptional beauty and harmony' but only 'when cultivated for its own sake', and still maintaining his reservations as to 'its applicability to mathematics'. In the final lines of his paper for the Irish Academy, Brouwer's last published paper, Boole is obviously singled out because of his contribution to symbolic logic. No doubt, 'the most captivating problems and surprising discoveries' referred-to include those raised by Intuitionist logicians; some reference to Heyting as 'the originator of all this' would not have been amiss.

> Fortunately, classical algebra of logic has its merits quite apart from the question of its applicability to mathematics. Not only as a formal image of the technique of common-sensical thinking has it reached a high degree of perfection, but also in itself, as an edifice of thought, it is a thing of exceptional harmony and beauty. Indeed, its successor, the sumptuous symbolic logic of the twentieth century which at present is continually raising the most captivating problems and making the most surprising and penetrating discoveries, likewise is for a great part cultivated for its own sake. Don't let us forget that it is Boole who has been the originator of all this. ([1955], p.116).

## Selected Contributions to the Formalization of Intuitionist Logic and Mathematics:

D. BRIDGES and E. BISHOP
[1985]    *Constructive Analysis.* (Springer Verlag, Berlin)

M. DUMMET
[1977]    *Elements of Intuitionism.* (Clarendon Press, Oxford)

A. HEYTING
[1930A]   Die formale Regeln der intuitionistische Logik. *Sitzungsberichte der preuszischen Akademie von Wissenschaften*, phys. math. Kl. 1930, pp. 42-56.
[1930B]   Die formale Regeln der intuitionistische Mathematik. *Sitzungsberichte der preuszischen Akademie von Wissenschaften*, phys. math. Kl. 1930, pp. 57-71, pp. 158-169.
[1930C]   Sur la logique intuitionniste. *Académie Royale de Belgique. Bulletin.* vol.16, pp. 957-963.
[1955]    *Les Fondements des Mathématiques, Intuitionisme, Théorie de la démonstration.* (Gautiers-Villars, Paris).
[1956]    *Intuitionism, an introduction.* (North-Holland, Amsterdam).
[1961]    Axiomatic Method and Intuitionism. In: *Essays on the Foundations of Mathematics dedicated to A.A.Fraenkel.* Jerusalem, 1961.

C. KLEENE and R. E. VESLEY
[1965]    *The Foundations of Intuitionistic Mathematics.* (North-Holland, Amsterdam).

A. N. KOLMOGOROV
[1925]     On the Principle of the Excluded Middle. Mame 32, pp. 646-667; [English translation in J. van Heyenoort, *From Frege to Gödel*. HUP, Cambridge Mass., 1967.]

G. KREISEL and A. S. TROELSTRA
[1970]     Formal systems for some branches of intuitionistic analysis. *Annals of Mathematical Logic* vol.1, pp. 229-387.

J. MYHILL
[1967]     Notes towards an axiomatization of Intuitionistic Analysis. *Logique et Analyse* vol. 35, pp. 280-297.
[1968]     Formal systems of Intuitionistic Analysis. In: *Logic, Methodology and Philosophy of Science* III (ed. B. van Rootselaar and J.F.Staal). (North-Holland, Amsterdam).

A. S. TROELSTRA
[1969]     *Principles of Intuitionism*. (Springer Verlag, Berlin).
[1977A]    *The Theory of Choice Sequences*. (Clarendon Press, Oxford).
[1980]     Intuitionistic extension of the reals. *Nieuw Archief voor Wiskunde*. (3) 28, pp. 63-113.
[1983A]    Analysing choice sequences. *Journal of Philosophical Logic*. vol.12, pp. 197-260.

A. S. TROELSTRA and D. VAN DALEN
[1988]     *Constructivism in Mathematics*. (North-Holland, Amsterdam).

D. VAN DALEN
[1978]     An interpretation of intuitionistic analysis. *Annals of Mathematical Logic*. vol.13, pp.1-43.

# CHAPTER VI
# BROUWER'S INTUITIONIST MATHEMATICS

## 6.1 Introduction

As an 'Intuitionist Manifesto' Brouwer's doctoral thesis is complete: it states his ideology, develops a coherent philosophy of mathematics and spells out the fundamental principles for reform of mathematical practice. However, the full implications of his constructive philosophy for the whole of mathematics were not immediately clear. Criticism in the thesis concentrates on contemporary trends in foundation theory, in particular logicism and on the excesses of Cantor's set theory. Classical mathematics is hardly affected. Brouwer's own mathematical work during the following years, his creation of modern topology, was built on the classical tradition and on Cantorian point-set theory. In a marginal note to his 1915 course on set theory he admits using even the Principle of the Excluded Middle, branded in 1908 as 'unreliable' ('...in my own work I have also often used the Principle of the Excluded Middle' ([BMS15], p.1). [1907] and [1908A] make some far-reaching observations about constructivity of sets and distinguish between 'mathematical sets' – discrete denumerable systems constructed on the basis of the Primordial Intuition – and 'mere methods', yet the point-sets of Brouwer's topological work in the years following are spatial manifolds, given or defined by property, and many of the tools used are the fundamental notions of point-sets in the style of Cantor and Schoenflies.

Not until about 1914 did Brouwer show awareness of the need for fullscale reform of classical mathematics, the 're-construction of mathematics based on Intuitionist principles'. The history of Brouwer's innovations often follows a pattern of critical investigation of contemporary authorities, identifying flaws and unproven conjectures and searching for new concepts and methods which avoid those pitfalls (cf. JOHNSON[1979]). The discovery of 'flaws' in the work of others leads to awareness of implications for his own work. Submitting his criticism of Schoenfliesian topology, (*Zur Analysis Situs*, [1910D]) Brouwer writes to Hilbert:

Last winter, when I had already completed my second contribution on

> finite continuous groups ready for submission to the *Mathematische Annalen*, I suddenly discovered that the Schoenflies investigations of the Analysis Situs of the plane, on which I had so fully relied, cannot be upheld in all their detail; this also called into question my own group-theoretical results.
> To clear up this matter I had to subject the relevant parts of Schoenflies's theory to a thorough examination and ascertain exactly which parts of his results we can trust and use in further construction. (Brouwer to Hilbert, 14.05.09, [DHI1]).

Again, a few years later, Brouwer's 'editorial' involvement in the re-edition of Schoenflies' *Bericht* alerted him to what he saw as fundamental deficiencies in set theory. After two frustrating years of attempting to correct and improve this 'standard work on set theory' he admits that it fails as a coherent theory based on and developing from one 'philosophische Grundanschauung' ([1914], p.78). Zermelo's axiomatic approach to the foundations of set theory had already been rejected in [1912A] for its complete reliance on language and presupposition of a Platonic universe of sets, partitioned on the basis of an axiom of comprehension. Brouwer's own theory of sets and the continuum sofar, although it did not compromise his constructive principles, proved inadequate as the foundation of major parts of mathematics. The notion of set as the activity and result of thinking-together discrete mathematical entities, was sufficient for the construction of the whole of 'separable mathematics'. By refusing to consider the continuum as a totality of points, Brouwer had been able to uphold this strict constructivity of sets. However, his characterization of the continuum as 'simply given and 'overlaid' on a heterogeneous framework of rationals was not a suitable basis of point-set topology, nor of analysis and function-theory. [In his 1916-Notes on Sets, Brouwer had to admit: 'A point-set is in fact a set of points defined by comprehension...' ([BMS15A], p.23)]

The *foundation of set-theory* became Brouwer's main preoccupation during the following years (1915-1917): the creation of a theory based on a concept of set which would accommodate the continuum and yet preserve the essential mathematical characteristic of being or having been constructed in time. His search for a new set-concept led to the introduction in 1917 of the notions of *spread* and *species*, dramatically described in later historical surveys as 'the Second Act of Intuitionism'; it is based on the 'new insight', the new 'recognition of Intuitionism' that free-choice sequences are legitimate mathematical objects and characterize the elements of the continuum. The proposed concepts of species and spread – still referred to as 'set' (Menge) – are a radical departure from the classical notion of set and supplement Brouwer's early restricted notion of 'mathematical set', in particular the notion of spread whose elements include free-choice sequences. As all Brouwer's intuitionist reforms, these new concepts were presented not as optional alternatives, but as *the* and only legitimate set-constructions in the treatment of the continuum. They were more than subtle philosophical interpretations justifying past practice; their adoption as the fundamental concepts of the mathematical continuum required complete and 'systematic reconstruction of set theory' and mathematical analysis. The need for mathematical reform was the more urgent as Brouwer grew more aware of the implications of his rejection of the Principle of the Excluded Middle. At the same time there was a growing confidence that the traditional instrument of expression and communication was capable of improvement.

The programme of systematic reconstruction of the whole of mathematics became part of Brouwer's grand design of 'intuitive-signific reform'. It was launched in 1917 with a paper presented to the Dutch Royal Academy and published in 1918 under the title *Foundations of Set-theory independent of the logical principle of the excluded middle* ([1918B] and [1919A]).

Part I, *General Set Theory*, introduces the basic notions of Brouwerian set theory: spread, species and their properties, cardinality and order.

Part II, *The theory of point-sets* applies the new set theory to Euclidean space and develops an intuitionist theory of point-sets with its own interpretation of continuum, interval, domain, content and measure.

The element of indeterminacy resulting from free-choice enabled Brouwer to give a 'natural' characterization of the continuum and provide proof of its global properties such as its non-denumerability. However, the essentially unfinished nature of free-choice sequences poses fundamental problems when used as elements of further construction, in particular in the definition of functions. Brouwer tried to solve these problems by distinguishing between the choice sequence proceeding with certain freedom ad infinitum and its constructed initial segment, and ensuring that analysis of functions and their properties relies wholly on the determined initial segments of the sequences generating real numbers. In a number of papers from 1923 onwards he developed an intuitionist function-theory based on these notions and principles.

The first paper, *The Foundations of the Theory of Functions independent of the logical principle of the excluded middle, Part One* [1923A], starts with the fundamental hypothesis that every function defined on the closed unit-interval is uniformly continuous, Brouwer's well-known Uniform Continuity Theorem. It is presented as following directly from the definitions of 'full function' and continuity. In [1924D] a proof is given, based on the Fundamental Theorem of Finite Sets ('finite Mengen'), later referred to as the 'Fan Theorem' ('If to every element $e$ of a finite set $M$ a natural number $\beta_e$ can be assigned then a natural number $z$ can be determined such that for every $e$, $\beta_e$ is completely determined by the first $z$ choices generating $e$.' ([1924D], p.192; see further 2.11.1 and 6.3.13). In one of his post-war lectures Brouwer speaks of it as: '... a wonderful theorem whose importance would justify to call it the fundamental theorem of intuitionism...' ([BMS66], App.10), the key to a simplified and elegant alternative to the classical theory of functions. Brouwer never succeeded in giving a direct proof of the Fan Theorem; in [1924D] and elsewhere it is a corollary of a more general hypothesis of securability, the 'Bar Theorem'. There is an element of regret in the quoted passage of [BMS66], which continues:

> ...the fundamental theorem of intuitionism, but whose absolutely rigorous proof till now has not been sufficiently simplified to allow it being reported here. So I shall only give its wording....

It is arguable whether failure to provide a satisfactory, 'simplified' proof of the Fundamental Theorem was the main cause of the loss of self-confidence which led Brouwer to abandon his programme of re-constructing mathematics. Brouwer set himself high standards, simplicity and elegance were the hall-mark of his topological work, as Hilbert testified*. Dissatisfaction with this particular solution no doubt was a contributory factor to Brouwer's loss of confidence, and the Fundamental Theorem marks the premature end of Brouwer's re-construction of Intuitionist analysis. This is reflected in the unpublished *Berliner Gastvorlesungen* and in particular in the uncompleted *Real Functions*; the manuscript ([BMS37]) ends abruptly after the introductory chapters on point-sets, functions and continuity and a proof of the Fundamental theorem based on the hypothesis of the Bar Theorem. The 'Table of Contents' attached

---

* e.g. Hilbert's letter to Korteweg of 06.02.11 [DJK79]: 'It is part of his character never to be satisfied with easy results...he always starts with specially hard and deep problems and does not stop until he has succeeded in finding a solution which completely satisfies him.'

to the manuscript (see app.11) summarizes the progress made in the programme of re-construction.

*Brouwer's Intuitionist Mathematics* forms a relatively modest part of the systematic re-construction of the whole of mathematics. Its significance lies in the insight it gives into what Brouwer describes as 'the Genesis of Mathematics': the constructive development of mathematical concepts in accordance with his Intuitionist philosophy. It shows how the fundamental intuitions determine concepts in further construction, and in turn, how all mathematical concepts and methods are constructively derived from Intuition. In his Intuitionist Mathematics Brouwer has put his philosophical theory into practice and so provided a model for all intuitionist mathematics or, in the preferred terminology of his later years, for 'modern mathematics'.

This chapter is intended as an introduction to some of the 'beginnings' of Brouwerian mathematics, in particular those which illustrate clearly the 'genetic' development of concepts or thought-constructions from fundamental intuitions and those which deviate most from the classical tradition. In no way does it claim to present a comprehensive survey of Brouwer's Intuitionist Mathematics.

[For further reading we refer the interested reader to Brouwer's own publications and his post-war lectures, some of which are included in the appendices to this monograph; we particularly recommend van Dalen's annotated publication of *Brouwer's Cambridge Lectures* (VAN DALEN [1981]. For post-Brouwer developments of Intuitionist-constructive mathematics consult the Bibliography entries under Bishop, Dummett, Heyting, Kleene and Vesley, Kreisel, Posy, Troelstra, van Dalen, van Rootselaar. For a comprehensive introduction to Intuitionist Mathematics to-date we recommend TROELSTRA AND VAN DALEN [1988] *Constructivism in Mathematics*.]

We start from the view that Brouwer's fundamental claim as to the nature and origin of mathematics should be taken seriously, in particular his consistent and ever repeated claims that mathematics is thought-construction on the basis of the Primordial Intuition alone. Analysis of his concept of Intuition (cf. 4.1) and his mathematical constructions reveals that these claims can be upheld; moreover, that the 'intuitive origin', rather than a spurious, post-factum justification, is the natural source of his new constructions and his new interpretation of classical concepts.

## 6.2 Brouwerian Separable Mathematics

Accounts of Intuitionist mathematics tend to concentrate on the implications of Intuitionist logic and proof-theory and on the new analysis resulting from the admission of free-choice sequences. Little or no attention is given to the Intuitionist construction of what Brouwer refers to as 'separable mathematics', the mathematics of discrete systems. Brouwer's own contribution consisted of a philosophical analysis of the human thought-processes generating discreteness, number and operations, and of an outline of the construction of discrete systems given in the first chapter of his dissertation. Belief in the 'Primordial Intuition' [PI] as the sufficient and necessary foundation of the whole of separable mathematics was re-affirmed in papers and lectures throughout his life; but no attempt was made at a systematic and detailed construction of discrete systems from 'the elements of Intuition'.

Comments in later years seem to suggest that one of Brouwer's reasons for this omission was the relative simplicity of such an exercise:

> It is introspectively realized, how this basic operation, continually displaying unaltered retention by memory succesively generates each natural number, the infinite sequence of all natural numbers, arbitrary finite sums of mathematical systems previously acquired, sums of fundamental sequences of mathematical systems previously acquired, finally a continually extending stock of mathematical systems corresponding to the 'separable' systems of classical mathematics. ([BMS49], p.4; App.7).

Moreover, the results of an Intuitionist construction of separable mathematics appear to be not very different from those of classical mathematics:

> It is introspectively realized how much of [classical Ed.] separable mathematics can be re-built in a slightly modified form by unlimited self-unfolding of the basic intuition.' ([1952B], p.141)

The lack of practical differences between the Brouwerian and the classical theorems of natural numbers, integers and rationals led post-Brouwer Intuitionists to dispense with the 'Primordial Intuition' as the the fundamental set-construction which generates the natural numbers and determines the nature and properties of number. Heyting, Bishop and their followers take the Kronecker view:

> Mathematics begins after the concepts of natural numbers and equality between natural numbers have been formed. (HEYTING[1956], p.15; cf. also BISHOP[1967], p.vii).

Troelstra dismisses the notion and theory of natural numbers, integers and rationals as 'unproblematic' (TROELSTRA[1977]).

Instead of probing deeper into the thought-process in search for the conception of number and its embryonic development, Heyting considers the natural number after its birth, complete with all its properties (and fully dressed):

> The notion of natural numbers does not come to us as a bare notion, but from the beginning it is clothed in properties which I can detect by simple examination. (HEYTING[1956], p.13).

Brouwer whole heartedly accepts the conceptual as the natural, 'home' domain of mathematics and conducts his search for the genesis of number and its properties in the processes of human reasoning. This 'conceptualism' sets him apart from other constructivists and intuitionists, even from Poincaré, for whom concepts and conceptual analysis are preliminaries in the process of mathematical creation. The Brouwer mathematical entity not only requires thought-activity as a preliminary, it *is* the constructive act of reasoning. Practising mathematics true to its nature, the Subject is naturally guided in his choice of constructions and their combinations, and safeguarded from error and irrelevance.

In Brouwer's interpretation of the conceptual genesis of mathematics the Primordial Intuition is identified as the most primitive act of mental construction; it not only generates and justifies number and each of its properties, it is the fundamental and characteristic construction of sets and of every mathematical system.

*6.2.1 The Primordial Intuition as Set-construction*

Brouwer's standard account of the primordial happening, generally criticized as metaphysical and mystical and disowned by Heyting and other neo-intuitionists as mathematically irrelevant, yet provides the basic ingredients for a constructive development of mathematics consistent with the Intuitionist interpretation of mathematics as thought-activity.

The 'genetic' analysis of the number-concept as a mental act linking past and present sensations seems close to Cantor's definition of set as: 'Any comprehension [Germ.: Zusammenfassung] into a whole $M$ of definite and separate objects $m$ of our intuition or our thought.' (CANTOR[1895], p.486). The most striking difference perhaps lies in the consistency of these definitions with the respective philosophies and mathematics of their authors. Cantor's 1895 characterization of sets conflicts with his Platonic view of the mathematical universe and would need drastic amendment to accommodate his new mathematical discoveries; to accommodate his infinity and transfinite, he had to compromise the human aspect of the comprehending intelligence and of the objects of mathematics and resort to a divine intelligence or 'absolute' as the generator and seat of mathematics. (cf. HALLETT[1984], 1.3-1.4). Brouwer's primordial set-construction is wholly in line with his philosophy of man and mathematics and remains the fundamental construction for the whole of his Intuitionist mathematics.

In accordance with the Brouwerian philosophy of mathematics as thought-activity the Primordial Intuition is an *act*; its only ingredients are the a-priori elements of all activity: the Subject and the dynamic time-continuum. The human nature of the Subject is recognized: mathematical activity, the life of the Subject-mathematician, proceeds along a half-open interval of time. The Primordial Intuition marks a definite beginning, it is the first act and has no mathematical antecedents. Unlike Cantor's definition of sets, which presupposes a manifold of things perceived as different (and Kant's synthesis of apprehension, 'the act of running through the manifold and then hold it together'), Brouwer's PI acknowledges the perception of 'difference' to be a binary relation, a vital element of which is the linking act of the Subject. Indeed, the concepts of two-ity, discreteness and union are so interdependent and equally indispensable that they can only be created in one single primordial act:

> ... a single a-priori Primordial Intuition which can be described as constancy in change or unity in diversity...the first constructive act which is the thinking-together of two discrete things. ([1907], p.179).

The standard account of the PI describes discreteness as generated by an act of the Subject 'splitting', effecting 'the falling apart of a life moment'. The ultimate and only a-priori of the perception of discreteness and two-ity is the time-existence of the Subject. Moreover, the distinctness of the 'parts' is perceived as 'past and present'; the maximal abstraction of Brouwer's genetic analysis results in the perception of 'before-after'. Although not explicitly mentioned among 'the irreducible elements of the Primordial Intuition', *order* is an intrinsic constituent of discreteness and the PI. The concept generated by the first constructive act has all the characteristics of the ordered pair. Indeed, the most fundamental mathematical concept, 'two-ity' or 'two-oneness' is the pure form of the ordered pair. Brouwer speaks of the: 'the empty form of the common substratum of all two-ities... this common substratum, this empty form which is the basic intuition of all mathematics.' ([1952B], p.141, et al.). It is an empty form also because in the PI there are as yet no mathematical entities 'previously acquired'. He remains vague about the nature of the empty form and does not himself refer to two-ity as ordered pair. His account of the primordial happening, however, supports this interpretation: 'two-ity' is one single, and the first mathematical construct; it consists in the order-relation abstracted from discreteness-in-time. In the pure empty form generated in the primordial happening the only individualizing characteristic of its members is their mutual relation of priority. Two-ity is : '... the two-membered sequence.' ([1933], p.45 ).

The numbers two and one emerge directly from the primordial happening, although Brouwer is not clear about the precise aspect of two-oneness with which each is to be identified. (see further 6.2.2). Thereafter the PI serves as the prototype of all further mathematical construction; applied to mathematical entities or systems 'previously acquired' it generates the ordinals and the whole of separable mathematics. In turn, each mathematical construct carries the genetic characteristics of 'two-ity', the essence of its generating construction:

> The basic operation of mathematical construction is the mental creation of two-ity of mathematical systems previously acquired and the consideration of this two-ity as a new mathematical system. ([BMS49], p.4; App. 7)

The Primordial Intuition, established as the justification of mathematics based on the sole a-priori of time, and as the characteristic construction of mathematics, becomes the criterion of mathematical legitimacy:

> From the present intuitionist point of view all mathematical sets deserving of this name, are those that can be constructed from the Primordial Intuition. ([1912A], p.13).

This remained Brouwer's consistent and most fundamental principle. It has been argued that his analysis of the Primordial Intuition may well provide a plausible philosophical justification of the construction of sets and discrete systems but adds little new to general constructive practice (as was indeed conceded by Brouwer; see quotation 6.2.1). To him, however, the conceptual limits of the PI constitute a more acceptable criterion of mathematical legitimacy than the arbitrary alternative of axiomatic systems. Further reflection and deeper understanding of the PI allowed him to widen the scope of legitimate set-construction and inspired the new concepts of his Second Act of Intuitionism.

Brouwer's pre-1916 definition of sets is restricted to separable systems and it carries the added condition that the construction be 'thought of as completed' (Du. 'af'; cf.[1907], p.10; p.148 ff.). Systems that cannot be so constructed are rejected as non-mathematical:

> It is wrong to call such a system a mathematical set, because it is not possible to construct it as completed from the mathematical Primordial Intuition. ([1908A], p.570).

Certain systems are disqualified as mathematical sets because they cannot be constructed as completed totalities, other classical sets lack any constructive basis and cannot even claim to be a mathematical system in a wider sense. Examples of the former are: the totality of definable points of the continuum, the totality of numbers of the second number-class... they are denumerably unfinished,... a *method* not a set. ([1908A], pp. 570-571).

The continuum is intuited as 'complete'; points of the continuum can be constructed into a system which is 'complete as a matrix, not as a set.' (ibid.).

[1914] introduces the notion of *well-constructed sets*. It is a first attempt to incorporate the continuous in a broader intuitionist set-concept and give an intuitionist interpretation to the 'perfect set' of his point-set topology ([1910B] and 1911B]). The well-constructed infinite set consists of a denumerable part – a fundamental sequence – and a non-denumerable part. The admission of 'choice' as the generator of elements of the non-denumerable part adds a new dimension to the construction of sets; in its sequential generation and in the domain of its choice, however, it still is based on the PI. (see further 6.3.4).

Before 1916 Brouwer's opposition to what he calls 'predicative definition' of mathematical sets or 'definition by comprehension', is absolute. Cantor's naive principle 'that every conceivable property creates a set' and definition of sets by axioms are rejected on the grounds of the mathematical irrelevance of language and logic and the presupposition of a Platonic universe complete with every conceivable set-partition:

> They all start from the pre-supposition of a world existing independently of the thinking individual and obeying the laws of classical logic, a world of mathematical things which could among themselves have the relation of a set to its elements...

Application of the axiom of comprehension leads the formalist to introduce various notions which are totally meaningless to the intuitionist such as 'the set whose elements are the points of space', the set whose elements are 'the continuous functions of one variable', 'the set whose elements are the discontinuous functions of one variable' etc. ([1912A], pp. 14-17).

Brouwer's *1916 Lecture Notes* mark a modification of his views on sets and hint at the possibility of a constructive interpretation of property, the notion of 'species' to be introduced in 1917 as part of the 'Second Act of Intuitionism'. In the margin he writes:

> Apart from these [mathematical sets] one could adopt the notion of predicative sets, defined by comprehension, but it is better to refer to them as species. ([BMS15A], p.1).

But again, both in its genesis as 'subconstruction' and in its attribution to 'mathematical entities previously acquired' the property or species is based on the Primordial Intuition, a 'modality of the Primordial Intuition'. The Second Act of Intuitionism is described as:

> The admission as modality of the self-unfolding of the Primordial Intuition of sequences of mathematical systems proceeding in complete freedom... the admission also, at every stage of the construction of mathematics, of properties to be attributed to mathematical thought-constructions already acquired as new mathematical thought-constructions under the name species. ([BMS32], p.7).

## 6.2.2 The Fundamental Sequence

In tracing the genesis of number to the Primordial Intuition of 'Two-ity' exclusively on the basis of time-awareness Brouwer made a profound and original contribution to solving the mystery of the concept of 'natural numbers'. Simple characterization of number in its construction as 'ordered pair', however, raises a number of questions such as e.g. to the nature of the number *one* and even the number *two*. Nor does Brouwer appear to be wholly consistent in his accounts of the construction of natural numbers. In [1908A] each number is clearly identified with the ordered pair constructed on its predecessor:

> ...the construction of the order-type $\omega$: if one thinks of the whole of the primordial Intuition as a new first, one can join in thought a new second, and this is called 'three' etc. (p.569).

[1912A] describes the construction of the number *three* as a process of splitting either element of two-ity and generating a three-membered sequence:

> This intuition of two-oneness not only creates the numbers *one* and *two* but also all finite ordinal numbers, since one of the elements of the two-oneness can be thought of as a new two-oneness and this process can be repeated as many times as one wishes... (p.12).

[1933] refers to the process at the level of 'mathematical viewing', generating 'temporal three-ity' by splitting one of the members of 'temporal two-ity':

> Of this temporal two-ity, born out of time-awareness, or this two membered time-sequence of phenomena, one of the elements can in turn and in the same way fall apart into two parts; in this way temporal three-ity or the three-element time sequence is born. (p.45).

[This process is recognized to be distinct from the pure-mathematical, i.e. 'achieved through mathematical abstraction which divests two-ity from all content, leaving only its empty form as the common substratum of all two-ities... which through self-unfolding produces first all the natural numbers...'. Unfortunately, Brouwers adds: 'we shall not concern ourselves here further with the manner of this construction.' ([1933], p.48).] [1954A] reverts to the original account, each natural number is 'successively generated' by means of 'the basic operation of mathematical construction, the mental creation of the two-ity of two mathematical systems previously acquired and the consideration of this two-ity as a new mathematical system.' (p.4; cf. also [BMS49], p.4).

### 6.2.2.1 The Order-type n

A coherent Brouwerian theory of ordinal numbers emerges from these cryptic accounts if the various notions introduced are identified and clear distinctions made. I suggest to distinguish between:
1. The Brouwer successor-construction or two-ity, the basic operation and constructive form of each ordinal number;
2. The whole of the finite sequence resulting from successive application of the basic operation, the 'empty form' of the ordered $n$-tuple, the Brouwer order-type;
3. The relative order-position of each of the components of the sequence: the pure ordinal.

The Brouwer successor-construction, the basic operation, is the construction of the ordered pair. If number is identified with the 'empty form', two is genetically the first number: the empty form of the ordered pair emerging from the Primordial Intuition. Subsequent application of the successor construction, using the so-created number two as a 'new first' and 'joining-in thought a new second' creates the number three. All further ordinals are generated in a similar way, each identified with the empty form of a sequence of nested ordered pairs.

The empty form of the ordered $n$-tuple abstracts from this binary composition and allows an extension by analogy to include the 'one-membered' sequence. [1933] identifies number, 'two-ity' and 'three-ity' with the form of the 'two-element' resp. 'three-element sequence' and similary any number, $n$-ity, with the empty form of the $n$-element sequence: 'sequences of any multiplicity created by the self-unfolding of the PI.' (p.46).

This notion of ordinal number seems close to that of Cantor's order-type, and in his earlier work Brouwer adopts this terminlogy. His accounts, however, place the origin of number in time-awareness, number is ultimately and exclusively abstracted from 'temporal sequences of phenomena', mathematical two-ity is 'the common substratum of all temporal two-ities'. Number, order-type, is subsequently related to other ordered mathematical systems and as such may be referred to as 'property of' or dimension of the 'ordered set' (cf. [1918B], p.14 where species are said 'to posses the ordinal number...').

Brouwer's order-type is the empty form of the whole sequence, considered as an ordered $n$-tuple or as nested ordered pairs. As such it is distinct from the relative order position of the elements of the sequence, which is genetically the more primitive aspect of the Brouwer ordinal number. The creation of 'two-ity' is the *primordial* intuition; yet Brouwer recognizes 'constructive elements which can be read off from the primordial intuition... elements which in the system of definitions must remain underivable.' ([1907], p.180). The mathematical abstraction of time-awareness in its must fundamental form of before-after is the relative order-position expressed by 'being-first' and 'being-second'. In [1908A] 'first' and 'second' are constructive elements in the genesis of 'two-ity' and clearly distinct from 'two-ity', 'the whole Primordial Intuition'. The primordial happening is described as: 'a *second* is being thought not on its own but with the *first* being held in memory. The first and the second are held together...' (p.569); and in the construction of three-ity 'the whole Primordial Intuition is thought-of as a new first' (ibid.), i.e. identified with the relative position of an element of the sequence.

'Ordinal' in this sense is the most primitive aspect of number. It is essentially nothing but the order-relation itself and we could speak of it as the *pure ordinal*.

As 'pure-ordinal' the number *one* has a definite interpretation in terms of the PI: it is the mathematical abstraction of 'before' in the PI, being-the-first-element of the primordial form of the ordered-pair. Circularity is avoided by considering the whole primordial happening as one single primitive process.

Similarly the pure-ordinal *two* is the abstraction of 'after' in the PI, the position of being the second element in the primordial form of the ordered pair.

In the form of sequences further constructed by means of the PI, *first* or *one* is the mathematical abstraction of 'before every other'; *second* or *two* the mathematical abstraction of the position 'after one' but 'before every other'; etc. These individual order-relations can be represented by words: first, second, third... or one, two, three..., or by symbols : 1, 2, 3,..., or by the conventional positional notation from left to right; the generalized relations of 'first-to' or 'second-to' resp. by the symbols '$<$' and '$>$'. (Cf. Brouwer's use of the terms 'left' and 'right' or 'higher in the sequence' rather than 'smaller' and 'greater' with their implication of magnitude.)

The interpretation of the order-relation among the ordinals as Brouwer order-type (i.e. the form of the whole sequence) is more complex and requires the notion of one-to-one correspondence. Between two successive numbers the relation could be interpreted directly in terms of the Brouwer successor-construction: the successor of $m$ is the ordered pair of which $m$ is the new 'first'. A more general order-relation including those between non-successive numbers relies on the notion of 'part-of' or 'sub-sequence' and requires the one-to-one correspondence construction (see further 6.2.3). A simpler interpretation of order-relation between order-types uses the order-relation of the pure-ordinals and the definition of ordered $n$-tuple as the sequence in which the $n$-th is the 'last' or highest position (e.g. in the ordered 5-tuple, the order-type 5, the fifth position or 5 in the ordering of the pure-ordinals is the last or greatest.)

The pure ordinals, the order positions of the members of the ordered $n$-tuple, and their verbal and symbolic representations form a natural sequence: first, second, third,...$n$-th and 1, 2, 3,...$n$; similarly, the finite order-types and their representations: one, two, three, ...$n$ and 1, 2, 3, ... $n$. Each of these sequences are again of order-type $n$.

### 6.2.2.2 *The Order-type* ω

The extension of the finite order-type $n$ to the infinite order-type $\omega$ is possible by a combination of the successor-construction and another, named 'element of the PI which must remain underivable': the 'and-so-forth', the freedom of the Subject, his ability to continue the construction infinitely, i.e. without restriction (in Dutch Brouwer uses the word 'onbepaald', literally 'without fence or barrier'; in the German [1918B] and [1925A]: 'unbegrenzt'; in later English papers: 'infinitely proceeding' e.g. [1952B], [1954A], [BMS49]; see further 6.3.1).

The notion of the open-ended sequence is fundamental in the analysis of the continuum and characterization of real number. Its abstraction, the empty form of the infinitely proceeding sequence, is the order-type $\omega$, constructed by means of the primordial operation and the 'and-so-forth'. The finite order-types arranged in order form the infinitely proceeding sequence, the ordered $\omega$-tuple, of natural numers: 'the order-type $\omega$ of the sequence of the finite order-types...'([1907], p. 144).

Surprisingly, in view of his deep concern with the nature and genesis of number, Brouwer shows little interest in the further treatment of natural numbers and number-theory. The continuum is uppermost in his mind and is the focus of his treatment of number right from the start. In the introduction of natural numbers in [1907] and [1908A] the emphasis is on the chasm between the discrete and the continuous, on the limitations of set-construction. The main argument is directed against Cantor's transfinite number-classes and the possibility of non-denumerable constructed sets. Mathematical sets are exclusively constructed by means of the PI; '... the possible systems in mathematics can therefore all be reduced to the above order types' and '...therefore there is only one infinite power, the denumerable; indeed, we can only construct denumerable finished discrete systems.' ([1908A], p.569).

### 6.2.2.3 *Brouwer's Generalization of the Concept Sequence*

The 'starting-point' of Brouwer's new set-theory and analysis in publications after 1916 is not the natural number as the pure-mental construct in any of its conceptual aspects, but its linguistic or symbolic representation, the numeral. His two major publications *Begründung der Mengenlehre* ([1918B], [1919A]) and *Zur Begründung der intuitionistischen Mathematik* ([1925A], [1926A], [1927A]) open with the statement:

> Set-theory is based on an unlimited sequence of signs, determined by a first sign and a law which for each of these signs derives the following sign. Among the many laws that can be used for this purpose the most suitable appears to be the one that generates the sequence $\zeta$ of the digit-complexes 1, 2, 3, 4, 5,... ([1918B],p.3; [1925A] instead of 'Ziffer-complexes' uses 'Nummer' (Brouwer's own quotation-marks Ed.))

The *fundamental sequence* is introduced here in its symbolic representation, its elements are not 'mathematical entities' but 'signs' generated mechanically and recursively. The sequence of numerals is one of them and singled out as 'geeignest', i.e. most suitable or useful. The use of the word 'Ziffercomplex' for each numeral is significant, it reinforces its linguistic, physical character referring as it does to its notational composition. The numeral is considered not as a single sign representing a mathematical entity, it is a 'group of digits'. The laws which govern its positional and decimal composition and the generation of further numerals are distinct from the intuitive construction of pure number, and they presuppose a.o. the notion of addition and its fundamental properties. Brouwer's *Notes on Sets* ([BMS15A]) clearly refers to the decimal composition and states that the laws of generation are 'well-known to us':

> We start from the sequence $\rho$ of the digit-groups [Dutch: 'cijfergroepen'] 1; 2; ...10; 11;...which can be continued without restriction according to a well-known law which enables us to determine for each digit-groep the 'next one'... (p.1).

One might be tempted to interpret this 'start from number-symbols' simply as a symptom of Brouwer's Signific optimism at the time, a move to a less extreme position as regards language and symbols, or towards the orthodox constructivist view on natural numbers as given and immediately clear. This interpretation, however, ignores the relatively minor role given here to the sequence of numerals and misses the import and basis of what Brouwer himself describes as 'the Second Act of Intuitionism: the 'new insight' or 'recognition' of a concept of sequence, wider and more fundamental than the one on which mathematical practice had traditionally been based. It recognizes the distinction between the pure sequential form or structure and the manner of determining the elements of a sequence. Moreover, it acknowledges the constructive a-priority of the sequential form, which does not emerge from abstraction and generalization, but was created with the construction of the pure ordinals from the PI. The construction of a sequence in essence is the creation of a new mathematical entity from existing ones, an act of 'combining' (Dutch 'samenvatting') certain mathematical entities in order, a process of selecting a mathematical entity for each order-position. An essential requirement – as for all mathematical construction – is that these 'building bricks' exist, are 'mathematical entities previously acquired'; in the case of infinitely proceeding sequences, moreover that the process be continued without restriction, is infinitely proceedable. As to the basis of the Subject's decision, Brouwer accepted that one single process constructed by the Subject, a 'law' or algorithm, can be used by him to determine each and every element of the sequence. His new 'insight' is the recognition that the concept of sequence allows other 'modalities', in particular the possibility that the decision on an element of the sequence is not wholly and uniquely determined by previous decisions, in other words that the Subject has left himself certain freedom:

> It allows as modality of the self-unfolding of the Primordial Intuition the combining of mathematical entities in finite sequences and of inductively pre-formed mathematical systems in lawlike unlimited proceeding sequences, but also the combining of mathematical entities in sequences proceeding indefinitely either in complete freedom or freedom subject to possibly varying degrees of restriction. ([BMS32], p.7).

Brouwer's extension of the notion of mathematical sequence helps to identify the most fundamental properties of sequence and leads to a more general abstract theory of sequences. The free-choice sequence also illustrates the ambivalent nature of the infinite sequence: its incompleteness as a mathematical object, and on the other hand the dynamic, growing aspect so characteristic of mathematical construction. But most important perhaps is Brouwer's recognition that the choice sequence can play a meaningful role in mathematics, characterizing the continuum and so forming the basis of a new analysis.

The elaborate definition of the Brouwer-set or 'spread' introduced in [1918B] (see further 6.3.10) is an attempt to isolate the choice process from the complexities of the real domain. The essential conditions of proceedability are stated in terms of 'dummy' elements. As such could serve any 'fundamental sequence' ('an ordered species of ordinal number $\omega$ '([1918B], p.14; see further 6.2.3 and 6.3.2)), wholly determined by a constructive law such as induction; even the conventional sequence of digit-complexes generated by positional notation:

> It seems that among the many rules that could be used for this purpose the

most suitable is the one which generates the sequence $\zeta$ of the digit complexes 1,2,3,4,5,... ([1918B], p.3).

The elements of the sequence $\zeta$, or of any other fundamental sequence, are used as elements or 'indexes' ([BMS32], p.7) in the ramification of choices.

The sequence $\zeta$, moreover, serves here as a paradigm of the classical notion of sequence, contrasting its complete determination with the freedom of the Subject creating the new infinitely proceeding choice-sequence.

Underlying both, however, their common, conceptually most general form and their genetic primitive is the empty form of the infinitely proceeding sequence, in essence nothing but the construction of the ordinals themselves, the most fundamental sequence.

### 6.2.3 Cardinality

Brouwer's pre-1916 notion of 'mathematical set', its identification with order-type, does not leave a major role to cardinality. Indeed, the term 'cardinal number' is hardly or ever used. [1908A] accepts the use of the word 'power' relating to non-denumerable systems but only as a façon de parler.

In Brouwer's genetic order of things absence of order is not a primitive condition in the construction of mathematical sets; it is the result of a deliberate act of the Subject scrambling the elements of an already constructed, ordered set. 'Number' ( Du: 'aantal') is established by a one-to-one correspondence with the whole or part of the sequence of ordinals; cf. Brouwer's version of the Fundamental Theorem of Arithmetic, the first page of [1907]:

> Let me break off [the sequence of written ordinals] for example at 23, and write under it the same broken sequence. Between the two rows there then exists a one-to-one correspondence. If I interchange two numbers of the top row the one-to-one correspondence still remains. By such interchanges I can arrange for a particular element of the top row to correspond to the element 1 of the bottom row; further, that a particular element of the remainder of the first row corresponds with the element 2 of the second row etc.. In other words, I can introduce an 'arbitrary order' in the elements of the first row, while the sequence of ordinals of the bottom row with which they correspond remains the same. This shows that an arbitrary collection of given symbols that has been counted once will produce 'the same number' if it is counted in a different order, i.e. the sequence of ordinals with which

they are brought into one-to-one correspondence will break off at the same number. ([1907], p.3).

The notions of one-to-one correspondence and equipotency ('having the same number') are introduced here and elsewhere without further comment. In Brouwerian genetic analysis equipotency is not a primitive notion, it is primarily a relation between completed sets and it presupposes number. The concept of 'correspondence', however, in its simplest form is immediately rooted in the PI, is a 'modality' of the primordial construction of the ordered pair: the Subject linking or mapping one mathematical entity to one mathematical entity. Consideration of these entities as elements of sets, the requirement that image-elements are determined by one pre-constructed common algorithm or law, and the one-to-one conditions of completeness and uniqueness are further qualifications which distinguish the one-to-one correspondence from other modalities of the primordial construction such as number, sequence and function.

With the broadening of the concept of set to include sets that are not necessarily ordered, the introduction of spreads and species in 1916, the notion of cardinality takes on a new respectability and importance. The 1916 *Notes* starts with a definition of cardinality:

> We say of two mathematical sets $M$ and $N$ (sets of points, numbers, curves etc.) that they have the same *number*, the same *cardinal number*, or the same *power*, and we write $M \sim N$, if we can construct a one-to-one relation between them, i.e. a law which with every element of the first set associates one element of the second set in such a way that elements of the second set are associated with one and only one element of the first. For example, the set of numbers between 0 and 1 have the same power as the set of numbers $> 0$; the required one-to-one relation is $x \cdot x' = 1$. ([BMS15A], p.1).

The opening section of [1918B] is entitled '*Cardinal Numbers*'. It introduces the new notions of Brouwer-set or spread and species and a complex range of power-relations.

It is interesting to note that the Fundamental Theorem of Arithmetic in [1907], restricted to countable 'mathematical sets' and based on the one-to-one correspondence with 'the sequence of ordinals', now becomes 'The Fundamental Property of Finite Species', based on the notion of cardinality and the one-to-one correspondence with 'the sequence of digit-complexes'. Unlike the [1907] treatment and later accounts of the PI, which concern the genesis of number and establish ordinal number as the most primitive concept, [1918B], as indicated by its title and as claimed in [1919D], is a 'systematic construction of Intuitionist Set-theory' ([1919D], p. 798)). Cardinal number and ordinal number are treated as properties of sets based on the notion of one-to-one correspondence between species, in the case of 'ordinal number' a one-to-one correspondence which preserves the natural order of ordered species.

## 6.2.4 Systems and Operations of Separable Mathematics

The development of 'separable' number systems and operations in Brouwer's [1907] outline, 'The Construction of Mathematics', follows the conventional pattern and practice of classical mathematics. Systems are extended and embedded and operations defined accordingly. Extensions are brought-about by construction on and with elements of an existing system and by the introduction of new elements, notions or constructive devices which are not part of the existing system. Classical mathematics introduces such new elements by recourse to axioms or to changes and additions to formal structures. In Brouwer's conceptual interpretation of mathematics as thought-construction on and with the PI, the justification of such extensions is the all-generating power of the PI, the 'basic' construction of the ordered pair, which in its various modalities 'generates each natural number... and finally a continually extending stock of mathematical systems corresponding to the 'separable' parts of classical mathematics'. ([BMS49], p.4; app.7). In this quoted passage Brouwer claims further that 'it is introspectively realized how this is to be done' (ibid.).

### 6.2.4.1 The Operations on Ordinal Numbers

[1907] defines the operations on the ordinals in terms of the basic acts of 'counting' and 'counting-on', the construction of a one-to-one correspondence between the elements of a particular ordinal number and part of the pre-constructed sequence of ordinals.

Addition is described as a 'shift-transformation' (Du: 'verschuivingstransformatie'), introducing the notion of 'free vector' in the space of ordinals.

Other operations are defined in terms of addition: multiplication as repeated addition, subtraction as 'the inverse' of addition and division as repeated subtraction.

The concept of 'inverse' is based on the antipodal character of the order-relation. The mutual relation between two different ordinals $m$ and $n$ is expressed in two ways $m < n$ and $n > m$ and generates in the space of ordinals two 'opposing directions', in the model of linear time: before-after, and in the conventional metaphor of linear space: lower-higher than or left-right of.

The natural direction in the successive generation of ordinal numbers is 'increasing', 'upwards', 'ascending' or 'to-the-right'. The construction of addition, 'counting-on' follows the natural direction of the sequence of ordinals. In the construction of subtraction the direction of counting is reversed, it is a count-down.

These operations are the basis of extending the number-system, generating the system of integers and rationals, and further the systems of irrational and complex numbers. In each case the new concept of number is: a sequence of constructions ultimately based on the PI, in 'language' represented by a combination of symbols of ordinals and operations defined on them. Cf. for example Brouwer's introduction of the irrationals in [1907]:

> In this way we can introduce step-by-step the traditional irrational numbers by writing them as agglomerates of symbols representing numbers constructed so far... (p.6).

In the context of the irrationals Brouwer further describes how one constructs order-relations between the 'new numbers' and those 'previously acquired' and 'the operations on the new numbers, operations which in turn generate new numbers.' ([1907], p.7).

### 6.2.4.2 The Integers

The construction of the system of the integers is based on the primordial concept of inverse direction, i.e. opposite to the natural ordering of the ordinals. In [1907] it is described in one line:

> We can continue the row of ordinals to the left with $0,-1,-2$, etc. ([1907], p.5).

It may be argued that the concept of negative numbers is more primitive than that of subtraction, which relies on the notion one-to-one correspondence, or even that the elements of the construction of integers are derivable directly from the Intuition of Time and linear time-space, the present moment modelling 'the origin' and dividing time-space into past and future, discrete points of time before and after.

In the simple form as quoted 'continuing the sequence of ordinals to the left', the construction of the system $Z$ of integers consists of linking the sequence of ordinals $N^+$, retaining the natural order of the ordinals, and a sequence $N^-$, in which the natural order of the ordinals is reversed, by means of a 'zero'-element, the construction of the ordered triple ($N^-$, 0, $N^+$). The zero-element 0 is the unique order-position $^-1 < 0 < {}^+1$, establishing one order-relation between elements of $N^-$ and $N^+$, and generating the ordered space of integers, the 'number-line' Each integer then is an ordered pair consisting of an ordinal number and an assignment to either $N^-$ or $N^+$.

The latter interpretation illustrates that the construction of the integer, unlike that of the rational number, does not necessarily involve *two* ordinal numbers and why the integers can be sequentially arranged in their natural ordering: $(...{}^-2, {}^-1, 0, {}^+1, {}^+2,...)$.

### 6.2.4.3 Rationals and the Order-type η

Brouwer's definition of rational number in [1907] conforms wholly to standard practice: rationals are ordered pairs of 'ordinals', i.e. elements of the extended set of ordinals:

> By a rational number we understand a pair of ordinal numbers, written as $a/b$; by putting $a/-b = -a/b$ we can ensure that the second number, the 'denominator' is always positive. We order them by putting $^a/_b \gtreqless {}^c/_d$ if $(a \times d) \gtreqless (b \times c)$. We order them between the ordinals by putting $^a/_1 = a$... etc. ([1907], pp. 5-6).

Everywhere-density of the rationals is simply stated and the existence of the rationals as an everywhere-dense metric linear space presumed for what Brouwer calls 'making the continuum measurable' (pp. 8-11). No attempt is made to derive these properties from the given definition of rational number; only vague indications are given of their justification directly in the Primordial Intuition and of a conceptual interpretation of rational number. I refer here in particular to the notion of *between*, used on p.7 to clarify everywhere-density of rationals and hailed on p.8 as the primordial concept characterizing the continuum.

The apparent anomaly between the discreteness of the rationals and a property which is characteristic of the continuum is part of the fundamental problem of the continuum: the relation of the discrete and the continuous, or in a more general and geometric context, the relation between two successive dimensions. The problem of the mathematical continuum and Brouwer's solutions will be discussed in detail in the following sections. We shall show in section 6.3 that the concept of the Brouwer-continuum, 'the Intuitive Continuum', is essentially that of open interval, i.e. linear space bounded by two points; further that in their primordial genesis and in the subsequent construction of points and intervals, the discrete and the continuous are, as Brouwer put it, 'inseparable complements' ([1907], p.8).

Allow me here to suggest that the construction of the rationals is to be characterized as the construction of a point-set on the mathematical continuum-interval; that the concept of rational number and the justification of its everywhere-dense and metric properties are based on the concept of interval and so founded in the P.I..

The construction of point-sets, in particular of the order-type $\eta$, is based on the primordial 'between', the continuum-interval as the potential for indefinite division into sub-intervals by insertion of points. Whereas in the construction of two-ity successive 'singularities' are isolated, considered in a vacuum, the 'between' is the source and sole justification of the construction of an order-type distinct from the succession of ordinals:

> The Primordial Intuition also makes possible the construction of the order-type $\eta$,... in this case it is experienced as a move between 'first-by-itself' and 'second-by-itself', which calls into being the notion of insertion. ([1908A], p.569).
> Singularities are joined by a between which is never exhausted by the insertion of new singularities. ([1907], p.8).

If, moreover, by abstraction of 'all quality'the continuum interval is conceived as a homogeneous substratum, i.e. indefinitely divisible into uniform parts, insertion of a point on the interval provides a conceptual interpretation of mathematical division beyond the limits of integral division defined as inverse multiplication, and so of rational number.

Brouwer's interpretation of the rationals as given in [1907] should be seen in the context of his views at that time on the 'intuitive continuum' and 'measure', (perhaps also his negative views on scientific measure). The [1907] 'intuitive continuum' is presented as 'measure-less' (p.8), i.e. without a ready-made everywhere-dense grid of points. Moreover, it is a non-metric topological space, even after it has been 'made measurable', i.e. after the construction of a point-grid of order-type $\eta$. At that time Brouwer was keen to show that Analysis and Function Theory 'should be started from Analysis Situs where length and content play no role' ([1909A], p.20). In [1907] he maintains against Russell that 'distance' is not a primitive notion (pp.110-112); his characterization of the mathematical operations on the continuum carefully avoids using the concept of 'size' and the metric properties of the rationals (the constructed and therefore denumerable point-grid of order-type $\eta$ is 'simply made to correspond with the system of binary fractions' (p.9)).

The most fundamental change in Brouwer's treatment of the continuum in 1917 was the recognition that 'interval' and 'content of interval' are primitive notions in the analysis of the continuum, that the concepts of the continuum as measurable and metrizable space are immediately derivable from the P.I. (see further 6.3 and 6.3.11). The concept of continuum as metrizable space provided a conceptual basis for the notion of rational number. The construction of the rationals, however is not further mentioned in Brouwer's writings, except as one of the 'separable' systems of mathematics which conform to the classical tradition.

## 6.3 The Brouwer Mathematical Continuum

### 6.3.1 Introduction

> To Brouwer we owe the new solution of the problem of the continuum whose provisional solution by Galilei and the founders of the differential and integral calculus was again destroyed from within by the process of history. (Weyl, *Über die neue Grundlagenkrise der Mathematik*, p.56).

When Brouwer entered the foundational debate at the beginning of the century the central issue and the root of the 'crisis in mathematics' was the age-old problem of the continuum, the mathematical abstract which underlies geometry and the theories of physical continua and the basis on which the whole of mathematical analysis is built. In Brouwer's historical reviews 'of the development of mathematical thought ([BMS49], [1952B], [1954A]) it is the issue which, more than any other, highlights the fundamental differences between the various philosophies of mathematics. The crisis in foundations was caused by the realization that the Kantian a-priori of space could not be upheld and the concept of a Euclidean continuum and its properties, abstracted from observation of physical space, could no longer serve as the basis and characterization of the mathematical continuum and real number. In the following attempts to find an alternative foundation for a rigorous treatment of number theory and analysis, the continuum lost its status as a fundamental, primitive concept.

They presented what Weyl calls 'an atomistic conception of the continuum': an arithmetical manifold of real numbers ('Zahlenmannigfaltigkeit' - Riemann, Weierstrass) based on a circular definition of real number; a special 'set' or totality of points, its continuity being characterized by the properties of 'perfect and connected' (CANTOR[1883]); the 'logico-linguistic method', which reduced the continuum to a logical combination of symbols representing discrete numbers or points, in its ultimate, formalist version a combination of symbols without any conceptual content. Even those referred to as the 'Pre-intuitionist School, Kronecker, Poincaré, Borel and Lebesgue' ([BMS32], p. 21) failed Brouwer's critical test of the continuum; for the construction of separable mathematics they insisted on an a-priori exactness independent of language and logic but they were not radical enough to apply the same principle in their treatment of the continuum:

> For the continuum, however, the pre-intuitionist school did not seek an intuitive origin, one which lies outside the domain of language and logic; on the contrary, they introduced elements alien to any form of intuition either by introducing some negative logical axiom which is neither evident nor intuitive, such as the completeness axiom, or by contentedly relying on an 'denumerable-unfinished' system of elements characterized by some genetic property related to the system of rationals. ([BMS32], pp. 2-3; cf.also [1930A], p.2 and [1952B], p.140).

The radical changes in Brouwer's treatment of the continuum during the period 1907-1917 reflect his mathematical development and his concern with different aspects of the continuum at the beginning and at the end of that period. Lack of 'continuity' is partly due to critics and commentators who tend to isolate Brouwer's twin interests and contributions in Topology and Foundations, describing them as 'two almost disjoint parts' (the editors of Brouwer's Collected Works). In fact, in the development of Brouwer's ideas on the continuum the topological and foundational can hardly be separated: his major contribution to the creation of Topology originated in his 'foundational' concern with the concepts of space and dimension; on the other hand, his work in point-set topology and the problems of dimension and measure highlighted the need for a reconsideration of the fundamental concepts of set and continuum and paved the way to what Weyl calls 'the Brouwer revolution'. Brouwer himself describes it as 'a new insight', one that sheds light on an aspect of the continuum so far left unresolved. It introduces into mathematics the elements of indeterminacy and freedom which Bergson had considered to be the exclusive prerogative of 'pure time' (cf. 4.1.3). It does not necessarily contradict Brouwer's earlier view of the continuum as 'completed' and 'given in Intuition'. Indeed, there is a consistency of purpose and principle throughout Brouwer's writings on the continuum. They are a search for an interpretation of continuum and real number which can form a sound and 'natural' basis for mathematical analysis, i.e. one that accords with his 'intuitive' and constructive view of mathematics. The fundamental posits of his theory of the continuum remain unchanged. They may be summarized as follows:

1. the 'sole a-priori element' of the continuum, as for all mathematics, is 'time';
2. the mathematical continuum is a concept, i.e. created by mathematical abstraction and existing in the human mind, the Subject;
3. moreover, it is a *primitive* concept, an element of the Primordial Intuition, directly abstracted from the awareness of time (often referred-to as the 'Intuition of the Continuum');
4. 'points' can be constructed 'on' a continuum, and have a role to play in the analysis of the continuum as defining the boundaries of intervals into which a continuum can be decomposed; they are not constituent parts of the continuum nor can the continuum be identified with a constructed totality of points.

The 'early papers' ([1907], [1908A], [1909A] and [1912A]) mainly concern the genesis of the continuum and the nature of the continuum as a whole. The relation between continuum and discrete points is described as 'complementary' and the constructions characteristic of points and continuum are found to be diametrically opposed. In particular, the synthetic construction of sets, combining discrete elements into a new and different kind of mathematical entity, is found to be a wholly inappropriate instrument for generating a continuum, whose characteristic construction is analysis, i.e. decomposition into homogeneous parts.

Brouwer's independent stand on the nature of sets and the continuum naturally resulted in his disagreement with current foundational thinking on almost every issue. These 'early papers' are therefore predominantly critical and negative, refuting the major conclusions of Cantorian set-theory on the power of the continuum, infinity and transfinite numbers as well as the formalist attempts to secure these results by means of logic.

At this stage Brouwer still assumes that the classical treatment of 'real number' can be justified and form a sound basis for analysis and function theory. Real analysis is identified with geometry of the one-dimensional mathematical continuum (cf. [1909A], p. 14), and the arithmetical operations on real number and their properties defined and justified as Lie-groups of transformations. As to real number itself, Brouwer seems reluctant to give a conceptual or 'geometric' interpretation. 'Real number' is identified with the sequence 'approaching' it (Du: 'benaderingsreeks') within a discrete scale on the continuum, or with the law which determines this approximation: the 'law of progression' (Du: 'voortschrijdingswet', e.g. [1907], p.150; cf. also [1912A] p. 22: 'The concept of 'real number between 0 and 1'... for the Intuitionist it means 'law for the construction of an elementary series of digits after the decimal point, built-up by means of a finite number of operations.'). His inevitable conclusion is that the power of any set of real numbers so constructed is denumerable, also on the general grounds that 'The intuitionist can only construct denumerable sets of mathematical objects' ([1912A], p.23; cf. also p.13). These 'real-numbers' form a kind of practical continuum à la Borel, or in Brouwer's later terminology a 'reduced continuum'.

In [1907] the measurability of this practical continuum is assumed to be secured by its everywhere-density and a vague association with the 'intuitive continuum'. The reasons for abandoning the practical continuum are given in Brouwer's historical reviews, where he criticizes Pre-intuitionism and by implication his own early interpretation:

> The Pre-intuitionist School... they seemed to have contended themselves with an ever-unfinished and ever-denumerable system of 'real numbers', generated by an ever-unfinished and ever-denumerable system of laws defining convergent infinite sequences of rational numbers. In doing so they seemed to have overlooked that such an ever-unfinished and ever-denumerable system of 'real numbers' is incapable of fulfilling the mathematical functions of the continuum for the simple reason that it cannot have a measure positively differing from zero. ([BMS49], pp. 2-3; also [1930A], p.3 and [1952B], p.140).

Significantly, the problems of measure and invariance of dimension seem to be an exclusive Pre-intuitionist pre-occupation at the time; the main contributions are due to Brouwer and prominent pre-intuitionists (pace Hausdorff). In Brouwer's case they led to a search for what one could call a 'real element of the continuum', one which could 'fulfill the mathematical functions of the continuum' and also qualify as a thought-construction derived from the Primordial Intuition alone.

The first step was his recognition that such a construction needs to contain the primordial elements of continuum and discrete point, and that his own notion of set-construction required modification. In [1914] Brouwer introduces the notion of *well-constructed infinite set*; it combines the synthetic construction of a collection of discrete points and the analytic construction of nested intervals:

> For the Intuitionist there exist only *well-constructed infinite sets*, which are composed of two parts: a part of the first kind, constructed as a single fundamental sequence, and a part of the second kind which is based on a fundamental sequence $f$ as Frechet-class, but whose elements are determined by a sequence of choices from elements of a finite set or fundamental sequence in such a way that to every sequence of choices there corresponds a sequence of nested intervals [Germ.'Teilgebiete'] of $f$ whose widths converge to zero, and that the final segments of every two sequences of intervals corresponding to different sequences of choices lie outside one another. ([1914], p.79).

The definition of well-constructed set already contains all the elements of Brouwer's 1917 characterization of 'element of the continuum' as a sequence of nested intervals and the changes in his set-theory to accommodate the new continuum. Two aspects of his previous stand on set-construction are modified: his insistence on a constructed law or algorithm as the only justification of the mathematical existence of infinite sets or sequences, and his rejection of any form of 'comprehension' as a defining principle of sets.

The choice of the Subject is now admitted as a legitimate constructive act and his freedom as the guarantor of the existence, i.e. the continuance of the convergent infinite sequence, and of the non-denumerability of the continuum; 'freedom' ensures that the limit is never reached, the essence of convergence, and generates sequences beyond the 'ever-denumerable set of fundamental sequences', i.e. those pre-determined by law.

Moreover, the newly introduced 'well-constructed set' in its composition and generating process cannot avoid the use of properties and classification of constructed mathematical entities according to property. In his 1916 *Notes* Brouwer still insists that classification by property does not have the status of mathematical set in the strict sense:

> Apart from these [sets] we can also introduce *pseudo-sets*, defined by comprehension, better referred-to as 'sorts' [Du. 'soorten', 'species' in Brouwer's later German and English translations] ([BMS15A], p.1).

These new ideas matured and became the basis of Brouwer's 'systematic re-construction of set-theory', completed in 1917 and published in March 1918. The fundamental 'insight', the 'Second Act of Intuitionism', is the recognition that the real-element-of-continuum is characterized by the continuum itself, and that the appropriate concepts and tools in the analysis of the continuum are *Spread*, the growing process of the convergent-infinite sequence in its widest sense and *Species*, the concept of mathematical property, both legitimate descendants of the Primordial Intuition:

> The Second Act of Intuitionism is the recognition that the 'self-unfolding' of the Primordial Intuition not only leads to the sequence of natural numbers but also to the spread-construction and to the foundation of analysis on the properties of the spread.
>
> ....and further admitting, as a modality of the self-unfolding of the Primordial Intuition at every stage of the construction of mathematics, properties to be posited for mathematical thought-constructions already acquired, as new mathematical thought-constructions under the name Species. ([BMS32], p.7).

## 6.3.2 The Genesis and Nature of the Intuitive Continuum

The key to understanding Brouwer's theory of the continuum at all its stages of development is his fundamental thesis of the Primordial Intuition: the primitive and complementary nature of the discrete and the continuous.

The notion of continuum is a primitive concept, i.e. it cannot be constructed from more elementary concepts or defined in more general terms. In particular the continuum is not built up from discrete points:

> The continuum as a whole is intuitively given; a construction of the continuum, an act which would create by means of the mathematical intuition 'all its points' is inconceivable and impossible. ([1907], p.62)

The discrete and the continuous are mutually complementary within the Primordial Intuition, one being the denial of the other and yet indispensable for the existence of the other:

> The Primordial Intuition...as the unity of the continuous and the discrete, the possibility of thinking-together singularities joined by a 'between' which is never exhausted by insertion of new singularities. Therefore, since in this Primordial Intuition the continuous and the discrete appear as inseparable complements, both of equal right and equally clear, it is impossible to

dispense with either as a primitive entity, to take one as self-sufficient and construct the other from it; for it is impossible to consider either of them in isolation. ([1907], p.8).

Not only are the discrete and the continuous 'inseparable complements' in their primordial genesis, they remain mutually indispensable in the definition of their characteristic properties. In particular the Brouwer continuum as open interval and its properties of infinite divisibility, order etc. can only be circumscribed by reference to its sole primitive partner, the discrete point.

In Brouwer's accounts of the primordial happening the mathematical continuum emerges from the Subject's observation of 'the move of time', 'change in all its forms', and the abstraction from 'all quality'. Unlike Bergson's intuitive continuum, 'pure duration', the Brouwer mathematical continuum is not time itself nor the concept of time; time is its a-priori condition. Brouwer's primordial act includes total abstraction of quality, which in Bergson's philosophy is the essence of time. Moreover, essential for Brouwer's perception of change is the existence of two distinct states or moments, 'a past and a present moment', emerging from the total abstraction of quality as two singularities or discrete points.

The Brouwer continuum is that what separates and yet joins the two discrete points: the 'between'. Conceived as successive, 'first' and 'second', they are the end-points of what is essentially the ordered domain of interval; and since they are not part of the 'between', the intuitive continuum is essentially the concept of *open interval* in its most abstract form, vaguely described as 'the matrix of points to be thought together' ([1907], p.9), or 'matrix of points or singularities' ([1908], p.570). The ability of mind to generate the 'between' is one of Brouwer's 'a-priori judgments':

> ...the ability to insert, i.e. the ability to see as a new element not only the totality of two elements composed into one [i.e. two-ity], but that which joins them: that which is not the totality nor the two elements. ([1907], p.119).

The mathematical continuum or open interval does not exist except as a pure concept or thought-abstract; any description has to remain metaphorical, using terms characteristic of its physical models. The most appropriate model is the time-continuum from which it has been abstracted. Convention, however, has favoured the spatial model, which is visual, more concrete and more easily captured in language and images. Discrete singularities are 'points' and the open interval the 'line-space' between them. The static nature of the spatial continuum, however, makes it less suitable as an illustrator of the dynamic aspects of the mathematical continuum such as order and progression.

### 6.3.2.1 The Construction of Intervals

The primordial act of construction on and with the continuum is 'analysis': splitting the interval or continuum into two parts, subintervals, homogeneous to each other and the whole interval, 'an operation which we could describe as 'bisection of the interval'' ([1907], p.9) or 'insertion of a point'. The new point $P_n$ is inserted in the interval $(P_1, P_2)$, i.e. it is the cut, the common limit which generates and separates the two open subintervals $(P_1, P_n)$ and $(P_n, P_2)$. Each of these subintervals is an interval in its own right: it has all the characteristics of its parent interval, in particular the capacity of being bisected.

The divisibility of interval into homogeneous subintervals is the basis of its 'infinite divisibility' and of the construction of the point-system of order type $\eta$. The homogeneity of open interval and subinterval further is the basis of an extension of the open interval into what is traditionally known as the 'infinite mathematical continuum', and of the isomorphism between them:

> The segment between two points of the continuum is equivalent (Du. 'gelijkwaardig') with the whole of the continuum; both form what is known as an open continuum. ([1907], p.11 footnote.)

The same footnote describes the 'cut' of an interval into two subintervals, and the inverse operation of joining them:

> An arbitrary point $P$ on an open continuum $\alpha\beta$ generates on its left and on its right two new open continua $\alpha\gamma$ and $\delta\beta$.
> Inversely, we build from two open continua $\alpha\gamma$ and $\delta\beta$ a new open continuum by inserting a new point $P$ and so joining $\gamma$ and $\delta$. (ibid.).

In this sense the mathematical continuum can be said to be 'constructed' from intervals:

> These [geometric] spaces are constructed from one or more parallelepipeds, being connected together... ([1909], p. 14).

(Cf. also Brouwer's topological papers on the invariance of dimension especially [1911D] and [1911E], where simply connected pieces, of which spaces are composed, are taken as the fundamental constructs.)

### 6.3.2.2 Relations and Ordering of Intervals

The act of insertion and the natural order of the continuum generate a relation of 'inclusion' and an order relation between subintervals.

The relation of 'inclusion' or 'sub-interval' is the fundamental relation of the continuum ('not the relation of set-and-element, but the relation of whole-and-part'), and of Brouwer's characterization of real-element-of-continuum as convergent sequence of nested intervals.

The order relation between (disjoint) subintervals is the natural order of the continuum abstracted from the progression of time:

> ...the Intuition of Time which is a one-dimensional continuum conceived by one single Subject in which two different time-segments differ absolutely... ([1909], p.13).

Again, since intervals are labelled and 'defined' by their endpoints, both relations are expressible in terms of order-relations between these points. Adopting the notation

$(P_x, P_y)$ : the open interval between $P_x$ and $P_y$ [ $P_x < P_y$ ];
$\subset$         : 'is a subinterval of';
$<$         : 'is before' or 'is smaller than';

we express the two relations as

$(P_x, P_y) \subset (P_a, P_b)$ iff $P_a \leq P_x$ and $P_y \leq P_b$.
$(P_x, P_y) < (P_a, P_b)$ iff $P_y \leq P_a$.

The relation of 'inclusion', even if expressible in term of order-relations between end-points, is not itself an order-relation in the strict sense. This leads to the problem of ordering the real-elements-of-continuum, which are defined by inclusion, especially when end-points are not pre-determined but generated by free-choice. (see further 6.3.11).

The main problem of the continuum, however, is the complementary partnership of points and interval, the necessary definition of the interval by means of endpoints which are themselves not part of the interval. Any analysis of the continuum is therefore dominated by a treatment of points, sometimes to the exclusion of the concept of interval. Even if 'interval' is acknowledged as the conceptual object of analysis, the fundamental element of continuum, there is still a tendency to avoid the elusive interval and substitute its more definite and simpler defining points. This approach is adopted in [1907] and [1909]: the continuous nature of elements-of-continuum is recognized, but the arithmetical operations and functions on them are defined and justified by transformations of their endpoints in terms of a point-grid super-imposed on the 'intuitive continuum'.

The 'well-constructed infinite set' of [1914] is Brouwer's first attempt to incorporate the notion of sub-interval and its characteristic property of having-content or 'width' (Germ. 'Breite') in the definition of element-of-continuum. Its characterization as an infinite sequence of nested open intervals whose widths converge to zero is the conceptual starting point of his new continuum theory of 1917. But as often in the systematic exposé of a mathematical theory, the grand idea gets somewhat lost behind the preliminaries, the groundwork of preparing tools and general devices, some of them so exciting and revolutionary that they become the focus of attention.

The characteristics of the general notion of the 'spread', in some respects restricting and in others extending the classical infinite sequence, are determined by the nature of the mathematical open interval. In particular 'free choice' is admitted as a legitimate and natural element in the generation of infinite sequences, not only because it extends the 'ever-denumerable' range of sequences determined by law: free-choice secures and highlights the necessarily unfinished character of the infinite sequence, and reflects the 'growing', dynamic aspect of the mathematical continuum, abstracted from the intuition of the movement of time. Above all, 'freedom' is the guarantee of continued proceedibility, which in the progressive selection of points is nothing but the nature of interval as the potential of indefinite division.

That the 'Intuitive Continuum' remained the conceptual basis of Brouwer's new continuum and real number is evident from his continued reference to the 'Intuitive Continuum' and the 'Intuitive Between' in later work; e.g. the opening lines of Chapter III of [BMS32], where he clearly claims to derive properties of the continuum from the primordial concept of the 'between':

> Together with the abstracted two-ity as the Primordial Intuition was created the 'between' that is never exhausted by the insertion of new singularities, not even by the insertion of infinite sequences of singularities. In Chapter II we recognized this 'intuitive between' as the matrix of point-cores' and as such added it to Intuitionist Mathematics under the name 'continuum'. Following-on, we shall now attempt to use the elements of order which are present in the intuitive 'between' and introduce them in mathematics to order the point-cores. ([BMS32], p.18; cf. 'The Cambridge Lectures' [BMS51], p.40, [BMS53], p.7, etc.).

### 6.3.3 Point-sets on the Continuum and the 'Made-measurable Continuum' (pre-1917)

#### 6.3.3.1 The 'Made-measurable Continuum'

Brouwer's theory of 'point-sets on the continuum' and 'the measurable continuum' of the period 1907-1912 is a critical phase in the development of his topological ideas and his new theory of the continuum. It is a compromise between his fundamental conviction that the whole of mathematics, including pure geometry and the mathematical continuum, is constructed from the Intuition of Time alone, and the contemporary practice of point-sets in the analysis of the continuum.

Awareness of the sole a-priori of time leads Brouwer to question the distinction between pure-geometrical space and the continuum:

> Singling out a special part of mathematics under the name 'geometry' certainly cannot be based on an a-priori distinction, and the question arises whether such a special name is even justified on the grounds of a pure-mathematical distinction.
> ... indeed the main operations with real numbers belong to geometry. ([1909A], pp.13-14; cf. also [1907], pp. 117-118).

In his search for a pure-geometry Brouwer in [1907] and [1909A] tries to exclude the 'spatial' primitive elements which had crept into Euclidean geometry, in particular the notion of distance or length:

> RUSSELL[1903] §37 tries to maintain 'distance on a straight line' as a primitive concept. He ignores the fact that one can very well conceive a continuum without being able to compare 'magnitudes' on it. ([1907], pp. 110-111; cf. also [1909A], p.20: '...If we succeed in starting the theory of functions from analysis situs... length and content can no longer play a role.') [This is one of the passages omitted from the 're-print' [1919B]. The CW version is a translation of [1919B] not the original [1909A].]

Indeed, one of the main themes of [1907] is the 'scientific', non-mathematical nature of the concept of 'measuring' and the notion of a metric. The 'intuitive continuum' is an abstraction of 'intuitive time':

> ...intuitive time which must be distinguished from 'scientific time', which is definitely a posteriori, appears only in experience, and as a one-dimensional coordinate can be introduced to catalogue phenomena. ([1907], p.99).
> ...the Intuition of Time, which is a measureless one-dimensional continuum... ([1909A], p.13).

[The fundamental reliance of euclidean geometry on the concept of 'distance' and measuring is reflected in the Dutch language (as indeed in Greek) where geometry is 'meetkunde': the science of measuring. Brouwer refers to this when he writes: 'Cartesian geometry becomes a *Euclidean geometry* when we define a distance between two points...'([1907], p.38).] Cantor's point-set topology seemed an ideal alternative to the 'metric' treatment of the continuum: the fundamental cantorian concepts of limitpoint, derived set and everywhere-dense set allowing a treatment of continuity without recourse to the notion of metric. The Lie characterization of arithmetical operations and functions in terms of transformation groups on sets of points seemed adequate for the purpose of analysis and function theory. Brouwer's problem in 1907 was the existence of pointsets and their relation to the continuum. Unlike Cantor's continuum, conceived as a given totality of points, Brouwer's intuitive continuum is a medium in which points can be inserted, a potential for 'cutting'. Points and point-sets are not ready-given in Intuition; as all mathematical entities they need to be generated before they can be used in further construction.

The construction of a suitable point-grid on the continuum, described as 'making the continuum measurable', was the first task Brouwer set himself in [1907]:

> We can only work with the continuum after we have made it measurable. (p.67).

'Making the continuum measurable' means constructing an everywheredense point-scale on the continuum and assigning numerical values to it. A brief sketch of the construction of the 'made-measurable continuum' is given in the introduction of [1907]:

> The binary scale arbitrarily constructed [on the continuum and of ordertype $\eta$] is not necessarily everywhere-dense, i.e. it does not necessarily penetrate into every segment of the continuum. But we agree to consider as contracted into a single point every segment not penetrated by the scale, i.e. we consider two points to be different only if their approximating binary fractions start to differ after a finite number of digits.
> If further we designate an arbitrary point of the constructed scale as the zero point, then the scale has made the continuum into a measurable continuum. ([1907], pp. 10-11).

The intuitive, measureless continuum, a one-dimensional non-metric space, has been made measurable, i.e. it is provided with an everywheredense reference-grid, a constructed and therefore denumerable set of points, a legitimate domain of one-one transformations. Although operations and functions are defined in terms of the points of the grid, the scale is not the continuum itself; it remains 'a point-set in the space...' ([1909A], p.14) or 'a point-set projected on the continuum' ([1907], p.66). The continuum is 'never exhausted' by the construction of a point-scale: every two points of the scale determine an interval with all the characteristics of the continuum, in particular its capacity for indefinite further division by insertion of new points.

Brouwer's [1907] interpretation of real number and real analysis was largely overtaken by his new theory of the continuum of 1917. However, the concepts of individualized point and point-set on the continuum, and the distinctions between constructed sets of points, definable points and possible points on the continuum, introduced in [1907] and [1908A], remained important elements of his intuitionist reconstruction of analysis and deserve some further comment.

*6.3.3.2 Point on the Continuum*

In his historical analysis ([1979]) Johnson describes Brouwer's famous proof of the invariance of dimension ([1911C]) as inspired by Poincaré's cut definition ('A continuum has $n$ dimensions when it is possible to divide it into several parts by means of one or more cuts which are themselves continua of $n-1$ dimensions.' POINCARÉ[1912], p.488). Poincaré had introduced his cut principle in 1903, significantly starting from the 'applied' three-dimensional space and defining other spaces by regressive recursion (*L'espace et ses trois dimensions*, POINCARÉ[1903]). Brouwer's 'pure' intuitive approach of 1907 starts from the one-dimensional mathematical continuum and builds spaces of higher dimensions upwards:

> Cartesian geometry can be constructed by taking $n$ open continua and defining a point of $^nR$ as the combination of $n$ points of these different continua. ([1907], p.38).

His proof of the invariance of dimension, however, was not based on the coordinate principle nor on group-theory. The concept of 'cut' clearly underlies the proof, in particular in the form given in [1913A] (Über den natürlichen Dimensionsbegriff). Defining dimensions of degree zero and infinity Brouwer further makes the cut definition inductive:

> The expression '$\pi$ has the dimension-degree zero' shall mean that $\pi$ has no continuum as a part; the expression '$\pi$ has an infinite general dimension' shall mean that neither zero nor any particular natural number can be the general dimension degree for $\pi$. ([1913A], p.147).

The specific reference to the dimension degree zero could be interpreted as a clue to the true origin of Brouwer's cut-concept: the primordial relation between the two primitives, point and continuum. In the Primordial Intuition the distinction between point and continuum is absolute, the archetype of dimensional difference. While the essence of the interval-continuum is the 'between', i.e. the infinite divisibility into homogeneous parts, 'point' is its atomic boundary, in line with the Aristotelian tradition 'point is that which has no part'; its sole function and essence is to 'define' the individual interval.

The first two points emerge from the primordial happening as the limits, the boundary points of the continuum conceived as open interval. Abstracted from two 'points in time', a particular past moment and the present moment, they occupy a definite 'position' in relation to the continuum, 'definiteness' being their established individual identity.

The construction of a point $P$ on the continuum is the act described above (6.3.2) as 'bisection of the interval', the 'insertion' or 'cut' generating two subintervals, or 'the common limit' of these two subintervals. According to these definitions point is essentially 'point on the continuum', it can never be entirely isolated from its relation to the continuum. In the Brouwerian 'inseparable complementary' partnership the construction of point on the continuum generates two subintervals and is itself defined by these intervals. The intervals are essential elements of the existence and individuality of the point; the operations described by Brouwer as 'contracting the interval into a point' (cf. the quoted definition of making the continuum measurable) and 'transformations reducing finite length and content to zero' (the above mentioned passage of [1909A] withdrawn in [1919B]) eliminate and destroy the interval and with it the individual point. Later, in the definition of the Spread, the contraction of interval into a point becomes the characteristic destruction of real-element-of-continuum; the generation of real-element-of-continuum is the continuing process of selecting boundary points of ever-decreasing nested intervals, never settling on one individual point thereby effecting the contraction of the interval.

Indeed, the constructed point on the continuum distinguishes itself from real-element-of-continuum by its definite position on the continuum, its 'complete determination'.

In Brouwer's interpretation of the continuum as a non-metric space the definite position of the point $P$ in the interval $(P_1, P_2)$ is not defined by 'distance', nor is it uniquely defined by the order relations $P_1 < P$ and

$P < P_2$. The only remaining basis of its definiteness, its individual identity, is the act of the Subject making a particular insertion in the interval at a certain point in time.

### 6.3.3.3 Sets of Constructed Points

The characteristic divisibility of the continuum and the possibility of extending the continuum interval allow the Subject to continue the process of insertion, constructing new points on the continuum. Consideration of the constructed individual points as a set further requires an act of the Subject linking them together. In Brouwer's early writings 'linking' is an immediate application of the Primordial Intuition (cf. 6.2.2): point-sets are 'sequences of points' (Dutch: 'puntenrijen') in the order of their construction or selection. The introduction of the notion of 'species' in 1916 allows a modified form of comprehension, a linking of points on the basis of a common property; even then the point-species is not defined as a totality of points but identified with the constructed property (see further 6.3.5).

Given Brouwer's early strict definition of 'mathematical set' one would expect the 'completed' set of individually constructed points to be necessarily finite. However, the denumerably infinite order-types $\omega$ and $\eta$ are accepted as legitimate 'individualized point sets' on the grounds of the primordial nature of inductive generation, quite distinct from other constructive procedures or 'mere methods':

> In the beginning of this chapter we were able to construct two kinds of linearly ordered sequences of points, the order-type $\omega$ of enumerated positive ordinals and the inverse type $\omega^*$ ; further the everywhere-dense denumerable sequence, the order-type $\eta$ of all rationals or those between 0 and 1...
> The mathematical intuition can create individually [individulalized] only quantities that are denumerable. ([1907], p.62);...
> In the first chapter we have seen that there exist only finite and denumerably infinite sets and continua; this was shown on the grounds of the intuitive truth that we cannot create other than finite sequences and on the basis of the clearly-thought 'and-so-forth' sequences of order-type $\omega$, but only those consisting of 'equal elements' (by the expression 'and-so-forth' is meant repetition of *one and the same* thing or operation, although this thing or operation may be defined in a rather complex way.) ([1907], pp. 142-143).

The basic denumerably infinite set is an inductive set, a sequence wholly defined by the indefinite repetition of a constructive act, which generates each and all its elements. Other 'infinite discrete individualized point-sets'can be constructed 'by alternate and subordinate combinations' of these basic sets and finite sets. ([1907], p.63).

### 6.3.3.4 Denumerably-unfinished Systems and Definable Points

In Brouwer's pre-1916 set-theory the finite and 'denumerably finished' constitute the full range of constructible point-sets. Analysing the claims of Cantorian set-theory, Brouwer acknowledges the possibility of other 'mathematical systems' and 'logical entities' combining points on the basis of 'comprehension', and examines their mathematical content. Cantor's diagonal procedure is accepted as a proof of the non-denumerability of possible points on the continuum, but not as a constructive basis for the continuum as a set of points. Such systems are introduced in [1907] as 'denumerably unfinished sets', although Brouwer hastens to add that the term 'set' here is used in a non-mathematical sense:

> By a *denumerably-unfinished* set we understand one of which only a denumerable part can be indicated as well-defined; on the other hand, having constructed such a denumerable collection one can by some previously defined mathematical process immediately derive from it new elements which are also deemed to be elements of the denumerably-unfinished set. But strictly mathematically speaking this set does not exist as a totality, nor does its power. We may, however, use these words as an arbitrary expression for a known intention. ([1907], p.148).

In [1908A] the denumerably-unfinished is not referred-to as 'set' (Menge), but as 'a mathematical system' and 'a method'; and constructibility is identified with denumerability:

> One could devise a method for building a mathematical system which from every given denumerable set belonging to the system derives a new element which also belongs to the system. But with such a method, as with all mathematics, one could still not build other than denumerable sets, the whole system can never be constructed since it cannot be denumerable. (p.569)
> ... a method, not a set. (p.571).

In both instances one of the examples given of denumerably-unfinished-systems is 'the totality of definable points of the continuum'. In the context of mathematics 'definable' means: to be generated by means of a constructive algorithm or law. The definable points of the continuum are the real numbers in the classical and pre-intuitionist sense:

> They [the pre-intuitionists] seem to have contended themselves with an ever-unfinished but ever-denumerable system of 'real-numbers' generated by an ever-unfinished and ever-denumerable system of laws... ([1952B], p.140).

Brouwer contends that the logical definition of the continuum as a *'totality* of points' can only refer to the mathematically definable points, and concludes that this logical continuum is denumerably-unfinished:

> One may introduce the logical entities of the totality of numbers of the second number-class and the totality of points of the continuum...
> But if one introduces the logical entity of *'totality* of points of the continuum', i.e. abandons the intuition of the continuum, one must define the points of the continuum, and this can only be done by definable laws of progression for approximating binary fractions. ([1907], p.150).

This is not a denial of the conceptual unity of the mathematical continuum, nor of a power of the continuum beyond the denumerably-unfinished: the mathematical continuum is thought-of 'as a whole' in the Primordial Intuition, a primitive concept, its unity in relation to its points is its inexhaustible potential for being 'cut', the basis of 'possible points'.

Brouwer's pre-1916 interpretation of mathematical point-sets may seem restrictive, his acknowledgment of the intuitive continuum as a primitive medium and the Free Subject as the sole generator of points leaves the way open to a much wider range of 'possible points' than those predetermined by law. In [1907] and [1908A] the limitations of 'law' are used to refute the Cantorian theory of powers of sets and the continuum. The positive aspects of 'choice' are not fully exploited until 1917, when 'indefiniteness' and 'free choice' become the fundamental elements in the characterization of the continuum.

### 6.3.4 *The New Set-theory 1917*

Brouwer's early stand on set-theory can be summarized as:
1. total rejection of the philosophical foundation of Cantorian set-theory and of attempts to found mathematical set-theory on logic or axioms;
2. recognition of the generating processes of the order-types $\omega$ and $\eta$ as the only legitimate constructions of 'mathematical sets'.

Opposition to platonism never weakened, nor did his challenge of what he called the 'belief' in the power of verbal definition generating mathematical elements, and the identification of 'property' with a totality of elements.

As to the constructive devices available to the mathematician, it was one of the fundamental tenets of Brouwer's philosophy that their legitimacy ultimately and solely depended on Intuition, which allows for new concepts to be derived from the Primordial Intuition in its 'free unfolding'. The combined effect of rejecting a logical set-theory and being restricted to the use of 'mathematical sets' had been felt as 'debilitating' and 'making mathematical analysis anaemic'. The Second Act of Intuitionism in 1918 'brought liberation from this situation' ([BMS32A], p.7).

The 'new insights' or 'Primordial recognitions of Intuitionism' ([BMS59B], p.3) modify Brouwer's early set-theory in two ways:
1. the notion of set as an effective process generating mathematical elements, i.e. sequences of order-type $\omega$ and $\eta$, is extended to procedures generating 'indefinitely proceeding sequences', including free choice;
2. a new entity 'mathematical species' is 'admitted', not so much a new mathematical concept as a new recognition of the legitimate role of 'mathematical property' in the construction of mathematics.

The extension of the notion of sequence has its major application in the analysis of the continuum, the concept mathematical property or 'species' is found to be a fundamental element of the whole construction of mathematics, 'at each stage of its development', and even to play a role in the primordial happening itself.

## 6.3.5 Mathematical Species

The change in Brouwer's views on the role of 'property' in mathematics was prompted by practical need and by growing awareness of its de-facto use in his own mathematical work. The use of 'property' or 'common property' as a generator of mathematical sets, 'predicative definition' or 'definition by comprehension' was outlawed in [1907], [1908A] and especially [1912A]. After his major investigation of point-set topology, however, and having embarked on a reconstruction of set-theory Brouwer had to admit in 1916:

> A point-set is in fact a set defined by comprehension; in this case a set of points which have the property that they can be approximated by means of a certain procedure. Further in this text we shall meet other point-sets defined by comprehension, e.g. interior boundary sets and measurable sets; and these have also led to important mathematical theories. ([BMS15A], p.23).

More fundamentally, in a construction of mathematics, restricted to progressively and sequentially building upwards there would be no room for hypothesis nor for 'absurdity'.

Moreover, the universality of mathematical concept, even within the subjective world of the mathematician, requires a constructive basis for 'sameness', the equivalence of mathematical entities acquired by various routes or at different times.

*The Brouwer 'sort' or 'species'*, introduced in 1917, is a constructive interpretation of 'property' which allows the mathematician to use hypothesis and generalization as legitimate constructive steps, and this without presuming the existence of a platonic universe partitioned by property. The cantorian Principle of Comprehension remains anathema:

> The axiom of Comprehension – on the basis of which all things which have a certain property unite into a set (even in the later, modified version given by Zermelo) – is inadmissible and useless; a legitimate basis of mathematics can only be found in a *constructive* definition of set. ([1919D], p.797).

The concept 'mathematical species' is invariably described in Brouwer's post-war papers as:

> ... properties supposable for mathematical entities previously acquired... mathematical entities for which the property 'holds' are called the elements of the species. ([1948C], p.1237, [1952B], p.142, [1954A], p.523, [BMS51], p.9, [BMS49], p.4, [BMS66], p.4).
>
> ... recognition at each stage of the development of mathematics of properties to be supposed for mathematical entities previously acquired. ([BMS52], p.2; also [1947], p.335)

In these texts the notion 'property' is not further defined (apart from the added 'conceptually completed' in [1925A]). In the Brouwerian universe of mathematics, however, it can only be a construction, and this is the interpretation given in [1907], [1908C] and [1923C] where property is 'a construction' or 'a system'. Even if for simplicity's sake the property is referred-to as residing in the mathematical entity or belonging to it, it must be understood that the Subject is its genitor and bearer.

The characteristic feature of 'property' is its being considered as a possible substructure, part of other constructions, in later definitions described as 'to be posited' or 'supposable for', in the earlier references more clearly as hypothesis:

> ... the conjectured fitting-in of one construction into another. ([1908C], p.5);

...properties of systems, i.e. the possibility of fitting systems into other systems. (1923B], p.1);
Within a constructed system it is often possible simply to fit-in new constructions quite independent from the original form of its construction... The so-called 'properties' of a given system are no more than the possibility of such building of new systems in certain agreement with an existing system. It is precisely this kind of *fitting-into a given system* of new systems which plays an important role in the construction of mathematics, often in the form of an investigation into the possibility or impossibility of such fitting-in... ([1907], pp.77-78; cf. also 5.8.2)

The main distinction between Brouwer species and Cantorian sets as well as the Brouwer-set or 'spread', is that species do not generate their elements: candidates for hypothesis are mathematical entities 'previously acquired', i.e. completed mathematical constructions, conceived in their entirety. Moreover, being-an-element-of the species is only established for each individual entity by 'a successful completed construction of fitting-in'; and in a similar way the 'absurdity' of it being an element. (cf. 5.8.2).

Neither is 'species' to be identified with the totality of its elements. Even if Brouwer conforms to standard set-theoretical terminology and uses terms like 'belongs-to', species is no more than the property itself, more like a Boolean elective operator, but one to be used by the Subject and on an existing universe of constructed mathematical entities.

### 6.3.6 *The Genesis of Species*

The genesis of species is analytical: it originates in the power of the Subject to abstract, isolate part of his construction, consider it by itself and as a part of its parent and other constructions. As such it is a fundamental element of the primordial process, and Brouwer can rightly claim it to be derived direcly from the P.I., 'a modality of the Primordial Intuition' ([BMS32], p.7).

Brouwer further claims that consideration of the isolated structure and the hypothesis of it being part of other constructed entities is a distinct constructive device, 'a new mathematical entity' ([1947], p.339; [1954A], p.2), and one that with the new concept of spread 'opens up a much wider field of development which includes analysis.' ([BMS49], p.4; cf. also the first version of [BMS32], p.7:

> .. the Second Act of Intuitionism, which recognizes in the self-unfolding of the Primordial Intuition not only the construction of natural numbers but

also the construction of spreads and the foundation of analysis on properties of spreads.)

The analytical origin of species and its status of 'a new mathematical entity', however, raise questions as to the nature of the existence of species. In the construction of a species $P$ and it elements Brouwer distinguishes the following phases of construction:
1. the construction of mathematical entities which form the domain of the species $P$;
2. the act of isolating an element of construction or a combination of such elements, and considering it as a 'property of', i.e. part of members of the domain, the 'conceptually completed' species $P$. We shall refer to it as the 'basic' or 'primary' species $P$;
3. testing the members of the domain for elementhood of $P$ and absurdity of elementhood of $P$.

We write $(P, x)$ for the hypothesis of $x$, a member of the domain, having the property $P$, the hypothesis that $P$ be 'fitted into' $x$, or be compatible with $x$. Testing the hypothesis may lead to the successful fitting-in of $P$ into $x$, confirmed in writing by '$x$ is proved to have property $P$', '$x$ is an element of $P$', '$P$ is compatible with $x$', or $\vdash (P, x)$. The attempt at fitting $P$ into $x$ may also lead to a proof that $x$ cannot have property $P$, that '$P$ is incompatible with $x$', or: $\vdash \sim(P, x)$, 'the assertion of an incompatibility' ([1954A], p.3). The attempt at fitting $P$ into $x$ may also remain unresolved. Both proof of elementhood, $\vdash (P, x)$, and absurdity of elementhood, $\vdash \sim(P, x)$ are binary constructions or relations, where $x$ is an existing mathematical entity and $P$ a 'supposable' property. [We leave aside here what constitutes a proof of elementhood or its absurdity, and the distinction between 'having been proved' and 'can be proved', i.e. 'a finite algorithm has been constructed leading to a proof of'. (see above 5.8.2.)].

It is one of Brouwer's fundamental theses that element-hood and absurdity of element-hood, although potentially contained in 'the mathematical entity previously acquired', are separate constructions to be joined to it; they only exist from the moment their construction has been completed, and mathematical entities only then 'acquire' the property.

One may add a further phase, and consider elements of the species $P$, the 'Brouwer-class' of mathematical entities which have been proved to possess the property $P$. Brouwer does not himself refer to it as a 'class' or 'totality', but he does speak of '*the elements of a species* [by which] we understand the mathematical entities previously acquired for which the property in question holds.' ([BMS51], p.8). In [BMS32] the Brouwer-class emerges as a new species, the property 'having been proved to possess property $P$', or 'the property of being element of $P$' ([BMS32], p.10). The new species, 'the element-species' or 'the property of having been proved to be elementof $P$', as well as 'the absurdity-species' or 'the property of element-hood of $P$ having been proved to be absurd', are distinct from the 'primary' species $P$, and 'of higher order' (see 6.3.7). In the definition of primary species, the existence of its elements and that of the totality of its elements do not play a role; indeed, the construction of element-hood and therefore the existence of the 'element' presupposes the existence of the property to be fitted and that of the particular candidate, a member of the domain of mathematical entities 'previously acquired'.

The existence of species, however, does pre-suppose the existence of a well-defined domain, and reference to this domain is an essential element in the definition of species. It is a 'dependent' existence, as is implied in the general concept 'property', and specifically referred-to in all Brouwer's definition of species by the condition 'supposable for...'. The property $P$ exists only as a construct supposable for the given domain, which Brouwer defines as all and only the mathematical constructions which genetically precede the construction of $P$.

The constructed species $P$ and its elements then become part of the domain in any subsequent species construction, resulting in a ramified cumulation of domains and species. (see further 6.3.7 The Hierarchy of Species).

By defining species in terms of its domain and independent of the existence of all its elements Brouwer avoided the pitfalls of self-reference and impredication. In this basic form the Brouwer species constitutes a principle of generalization. It allows the Subject to treat 'common property' as a definite, constructed mathematical entity, irrespective of whether all its elements have been or can be constructed. As legitimate mathematical objects common properties can be used in further constructions, including the construction of new species, i.e. properties of properties or species of species, reflected in the 'general statements' of analysis and function-theory.

The importance Brouwer attached to his species-principle and its role can be gauged from his comments on Weyl's failure to appreciate the distinction between Brouwer-sets and species, or his refusal to accept the species-principle and its consequences. In his otherwise enthusiastic support of Brouwer's *Foundations of Set Theory* Weyl had expressed his reservations as to the notion species-of-species and the legitimacy of generalization on the basis of species; indeed he failed to acknowledge the distinction between 'Menge' and 'Spezies':

> We shall wholly banish from our mind sets of functions and sets of sets. [Mengen von Funktionen und Mengen von Mengen sollen wir uns ganz aus dem Sinne Schlagen.] There is, therefore, in our analysis no room for a general set-theory, nor for general statements about functions.
> ...In my radical conclusions, as far as I understand, I do not wholly concur with Brouwer. (WEYL[1921], p.70)

In a private note Brouwer comments:

> p.71 ...*Limiting Arithmetic and Analysis to general statements about numbers and free-becoming sequences*. This restriction of mathematics to mathematical entities and species of the lowest order is totally unjustified. This obviously is a reference to his p.70, where he dismisses my theory of species as meaningless [Germ. 'sinnlos'; Brouwer's misunderstanding of the expression 'sich aus dem Sinne Schlagen'? Ed.], and it shows that in the end Weyl only half understands what Intuitionism is about.

Heyting, on the other hand, accepted Brouwer's species-principle and declared it to be the constructive basis of axiomatics and of the axiomatization of projective geometry of his doctoral thesis:

> We take the view that Professor Brouwer's principle of the *Begründung der Mengenlehre* I, p.4, by which from species of order $n$ a species of order $n+1$ is formed, has a clear meaning. Axiomatics is an application of this principle... (HEYTING[1925], p.2).

### 6.3.7 The Hierarchy of Species

In Brouwer's set-theoretical foundations of the years 1917-1927 the species-principle plays a prominent role. It is used to develop a general hierarchy of species constructions, genetic levels of abstraction and generalization, intended to specify the 'previously-acquired' and the constructive steps in the make-up of real numbers and functions.

In all Brouwer's accounts the first species 'construction' follows the constructive generation of 'mathematical entities': the individuals of the species domain are the fundamental generating procedures and their elements. For example in [1918B]:

Sets [Germ. 'Menge'] and elements of sets are called *mathematical entities*. By a *species of the first order* we understand a property which only a mathematical entity can possess; such a mathematical entity is then called *an element of a first-order species*. ([1918B], p.3; [1925A], p.245).

Species of higher order are then inductively defined:

> By *a species of the second order* we understand a property which only a mathematical entity or a species of first order can possess; this mathematical entity and the species of first order are then called *elements of a second-order species*.
>
> In a similar way we define *Species of order n...* (ibid.)

The condition restricting the domain of $P^n$ (the species $P$ of order $n$) to mathematical entities and species of lower order, ensures that 'species cannot be elements of themselves' (cf. also [BMS51] p. 8); on the other hand, since the species $P^n$ and its domain are included in the domain of species $P^{n+1}$, a species of order $n$ satisfies the conditions for a species of order $n + 1$ and, as Brouwer claims in the *Berliner Gastvorlesungen*:

> Any species of order $n$ is always also a species of order $n + 1$. ([BMS32], p. 9).

Brouwer's statement in [1918B] and [1925A] that 'spreads form a special case of first-order species' should be interpreted as a simple extension of this principle to the domain of the species of order 1.

The status of 'element-hood' in the genetic hierarchy of properties is at first left somewhat vague. In [1918B] species is defined 'intensionally', as property, but interpreted 'extensionally', i.e. in terms of its elements, in the definitions of congruence of species, subspecies, union and intersection. For example:

> The species $M$ is called a *subspecies* of the species $N$ if every element of $M$ is also an element of $N$. ([1918B], p.4)

[1925A] introduces the notion of equivalence of spreads and species in terms of their elements (not mentioned in [1918B]) and acknowledges the need for the construction of elementhood, i.e. the existence of elements, prior to the further construction of relations and combinations of species. For example:

> The species $M$ is called a *subspecies* of the species $N$ if for every element of $M$ there exists an equivalent element of $N$. ([1925A], p.246).

In the *Berliner Gastvorlesungen* and later papers Brouwer goes one step further and refers to element-hood as a 'property', for example:

> The property of being an element of the species $M$ as well as of the species $N$ is called the intersection of $M$ and $N$. ([BMS32], p.10; cf. also [BMS51], p.9).

The constructions of element-species and equivalence – as of any property of species expressed in terms of its elements – constitute a species of higher order. This conclusion was clearly drawn in [BMS32] w.r.t. equivalence at the first level of species construction. The construction of equivalence-species of spread-elements is introduced as an intermediate stage, 'a species of order zero', a stage of abstraction and generalization which follows the Subject's act of generating individual entities in time. Application of the species-principle to a generated spread-element results in what one might call a Brouwerian equivalence-class or rather equivalence-property, 'the spread-species':

> The property of being equivalent with the element generated by the spread $S$ we call the *spread-species S*, often also the *spread S* for short. A mathematical thought-construction which at this stage has already been generated and which has the property concerned is called an *element* of the spread-species $S$. Spread-species are also called *species of order zero*. ([BMS32], p.8)

It should be noted at this stage in the genetic hierarchy of mathematical construction the species-principle is applied to a particular spread-element and produces the property of equivalence to that spread-element, (e.g. the abstract 'two-ity' supposable for all constructed two-ities).

Introduction of the intermediate stage could be interpreted to be due to Brouwer's increasing awareness of the fundamental role of the species principle 'at every stage of the construction of mathematics', in particular in the creation of abstract mathematical concepts, which is based on the property of equivalence. At the fundamental stage, the zero level, the construction of the equivalence property isolates from a generated entity all but its individual identity, its contingent construction by the Subject at a certain moment of time. As such it provides a link between the contingency of the Subject's mathematical activity and the universality of his mathematical concepts.

In a letter to Heyting Brouwer speaks of his new awareness of the distinction between mathematical entities and their equivalence species:

> When I was giving my Berlin lectures it became evident to me that some

amendments were needed in my 'Zur Begründung der intuitionistischen Mathematik'; in particular I then introduced for every property as an added species 'the identity with an arbitrary thing which posseses the property in question'. I started from the 'species of order zero, i.e. either a given set-element or the identy with an arbitrary element of a given set.
Perhaps it is a better way of treatment to introduce along with the things themselves the 'species of identical things' and then mainly use the latter, in the same way that topological set-theory considers point-cores rather than the points themselves. (Brouwer to Heyting, 17.07.28).*

At this level, and all further levels, the construction of element-species and equivalence-species provides a justification of an extensional interpretation of species and extensional equivalence (recognized as necessary on practical grounds: '...since in mathematics we are only concerned with properties which are valid for mathematical entities together with their equivalents.' ([BMS32], p.9a):

> Two species (and in particular two spreads are called *equivalent* or *identical* if for every element of one an element of the other can be indicated which is equivalent with it. ([BMS32], p.9a; cf. also [BMS51], p.9 and [1954A], p.6).

Equivalence so expressed, i.e. in terms of 'having the same elements', forms the basis of other relations between species and a 'lateral' division of species (see further: The Theory of Species, 6.3.8).

The importance Brouwer at the time attached to the hierarchical structure of species and equivalence can be gauged from the elaborate detail in which he spells out the levels or 'orders' of species-construction. For this reason we quote the passage of [BMS32] in full:

> The property of equivalence with an element of the spread $S$ we call *spread-species S*, often *the spread S* for short.
> A mathematical thought-construction which has at this stage already been introduced and which has the above property is called *an element* of the species $S$.
> Two spreads, resp. two spread-species, are called equivalent or identical if for every element of one an equivalent element of the other can be indicated.

---

* In the same letter Brouwer announces the publication of his Berlin Lectures [BMS32] 'shortly'. The publication of [BMS32] did not materialize, but Brouwer made amendments in all his off-prints of [1925A, 1926A and 1927A], including this distinction and his hierarchy of species. These 'corrections' are given by Heyting in *Brouwers Collected Works* I, pp. 590-591.

Spread-species are also called *species of order zero*.

By *a species of order one* we understand a property which only spreads, spread-species and spread-elements can possess, and this only together with all mathematical thought-constructions introduced before with which they are equivalent.

Two species of order one are *equivalent* or *identical* if they have the same elements.

By a *species of order two* we understand a property which only spreads, spread-species, spread-elements, and species of order zero and of order one can possess, and this only together with all mathematical though-constructions introduced before with which they are equivalent, in which case they are called *elements* of this species of order two.

Two species of order two are *equivalent* or *identical* if they have the same elements.

By a *species of order three* we understand a property, which only spreads, spread-elements and species of order zero, of order one and of order two can possess, and this together with all mathematical thought-constructions introduced before with which they are equivalent, in which case they are called *elements* of this species of order three.

Two species of order three are called *equivalent* or *identical* if they have the same elements.

In a similar way we define *species of order n* and their equivalence for every natural number $n$, in which case a species of order $n$ is always also a species of order $n + 1$.

Two species are called *equivalent* or *identical* if they have the same elements. ([BMS32], pp. 8-9).

The distinction between different orders of species construction is not further referred-to in the *Berliner Gastvorlesungen*. *The Cambridge Lectures* and Brouwer's other post-war papers no longer mention the detailed hierarchy of species construction and equivalence species. They revert to the more general requirement that legitimate candidates for elementhood of species are only 'mathematical entities previously required'; and the construction of equivalence-species becomes an added condition of elementhood:

> ...mathematical species, i.e. properties supposable for mathematical entities previously acquired, satisfying the condition that if they hold for a certain mathematical entity, they also hold for all mathematical entities which have been defined to be 'equal' to it, definitions of equality having to satisfy the conditions of symmetry, reflexivity and transitivity. ([BMS51], p.8; also [1952B], p.142, [1954A], p.2, [BMS49] (App.7), p.4 etc.).

Brouwer's hierarchy of species construction is an attempt to analyse and justify mathematical construction based on 'comprehension', identifying its essential components and placing them in their 'genetic' order.

Most significant is his interpretation of species and its non-generative role. The species presupposes the existence of its domain and is 'conceptually complete' as 'property supposable for'; it does not produce 'elements' and does not itself partition the mathematical universe, not even in the sense of domain of mathematical entities previously acquired. Such partitioning or 'splitting' of the domain can only be the result of further constructive activity of the Subject, proving element-hood and absurdity of element-hood, and in some cases may prove to be complementary. (see further 6.3.9).

The principle of species-construction, however, allows both 'proven element-hood' and 'absurdity of element-hood' to be considered as (higher-order) species and therefore as legitimate mathematical thought-constructions. Further application of the species-principle, moreover, leads to the species 'absurdity-of-absurdity of element-hood of a species $P$, distinct from 'element-hood of $P$', but a legitimate mathematical thought-construction.

## 6.3.8 The Theory of Species

The binary relations 'proven element-hood' and 'absurdity of element-hood' are the fundamental constructions on which all relations between species and 'operations' on species are based. Relations and operations are therefore defined extensionally, which accounts for a certain measure of agreement with traditional set-theory. For example, definitions of equivalence of species, subspecies, union and intersection appear to be little different from their classical counterparts:

> Two species are called *equivalent* or *identical* if for every element of one an element of the other can be indicated which is equivalent.
> ...A species $M$ is called a *subspecies* of the species $N$ if ever element of $M$ is also an element of $N$.
> ...The property of being an element of the species $M$ as well as of the species $N$ is called the *intersection* of $M$ and $N$ and is denoted by $\mathscr{D}(M, N)$.
> ...The property of being an element of either the species $M$ or the species $N$ is called the *union* of $M$ and $N$, and is denoted by $\mathscr{V}(M, N)$. ([BMS32], pp. 9-10; cf. also [BMS51], p.9 and [1954A], p.6).

However, the Brouwer interpretation of elementhood and absurdity gives these definitions a special meaning and leads to considerable divergence from classical set-theory. The most conspicuous result is his rejection of the universal complementarity of species (see further 6.3.9). The condition of proven elementhood and absurdity also leads to a wider range of distinctions and relations between species and to a stricter interpretation of species-combinations.

### 6.3.8.1 Union and Intersection of Species

The notions 'and' and 'or' in the above-quoted definitions of union and intersection are not further defined; but one could readily give a constructive interpretation of intersection and union as 'species-constructions', similar to the Boolean 'product' and 'sum' of elective operators.

The significance of *proven* element-hood of the species $M \cup N$ is that proof of element-hood of $M \cup N$ requires in all cases a proof of element-hood of $M$ or a proof of element-hood of $N$.

Another implication, mentioned in [BMS32], is that species are not necessarily closed under the 'operations' of union and intersection. The counterexamples given to show that intersection of spreads and union of individualized spreads do not necessarily constitute spreads, resp. individualized spreads, rely on the impossibility of 'testing', i.e. deciding on elementhood, which results from definition of spreads by 'fleeing property':

> To show that the intersection of two spreads does not necessarily have to be a spread, we start from the fleeing property $f$, define $R$ as the spread whose only element is the indefinitely proceeding digit-sequence consisting exclusively of 0's, and $S$ as the spread whose only element is the indefinitely proceeding sequence of which the $n$-th element is the digit 1 if $n$ has the property $f$ and the digit 0 otherwise.
> 
> $\mathscr{D}(R, S)$ in this case would itself be a spread only if the question of the existence of an element of $\mathscr{D}(R, S)$, i.e. the *absurdity of the existence* of a critical number for the property $f$ could be made to be equivalent to the question of the existence of an element of a given spread. The latter is again a question of the existence of a critical number for a given fleeing property.

> ...That the union of two individualized spreads in the restricted sense, need not be an individualized spread even in the wider sense, is clear when we again consider the above spreads $R$ and $S$, each of which possessing only one element. For, if there were an individualized spread $U$ in the wider sense equivalent to $\mathscr{V}(R, S)$ then, after an initial sequence of $v-1$ unhemmed choices and as a result of the spreadlaw of $U$, two unhemmed $v$-th choices could occur for at most one natural number $v$, and for all other natural

numbers $v$ in every case only one unhemmed $v$-th choice.

In addition, the spread-law of $U$ would have to make the v-th choice dependent on whether or not the first $\mathscr{D}(v) \geq v$ natural numbers possess the fleeing property $f$, and for only the smallest value $v_1$ of $v$, for which $\mathscr{D}_{v_1} \geq \kappa_f$ would have to indicate two unhemmed $v$-th choices. If $U = \mathscr{D}(R, S)$, then because of the main theorem of closed spreads there would exist a natural number $m$ such that for an arbitrary element of $U$ it would be certain at the $m$-th choice whether it belonged to $R$ or to $S$. But this would require at the $m$-th choice the certainty that for $v > m$ after an initial sequence of $v-1$ unhemmed choices there could never be the possibility of two unhemmed $v$-th choices. This would only be possible if the existence of $\kappa_f$ were certain or absurd. ([BMS32], p. 11a).

### 6.3.8.2 The Empty Species

The notion *'empty species'* is first mentioned by Brouwer in [BMS32], and is there defined as:

> A species which cannot have an element is called *empty*. Two different species whose intersection is empty are called disjoint. ([BMS32], p. 10; Du. 'uiteenliggend', translated by Brouwer on p.11a as 'exterior to each other', in [BMS51] as 'disjoint'; cf. also [1954A], p.6).

It is significant that the notions 'empty species' and 'disjoint species' are introduced in one paragraph. We argued above (5.8.2) that the concept of 'absurdity' is a primitive binary relation, an 'incompatibility'. The empty species is essentially the intersection of two incompatible species; as 'property supposable for' it can exist only as the hypothesis of the conjunction of incompatible constructions, absurdity itself.

### 6.3.8.3 Relations between Species

The Brouwer interpretation of 'element' and 'mathematical absurdity' and the nature of his fundamental mathematical entities, i.e. spreads, give rise to a number of distinctions and further relations between species. They are introduced and developed in his work on general mathematical species during the period 1917-1927, in particular [1918B], [1925A] and [BMS32].

Application of his species-theory to 'point-sets' ([1919A]) and the introduction of 'absurdity' as a logical operator ([1923C]) led to a further range of relations and distinctions, and to a proliferation of order-relations, power-relations and denumerabilities. ( Cf. e.g. the logical relations between point-species of [1923C], p.880): 'apartness', 'separation', 'agreement', 'deviation', 'detachment', 'congruence', 'coincidence', 'intertwinement', 'dissociation', and in addition 'sharp difference' in [1930A], p.10; the power-relations of [1925A], p. 255:'over-associated', 'of-greater-range', 'of-equal-range', 'overlaid', 'of-greater-width', 'of-equal-width', 'super-imposed', 'of-greater-extension', 'of-equal-extension', 'covered', 'of-greater-weight', 'of-equal-weight'; and the relative denumerabilities, almost impossible to translate from the German: 'abzählbar', 'nachzählbar', 'auszählbar', 'überzählbar', 'durchzählbar', 'aufzählbar' ([1925A] p.255), and in addition 'überabzählbar' in [1930A], p.5).

The pure-mathematical relations between species are: equivalence and its various alternatives, based on common element-hood and absurdity of common element-hood. The two extremes, expressing either common element-hood or absurdity of common element-hood for *all* elements of the two species, are *equivalence* and *disjointness*. Definitions of equivalence and disjointness of species are given in [1918B] and remain practically unchanged, e.g.:

> Two species are called *equivalent* if for every element of one an element of the other can be indicated which is equivalent to it. ([BMS32], p.9a; cf. also [BMS51], p.9);

in other words: two species $M$ and $N$ are equivalent if $M$ is a subspecies of $N$ and $N$ a subspecies of $M$.

> Two species $M$ and $N$ are said to be *disjoint* (Germ. 'elementefremd') when no element can exist which belongs to $M$ as well as $N$. ([1918B], p.4).

Other relations express a partial sameness or difference, based on the various possibilities of common element-hood or its absurdity for some of the elements.

### 6.3.8.4 Difference, Deviation and Apartness

The most general form of difference of species is the simple absurdity of their equivalence. In this general form it is introduced in [1925A] as '*difference*':

> Two species $M$ and $N$ are said to be different [Germ.'verschieden'] if the absurdity of their equivalence has been ascertained. (p.246; cf. also [BMS32], p. 10a).

The case when evidence of difference is provided by the existence of an element of one species whose elementhood of the other species is proved to be absurd is referred to as *'deviation'*. [1925A] defines a relation of deviation for general species:

> We say that the species $M$ stands out [Germ.'herausragt'] from the species $N$ if $N$ possesses and element which is different from each element of $M$. ([1925A], p.246).
> Cf. the definition of [BMS51] p.9:
> If the species $M$ possesses an element which cannot possibly belong to the species $N$, or what is the same, is different from each element of $N$, we shall say that $M$ *deviates* from $N$. (also [BMS32] p.9a).

The term 'deviation' (Du. 'afwijking', Germ. 'Abweichung') is first used in [1923C] where it refers to a relation between 'points' and 'point-species'. Allow me here to quote definitions of point, point-core and coincidence (a fuller treatment is given in section 6.3.11):

*Points*, or Cartesian points are defined as 'indefinitely proceeding sequences of nested plane-intervals, or λ-squares' (cf.[BMS32], p. 16);

*coincidence of points*: 'if $p'$ and $p''$ represent the points $\kappa'_1, \kappa'_2, \kappa'_3,\ldots$ and $\kappa''_1, \kappa''_2, \kappa''_3,\ldots$, we say that $p'$ and $p''$ are *coincident* if there is a certainty that, for arbitrary $\mu$ and $\nu$, $\kappa'_\mu$ and $\kappa''_\nu$ overlap [i.e. cover one another entirely or partially]'. ([BMS32], p.16, also [BMS51], p.32);

a *point-core* or Cartesian point-core $P$ is defined as 'a species of Cartesian points coincident with a point $p$' (ibid.);

*coincidence of point-species*: ' Two species $Q$ and $R$ are said to be coincident if every point of $Q$ coincides with a point of $R$ and every point of $R$ coincides with a point of $Q$.' ([1923C], p.880).

Deviation of points is proven 'difference', or 'the absurdity of coincidence of points':

> The points $p'$ and $p''$ of the Cartesian plane are said to deviate from one another if they cannot possibly coincide. ([BMS32], p.16a; cf. [BMS51], p.32).

Deviation of point-species is then defined in terms of deviation of points:

> Two point-species $Q$ and $R$ are said to be deviating from one another if one of the point-species possesses a point which deviates from the other point-species [i.e. from every point of the other species]. ([1923C], p. 879).

A particular case of 'deviation' of points and point-species had already been introduced in [1919A] as *'local distinctness'*, and was later referred to as *'apartness'* (Du. 'verwijdering', Germ. 'Entfernung'). Apartness in its simplest form is disjointness of plane intervals ([BMS51], p.32). The apartness relation of two points $p'$ and $p''$ is defined as the (proven) disjointness of a $\lambda$-square of $p'$ and a $\lambda$-square of $p''$, and therefore of all their subsequent $\lambda$-squares:

> The points $p'$ and $p''$ are said to *lie apart* if a square of $p'$ and a square of $p''$ can be indicated which lie outside one another. ([BMS32A], p.16; cf. [1923C], p.878; also [BMS51], p.32 which uses 'separation' instead of apartness, a term in [1923C] reserved for 'absurdity-of-absurdity of apartness').

Apartness of species is then again defined in terms of apartness of points:

> Two species $Q$ and $R$ are said to lie apart if one species contains a point which lies apart from every point of the other species ([1923C], p.880) ...i.e. 'if it lies apart from every point of the other species'([1923C], p.879).

Difference, deviation and apartness of species are each based on the absurdity of an equivalence aspect in the make-up of species and point-species: difference of species is the absurdity of general equivalence of species, 'deviation of species' is based on the absurdity of equivalence of their elements (coincidence of their points), 'apartness' on the absurdity of 'concidence' or 'overlapping' of the $\lambda$-squares of their points. These aspects of equivalence and their absurdities constitute what Brouwer calls the 'fundamental relations' of point-species ([1923C]).

Other relations between species are generated in [1923C] by applying the logical operators, the 'predicates' absurdity and absurdity-of-absurdity to these fundamental relations:

> Applying the predicates absurdity and absurdity-of-absurdity to coincidence of species we generate resp. dissociation and intertwinement
> ...to deviation, we generate resp. congruence and detachment
> ...to apartness, we generate resp. agreement and separation. (p.880).

[1923C] is a logical exercise, it examines what *distinct* relations can be generated by absurdity-operators on the basis of the principle of the Brouwer-logic that 'absurdity-of-aburdity-of-absurdity is equivalent with absurdity', in other words that 'a finite sequence of absurdity predicates can be reduced either to absurdity-of-absurdity or to absurdity' (p.878). Although application of the logical operators is justified by the corresponding mathematical species construction and the relations so generated reflect mathematical species, only few play a role in Brouwer's Intuitionist mathematics. Significantly, those that are used, such as 'congruence' and 'agreement', had already appeared in his mathematical papers prior to the publication of his logical calculus of absurdities.

### 6.3.8.5 Congruence and Agreement

*Congruence* and *agreement* are both defined as 'absurdity-of-absurdity' of an equivalence relation. The distinctness from their corresponding equivalence examplifies Brouwer's fundamental thesis on double negation; instances of species that are congruent or concordant but not equivalent provide counterexamples refuting the PEM.
Congruence is defined as absurdity of mutual deviation:

> Two species $M$ and $N$ are said to be *congruent* if no element of $M$ can exist which is different from every element of $N$ nor an element of $N$ which is different from every element of $M$. ([1918B], p.4).
> ...Two species $M$ and $N$ are said to be congruent if neither can deviate from the other. ([1925A], p.246; also [BMS32], p.9 and [BMS51], p.9).

A special case of congruence, referred-to as 'semi-equivalence' or '*half-identity*', is introduced in [1918B]:

> [$M$ and $N$ are congruent species] If, moreover, every element of $M$ is also an element of $N$ then $M$ and $N$ are called *half-identical*. ([1918B], p.4);

in other words, if every element of $M$ is an element of $N$, and no element of $N$ can deviate from $M$, i.e. proved to be different from every element of $M$. Half-identity is obviously a one-way relation, and is so described in [BMS32]:

> A subspecies $M$ of $N$ which is congruent with $N$ is called *half-identical* with $N$. ([BMS32], p.10; also [1942A], p. 323 where the printing-error of [1925A] is corrected.)

*Agreement* appears first in [1919A] as 'local agreement', the absurdity of 'local distinctness', later as 'agreement' or 'concordance', the absurdity of apartness:

> Two point-core species of the Cartesian plane are said to be *locally concordant* or simply *concordant* if neither can contain a point-core which lies apart from the other. ([BMS32], p.17a; also [BMS51], p.33).

Congruence and concordance are important relations in Brouwer's alternative analysis (see further 6.3.11).
Examples of concordant and of congruent species which are not necessarily equivalent, are given a.o. in [1925A], [BMS32] and [BMS51]. We quote

the example of congruent but not equivalent species given in the unpublished [BMS32]:

> Let $P$ be the species of indefinitely proceeding sequences of digits 0 and 1, $M$ the species of those digit-sequences of $P$ of which it is known either how many 0-digits precede the first occurrence of the digit 1, or that the digit 1 does not occur in the sequence, and let $N$ be the species of those digit-sequences of which it is known either how many 1-digits precede the first occurrence of the digit 0 or that the digit 0 does not occur in the sequence. Then $M$, $N$ and $P$ are all mutually congruent. ([BMS32], p. 10a).

### 6.3.9 The Complementarity of Species

The space in Brouwer's writing given to the treatment of 'Complementarity' is somewhat disproportionate to the restricted interpretation allowable within his species theory and the relatively minor role it plays in his intuitionist mathematics. Most of it is critical, analysing and refuting the classical thesis of complementarity of sets, identified with the logical Principle of the Excluded Middle. His most fundamental objections were raised in his 'philosophical writings' ([1907], [1908C], [1912A], [1933], a.o.); they concern the Platonistic conception of sets, the presumption of a universal mathematical domain, automatically partitioned by any one property into two complementary sets.

Having found a valid, and in his views the only valid interpretation of 'universe' and 'set', he now examines various aspects of 'complementarity' in the more restricted setting of 'domain of mathematical entities previously acquired' and the constructed species of 'element-hood' and 'absurdity' of one single property. In this interpretation a partitioning of the domain is brought about by any aspect of species construction between the mathematical entities for which the construction has been completed, i.e. has been proved or can be proved (i.e. an algorithm is known which will lead to the proof), and the residual part of the domain for which a proof does not or does not yet exist; in the case of element-species $P$: a partitioning between those mathematical entities which have been proved to be elements and the residue of the domain for which element-hood (as yet) has not been or cannot be proved. Likewise, the absurdity-of-$P$ partitions the domain into those elements for which the absurdity of $P$ has been proved and a residue for which no such proof exists; similarly, a partition is generated by the construction 'absurdity-of-absurdity-of $P$' etc..

In Brouwer's view the various aspects of classical complementarity arise from the confusion of the 'residue' resulting from a species construction and its corresponding absurdity-species. The classical principle of complementarity, referred-to by Brouwer as the '*Complete Principle of Judgeability*' or 'the Complete Principle of the Excluded Third', asserts that the residue resulting from the species of proven elementhood is identical with the absurdity-species of the same property, in other words, that there is no residue after the combined construction of element-species and absurdity-species of a single property:

> We formulate the complete principle of the excluded third as follows:
> If $a$, $b$ and $c$ are species of mathematical entities, if further both $a$ and $b$ form part of $c$, and if $b$ consists of those elements of $c$ which cannot belong to $a$, then $c$ is identical with the union of $a$ and $b$. ([1948C], p.1245; also [1948B], p.1240).

The Complete PEM is a universal assertion about any domain and for any property, 'a species of assertions' ([1954A], p.3). In its generality and within the Brouwer interpretation of 'element-hood' and 'absurdity' it is an obvious mis-statement of fact and, as Brouwer claims, even 'contradictory' (ibid.). His counterexamples of unsolved mathematical problems, undecidable properties and 'fleeing properties' provide the evidence (see 5.9.4.).

The same confusion of 'residue' and 'absurdity' in the case of the 'absurdity-species' leads to what Brouwer calls '*the Complete Principle of Testability*', which asserts that the residue of the 'absurdity-species' of a property is identical with the absurdity-of-absurdity species of that property, in other words, that the residue of the combined construction of the absurdity-species of a property and the absurdity-of-absurdity species of that property is empty:

> If $a$, $b$, $d$ and $c$ are species of mathematical entities, if each of the species $a$, $b$, and $d$ form part of $c$, if $b$ consists of elements of $c$ which cannot belong to $a$, and $d$ of the elements of $c$ which cannot belong to $b$, then $c$ is identical with the union of $b$ and $d$. ([1948C], p.1245).

[1948C] and [BMS51] describe the Principle of Testability as a corollary of the Principle of Judgeability, based as it is on the apparently weaker supposition that 'every mathematical assertion can be tested, i.e. that for every mathematical entity either the absurdity of a property can be proved or the absurdity of its absurdity (the Simple Principle of Testability).'

(ibid.). The counterexamples given in [1948C] and [BMS51] are refutations of the Complete Principle of Testability as well of Judgeability.

The Principle of Testability is an assertion about the absurdity-species and absurdity-of-absurdity-species of a property, it is not immediately concerned with the element-species of that property. The relation between element-species and absurdity-of-absurdity-species, resulting from the Complete Principle of Judgeability is expressed in *the Principle of Reciprocity of Complementarity*, which asserts the identity of element-species and its absurdity-of-absurdity species, in logical terms the equivalence of truth and non-contradictority. It is first mentioned in [1923C] as a logical principle, 'a special case of the logical Principle of the Excluded Middle' (in the German 1925 version as 'a corollary of the PEM'):

> The Principle of Reciprocity of complementary species, which for an arbitrary mathematical system derives the truth of a property from the absurdity of the absurdity of that property. ([1923C], p.877).

The Principle of Reciprocity of Complementarity as a universal principle is rejected and refuted on the evidence of counterexamples. The same paper, however, makes an important exception and establishes an intuitionist alternative to the classical principle of complementarity:

> Absurdity-of-absurdity is not equivalent with correctness...(p.877), but ...absurdity-of-absurdity-of-absurdity is equivalent with absurdity. (p.878).

The latter principle is given as a theorem; the proof consists of two parts and is a deduction apparently based on the classical logical principles $(a \rightarrow b) \rightarrow (\neg b \rightarrow \neg a)$ and $a \rightarrow \neg\neg a$ (quoted in full in 5.10). The same proof is almost literally repeated in [BMS51], p.12 and [1948C], p.1245.

Significantly, Brouwer uses the terms 'predicates' ([1923C]) and 'assertions' ([BMS51] and [1948C]) and emphasizes the logical nature of the principle ('a theory which with some right may be called intuitionist mathematical logic, and to which belongs a theory of the principle of the excluded third' ([1954A], p.3). [1948C] insists that such logical principles are 'assertions about assertions, only then realized, i.e. only then convey truths, if these truths have been experienced [i.e. mathematically constructed]' (p.1245).

In the mathematical reality of Brouwer's species construction, however, the complementarity-principle plays a relatively minor role and no direct attempt is made in Brouwer's 'mathematical' papers to translate the proof of his modified principle of complementarity in terms of species constructions. An interpretation of the logical proof of $\neg\neg\neg a \to \neg a$ in terms of species constructions and based on the intuitive principles of contradiction and what we have referred-to as 'conclusiveness' (5.8.1) is readily given.

However, no justification is given of the substitution of a negative property in the implication $a \to \neg\neg a$, nor does Brouwer restrict his modified complementarity principle to what he later described as 'essentially negative properties'.

A modified theory of complementarity of mathematical species was developed in [1925A], [BMS32] and [BMS51], prompted by the needs of the Intuitionist analysis of the continuum (cf. e.g. his proof of the 'connectedness' of the intuitionist continuum in [1930A], p.10; cf. also [BMS51], p.81 ff.). Its purpose is to specify the conditions for the construction of a species to determine a unique 'complementary' species, i.e. such that the 'residue domain' constitutes a uniquely defined species.

Complementarity is considered in a restricted setting of domain and species. The two species under consideration are simply defined as 'disjoint'. Neither is considered as the element-species of a property of which the other is the absurdity-species. Both species are given and their mutual relation is defined 'negatively' in the sense of disjointness, not necessarily as the unique 'absurdity' of the other.

As to the 'domain', Brouwer considers two cases, which give rise to two distinct forms of complementarity: the case when the union of the two species concerned is identical with the domain and the case when the union is 'half-identical' with the domain.

In the latter case the domain is said to be 'composed' of the two species, and the two species are called 'complementary' ('conjugate' in [BMS51]):

> If $M_1$ and $M_2$ are disjoint subspecies of the species $N$ and if $\mathscr{V}(M_1, M_2)$ is half-identical with $N$, we say that $N$ is *composed of* $M_1$ and $M_2$ and that $M_1$ and $M_2$ are *mutually complementary species in* $N$.
> An example of complementary species are the species of numbers which, used as exponents of the Fermat equation, make this equation respectively solvable and unsolvable. ([BMS32], pp.10-11; cf. also [1925A], p.247 and [BMS51], p.25).

Given a domain $N$ and a species $M_1$, it is always possible to construct a 'complementary conjugate' species $M_2$ such that the conditions of half-identity of $\mathscr{V}(M_1, M_2)$ and $N$ are satisfied. But, as Brouwer insists, such a 'complementary' species is not necessarily uniquely defined:

Within a given $N$ one can for every $M_1$ indicate a complementary species $M_2$, but it is out of the question [Du. 'er is geen sprake van'] that such $M_1$ and $M_2$ for a given $N$ should be defined by each other. ([BMS32], p.11; also [BMS51], p. 25 where Brouwer gives the example of the species of irrational numbers within the species of reals, and as its corresponding 'complementary' species 'the species of those real numbers whose rationality is non-contradictory as well as the species of rational numbers').

The uniqueness of the complementary species is not in question if the union of the two species concerned is identical with the domain, if, as Brouwer puts it, the domain is 'split into conjugate splitting species'. But Brouwer questions the general possibility of 'splitting' a given domain on the basis of a given property, i.e. defining the residue domain as a species:

> If $K_1$ and $K_2$ are disjoint subspecies of the species $N$ and if $\mathscr{V}(K_1, K_2)$ is identical with $N$, we say that $N$ is *split* into $K_1$ and $K_2$ and we call $K_1$ and $K_2$ *conjugate splitting species* in $N$, and we call $K_1$ as well as $K_2$ a *removable* subspecies of $N$.
> 
> For example, the species of numbers consisting resp. of at most five and more than five digits are conjugate splitting species in $A$.
> 
> But it is out of the question that for a given $N$ and for any $K_1$ one should be able to indicate a conjugate splitting species $K_2$. There are even species which do not contain any removable real subspecies. ([BMS32], p.11; also [BMS51], p.25 which quotes 'the species of the prime numbers and the species of composite numbers' as an example of conjugate splitting species.)

Genuine and full complementarity is – by definition – found in the case referred-to in [BMS51] as 'directly conjugate species', conjugate species which are defined not only as disjoint but as each other's absurdity-species, when in other words 'element-species' is equivalent with its absurdity-of-absurdity species:

> If $V$ and $W$ are conjugate subspecies of $P$, and if in addition $V$ consists of those elements of $P$ which cannot belong to $W$, and $W$ of those elements of $P$ which cannot belong to $V$, we shall say that $P$ is *directly composed* of $V$ and $W$, and that $V$ and $W$ are *directly conjugate subspecies* of $P$. ([BMS51], p.25).

Even though 'direct conjugation' seems to be given as a general case, definition of two species by mutual absurdity ultimately requires a 'primary property' which, moreover, is to be equivalent with the absurdity of its absurdity, and which on Brouwer's presupposition can only be a negative property, the 'example' quoted:

> Thus, for instance, the conjugate species $D$ and $E$ consisting of those elements of $P$ for which a certain negative property $\varphi$ is true and absurd respectively are directly conjugate species of $P$. ([BMS51], p.25).

### 6.3.10 Spreads and Choice-sequences

> Liberation from this situation was brought by the *second act of intuitionism*, the admission as a modality of the self-unfolding of the Primordial Intuition of Mathematics not only of the assemblage [Du. 'samenvatting'] of finite sequences of mathematical systems and lawlike indefinitely proceeding sequences of inductively pre-formed mathematical systems, but also of the assemblage of sequences of mathematical systems proceeding indefinitely in complete freedom or in freedom subject to possibly changing restrictions. ([BMS32], p.7).

The broadening of the concept 'sequence' was considered by Brouwer to be the most radical proposal of his reform of the fundamental concepts of classical analysis and, as the introduction of the concept 'species', a radical change from his own earlier views, 'a new insight'.

The sequential nature of the generative processes of mathematics had already been recognized in his early work and was firmly based on the Primordial Intuition of Time. The fundamental concept of sequence had further led to a notion of infinity as step-wise proceedibility, i.e. the denumerably infinite. The classical infinite sequence' had been acknowledged as a legitimate mathematical construction: although proceeding indefinitely it was considered to be a completed construction as algorithm, the constructed procedure for the generation of its elements. Definition as algorithm, however, restricts the use of the infinite sequence, in particular its use as a constructive device in the analysis of the continuum. Brouwer's argument against the classical analysis of the continuum as a totality of infinite sequences of rational points was based on the denumerability of such algorithms. The classical system of real numbers and Brouwer's own 'everywhere-dense scale' were no more than an alien grid superimposed on the continuum for practical purposes; in no way could they be considered as a constructive analysis of the mathematical continuum. Brouwer describes his doubts about the merits of such analysis in the lines preceding his announcement of the Second Act of Intuitionism, the 'situation from which liberation was brought':

At this point one may well raise the question whether analysis should be given any significance beyond that of an empirically efficient method of calculation in mechanics and the physical sciences without any deeper content in itself. ([BMS32], p.7).

The admission of 'free choice' as a legitimate mathematical act and infinite sequences determined by choice as genuine mathematical constructions completed Brouwer's new design of real number. While the characterization of element-of-the-continuum as convergent sequence of nested intervals guaranteed the measurability of the continuum, extending the range of infinite sequences beyond the limits of algorithm secured its non-de-numerability. Moreover, and wholly in line with Brouwer's philosophy, the element of choice highlights the active role of the Subject, assigning elements, and the positive growing aspect of the mathematical infinite.

Consideration of choice and choice-sequences as mathematical concepts was not altogether new. References to 'choice' and 'choice-sequence' can be found in Brouwer's earlier work and indeed in that of others. The words 'keuze' (choice) and 'willekeurig' (chosen at will) are frequently used in [1907].* 'Arbitrary choice' is the basis of the construction of 'a wholly freely constructed everywhere-dense scale of points on the continuum' ([1907], p.10, pp.64 ff.; cf. also [1908A], p.570). Choice is further used to prove the impossibility of constructing the continuum as a totality of points. Choice and arbitrariness are identified with absence of law, while law is recognized as the only justification of infinite sequences as 'completed constructions'. Extensionally, considered as an ordered totality of elements, divorced from the generating process, every 'infinite sequence' exists only as a finite initial segment:

> The sequence converging to a definite point, however, can never be thought of as completed, and we must therefore consider it as partly unknown. ([1907], p.10).

In particular, sequences determined by free choice, which lack the math-

---

\* e.g. [1907], pp. 8-9.
  Given Brouwer's precise, somewhat puritanical use of language, the Dutch word 'willekeurig' should be interpreted in its original meaning, i.e. resulting from choice of will.
  The translation of this passage of *the Collected Works* , 'willekeurige denkbare...' into 'arbitrarily given', interprets 'willekeurige' as an adverb qualifying 'denkbare' - which in Dutch would have required 'willekeurig'-; the 'given' is an unfortunate translation of Brouwer's 'denkbare', i.e. 'thinkable'.

ematical existence of constructed law, only exist as the ordered totality of their completed intitial segments; there is no justification for the existence of the infinite choice sequence as a completed totality of all its elements as is, for example, presumed in the Axiom of Choice. Neither can Brouwer at this stage conceive the convergent choice-sequence in its uncompleted, growing form as sufficiently well-defined to individualize a point-on-the-continuum and serve as an argument of real functions.

The notion of 'free-choice sequence' was mooted by Borel, in particular in his paper (BOREL[1909]) delivered to the 1908 International Congress of Mathematicians, attended by Brouwer. Borel's view that the uncountable infinity of choice sequences was needed to justify the 'arithmetic continuum' and his rejection of choice-sequences as not sufficiently defined to be legitimate mathematical objects were intitially shared by Brouwer. As late as 1912 he dismissed the notion of choice sequences defining real numbers as 'formalist', contrasting it with the intuitionist identification of real number and constructed, finite law:

> The Formalist creation of 'the set of all real numbers' is devoid of any meaning, whether one thinks of the Formalist real number defined by a fundamental sequence of freely chosen numerals or the Intuitionist real number determined by a finite law of generation. ([1912A], p.22).

But like Borel he accepts the continuum power of the hypothetical set of free-choice sequences:

> The Intuitionist can only construct denumerable mathematical sets, and if he, on the basis of the intuition of the linear continuum, admits a fundamental sequence of free choices as an element of construction, he proves quite easily that the non-denumerable sets so constructed have no other power than that of the continuum. ([1912A], p.23).

The *Review of Schoenflies and Hahn* ([1914]), and the *1916 Notes* ([BMS15A]) mark a fundamental change in Brouwer's conception of real number and the role of choice sequences. Definition of real-element-of-continuum is no longer restricted to 'finite law of generation', choice is now accepted as a constructive device in its characterization as *well-constructed infinite set*:

> For the Intuitionist there exist only *well-constructed infinite sets*, which are composed of two parts: a part of the first kind, constructed as a single fundamental sequence, and a part of the second kind which is based on a fundamental sequence $f$ as Frechet-class, but whose elements are deter-

mined by a sequence of choices from elements of a finite set or fundamental sequence in such a way that to every sequence of choices there corresponds a sequence of nested intervals [Germ.'Teilgebiete'] of $f$ whose widths converge to zero, and that the final segments of every two sequences of intervals corresponding to different sequences of choices lie outside one another. ([1914], p.79).

The 'well-constructed infinite set' retains the essential characteristics of Brouwer's original set-concept: it is identified with the constructive procedure for generating elements. Such constructive procedures, however, are now recognized to leave sufficient scope for 'free acts of the Subject' and so allow full representation of points of continuous space.

Reference to 'final segments lying outside one another' is the first mention of the 'apartness' relation and an indication of Brouwer's solution to the problem of insufficient determination of point-of-the-continuum by choice sequence.

The problem and its solution are clearly stated in an added note on page 1 of [BMS15]. The question is whether the infinite choice sequence, which can only claim mathematical existence in its incomplete, growing form, is sufficiently defined to determine an element of the continuum uniquely, and can be used as an argument of real functions. Brouwer's answer is that the mathematician 'can work very well' with such a sequence on the basis of its completed initial segments and a 'law' which assigns a definite mathematical entity to each initial segment:

> A mathematical thing is either:
> an element of a previously constructed fundamental sequence $\mathscr{F}$ (which is governed by induction, such as the sequence $\rho$);
> a finite sequence;
> or a fundamental sequence $f$ of freely chosen elements of $\mathscr{F}$ (which is never completed and is not governed by induction).
> With such a sequence [of the latter type] one can work very well as long as, when relating the *finite thing d* or the *fundamental sequence r* to this sequence one can *at every stage* work with a *suitable* initial segment of $f$ ($r$ is therefore in general never completed).
> If the fundamental sequence $f$ is to represent a point, it must lead to a *point*, i.e. converge to a limiting value for the distances to elements of the Frechet V class.
> A set then is a *law* by means of which a $d$ or an $r$ is derived from $f$; such an $r$ may then also contain as elements relation-symbols (such as e.g. the ordering symbols), so that the law may lead e.g. to a well-ordered set or other ordered sets or to functions. ([BMS15A], p.1)

Brouwer's introduction in 1917 of Spreads and of species ([1918B]) comes after a long process of reflection on the nature of element-of-the-continuum, development of new concepts and study of their implications.

In the best mathematical tradition the 'theory of spreads' is introduced as a generalized structure rather than the model that inspired it and the particular purpose Brouwer had in mind. The general theory of spreads and choice-sequences was further developed by Brouwer and by his 'successors' (in [1923A], [1924D], [1925A], [1927A], [1942A], [1942B],[1942C], [1954A], [BMS32], [BMS37], [BMS45], and [BMS51]. From the work of his successors we mention HEYTING[1956] and [1981], DUMMETT[1977], KLEENE AND VESLEY[1965], KREISEL[1958], [1968], KREISEL AND TROELSTRA[1970], POSY[1975], VAN DALEN[1974], [1975], TROELSTRA[1968], [1969A], [1969B], and in particular TROELSTRA[1977A], [1982] and [1983]. A survey of development to-date is given in TROELSTRA AND VAN DALEN[1988].)

In line with the express purpose of this chapter, I shall confine myself to an introduction of the notion 'spread' as it developed in Brouwer's search for a characterization of elements of the continuum and real function, examine the 'new' fundamental concepts and Brouwer's claim that they are directly rooted in Intuition. This 'development' is more clearly reflected in the treatment given in Brouwer's later publications and manuscripts where the general theory of spreads and species is preceded by an introduction of 'point-species' and 'point-spreads' ( cf. e.g. [BMS51]).
Our main reference will be the introductory chapter of Brouwer's incomplete and un-published book on *Real Functions* ([BMS37]) (a full translation of the first 7 pages of the manuscript is included in this monograph as 'appendix 11').

### 6.3.11 Real Elements of the Continuum

The new Brouwer concept of real-element-of-continuum can be simply defined as 'species of coincident points' or 'equivalence species of infinite sequences of nested intervals whose lengths converge to zero'. So described it may appear to be little different from the classical concept of real number. The definite distinct identity of the Brouwer real number or 'point-core' is due to his interpretation of the ingredient notions used in this definition; traditional terms such as interval, point, sequence, law, convergence, almost all need considerable qualification.

### 6.3.11.1 Interval

'Interval' remains the fundamental primitive in the construction of real number: the 'intuitive between' identified with the continuum itself and characterized as the potential for infinite division by insertion of discrete points. Brouwer's earlier view, tracing the genesis of interval directly to the Primordial Intuition, is almost literally repeated in the major works of later years:

> Together with the abstracted two-ity as the Primordial Intuition of mathematics was created the 'between' which is never exhausted by the insertion of new singularities nor by the insertion of infinite sequences of new singularities. In Chapter 2 we recognized this 'intuitive between' as the 'matrix of point-cores' and as such joined it to intuitionist mathematics under the name 'continuum'. ([BMS32], p.18; cf. [BMS51], p.42, also section 6.3.2).

The dual aspects of the Primordial Intuition, the discrete and the continuous, justify the construction of a rational grid (cf. 6.3.3), generating a denumerable system of rational intervals of the continuum and of $n$-dimensional space. The preferred prototype in [BMS37], [BMS32] and [BMS51] is two-dimensional space, the domain of real functions, with its obvious lack of 'natural order'. Intervals are introduced as 'squares', and a sequential binary subdivision of the unit-square under the name $\kappa$-squares and $\lambda$-squares.

### 6.3.11.2 Real Point or Real Number

In Brouwer's conceptual genesis of real element-of-the-continuum the first construction is the generation of 'point' as an infinite sequence of nested intervals whose 'lenghts' converge to zero. The subsequent construction of 'species of coincident points' or 'point-core' may well represent the immediate object of his analysis and function theory and correspond more closely to the classical notion of real point, to Brouwer the fundamental element, most characteristic of the continuum, is the point generated as a sequence of nested intervals. Confusingly, the term 'real number', hitherto reserved for the classical 'law of progression' (cf.6.3), is now the name given to this new 'element'*:

> We call such an indefinite [Germ. 'unbegrenzt'] sequence of nested $\lambda$-intervals the *point P* or the *real number P*. ([BMS37], p.3)

---

\* In [1921] footnote p.955 Brouwer remarks on this change of meaning of the expression 'real number', which was 'considerably narrower' in his earlier work; 'real number' is the name chosen for his new, wider concept 'because of its expressiveness'.

'Nesting' (Germ. 'Einschachtelung') is decribed as the next interval lying strictly within in its predecessor, 'strictly' being interpreted as having no common 'side-points'. The definition of $\lambda$-intervals (see [BMS37], pp.1-3; app. 11, pp.471-472) ensures that their lengths converge to zero.

The implications of this new definition of point are far-reaching.
First of all, 'point' is not to be confused or identified with the discrete, zero-dimensional real point of classical mathematics or with Brouwer's own discrete 'singularity' or 'unit' inserted in the continuum (cf. 6.3.3). (To avoid confusion we shall refer to the new Brouwer-point as 'Brouwer real-point' or 'real point'.)

To dispel any doubt about the nature of 'real point' Brouwer adds:

> We must stress that for us the point P *is the sequence* $\lambda_{v_1}, \lambda_{v_2}, \lambda_{v_3},\ldots$ itself, not something like 'the limiting point to which according to the classical view the $\lambda$-intervals converge or something to be defined as the unique accumulation point of midpoints of these intervals'.
> Every one of these $\lambda$-intervals is therefore part of the point P.
> We must further point out that every approximating procedure can be reduced to such a nesting of $\lambda$-intervals. (ibid.).

Moreover, at any stage of the construction of 'real point' the last constructed element of the initial segment is an interval, i.e. a homogeneous part of the the space under consideration, in one-dimensional space the measurable element interval of the mathematical linear continuum, in two-dimensional space the element square of the plane and of the continuous function.

### 6.3.11.3 Law and Choice

A construction such as the $\lambda$-interval system and the requirement of nesting are sufficient to ensure that successive generation of element intervals results in a sequence which is convergent and proceedable indefinitely, i.e. in a 'real point'.

The definition of real number does not require an inductive law of progression, i.e. a rule which pre-determines every element-interval uniquely. Apart from the limits of 'nesting', the manner of determining each element is subject to no other restriction than that of all mathematics, and allows 'free choice':

> An indefinitely proceedable sequence is in general not a fundamental sequence (cf.§3), since in the process of its generation free choice of elements is not excluded. (ibid.).

The admission of free-choice sequences follows a fundamental change in Brouwer's views on the nature of the 'infinite sequence'. Hitherto the infinite sequence was considered to be a legitimate mathematical object only as the completely constructed algorithm for the generation of its elements; considered extensionally, i.e. as a sequence of generated elements, it was essentially incomplete and therefore not acceptable as a real mathematical object. Brouwer now views the infinite sequence as an ongoing process in time, split by the present moment into a completed but finite part and the proceedability of the future remainder. This split and the distinction between the definite and existent results of past acts and the possible outcome of future constructions are epitomized in the ambivalent nature of 'free choice'. Free choice in respect of future acts requires the availability of alternatives and is absence of complete pre-determination; the result of a past act of free choice or decision is one definite object singled out from those that were available.

The infinite choice-sequence, considered extensionally, in the process of construction, consists of a completed finite sequence of uniquely determined elements, an 'initial segment', and a yet-to-be-constructed 'tail'. As such it is not different from other infinite sequences, including 'fundamental sequences', i.e.those determined by 'law'.

In the process of constructing a sequence of nested intervals the last element of any 'initial segment' is an interval. But whereas the next element of a fundamental sequence is to be uniquely determined by a pre-constructed law, the 'choice' of the next element of a free-choice sequence is restricted only by the conditions of 'nesting' and convergence. Within these restrictions the domain of choice is an interval with its infinite potential of insertion of sub-intervals.

In all cases the mathematical existence of the to-be-constructed tail is the whole of constructed procedures which govern the generation of elements. This complex of procedures for the generation of 'indefinitely proceedable sequences' in its most general form is enshrined in Brouwer's new definition of 'set' or 'spread', and the new Brouwer-set identified with 'law': 'A set is a law...' (cf. any of Brouwer's definitions of set or spread after 1917).

*6.3.11.4 Species of Coincident Points or Point-cores*

Identification with a unique sequence of intervals defines the individuality of each real point. Real points or real numbers are equal or identical only if their corresponding elements are the same (op.cit., p.4). Such individual identity is obviously inadequate as a basis of analysis and function theory. Brouwer's new device of species-construction, however, enabled him to widen the concept 'real point' through the construction of an equivalence-species or 'coincidence' relation:

> Two points $P' = \lambda'_{\mu_1}, \lambda'_{\mu_2}, \lambda'_{\mu_3},\ldots$ and $P'' = \lambda''_{\nu_1}, \lambda''_{\nu_2}, \lambda''_{\nu_3},\ldots$ *coincide* if every $\lambda'$-square *strictly contains* a $\lambda''$-square and vice versa. (If there is a $\lambda$-square of $P''$ in every $\lambda$-square of $P'$, it follows that there is a $\lambda$-square of $P'$ in every $\lambda$-square of $P''$.)...
> If $P'$ is coincident with both $P''$ and $P'''$, then $P''$ is also coincident with $P'''$. (op.cit. p.4)

The property of coincidence with a given real point $P$ is an equivalence-species, and ranks as the 'fundamental species', or 'species of order zero' (cf. 6.3.7), in later work it is also referred-to as 'point-core' or 'limiting number-core' (e.g. [1954A], p.4). The point $P$ and every point equivalent with $P$ are elements of the species:

> For example, if $Q$ is the property of coincidence with $P$, then $P$ itself is an element of $Q$. (op.cit. p.5).

It should be noted that coincidence with a given point $P$, although well-defined, does not itself partition the continuous space, not even in the restricted Brouwer domain of 'points' constructed or in the process of construction; elements of the coincidence-species are only those previously constructed points which have been proved to be coincident with $P$ ('Mathematical thought-constructions which at this stage have already-been generated and have the property' ([BMS32], p.8; for full quotation see 6.3.7)). Neither is the 'residue-domain', i.e. other points of the space, constructed or in the process of construction, to be identified with the species defined by the property of 'apartness from $P$' or 'local-distinctness from $P$'. Apartness of $P'$ from $P''$ requires the proven existence of disjoint intervals:

> Two points $P'$ and $P''$ are *locally distinct* if two squares $\lambda'_{\mu_r}$ and $\lambda''_{\nu_\kappa}$ can be indicated which lie strictly outside one-another. ([BMS37], p.4).

The relations of coincidence and apartness are mutually exclusive, but their complementarity is an assumption of the Principle of the Excluded Middle:

> The two relation ($\alpha$) $P'$ coincides with $P''$, and ($\beta$) $P'$ is locally distinct from

$P''$, are mutually exclusive, i.e. if ($\alpha$) has been proved then ($\beta$) is absurd and vice versa. But if ($\alpha$) has not been proved then it does not in any way follow that ($\beta$) must be true; here too *a third is not to be excluded*. (ibid.)

Characterization of 'real point' by element-intervals and the general requirement of 'nesting' does not generate a 'natural' ordering of real-points and point-cores (unlike the singularities or discrete insertions made in the 'intuitive continuum'; see 6.3.3). Brouwer ends his section on real points with the reminder:

> Finally, we particularly wish to point out that we do not consider the traditional concept of the continuum to be consistent. For us the totality of real numbers is not an ordered set; and the apparent intuitiveness of the classical continuous infill of the number-axes by 'real numbers' is in our system off-set by the *range of choice* available when selecting $\lambda$-intervals of a *becoming* sequence of intervals. (op.cit. p.5).

### 6.3.11.5 *Point-set or Point-spread*

[In his German papers Brouwer invariably and as late as 1942 uses the word 'Menge' (Eng.'set'); the word 'spreiding' (Eng.'spread') appears for the first time in the Dutch version of [BMS32]. To avoid repetition, I shall in this and following sections simply use the term 'spread', except in quotations, where a literal translation of the original text is given.]

A Brouwer Point-Set or Point-Spread is a procedure for generating points, i.e. indefinitely proceedable sequences of nested intervals as defined above. For example, the procedure for generating 'points' of the continuum is a point-spread, 'the spread $C$' ([1918B], p.9); other spreads are: the procedure for generating points of the plane, the procedure generating points of the unit interval, etc..

The sequences generated by a Spread are the 'elements of the Spread'.

The Spread is often likened to 'a subtree of a universal tree', and the growing, branching tree is a favourite image with Brouwer, used to illustrate the continuing, free growth of sequential elements. (Cf. for example [1907], p.65, where the tree (the only illustration in [1907]) represents the freely constructed everywhere-dense scale on the unit-interval, and [BMS37], pp. 6ff. (app.11, p.475) where the tree-diagram illustrates the generation of points. Much of Brouwer's spread-terminology derives from the tree-image.)

In as much as 'tree' is a living organism, the source of its characteristic growth, it represents the spread itself. Its outward and above-ground manifestations, its trunk, 'nodes' and branches, represent the results of the Spread in action: the completed initial segments of spread-elements, i.e. indefinitely proceedable sequences. As a living tree it has the potential for further growth, branching out at the top 'nodes' in various directions and proceeding indefinitely.

Each spread-element or IPS then is represented by an upward sequence of 'nodes' (referred-to later by Brouwer as an 'arrow'). The elements are united by the Spread as their common genitor.

Brouwer usually avoids reference to a 'universal tree', with its implication of an existing universe of all possible IPS's, or to the tree as a whole, representing the spread as the totality of all its elements. His definition of Spread concentrates on the generation of 'node', the constructive processes and restrictions which govern the selection of the next element of the IPS, in the case of point-spread the next interval.

Having constructed an initial segment, a finite sequence of $n-1$ nodes, the Subject is faced with a number of options, intervals, which are not 'given' but constructed by him. The options are denumerable and are labelled by the Subject, using the elements of a 'fundamental sequence' such as 1, 2, 3, ....

Some of these options will be intervals which satisfy the conditions of 'nestedness' and the specific requirements of the Spread in question; one such option, if selected by the Subject, then becomes the $n$-th node of the IPS. Any options at the $n$-th node which do not satisfy these conditions are 'blind alleys': the initial segment with such a choice as $n$-th term is disqualified, or in Brouwer's terminology 'destroyed' or 'sterilized'.

[At this stage and in the context of 'points' the Brouwer Set or Spread is by definition the generalized procedure for generating *Infinite* sequences. In [1925A] the notion of spread is extended to generate finite as well as infinite sequences, adding a third possibility to the two alternatives of 'continuation' and 'destruction': 'the possibility that after a certain choice any further choice will, instead of a sequence of numbers, produce 'nothing.' '([1925A], p.244; see further 6.3.12).]

'Indefinite proceedability' is by definition a condition of the IPS, and requires that at the $n$-th node there is at least one option which will satisfy the spread-conditions and allows the process to continue.

The concept IPS abstracts from the method by which individual elements are finally decided. The definition of Point-Spread therefore restricts itself to formulating the ingredient-constructions and the constructive processes, 'laws' which circumscribe the domain of choice: the construction of the fundamental sequence, the construction of an interval-system such as the $\lambda$-intervals together with 'laws', procedures for labelling intervals and for guaranteeing 'nestedness' and proceedability, as well as the restrictions characteristic of the individual spread. Since the new 'Set' or 'Spread' is a fundamental concept, all those constructions and laws are brought together in one complex definition, and in Brouwer's earlier accounts, in one single sentence, usually accompanied by the apology that 'This long-winded definition regrettably cannot be replaced by a shorter one.' ([BMS37], p.6; cf.also [1925A], p.244 a.o.).

[BMS37] defines *Point-Set* as follows:

> 'By a *point-set* we understand a *law* by means of which with the numbers $n_1, n_2, n_3, \ldots$ freely and in sequence selected from the sequence of natural numbers 1, 2, 3, ... either:
> 1) a $\lambda$-square is associated in such a way that of two subsequent $\lambda$-squares the first lies strictly within the second, or
> 2) *nothing* is associated with the chosen first number $n_1$, or, if $\lambda$-squares have been associated with $n_1, n_2, \ldots, n_{h-1}$ ($h \geq 2$), nothing is associated with the choice of the $h$-th number $n_h$, and at the same time all intervals generated so far are destroyed and the process is terminated. In the last case, however, there must be at least one $n'_h$ different from $n_h$, with which, when chosen, the law does associate a $\lambda$-square.' (p.5).

For 'further clarification' [BMS37] describes the genesis of a point of a point-spread, and illustrates the general spread-procedure by means of a diagram. We refer the reader to the relevant passage on p.6 as given in appendix 11, p.475.

Point-Spread is the procedure by means of which the Subject generates new mathematical entities, i.e. points. Ontologically, the Spread precedes the points it generates; element-hood-of-Spread is the birth of the point itself, unlike element-hood-of-Species which is a proof of the property for an existing mathematical entity, 'previously acquired':

> Unlike the Species, the Spread is *primary* with respect to its elements. ([BMS32B], p.9).

In as much as 'having a common genitor $M$' may be considered to be a property of points generated or in the process of generation, the spread $M$ can be said to be a species, 'a special case of species':

A point-set *M* is always also a point-species, but a point-species is not necessarily a a point-set.
The common origin [Germ. 'Entstehungsart'] of interval-sequences generated by the law *M* is also referred to in short as *M*. ([BMS37], p.7)

Intuitionist logicians have been puzzled by the considerable variations in Brouwer's definition of point-spread and its generalization, most of which seem irrelevant in the construction of Intuitionist mathematics. It seems to me that Brouwer's distinctions, here and elsewhere, arise from his habitual reflection on and analysis of fundamental concepts, and this often without regard to their immediate usefulness. Other instances of such elaborate differentiation are: his 'levels' of mathematical construction ([1907], pp.173-175), his power relations and 'numeracies' ([1918B] and [1925A]), his hierarchy of species (cf. 6.3.7), and the relations between points and point-species generated by his Calculus of Absurdities ([1923C]), etc. In a letter to Heyting he gives the reason for developing the latter: they only serve as an example of relations that *can* be developed:

> The purpose of my paper 'Intuitionist Splitting' is only to indicate relations that can be derived from some definitely given fundamental relations by means of the operation of stating absurdity. Ways of extending these tables are legion. (Brouwer to Heyting, 26.06.24).

Generating distinctions within concepts is part of mathematics 'left to free unfolding'. The lack of purpose or usefulness is claimed by Brouwer to be an essential characteristic of mathematics; pure mathematics is practised 'playfully, i.e. without purpose' ([1948C] and [1950C]).

As to the variations in his definition of point-spread, Brouwer was convinced that his new fundamental concepts Species and Spread were the twin pillars of an alternative set-theory on which the whole of Analysis could be constructed, in particular the point-spread, a general procedure for generating sequences of all kinds which was based on the intuition of the continuum as the 'between', its infinite potential of insertion. Generalization of this natural concept and further analysis of all its aspects, no doubt, would lead to deeper insight and produce the essential ingredients of Analysis and Function Theory.

### 6.3.11.6 *Fundamental Sequences of Intervals and Finite Point-spreads*

The definitions of 'point' and 'point-spread' are generalizations of infinite sequence of nested intervals and the procedure of generating such a sequence; they cover free-choice sequences as well fundamental or inductive sequences, i.e. sequences of which each term is uniquely predetermined.

The law generating a fundamental sequence of nested $\lambda$-intervals is a special point-spread:

> A *Fundamental Sequence* of points is a point-set which assigns a $\lambda$-square to every number $n_1$ selected as the first choice, and where at every subsequent choice $v$ all selected numbers $n_v$ except one lead to the 'destruction of the preceding' (as shown in the following diagram).
> A Fundamental sequence of nested $\lambda$-intervals is a point but not every point a fundamental squence. ([BMS37], p.7; for full quotation and diagram see app.11, p.476).

The spread generating a fundamental sequence of $\lambda$-intervals is a law which restricts the choice of each successive term to one single $\lambda$-interval.

The point-spread which restricts the domain of choice at each stage to a finite number of $\lambda$-intervals is referred to by Brouwer as *Finite Set, Finite Spread* and later as *Fan*. [It should be noted that 'finiteness' does not refer to the elements of the spread, which are points, i.e. infinite sequences, nor to the number of elements of the spread. To avoid confusion, Heyting and modern Intuitionists adopted the term 'finitary', introduced by Kleene.]

Finite spreads play an important role in Brouwer's theory of functions, in particular in the proof of his Fundamental Theorem of Spreads (see 6.3.13). They are first introduced in [BMS37] (dating from around 1923), which gives a brief definition and states their fundamental property:

> If, at every stage of the construction of a choice sequence only a finite number of options from the numbers 1,2,3,... allow the continuation of the generating process, we speak of a *finite* point-set. In this case, for every $n_\alpha$ there exists a maximum $m_\alpha$, such that the choice of $n_\alpha > m_\alpha$ leads to the arresting [Germ. 'Hemmung'] of the generating process. (ibid.)

## 6.3.12 The General Spread Concept

The definition of point-spread distinguishes between two phases in the process of 'deciding the next interval':
1. the indefinitely proceedable process of sequential choice from an ordered finite or denumerably infinite domain of labels or 'indices', and
2. the association of a particular $\lambda$-interval with the selected finite sequence of indices.

A corresponding distinction is made within the 'law' which constitutes the spread, the pre-constructed procedures which govern and limit this

choice and which determine the individual characteristics of each spread, between:
1. the rules which govern the formation of initial segments of indefinitely proceedable sequences, the spread-law proper, and
2. a law, referred-to by Heyting as 'the complementary law', which in the case of a point-spread assigns a $\lambda$-interval ('or nothing') to each initial segment.

Abstraction from the nature of the elements assigned by the 'complementary law' leads to the general spread concept. In this form the Spread was launched in [1918B]. The elements assigned are simply 'symbols' (Germ. 'Zeichen'), or 'a 'figure', i.e. a mathematical thought-construction [Du. 'denkbaarheid']'.

Brouwer's definitions of Spread in [1919B], [1919D], [1925A], [BMS37] and [BMS32] are substantially the same; minor modifications, especially after 1925, concern the respective roles of free choice and restriction of this freedom in the process of generating IPSs. In the early definitions, [1918B], [1919D] and the printed version of [1925A], the only restrictions are 'laws', i.e. restrictions constructed in advance of the generation of the IPS and applying to every stage of its generation; and Spread is identified with 'Law': ('A set is a law...').

During the following years Brouwer considered freely imposed restrictions, operating at various stages of the construction of the IPS and at varying degrees, and the notion 'Spread' covers the whole process of constructing IPS, in which both the free activity of the Subject and his self-imposed restrictions play a role; instead of the Spread being identified with law, the Spread is now 'based on a law' ([BMS32], p.7).

These changes are most marked in the numerous corrections inserted by Brouwer in his copies of [1925A] and in [BMS32]. The innovations introduced are: a greater prominence of the role of Species, a hierarchy of levels of species-construction (see 6.3.7), and the use of 'choice' and 'restriction' as operators at various stages of the construction of the IPS. Compare e.g. the amended and added footnote to the definition of Spread of [1925A], p.245:

> This includes as a characteristic the freedom to proceed, a freedom which can be narrowed arbitrarily (possibly even to the point of complete determination or in accordance with a spread-law). The arbitrariness of this 'narrowing clause' adjoined to a finite sequence of choices — always with the condition that proceedability is safeguarded — adds a new element of willfulness to this choice-sequence and therefore also to its continuations.

> One can further adjoin to the spread an arbitrary well-ordered species of such restrictions (and therefore one may join for example to a finite sequence of choices a clause which restricts the freedom to add restrictions to further choices).

and [BMS32], p.7:

> The freedom to proceed with the choice sequence can after every choice be arbitrarily restricted (possibly in a dependence, imposed by the choosing mathematician, on the events in his mathematical thought-world), which may lead e.g. to complete determination or to restriction by a spread-law. The arbitrary nature of this restriction at every new choice of the freedom to proceed, which is allowed provided the possibility to proceed is retained, is an essential element of the free-becoming of the element of the spread, as is the possibility to link to every choice a restriction of the freedom to make further restrictions of freedom etc. (cf. also [1942A], p.323).

In his latest work, however, such higher-order restrictions were discarded for the sake of simplicity:

> In former publications I have sometimes admitted restrictions of freedom with regard also to future restrictions of freedom. However, this admission is not justified by close introspection and more over would endanger the simplicity and rigour of further developments. ([1952B], p.142).

Arbitrary high-order restrictions were investigated by Troelstra and found to be 'so dazzlingly complex as to make the whole theory unviable' (cf. TROELSTRA[1982]).

[In the quoted passage Brouwer retracts even his previous admission of 'restrictions to freedom to make further restrictions' and this on the ground of 'close introspection'. I venture to suggest that his reference to 'close introspection' points to a further examination of the *concepts* 'free choice' and 'restrictions of freedom'. One of the requirements of free choice, as analysed above (6.3.11.3), is the avaibility of a 'domain of choice', determined by past decisions of the Subject. Freedom, as Brouwer pointed out, includes the freedom of the Subject to impose restrictions on the domain of his future choices. It must be understood that such decisions to be meaningful should be irrevocable.

In respect of a particular act of choice all decisions taken at earlier stages can therefore only lead to further restriction of the domain, in the case of a spread, initially defined by the Subject to be a fundamental sequence of mathematical entities. Indeed, the process of chosing itself can be described as a narrowing of the domain of one's choice by restricting decisions, ultimately to the one selected object. It requires the power of the chosing Subject to restrict his domain of choice to be unlimited except in respect of one element of the domain. In other words 'restricting one's freedom to make restrictions' is incompatible with choice itself.]

The definition of 'spread' given in the final version of [BMS32] incorporates various developments and reflects Brouwer's view at the end of his 'creative period'. We shall leave aside the issue of 'higher-order restrictions' and consider the general spread concept as it is defined in [BMS32] (for a full translation of various versions of the relevant passage see appendix 12).

> First we determine an indefinitely proceeding sequence of symbols by a first symbol and a law which derives from each of these symbols the next one. One could choose e.g. the sequence $\zeta$ formed by the succession of ordinal numbers or 'numerals' 1, 2, 3, ... [left out in the final version].

> ...A species whose admission in intuitionist mathematics has far-reaching consequences and which in particular makes possible a re-creation of analysis is the *Spread*. The Spread is based on a spread-law, according to which every time when an arbitrary (natural) number is chosen as 'index' each of these choices either produces a 'figure', i.e. a mathematical entity [Du. 'denkbaarheid', thought-construction] (as such also 'nothing' can serve), or results in the the process being arrested together with the definite destruction of what the process had produced so far. To this must be added the condition that, for every $n > 1$, after every sequence of $n - 1$ choices at least one number can be indicated which, when chosen as the $n$-th index, will *not* result in the the process being arrested. The columns of 'figures' which in this way are generated by indefinitely [Du.'onbegrensd',i.e. 'unlimitedly'] proceeding choice-sequences are the elements of a species which is called '*spread*'. ([BMS32], p.7).

The spread-process as defined here consists of the following consecutive constructions, i.e. in their 'genetic' order:
1. the preliminary constructions of the 'fundamental sequence' and of 'mathematical entities';
2. the construction of the 'spread-law', which combines what we described above as the spread-law proper and the 'complementary law';
3. the generation of spread-elements by acts of free-choice in accordance with the spread-law;

4. the 'assemblage' of constructed spread-elements into the spread-species.

1) The construction of the 'fundamental sequence' is invariably mentioned as a preliminary to the spread-construction. Although based on the concept of natural numbers and ultimately on the Primordial Intuition, the 'fundamental sequence' here is an ordered, inductively defined set of symbols. Brouwer deliberately uses words such as 'Zeichen' (symbols), 'Ziffernkomplexe' (digit-groups), and 'Nummern' (numerals) as distinct from 'Zahlen' (numbers). It is also clear from [BMS15A] that the 'law of construction' refers to the decimal composition of the digit-groups ('We start from the sequence $\rho$ of the digit-groups 1,2,...10,11,... which can be continued without restriction according to a well-known law which enables us to derive from each digit-group the next one...' ; see 6.2.3).
For the purpose of 'labelling' and as 'indices' the sequence of physical symbols is sufficient and 'the most suitable' (Germ. 'geeignest'), since they are clearly distinct from the 'mathematical entities'.

The mathematical entities to be assigned are: 'mathematical thought-constructions', such as the $\lambda$-intervals in the construction of point-spreads. They must have been constructed before they can serve as elements of an IPS:

> ... to each node an assignment is made of a mathematical entity *previously acquired*, ... ([BMS37], p.14).

Unlike the fundamental sequence, the domain of mathematical entities to be assigned is not necessarily ordered.

2) The 'spread-law' in the sense of [BMS32] comprises:
i. a law associating 'labels' with mathematical entities for every stage of the generating process (the 'Complementary Law').
In *Zum freien Werden von Mengen und Funktionen* [1942A] Brouwer extends the notion of Spread by allowing free choice sequences to determine this assignment, leading to the notion of 'higher-order spreads'. (see further TROELSTRA[1969A], 11.6);
ii. a law of continuation of the sequence of 'labels', determining at each stage the scope of choice available to the Subject. In general this law determines at the $n$-th stage whether the choice of a particular label $a_n$ allows the process to continue or not. Such a choice is either:
a) successful, i.e. it results in the construction of a finite initial segment

with $a_n$ as its last element, to which then a mathematical entity is assigned by the complementary law; or

b) the choice of label $a_n$ is deemed impossible. As in the construction of absurdity, the Subject finds the chosen avenue blocked (see 5.8.3). Indeed, Brouwer uses the same expression: progression is 'checked' or 'arrested' (Du. 'gestuit', Germ. 'gehemmt'; in later English publications 'sterilized'). The sequence $a_1, a_2,...,a_n$ is ruled impossible as an initial segment, $a_n$ is eliminated together with its predecessors, 'the definite destruction of what the process had produced so far.'

The law specifies which choices are compatible with the characteristics of a particular spread such as e.g. the continuum and the unit interval. Incompatibility or impossibility of certain choices may in this case be due e.g. to the condition of 'nestedness' not being met, etc..

[In [1926B] Brouwer introduces an extension of the general spread concept by allowing the Subject to change the 'law' during the generating process, i.e. replacing the pre-constructed law by 'an infinite sequence of 'decisions' which determine for each non-arrested finite choice-sequence whether it will remain non-arrested or will be arrested... The concept spread is hereby extended, the decision as to 'arrestedness' or 'non-arrestedness' then no longer needs to be wholly fixed in advance by one law for all finite choice-sequences'. (p.634).]

For Indefinitely Proceedable Sequences, moreover, the Spread Law includes by definition the condition that at every stage there is at least one option 'which, chosen as the $n$-th index, will not result in the process being arrested'.

In a footnote Brouwer further adds that this condition is distinct from the requirement 'that for every $n$ at least one initial segment of $n$ indices can be indicated which is not arrested, in which case we shall speak of a *spread-haze law*, and we will call the species of sequences of figures so generated a *spread-haze*'. ([BMS37], p.7). [In [BMS51] the spread, as distinct from the spread-haze, is referred-to as 'spread direction'.]

Generalization of Spread as a procedure for generating finite as well as infinite sequences requires a third possible outcome of choice at the $n$-th stage: the simple 'ending' of the process.

In [1918B], [1919D], [BMS32] and [BMS51] such 'ending' is described as a special case of successful choice, the chosen index $a_n$ is not ruled out as impossible and its predecessors not eliminated, but $a_n$ and subsequent choices are assigned to 'nothing', which is given the status of 'mathemat-

ical entity' ('...is assigned to a mathematical thought-construction (as such 'nothing' can serve)...'; cf. also [1928B], p.380 footnote 1)).
[1925A] and the German version of [BMS32] include 'ending of the process' in the definition of spread:

> Each of these choices either produces a definite sequence of symbols with or without ending of the process, or cause the process to be arrested...

In a footnote 'ending of the process' and 'assignment to 'nothing'' are then simply stated to be equivalent:

> In this definition of set one may obviously replace the possibility of ending the process by the possibility that after a certain choice every further choice, instead of being assigned to a sequence of symbols, is assigned to 'nothing'. ([1925A], p.244).

Although the process of choosing indices is proceedable indefinitely, the sequence of corresponding mathematical entities can be thought of as 'having been completed', consisting as it does of an initial segment of $n - 1$ elements, and a determinate tail of 'nothing'.

Examples of such spreads are:

> Laws generating finite sequences of symbols and infinite sequences such as the sequence $\zeta$, they are special cases of spreads; in particular the spread which we shall call $A$ and which successively allows every first choice together with the ending of the process. ([BMS32], p.8).

3) The activity of free decision within the restrictions of 'law' and those imposed by other previous decisions and assigning to each of these choices a 'figure', i.e. a mathematical entity, generates 'a column of figures'.
If the process is not 'ended' the column of figures is an indefinitely proceeding sequence of mathematical entities (other than 'nothing'), remaining in the process of construction or 'becoming'.

In the original version of [BMS32] the column or sequence is simply called 'spread-element':

> The sequence of figures produced in this way by an indefinitely proceeding sequence of choices, in general not representable as completed, is called an element of the spread. (p.7).

4) Changes in the text here and elsewhere (dating from around 1927) are evidence of the increasing importance Brouwer attached to the role of 'species' in the construction of relations (see also 6.3.5 ff.). In the amended version of [BMS32] the column of figures are 'elements of a *species* which

is called spread'. Introduction of the species-construction at this stage suggests a distinction between the generating process and the (permanent) relation resulting from it, a property residing in the sequence produced ('the common origin of interval-sequences generated by the law' ([BMS37], p.7 quoted above)). Such a property is referred-to as 'spread-species', but confusingly 'also as *spread* for short'.

The spread-species also includes elements, i.e. sequences, previously and otherwise acquired, which are equivalent to any elements generated by the spread-law:

> The infinite columns of figures generated according to the spread-law by infinitely proceeding sequences of choices are by virtue of their genesis – together with all sequences equivalent with them – the elements of a species. This species is called a spread. ([BMS52], p.3; app.6).

Interpretation of spread as element-species allows an extensional use of spread as e.g. in the construction of relations between spreads such as 'equivalence' and 'sub-spreads':

> Two sets are equivalent or identical if one is certain that for every element of one set an equal element of the other set can be indicated.
> The set $M$ is called a subset of the set $N$ if for every element of $M$ there exists an equal element of $N$. ([1925A], p.245).

The theory of relations between spreads, their powers etc. are part of the general theory of species and are treated there. (see also 6.3.7-8).

While the spread-concept, in particular the point-spread as the universal generator of all points of the continuum, provides a comprehensive characterization of the intuitive continuum, the construction of the spread-species, the 'assemblage' (Du. 'samenvatting') of generated elements on the basis of property, allows subsequent de-composition and is the basis of mathematical analysis of the continuum. The construction of spread-species and further constructions of point-species, however, do not necessarily carry all the characteristics of the continuum, unlike the continuum-spread which generates the continuum with all its characteristics. For example:

> Content and quantity are attributes of sets not of set-species. The set concept, in contrast with the species concept, is primary with respect to its elements, and is an indispensable fundamental notion of the concept quantity. (note in the margin of [BMS32B], p.9).

### 6.3.13 Continuity and Uniform Continuity of Spread-Functions

#### 6.3.13.1 Finite Spreads

In the general definition of Spread the domain of choice is a denumerably infinite sequence of indices. The spread defined by such a domain and by completely free choice from this domain, i.e. without restrictions leading to 'ending' or 'arresting' of the process, generates and characterizes the whole, intuitive continuum. The 'spread $C$', so defined, is the simplest example of a spread ([1925A], p.244 footnote 2).

Such complete freedom or 'lawlessness', however, is inadequate as a basis for 'precision analysis of the continuum' and function theory, it does not allow any construction of property or relation beyond that of individual identity. Effective restrictions are needed which can secure conditions such as convergence and define the construction of points, point-cores, intervals and functions.

The spreadlaw which restricts the domain of choice at every stage to a finite number of indices, i.e. defines a finite spread, characterizes the closed continuum interval. The relative simplicity of the finite spread, in Brouwer's view, further facilitates the proof of the 'Fundamental Theorem' of Brouwer's intuitionist analysis, which in turn yields proofs of the characteristic properties of the continuum and real functions, in particular the uniform continuity of all full functions.

The general notion of Finite Spread is introduced in the context of the proof of the Uniform Continuity Theorem, first in [1923A]. We quote the definition given in [1925A]:

> If for every $n$ in [the fundamental sequence] $\zeta$ there is a definite number $k_n$ such that whenever a number is chosen as the $n$-th choice which in $\zeta$ lies higher than $k_n$ the process is arrested, then the set is said to be finite. In particular, indefinitely proceeding sequences of single-digit numbers form the elements of a finite set. (p.245).

It should be noted that in defining 'finite' Brouwer uses his strong negative: the choice of any index beyond a certain element in the ordered sequence of indices is impossible, i.e. leads to the generating process being arrested.

According to the general definition the value of $k_n$ can be different for different $n$; the example given of the spread generating sequences of single-digit numbers is a special case of what Brouwer later called 'pure $n$-finite spreads', where the domain of choice at every stage is uniformly restricted to a given number of indices. ([1930A], p.5).

### 6.3.13.2 *The Brouwer Continuity Hypothesis or the Fundamental Hypothesis for Real Functions*

Brouwer's Uniform Continuity Theorem is perhaps the most radical and controversial conclusion of his Intuitionist set theory. The search for a simple and rigorous proof, based on the notions of spread and species, dominated his work during the years following the publication of *The Foundations of Set Theory*. At various stages a proof was claimed to have been given, a corollary of the 'Fundamental Theorem of Finite Spreads' or 'Fan theorem', which in turn was based on a complex and revolutionary proof of the 'Bar Theorem' (cf. Chapter II). But as late as 1952 Brouwer had to admit that a simple proof of the 'Fundamental Theorem' had eluded him ('For Fans a wonderful theorem holds whose importance would justify to call it the Fundamental Theorem of Intuitionism but whose absolutely rigorous proof has not sufficiently been simplified...' ([BMS66], p.6; app.10)).

Yet, continuity and uniform continuity of functions defined on a closed continuum interval were recognized by Brouwer to be essential characteristics, directly evident from the nature of the continuum and contained in the concept of 'function defined on a continuum'. It is an 'insight' which underlies his *Foundations of Set-Theory* ( [1918B], [1919A]), and is clearly stated in Weyl's 1920 Zürich lectures in defense of Brouwer's continuum and in Brouwer's approving comment added as editorial notes in the margin of Weyl's manuscript:

> [WEYL] 'It is clear that one cannot explain the concept 'continuous function in a bounded interval' without including 'uniform continuity' and 'boundedness' in the definition. Above all, there cannot be any function in a continuum other than continuous functions. When the Old Analysis introduced 'discontinuous functions it showed most clearly how far it had departed from a clear understanding of the essence of the continuum. What is nowadays called a discontinuous function is in reality no more than a number of functions in separate continua...'
> [BROUWER] 'Better to say 'the function is not everywhere defined.'
> [WEYL] 'Take for example the continua $C$, $C^+$ ($x > 0$) and $C^-$ ($x < 0$)... If we consider the two functions $+1$ in $C^+$ and $-1$ in $C^-$ then *there does not exist* a function defined for the whole of $C$ equivalent with the one value for $C^+$ and the other value for $C^-$.'
> [BROUWER] 'Very true! Underline,because this is the main and most important point.' (BA[HWE], p.47; cf. [1921], p.76).

For Brouwer continuity is implied in the essential requirement for all functions that they be defined, i.e. that the function-rule effectively assigns a value to each element of the continuum, in terms of Brouwer's characterization of real-point, the condition that a definite value be assigned to any Indefinitely Proceeding Sequence generated by the continuum spread.

The fundamental problem that faced Brouwer is the apparent conflict between effective determination of the function value and the incompleteness, the indeterminacy of the IPS. His analysis and constructive interpretation of the infinite sequence both put the problem into focus and pointed to a solution: If a definite value is to be assigned to an IPS it cannot be dependent on the 'incomplete part'; the basis of such assignment is to be sought in the only aspect of the IPS which can claim mathematical existence: its completed finite initial segments. A solution was hinted at as early as 1916, when Brouwer remarked in his *Notes* 'that with such a sequence [IPS] one can work very well, as long as when relating the finite thing... to this sequence, one can work at every stage with a suitable initial segment.' ([BMS15A], p.1; for full quotation see 6.3.10).

The assertion that the function assignment must be wholly dependent on an initial segment of the IPS became the fundamental hypothesis of Brouwer's analysis and function theory; we shall refer to it as *The Fundamental Hypothesis for Real Functions* or *the Brouwer Continuity Hypothesis*. It claims that for a function $f$ to be defined for a real point $\xi$, the value $f(\xi)$ must be determined by a finite, completed initial segment of the sequence of choices generating $\xi$. It is first stated in [1918B] where it is used as part of Brouwer's proof of the non-denumerability of the continuum:

> A law which assigns a natural number $h$ to each element $g$ of $C$ must have determined $h$ completely after a certain initial segment $\alpha$ of the sequence $g$ has become known... (p.13).

The underlying intuitive argument is based on the 'insights' that:
1. a real point of the continuum is characterized by a convergent sequence of nested intervals generated by an indefinitely proceeding sequence of choices;
2. a real function is a law, an effective procedure or algorithm which, when applied to a real point, generates a number, i.e. an integer, a rational number or 'a definite point';
3. only the 'previously acquired' are legitimate elements of a mathematical construction or calculation, and the result of a mathematical con-

struction can therefore only be dependent on such completed constructs.

The 'intuitive argument' further presupposes that real functions exist and can effectively operate on real points as defined.

'Absolutely rigorous proof' of the Continuity Hypothesis, as the proof of any general property, requires a construction which establishes the conjectured property to be inherent in the given construction, in this case the Indefinitely Proceeding Sequence generated by the spread. In accordance with good mathematical practice Brouwer started his search for such a proof by considering the special, simpler case which retains all elements relevant to the general case. Simplification was achieved by restricting both the domain and range of the function: restricting the domain to the closed continuum interval, the range of the function to the system of integers. Earlier investigations had shown that every closed interval 'coincides with a finite spread' ([1919A], p.14); the relative simplicity of the finite spread suggests a corresponding simplification of proof of the Hypothesis for functions defined on a closed interval. Another result, the proof of the non-denumerability of the continuum ([1918B], p.13) pointed to a possible proof of the Fundamental Hypothesis and the Uniform Continuity Theorem when restricted to integral-valued functions.

### 6.3.13.3 *The Fundamental Theorem of Finite Spreads*

Uniform continuity of all 'full functions' was first claimed and a sketch of proof given in [1923A]. In the detailed proof of [1924D] Uniform Continuity is a corollary of what Brouwer considers to be the fundamental theorem of Intuitionist Analysis:

> ... a wonderful theorem whose importance justifies it being called the Fundamental Theorem of Intuitionism. ([BMS66], p.6; app.10).

It is referred to as such in [BMS37] and [BMS32], later also as *the Bunch Theorem* ([BMS51]) and *the Fan Theorem* ([1952B]).

The *Fundamental Theorem of Finite Spreads* asserts the Continuum Hypothesis for integral-valued functions defined on a finite spread and includes uniform continuity:

> If to each element $e$ of a finite set $M$ a natural number $\beta_e$ is assigned then a natural number $z$ can be indicated such that $\beta_e$ is completely determined by the first $z$ choices generating $e$. ([1924D], p.192; cf. also [BMS32A], p.8, app.12; [1927B], p.66; [BMS37], p.111; [BMS66], p.10, app.9; a.o.).

Brouwer's proofs ([1924D], [BMS37] and [1927B]) concentrate on the species of finite initial segments generated by the spread and its subspecies, and on the relations between the elements of these species.

The species of arrested as well as non-arrested finite choice sequences generated by the spread $M$ is referred to as $\mu$. The species $\mu$ is numerable (Germ. 'zählbar').

Subspecies of $\mu$ are defined by the property of arrestedness and by a property defined in terms of continuations of the initial segments and their function values, the property that all continuations of one initial segment are assigned by the given function to one and the same natural number. In particular $\mu_1$ is the subspecies of non-arrested finite choice sequences generated by $M$ such that all continuations of each one are assigned to one natural number; in other words:

> ...such that all elements of $M$ (i.e. infinite choice sequences) issuing from any one of these finite choice sequences are assigned to one natural number $\beta$. ([BMS37], p.108)

Moreover, each element of $\mu_1$ is a minimal segment, i.e. if $a_1, a_2, ... a_n$ is an element of $\mu_1$ no proper segment of $a_1, a_2, ..., a_n$ is an element of $\mu_1$:

> We can therefore speak of an element of $\mu_1$ only if by the algorithm of the assignment rule the finite natural number $\beta$ has been wholly established and its determination is not left to further choices, while none of the finite initial segments of the element have this property. (ibid.)

The species $\mu_1$ is numerable and, Brouwer claims, a 'separable' or 'removable' subspecies of $\mu$, i.e. $\mu$ is 'split' into $\mu_1$ and its 'direct conjugate' (see 6.3.9).

Another species subdivision of $\mu$ is generated by a property referred to as 'securedness'. An element of $\mu$ is 'secured' if it is either arrested or has a finite initial segment which is an element of $\mu_1$.

The Fundamental Theorem is based on the hypothesis that every element of a spread $M$ ('i.e. every infinite sequence generated by $M$') is 'securable', i.e. can be shown to have a finite initial segment which is an element of $\mu_1$. It asserts that each element of $\mu$ can be reached 'inductively' starting from an element in $\mu_1$.

The first stage of Brouwer's proof of the Fundamental Theorem consists of a proof of what he later ([1954A]) refers-to as *the Bar Theorem* :

> If to each element of a set $M$ a natural number $\beta$ is assigned, then $M$ is split

by this assignment into a well-ordered species $S$ of part-sets $M_\alpha$, each of which is determined by a finite number of choices, and to each element of one $M_\alpha$ the same natural number $\beta_\alpha$ is assigned. ([BMS37], p.110; [1924D], p.191; cf. also [1927B], p.65; [BMS51], p.78 and [1954A], p.14).

In other words, if a function $f$ assigns each element of a spread $M$ to a natural number, then $M$ is split into sub-spreads $M_\alpha$ so that all elements of one $M_\alpha$ have the same initial segment and are all assigned by $f$ to the same natural number $\beta_\alpha$.

Again, Brouwer's conviction of the truth of the Bar hypothesis precedes his proofs, as is clear from a remark in [1927B] which also throws some light on his own appraisal of the new methods introduced in these proofs:

> Profound intuitionist reflection reveals that this securability is nothing but the kind of property defined by the condition that it holds for every element of $\mu_1$ and for every arrested element of $\mu$, further that it holds for an arbitrary $\mathscr{F}_{sn_1\cdots n_r}$ if it has been established for $\mathscr{F}_{sn_1\cdots n_r v}$ for all $v$. This implies the well-ordering property of any arbitrary $\mathscr{F}_{sn_1\cdots n_r}$. Nevertheless, the proof of this property as given in the text seems to me to be of some interest because of the assertions made in the course of it. ([1927B], p.63).

The major innovations Brouwer introduced in his proof of the Bar Theorem are the proof itself, which is metamathematical even in the Brouwer sense, i.e. based on reflective analysis of the structure of possible proofs, and a new principle of induction, known as *Bar Induction*, defined in the above quotation and described as 'evident from profound intuitionist reflection'.

'Bar Induction' is a form of regressive induction, appropriate to and inspired by the branching-tree structure of choice sequences and in particular by the properties defining the species $\mu_1$. As such it could be claimed by Brouwer to be clear by intuitionist 'insight'.

Bar Induction and the Bar Theorem have been the subject of detailed research and frequent discussions among mathematical logicians. The most illuminating are the expositions by Kleene in KLEENE AND VESLEY[1965], Dummett in DUMMETT[1977] and Martino and Giaretta in MARTINO AND GIARETTA [1979]. Post-Brouwer developments are not within the scope of this book; we refer the interested reader to the quoted books or papers, in particular to Dummet's exposition on Bar-induction and Kleene's proof of the Bar Theorem, which follows Brouwer's argument closely. For Brouwer's own proof we refer the reader to the version of

[BMS37] given in app.11. p.477ff., which is similar to that of [1924D] but contains some of the elements of [1927B].

### 6.3.13.4 The Uniform-Continuity Theorem

The Bar-Theorem and the Fundamental Theorem on Finite Spreads were conceived in 1924 as lemmas of the Uniform Continuity Theorem; they are referred-to simply as 'Theorem 1' and 'Theorem 2' in the paper entitled *Proof that every full function is uniformly continuous* ([1924D]).
In [BMS37] and [BMS32B] the Uniform Continuity Theorem is one of the 'applications' of the Fundamental Theorem, 'the most important application' being the Intuitionist alternative to the Heine-Borel Covering Theorem ('If a law $w$ assigns to each element $e$ of a compact point-core species $Q$ a neighbourhood of $e$, then a finite number of these neighbourhoods can be indicated such that for an arbitrary pointcore of $Q$ at least one of them is a neighbourhood of that point-core.' ([BMS32B], p.56)). The 'unsplittability of the continuum' is the 'second application', followed by the Uniform Continuity of every full function as 'third application', and the Uniform Continuity of the limit function of convergent full functions as 'fourth application'. In [BMS51] the Uniform Continuity Theorem of every Full Function is re-established as the first and most important application.

Brouwer's Uniform-Continuity Theorem asserts that 'Every full function is uniformly continuous'.
A real function is defined as a law which assigns to every point-core $\xi$ one and only one point-core $\eta = f(\xi)$.
The species $D$ of point-cores $\xi$ for which the function is defined is the 'domain of definition' of the function.
A 'full function' is a function whose domain of definition is the closed unit-interval ('including the end-points 0 and 1').
Continuity and uniform continuity are defined in their standard metric form, although Brouwer in [BMS37] stresses that this is done only for simplicity's sake:

> One should note that the above definition of continuity is given in metric form only for the sake of simplicity, the concept [Germ. 'Inhalt'] continuity is independent of this form. (p.104)

The proof of the Uniform-Continuity Theorem as a corollary of the Fundamental Theorem was given in [1924D] and with minor modifications again in [BMS37], [1927B], [BMS32], [BMS51] and [1954A].

For Brouwer's own proof we again refer the reader to the version of [BMS37] given in app.11, pp.479-480, which conforms to his earlier terminology and notation as introduced in our section 6.3.10 ff. For a short proof in modern terminology, which derives uniform continuity from 'the Intuitionist Compactness Theorem' i.e. the Fan Theorem, we recommend Troelstra *'Aspects of Constructive Mathematics'*. (TROELSTRA[1977B]).

# APPENDICES

⟦The following appendices are a selection mainly from the manuscripts of Brouwer's private notes, lecture notes, letters and other unpublished papers. In this selection I have been guided by the importance and relevance of the work to the subject of this book: the genesis of Brouwer's Intuitionism, and by the inaccessibility of the work to wider readership. These are also the reasons why I have included an English translation of a published paper, BR[1933] (appendix 5), perhaps the most comprehensive and mature statement of Brouwer's views on the nature of mathematics and language, but as a Dutch publication not accessible to a wider public.

The English translations are all my own. Recognizing that every translation necessarily includes an element of personal interpretation by the translator, I have included the original Dutch text in all cases where the Dutch original has not already been published.

The numbers in the margin refer to the page-numbers of the MS. Ed⟧

## Appendix 1

### Geloofsbelijdenis van L.E.J. Brouwer [BMS 1A]

⟦Made by Brouwer on 3 April 1898 before the congregation of the Remonstrant Church at Haarlem and its minister Ds B. Tideman. It is written as a response to a number of questions ('punten') put to candidate members. Ed.⟧)

1   Punt 14: Wat is de grondslag van mijn vertrouwen op God?
    Dit is voor mij het hoofdpunt der belijdenis, en het eenige dat eigenlijk de naam mag dragen van "geloofsbelijdenis van een persoon".
    Dat ik Godsgeloof heb, heeft zijn oorsprong voor geen deel in een verstandelijke overweging in die zin dat ik uit verschillende verschijnselen, die ik om mij heen waarneem, zou zien de openbaring van een 'hoogere macht', maar juist in de volkomen machteloosheid van mijn verstand. Immers het eenige ware voor mij is mijn eigen ikheid van het oogenblik,

omgeven door een schat van voorstellingen waaraan de ikheid gelooft, en die haar doen leven. Een vraag of die voorstellingen 'waar' zijn heeft geen zin, voor mijn ikheid bestaan alleen de voorstellingen en zijn als zodanig reëel; van een tweede, onafhankelijk van mijn ikheid, daaraan beantwoordende realiteit is geen sprake. Mijn leven nu is mijn overtuiging van mijn ikheid en is mijn gelooven in mijn voorstellingen en daaraan knoopt zich direct vast het geloof in Dat, wat de oorsprong van mijn ikheid is en dat mij mijne voorstellingen geeft, onafhankelijk van mij, dus iets dat, evenals ik, leeft en boven mij staat, en dat is mijn God.

2   In de vele woorden die ik hiervoor gebruikte, moet geenszins gelezen worden een verstandelijke afleiding van het Godsbestaan, want dit Godsgeloof is de onderste grond, waarvan kan worden afgeleid, maar die niet zelf wordt afgeleid. Het Godsgeloof is een direct spontaan gevoel in mij.

Nu geloof ik wel, dat dit Godsgeloof van eenigszins anderen aard is, als het gewone, en wel voornamelijk hierom, dat het mijne steunt op een wereldbeschouwing, die alleen mijzelf en mijn God als levende wezens erkent, waarvan ik mijzelf ken en mijn God, mijn meester, voel.

De voorstellingen verder, die mij gegeven zijn, bevatten in zich b.v. ook, dat er andere ikheden zouden zijn, ook met voorstellingen, maar deze zijn niet reëel, zij zijn de deelen van mijne voorstellingen, dus van mij. Mijne voorstellingen zijn mijn leven, zoo leef ik op dit oogenblik in de voorstelling, dat ik denk over mijn leven, en een belijdenis schrijf, maar in dat leven vind ik mijn God niet, mijn God staat onder* en buiten mijn leven, slechts het feit, dat ik leef, doet mij mijn God voelen, niet in het hoe ik leef vind ik mijn God. (* onder, d.w.z. is de 'Urheber' van mijn leven.)

3   Deze opvatting sluit in zich mijn onsterfelijkheid, of liever kan niet denken aan sterfelijkheid. Immers 't begrip tijd, zowel als ruimte behooren tot mijn voorstellingen, maar mijn ikheid is daar volkomen los van. Mijn verhouding tot mijn God is een afhankelijk vertrouwen op hem, die mij doet leven.

Maar het leven dat mijn God mij laat leven, kan zijn denken over dingen die ik waarneem en toestanden, die ik om mij heen zie, en opinies hebben over verschillende kwesties, ook zgn. godsdienstige kwesties, maar dan zijn het voorstellingen, mij door God gegeven, die er buiten en boven staat; zij kunnen mijn God niet omvatten, want zij zijn uit Hem voortgekomen. Tot mijn eigenlijke Godsdienst behoort alleen het mijn God voelen.

Maar de taal is hier een te onhandig instrument. Het Godsgevoel en Godsvertrouwen is niet een bewuste gedachte dus een voorstelling, want dan zou het weer buiten God zelf staan, maar het is iets dat, daar het boven de gedachte staat, zich niet laat denken, nog minder neerschrijven, het is iets, dat vast is aan de onbewuste ikheid, die bewust wordt m.a.w. voorstellingen krijgt, maar afgescheiden van die voorstellingen. Wel kan er een beeld van ontstaan in een bewuste gedachte, maar niet dan zeer vaag.

4   Deze mijne opvatting over het eerste punt der belijdenis maakt de bespreking van vele andere punten overbodig.
Punt 1-13: Vooreerst een historisch godsdienstoverzicht kan bij mij niets in zich bevatten dat mij tot mijn overtuiging heeft geleid en is dus hier niet op zijn plaats.
Punt 15: Bezwaren tegen de erkenning van een godsbestuur in deze zin dat ik dat niet zou weten te rijmen met verschillende dingen, die ik om mij heen waarneem, bestaan voor mij niet. Immers mijne waarnemingen behooren tot mijn voorstellingen, en geen van deze kan wegens haren aard een bezwaar zijn, zij allen zijn in hun bestaan het bewijs van God's bestaan.
Punt 16 -17: Het kenmerkende van mijn levensopvatting tegenover die van anderen is ook reeds gezegd. Ik vat het leven op noch als een last die ik te dragen heb, noch als een taak dien ik te vervullen heb; neen, mijn leven is een voldongen feit, waarover ik geen meening kan zeggen. Immers daartoe zou ik het objectief van buiten moeten kunnen beschouwen, maar dat kan ik niet, voor mij is het leven het groote enige Het dat ik geen eigenschap kan geven omdat er niets mee kan worden vergeleken.
    Deze opvatting sluit volstrekt niet in, dat mijn leven een dof, blind, willoos mij laten gaan zou zijn. Het leven dat mijn God mij laat leven, kan rijk zijn aan hoop, angst en streven vol van hartstochtelijke najaging van idealen, en mijn eigen vrije wil kan sterk zijn; dit alles toch behoort tot de voorstellingen die mij gegeven kunnen worden.

5   Punt 18 -19: Van mijn onbepaald vertrouwen op God, en overtuiging van mijn onsterfelijkheid sprak ik reeds bij punt 14.
Punt 20: Onder de voorstellingen die mijn God mij geeft behooren die, welke mij intens doen voelen zijn bestaan op sommige oogenblikken waarop dan volgt eenkrachtig zelfvertrouwen, en blijde levensmoed. Telkens wanneer dat bewustzijn zich krachtig aan mij opdringt, dat geheel mijn innerlijk leven aanblaast, kan ik spreken van liefde tot mijn God. Zulke oogenblikken van contact dragen bij mij niet het karakter van gebed, omdat mijn wenschen en verdrietelijkheden er geen rol bij spelen, maar integendeel dan juist voor mij geheel verdwenen zijn.

Tot zoover heb ik de punten in verband met mijn godsdienstige overtuiging kunnen brengen. Het overblijvende heeft voor mij een geheel andere beteekenis.

Punt 21: Mijn God n.l. heeft mij ook het streven gegeven om mijn leven, d.w.z. mijn voorstellingen zo schoon mogelijk te maken; daaruit vloeit voort dat ik in de, deel van mij zijnde, mij omgevende wereld, getroffen word door de walgelijkheden, en die wil trachten weg te nemen, ook wat betreft de menschenwereld. Liefde voor mijn naaste zal ik dit ternauwernood kunnen noemen, immers aan de meeste menschen heb ik het land, bijna nergens vind ik mijn eigen gedachten en zieleleven terug: de menschenschimmen om mij zijn voor mij het leelijkste deel van mijn voorstellingenwereld. In theorie zal ik mij ook nooit opofferen voor een ander; mij worden echter door mijn God gevoelens als medelijden gegeven die mij soms dwingen in die richting te werken.

Punt 22, 23, 24: Door het onbewuste streven naar mooier maken van mijn voorstellingen heb ik natuurlijk opinies over het al of niet nuttig zijn van instellingen in de menschenwereld, dus kan ik ook over deze punten schrijven, ook al raak ik hoe langer hoe verder van mijn religieuze overtuiging.

Ik vind een kerk goed, omdat zij, ook al hooren wij er niet onze eigen overtuiging, onze gedachten op gebieden kan leiden waar door eigen werken en denken geluk kan worden gevonden. Kerkelijke heerschappij en dogmatiek zijn natuurlijk gedegenereerde toestanden. Godsdienstige vormen vind ik goed voor de domme menigte, om door een kerk, die heerschen wil, in eerbiedig niet begrijpen onder de duim te worden gehouden. Nog eens, ik vind de kerk goed, als degene die ons wijst op onze taak, om in een godsdienstige overtuiging een staf te vinden, om ons leven mee door te gaan.

Punt 25: En dit is de belijdenis van mijn godsdienstige gevoelens en overtuigingen, waarvan ik mij thans voor 't eerst rekenschap heb gegeven, en ze heb geordend en geschift, ook al heeft de eenheid en kracht waarschijnlijk geleden door een verdeeling over punten, die niet de mijne was.

30 Maart 1898

### Profession of Faith by L.E.J.Brouwer (translation)

Point 14: What is the foundation of my faith in God?

This to me is the main point of a profession of faith, and the only part that can really be called the 'profession of faith of a person'.

That I believe in a God is due, not so much to some rational consideration in the sense that in the phenomena around me I would see the revelation of some 'higher power', but to the complete helplessness of my intellect. For the only truth to me is my own ego of this moment, surrounded by a host of images in which the ego believes and which causes the ego to live. There is no sense in the question whether these images are 'true'; they only exist for my ego and as such they are real. There is no such thing as a second reality, independent of my ego and corresponding to these images. My life of this moment is my conviction of my ego and my belief in my images. This is directly bound up with the belief in That which is the origin of my ego and which gives me my images; which is, therefore, independent of me, something that like myself is alive and which is higher than me; that something is my God.

2  In the many words which I am using to describe this, one should not see an attempt at rationally deriving the existence of God, this belief in God is the deepest foundation of any further reasoning, but it cannot itself be proved. Belief in God is a direct, spontaneous sensation in me.

I do accept, however, that this belief in God is taken here in a somewhat unusual sense, mainly because my belief is founded on a philosophy which recognizes only myself and my God as the sole living beings, myself as the one I know and my God, my master, as the one I feel.

The images given to me contain in themselves among other things also the possibility of the existence of other egos with their own images; but these are not real, they are part of my images and therefore of me. My images are my life; at this very moment, for example, I live in the image that I am thinking about my life and write my profession of faith. But I do not find my God in this life: my God is under* and outside my life; only the fact of my living makes me feel the existence of God, I do not find Him in the way I live.

(*   Under , i.e. He is the originator of my life.)

3  This view includes my immortality, or rather it excludes mortality. For time as well as space are part of my images of which my ego is completely independent. My relation to my God is a dependent trust in Him who makes me live.

But the life that my God allows me to live can be: thinking about the things I perceive and the conditions I observe around me, having an opinion about different questions including the so-called religious questions; but these are images given to me by God, who stands outside and above them, they cannot capture my God because they originate in Him. My real religion is only my feeling of my God.

But language is here too clumsy an instrument. The feeling of God and trust in God is not a conscious thought, i.e. an image, otherwise it would again remain outside God; it is something that transcends thought, that cannot be thought, let alone be described; it is something connected with the unconscious ego at the moment of it becoming conscious i.e. receives images, but separate from these images. It is possible that an image of it grows in some conscious thought, but only very vaguely.

4 This, my view on the first point of the profession, makes discussion of many other points superfluous.

Point 1 - 13: First of all I don't think that a historical survey of religion could contain anything that led me to my conviction, and therefore it is here out of place.

Point 15: As to objections against acceptance of a divine rule in the sense that I would be unable to reconcile this with various things and events which I see around me, they do not exist as far as I am concerned. My perceptions are part of my images and none of these by their very nature can constitute an objection: their very existence is a proof of the existence of God.

Point 16 - 17: I have also already given the essence of my views on and attitude to others. I consider life to be neither a burden to be carried nor a task to be fulfilled. No, my life is a given fact on which I cannot express an opinion; this would require an objective view from the outside and that is impossible. To me life is the great and only *It* to which I cannot attribute any properties because there is nothing to which it could be compared.

This view does not in any way imply that my life were some dull, blind and will-less letting-myself-go. The life which my God allows me to live can be rich in hope, fear and aspiration, full of the passionate pursuit of ideals, and my own free will can be strong. Isn't all this part of my images that may be given to me?

5 Points 18 - 19: I have already spoken in point 14 of my trust in God and my conviction of my immortality.

Point 20: Among the images which my God has given me there are those which make me sometimes intensely aware of His existence; such moments then lead to strong self-confidence and a new zest for existence. Whenever this awareness forces itself upon me and sets my inner life alight, I can speak of love of my God. Such moments of contact in my case do not take the form of prayer, since my wishes and vexations do not play a part in it; in fact these have at such moments completely vanished from my consciousness.

So far I have been able to link the points of the questionnaire to my own religious convictions. The remaining points to me have a completely different meaning.

Point 21: My God has also given me the ambition to make my life, i.e. my images, as beautiful as possible. It follows that in the world surrounding me and part of me I am struck by its loathsomeness and that I want to remove this, also from the human world. I can hardly call this love-of-my-neighbour, for I don't care two-pence for most people. Hardly anywhere in human society do I recognize my own thoughts and inner life. The human spectres around me are the ugliest part of my world of images. In theory, therefore, I shall never sacrifice myself for someone else, although my God also gives me feelings of pity which make me sometimes move in that direction.

6 Points 22, 23, 24: Because of this unconscious tendency to making my images more beautiful, I do, of course, have opinions on the usefulness of the institutions of human society and therefore I can also write about these points, even if I move further and further away from my religious convictions.

I approve of a church because, even if we don't hear our own convictions preached, it may guide our thoughts in areas where we may find happiness by our own work and thought. Of course, church domination and dogma are degenerate situations. I consider religious forms to be good for the stupid masses, to be kept in reverent ignorance under the thumb of a power-thirsty church. Again, I approve of a church as a guide, showing us our task, providing a supporting staff on our path through life.

Point 25: This is the profession of my religious feelings and convictions, of which I now for the first time render an ordered account. Unity and force have probably suffered because of the division in points which was not mine.

30th March 1898

## Appendix 2

### Student Notebook [BMS 1B]
⟦a collection of 'thoughts'⟧

1  Waarzeggerij? Och, wat is 'toekomst zien' anders dan 'toekomst willen'? Pas op voor kwade toekomst en visie zoo goed als voor kwade wil.

   Dat iets gaat groeien en assimileren is, dat het uit een gevoel van onvolledigheid, van excentriciteit, moet gaan parasiteeren op anderen.

   Is het individu eenmaal door de gemeenschap aan zijn natuurlijk milieu onttrokken, dan wordt hij in de eenzaamheid zoo weerloos en direct gegrepen door suggesties.

   Bij het leven in de gemeenschap is het eenige, wat met 'trots' is vol te houden leven in een specialiseering, waarop je je toelegt; de spanning der te velen in de samenleving perst ieder in zijn specialiseering, en harmonische menschen krijgen hoe langer hoe kleinere waarschijnlijkheid.

2  Iemands 'talent' of specialiteit is dus het zwakste (minst weerstand biedende tegen druk) deel van zijn organisme.

   Helderziendheid geeft een mensch niets in den strijd om het bestaan. Daarom geldt het niet, de wereld te kennen, maar slag te hebben haar te verknoeien.

   Fatsoen is: op het oorlogsterrein neutraal doen, d.i. doen niet volgens persoonlijke instincten (want die treden in den strijd), maar volgens wiskundige wetten (die voorondersteld worden, algemeen te zijn aangenomen.)

   De uitloopers van het beschavingssysteem, als archeologie, behoeven in zichzelf niet zuiver te zijn; immers ze zijn er slechts om het beschavingssysteem af te ronden, door ook wat te doen aan de consequenties die het bevangen verstand ervan ziet.

3  Als levensdoel zou kunnen worden gezien: Afschaffing en verlossing van alle wiskunde.

   De veelgeprezen kunst heeft het nog niet verder gebracht, dan tot kinder-romantiek; het zien der droeve lijnen is nog niet bereikt dan als vaag pessimisme.

Je gevoel verintellectualiseeren mag je alleen ter bestrijding van de suggesties der anderen.

4    Ze verkondigen hun oppervlakkige ideëen alleen, zoolang hun eigenbelang hen drijft, al of niet door valsche hoop gepaaid.

Niets van kunst of wetenschap, dat waar is, heeft waarde (ruilwaarde).

Het plebs weet niet, dat de taalwaarheid, de verstandhoudingswaarheid, niet bestaat; vandaar dat hun passie zich manifesteert in: willen verdraaien van de waarheid.

Verstandhouding tusschen 2 menschen is gradueel; maar wiskundige verstandhouding is iets van wel of niet, zooals slapen iets is van wel of niet.

De meesten worden socialist, om in een voor hen geschikt milieu te komen, en daar relaties aan te knoopen.

5    Dat de wiskunde en haar toepassing zondig zijn volgt uit de direct als zondig gevoelde tijdsintuitie.

Het wiskundig willen bekijken is tevens zich willen afsluiten van als bijkomstigheden gewilde dingen.

Schilders kruipen bij boeren als parasieten; aan de boeren hun bescherming tegen het natuurgewetenwillende opdragen.

De volgroeiing van het individu (zooals van een boom) is het langzamerhand zich ompantseren tegen de bedorven omgeving; was de omgeving zuiver, er ware geen groei, en ook geen dood.

6    Pas in de renaissance breekt de werkelijkheid der materieele vormen door als iets wezenlijks; te voren trachtte men er juist boven uit te leven in de kunst.

Het gewone is gewild door Tintoretto, en doet wonderlijk aan. Het dramatisch extract uit het gewone geven Carpaccio en Montagna, en nog betere consequentie de heroische kunstenaars.
Giotto is vrij van het gewone. Altichieri en andere burgerlijke stakkers

willen het gewone verhoogen tot het burgerlijke mooie (bijzonder mooie gewaden of bijzonder flinke kerels), maar vallen telkens weer in het zeer gewone terug, en worden dan belachelijk.

7 Wetenschap is te veroordelen en om de primaire tijdsintuitie en, omdat haar toepassing is: werken op de toekomst.

In het leven alleen blijft de man alleen; maar in het werken tusschen de menschen behooren man en vrouw een te zijn.

Wil je in je hart een practisch ideaal nastreven, dan kan het alleen zijn het vernielen van alle menschelijke kennis over de wereld.

Heb eens een korte tijd gehandeld en retireer dan, dan merk je wat een vereenzijdiging er voor noodig was.

Rijken en armen zijn elkaar completeerende elementen van een in de wereld gevallen zonde. Wie zich daaraan weet te onttrekken, wordt noch rijk noch arm.

8 'Idealen' of speciale optimismen zijn begeleidingen van speciale zonderichtingen van enkele menschen of van in zondegewrongen richting gehouden agglomeraten van menschen.

In de wereld der specialiseering is plaats voor enkele wilden, die direct de vrijheid willen houden, maar dat alleen kunnen doen door met hun vrij gedrag handel te drijven: zoo wordt hun vrijheid toch steeds aangebotst en verwart onherstelbaar, tot een meer waarschijnlijke toestand van specialiseering.
Maar wie nederig begint met het speciale, en wacht tot gegeven doorgangen komen, die kan zich dan eraan laten ontketenen.

9 Er is geen plaats voor alle menschen, zoo allen hun algemeene leven wilden leven; de uitkomst zit in specialiseering: die specialiseering wordt gericht en getroost in een evenwijdige wegrichting voor allen door samen komen in een tempel.

Het handelsverkeer (van geven en nemen) gaat over wiskundig omgrepen dingen der wereld; het niet wiskundig omgrepene omvat wel veel meer, maar treedt niet in geregelde ruilhandel, wordt tuchteloos en veronachtzaamd gegeven en genomen.

De wereld verbeteren of mooi maken? Zijn wij menschen dan niet de verrottingsbacterien der aarde, die het ontbindingsproces waarvoor het tijd was, bemiddelen en bespoedigen?

10   Enthusiasme is leven in onbereikte bereikbaarheid.

Men scheide tusschen hen, die zoo goed mogelijk willen zijn en hen, die zoo goed mogelijke waar (dus, daar ieder zijn waar als goed aanprijst, zoo eerlijk mogelijke waar) trachten te leveren. Van Eeden b.v. levert zeer goede waar, maar is zelf zeer slecht.

Wie de menschen als een lagere diersoort dan zichzelf ziet, duelleert niet meer; zoo min als men in den strijd tegen de dieren toestemt een in onderlinge verstandhouding daarom gekozen wapen te gebruiken.

De overwinning der listigen berust hierop, dat de naieven nog enkele primitieve paradijsinstincten (b.v. vatbaarheid voor vriendelijkheid) hebben, die de listigen weten op te merken en te leiden.

11   Altijd door lijden, en altijd door je dwingen.

Je zou jezelf te veel vermoorden, als je tegen iedereen eerlijk wou zijn. En als iemand je oneerlijkheid jegens hem opmerkt, en daarover gepiqueerd is, dan heb je dat aan te nemen als een van de als gevolg van je wezen noodwendig voorkomende gebeurtenissen.

De menschen, die altijd bezig zijn met met zelfverdediging, hetzij door loon, hetzij door plicht-proclameering, verliezen daardoor het vermogen, om werkelijke innigheid met iemand te voelen.

12   Leer je teveel wetenschap, dat is gemeenschappelijk denken, dan verleer je daardoor het denken voor jezelf in strijd met de anderen.

Dat volk presteert het meest, dat als geheel naar buiten perfide is, een handelsman met streken, maar in zich een strenge regelmaat en onderlinge plichtsvervulling handhaaft. Dat laatste is echter noodig, en, daar een natuurlijke leefwijze in 't algemeen de plichtsbetrachting en het vermogen tot plichtmatige prestaties verzwakt, hoewel misschien het vermogen tot handhaving door streken versterkt, moet een volk, dat natuurlijk leeft, het afleggen; kan echter in de beschavingsperiode, die nog het dichtst bij de natuurlijke ligt, misschien andere volken overwinnen.

13   Het net van categorieen uit zijn milieu, dat ieder mensch omstrikt, zie je het helderst, als je je aan iemand, wiens taal je niet of weinig kent, verstaanbaar moet maken en hem aankijkt, terwijl je woorden stamelt, en hij tracht te begrijpen.

De wiskundige systemen in gedichten en geschreven religie zijn zoo ongeveer niets dan coordineeringen.

Immanente waarheid is het zien van de wereld als verleiding, dus passeren van den schijn van wetten en begeerenswaardheden er in.

Wanneer men 'reine', 'doelloze' ervaringswetenschappen bedrijft, heeft men eigenlijk evengoed een concreet doel n.l. het zeker zijn, zeker gaan van een beloop van verschijnselen.

14   De degeneratie van het practisch wiskundig doen kunnen zijn zucht tot zien van wiskundige systemen (theoretiseerders) en zucht tot verwerkeliking van wiskundige systemen (knutselaars, agitatoren en intriganten). Het gewone organiseerend talent hoeft men nog geen degeneratie te noemen.

Het gevoel van waardigheid en geëerd te zijn, is analoog aan het foutieve gevoel van gemak, van macht, van gewapend zijn, en andere, die tot inslapen verlokken.

Het is een practische kunst, een ander je meening, dus je wil en wensch te doen gevoelen, zonder het wiskundig systeem er van uit te spreken, en hem daardoor gelegenheid en vat te geven, zich tegen dien wil in den vorm van een wiskundig systeem te verdedigen, b.v. door het aan anderen over te vertellen.
Men kieze dus zijn woorden zoo, dat ze met voldoende zekerheid hun beteekenis alleen voor den toegesprokene hebben.

15   Men retireert zijn persoonlijkheid, die vereenzijdiging, om haar te kunnen verdedigen, want door concentratie alleen wordt zijn reaktionsfähig. Echter kan men veel gemakkelijker het leven houden, door vermenging met de omgeving toe te laten, d.w.z. prijsgeven der persoonlijkheid; zoo n.l. worden de potentiaalverschillen zwak, dus ongevaarlijk.

Alle natuurregels zouden kunnen worden gebroken; ze bestaan door traagheid, gebrek aan helderziendheid, om het brekingsmiddel te vinden.

16      Smart is onmacht, in zijn hoofd het rust gevende wiskundige systeem te vinden. Alleen hij die rust wil kent dus smart.

## Student Notebook (translation)

1       Sooth-saying? What else is 'seeing the future' if not 'wanting the future'? Beware of evil visions of the future as much as of evil desires.

That something begins to grow and assimilate is due to a sense of insufficiency, of excentricity, which forces it to be a parasite on others.

Once an individual has been taken out of his natural environment he is totally defenceless, and is immediately affected by the suggestions of others.

In life in a community the only thing that can be maintained with 'pride' is living in a chosen specialism; the tension of 'too-many' in society forces everyone into his specialism; harmonious people become increasingly scarce.

2       One's 'talent' or speciality is therefore the weakest (i.e. offers least resistance against pressures) part of one's organism.

Clairvoyance does not offer anything in the struggle for existence. That's why the rule should not be to know the world but to have a knack of spoiling it.

Good manners is acting neutrally on the battlefield, i.e. acting not according to one's instincts (for these would join in the battle) but according to mathematical laws (which are supposed to have been universally accepted).

The off-shoots of our system of civilization such as archaeology do not need to be pure-in-themselves; for they only exist to round off the system by also doing something to the results which the enchanted intellect sees of it.

3       One's aim in life could be the abolition of and the deliverance from all mathematics.

This much praised Art has not gone much beyond child-romanticism; seeing the sad lines has not been achieved except as some vague pessimism.

Intellectualizing one's feelings is only permitted in the fight against the suggestions of others.

4   They only preach their superficial ideas as long as they are driven by self-interest, maybe fed on false hopes. Nothing in art or science that is true has value (i.e. commercial value).

The riff-raff do not know that truth of language, i.e. truth in human communication, does not exist; that's why their passion is to try and twist the truth.

Understanding between two people is a matter of degree, but mathematical understanding is like being asleep: you either are or you are not.

Most people become socialist in order to find a way into a milieu that suits them and to make contacts.

5   Mathematics and its application are sinful because of the intuition of time which is directly experienced as sinful.

The will to view things mathematically includes the will to shut oneself off from things which one wants to be accidental circumstances.

Painters scrounge on farmers like parasites, trying to offer them protection against their natural conscience.

The growth of the individual (like that of a tree) is slowly forming an armour round oneself against a decaying environment. Had the environment been pure, there would not have been any growth nor death.

6   Only during the Renaissance does the reality of material form break through as something essential; before that time one rather tried to rise above it in art.

The 'ordinary' is wanted by Tintoretto, and it strikes one as magical. Carpaccio and Montagna give a dramatic extract of the ordinary, and the heroic artists an even better result. Giotto lacks the ordinary. Altichieri and other poor bourgeois devils try to raise the ordinary to the level of the bourgeois 'nice' (specially pretty garments or specially sturdy fellows), but they always backslide into the very ordinary and then become quite ridiculous.

7    Science must be condemned because of the primordial intuition of time and because of its application: influencing the future.

In the life of solitude man stands alone, but in the interaction between human beings man and woman should stand united.

If you wish to pursue a practical ideal, it can only be the destruction of all human knowledge over the whole world.

When you have been active even for a short time and then withdraw into yourself, you notice how much one-sidedness such action demand.

Rich and poor are the mutually complementing elements of sin visiting the world. Those who are able to draw back from it are neither rich nor poor.

8    'Ideals' or special optimisms are accompaniments of special sinful tendencies of some individuals or the sinful direction forced upon agglomerations of human beings.

In the world of specialization there is room for a few wild individuals, those whose immediate aim is to keep their freedom but who can only do that by bartering their free behaviour; in this way their freedom still receives a pounding and suffers irreparable damage, resulting in a more likely state of specialization.
But he who humbly starts with the special and waits for certain breakthroughs, he will be able to free himself from its fetters.

9    There is no room for all human beings if they all want to live an 'all-round' life. The only solution is specialization; such specialization is directed and calmed into a parallel movement for all, coming-together in a temple.

Commercial intercourse (of give and take) deals with the mathematically-conceived things of this world; what has not been mathematically conceived comprises far more, but it is not the subject of barter, one gives and takes without any restraint or care.

Improving the world or making it beautiful? Aren't we all just the bacteria rotting down the earth, causing and speeding-up its timely process of decomposition?

10    Enthusiasm is living in unrealized potential [literally: attainability that is never attained].

One should carefully distinguish between those who wish to be as good as possible, and those who try to deliver wares as good as possible (and since everybody praises his own ware as good, it should be as honest as possible). Van Eeden e.g. delivers very good wares but is himself thoroughly bad.

He who considers human beings as an animal species inferior to himself, no longer fights duels, not any more than that in the battle against animals one comes to a mutual agreement on the choice of weapon.

The victory of the cunning is based on the primitive instincts of paradise still found in the naive (e.g. their responsiveness to kindness); these are noticed and used by the cunning.

11    Ever suffering and ever forcing yourself.

You would murder too much of yourself, if you were to be honest with everybody. If anyone notices your dishonesty towards him and takes offence, you'll have to accept it as one of those things that inevitably follow from your human nature.

People who are always busy defending themselves, quoting as excuse their position or their duty, thereby lose the ability to feel really deeply for someone else.

12    Most is achieved by a society which is united in their perfidy towards the outside world, but which for itself maintains a strict code of conduct and a sense of duty among its members. The latter is absolutely necessary; but since a natural life-style usually weakens one's sense of duty and one's ability to act dutifully − although perhaps the power of self-preservation grows by cunningness − a society which lives naturally must go to the wall. Maybe during the period of civilization closest to nature, it may succeed in conquering other peoples.

13    The net of categories in which society ensnares all its members is most obvious when you have to make yourself understood by someone whose language you don't know or hardly know, when you look at him, stammer a few words and he tries to understand you.

The mathematical systems in poetry and written religion are roughly speaking no more than coordinate systems.

Immanent truth is seeing the world as a temptation, i.e. seeing through the outer appearance of its laws and its attractions.

Even when one is engaged in 'pure', 'purpose-less' experience-based science, one still has a concrete aim: one wants to be sure, and make sure that one's theory works.

14  The degeneration of practical mathematical acting could be a desire to see mathematical systems (theorizers), or a passion to see mathematical systems realized (tinkerers, agitators and schemers). The ordinary organizing talent need not necessarily be a degeneration.

The feeling of dignity and being honoured, is analogous to the wrong feelings of comfort, of power, of being armed, and other feelings that tempt you to fall asleep.

It is a matter of practical cunning, to be able to let someone else know your opinion, and therefore your will and wishes, without expressing the mathematical system which underlies it; he then will not have the opportunity and chance to defend himself against that will in the form of a mathematical system, e.g. by speaking to others about it.
One should therefore choose one's words carefully, so that they can be understood with sufficient certainty by the listener alone.

15  The purpose of withdrawing one's personality (one-sidedness) is defence, since concentration alone makes it 'reaktionsfähig'. But it is much easier to save one's life by allowing it to blend-in with the environment, i.e. by surrendering one's personality; in this way potential differences become weakened and therefore harmless.

All natural laws could be broken; they continue to exist because of our inertia, our inability to see clearly and find the means of breaking them.

16  Grief is weakness, the inability to find in one's mind the mathematical system that secures tranquility. Therefore only those who seek tranquility suffer grief.

# Wiskunde en Ervaring

~~Het~~ leven, (na) ontstaan (te zijn) als vereenzijdiging ~~van~~ de natuur, ~~[illegible]~~ ~~maar~~ uitte ~~[illegible]~~ zijn bestaan in een verinnerlijking; een doordringing van de natuur met zichzelf, in terugdringing van andere ~~[illegible]~~ vereenzijdigingen.(1)

De verinnerlijking der menschheid, het dienstbaar maken van de omgeving aan de ontplooiing der menschelijkheid, verschijnt ons(2) als een rechtlijnig en regelmatig worden van de natuur; waardoor al het andere leven wordt teruggedrongen of aan ~~[illegible]~~ de menschheid aangepast.(3)

Wat is nu het essentieele van de menschelijke verinnerlijking, dat zooveel machtiger blijkt, dan de brutale assimilatie of vernieling, die van andere schepselen uitgaat? De rechtlijnigheid en regelmatigheid komt b.v. ook bij de bijen voor, maar daar brengt ze geenerlei bijzondere macht. Maar de mensch heeft een vermogen, dat al zijn wisselwerkingen met de natuur begeleidt, het vermogen n.l. tot objectiveering der wereld, het zien in de wereld van herhalingen, van volgreeksen, het zien in de wereld van causale systemen in den tijd.(4)

---

(1). Dàt het zich verinnerlijkt, en niet te sterven ligt, wordt als genie van wijsheid, als genie van verband met het Al, door religie gevoeld. — Overigens drijft het zich verinnerlijken, het willen vernietigen of willen heerschen, meteen af van alle voeding ([illegible]) uit de natuur. — Wie heerscht, is reeds gevloekt, en het zijn gevloekte eigenschappen, die helpen tot heerschen.

(2), n.l. als wij intellectueel, met wiskundig-causalen blik de wereld bekijken.

(3). Door het aanpassen van het milieu het steeds verder afvoert van den natuurtoestand, zooals die oorspronkelijk de menschheid droeg, wordt elk overwonnen en aangepast milieu ten slotte voor de menschheid zelf onhoudbaar.

(4). Dat zien is intusschen niets, dan een daad van verinnerlijking; van een bestaan der objectieve verschijnselen der natuur, in den zelfden zin als (van) een bestaan der natuur zonder meer is geen sprake: het zien gaat uit van den bekijker, is een ~~[illegible]~~ (uitwerking) van den bekijker alleen, buiten de natuur om, die zelf bestaat ~~[illegible]~~ voor het subject buiten zijn wil om.

**Appendix 3**

**The Rejected Parts of Brouwer's Dissertation** [BMS 3B]

⟦Of the original (Dutch) manuscript of the dissertation only 22 pages have been preserved; they are the pages wholly or partly rejected by Brouwer's supervisor, Professor D.J. Korteweg.

The passages, crossed-out by Korteweg and not published in BR[1907], are given here in my English translation.

The number in the margin refers to the page-number of the manuscript. A line is used to separate the text of each page from the footnotes and a dotted line to indicate the end of the page.

Korteweg's comments are marked DJK and placed between double square brackets ⟦...DJK⟧; editorial comment is marked ED and placed between double square brackets ⟦...ED⟧. For further editorial comment see VAN STIGT[1979]. ED⟧

CHAPTER 2

Mathematics and Experience

1     All human life originated in a one-sided constriction of nature [*] and has protracted its existence in an 'externalization,' man impregnating nature with the human self and repressing other one-sided developments [1].

This externalization by man, making his environment subservient to the full development of his humanity, appears to us [2] as a process whereby nature itself becomes linear and regular and all other life repressed or adapted to mankind [3].

What then is the nature of this human externalization which evidently is so much more powerful than the brute assimilation and destruction practised by other creatures? We feel linearity and regularity, for example, also in bees; there it does not result in any sort of special power. But man has the faculty, accompanying all his interactions with nature, of objectifying the world, of seeing in the world causal systems in time [4].

---

[*] ⟦literally, 'making one-sided'--human concentration on one single aspect of nature and adapting nature accordingly.ED⟧

[1] This externalization of life and the holding off of death, from the point of view of religion, reflects a lack of wisdom and the absence of a bond with the universe. Moreover, this externalization, the will to destroy and rule, immediately obstructs any nourishing of the heart by nature. Those who rule are already damned and damned are those qualities that promote man's rule.

[2] If we view the world intellectually, i.e., with a mathematical causal eye.

[3] Since the adaptation of the environment leads human life further and further away from the natural state which originally supported him, this conquered and adapted environment will ultimately become intolerable to mankind.

[4] This 'seeing', however, is a human act of externalization: there is no real *existence* of objective natural phenomena as can be ascribed to nature itself: the seeing originates in man, is an expression of man's will alone, independent of nature which itself exists independent of man's will.

........................

2    The primordial phenomenon is simply the intuition of time in which repetition of 'thing in time and again thing' is possible, but in which (and this is a phenomenon outside mathematics) a sensation can fall apart in component qualities, so that a single moment can be lived through as a sequence of qualitatively different things. One can, however, restrict oneself to the mere sensation of these sequences as such, independent of the emotional content, i.e., independent of the various degrees to which objects perceived in the world outside are to be feared or desired. (The attention is reduced to an intellectual observation.) The human tactics of 'acting purposively' then consists in replacing the end by the means (a later occurrence in the intellectually observed sequence by an earlier occurrence) when the human instinct feels that chance favours the means. However, since the link between end and means is observed in the intellect without the control of more central instincts (a restriction which will make the process even more intensive and more generally applicable), the reliability of the human conviction that the parts of the sequence belong together in reality is far from absolute and can constantly be disproved; this is experienced by the intellect as a discovery 'that the rule no longer applies'.

In general, however, these tactics, i.e. of observing sequences and then jumping from the end to the means, prove effective and are the source of human power [1]. Indeed, if this faculty did not achieve its end it would not exist, as lion's paws would not exist if they failed of their purpose. It is possible to discover regularity in a limited domain of phenomena independently of other moments and other phenomena, which therefore can remain completely concealed from the intellectual observation. In this way one succeeds in getting on nature's weak side and in disarming enemies in an essential sphere of life.

---

[1] There is a distinction between this form of replacing end by means, which is based on a purely active intellectual observation − not merely waiting for, but actively seeking such sequences − and aiming for something which is different from the end itself but is associated with this end, and that therefore appears as a phase of the end itself; the latter is an animal as well as a human phenomenon.

........................

3   To maintain the certainty of observed regularity as long as possible, one tries to isolate systems, i.e. to exclude observations which disturb this regularity. In this way man makes in nature far more regularity than had originally and spontaneously occurred in it. He wants this regularity because it strengthens him in his struggle for life as it enables him to predict the future and take his measures.

The process of objectifying the world through the primordial intuition of 'repetition in time' and 'following in time' gains in generality by the construction of mathematics from the same primordial intuition, without reference to direct applicability. In this way man has a ready-made supply of unreal causal sequences at his disposal, just waiting for an opportunity to be projected into reality. One should bear in mind that in mathematical systems with no time coordinate, all relations in practical applications clearly become causal relations in time; e.g. Euclidean geometry when applied to reality shows a causal connection between the results of different measurements made by means of the group of rigid bodies. Needless to say, in the application of a mathematical system, in general, only a fraction of the elements and substructures find their correspondence in reality; the remainder plays the role of an unreal 'physical hypothesis'. Similarly, even with a limited development of method, the observed sequences no longer consist exclusively of phenomena evoked by man himself (acts without any direct instinctive aim, but carried out solely to complete the causal system into a more manageable one). The simplest example is the sound-image [1] of number as a result of counting, or the sound-image [1] of number as a result of measuring [2].

[1] or written symbol

[2] This example shows how infinitely many causal quences can be brought together under the viewpoint of one single law of causality on the basis of a mapping of number-sequences onto one-another through *mathematical induction*.

........................

4  The strategy of objectifying the world forces man even more to eliminate the 'deviating' influences and thereby to abnormalize his environment. The nature of the phenomena within a certain domain changes not only through the elimination of the influences which deviate in this domain itself, but also through the degeneration of the environment of these phenomena because of the removal of influences which deviate with respect to another quite different group of phenomena.

In the case of a *negative* application of this science (i.e. by a process of destructive removal) this one-sided constriction often causes no further trouble, but destruction remains destruction, although the element to be destroyed was originally thought not to belong. However, if this science is positively applied (i.e., by adding or by forcing into the right form) the influences which disturb the intellectual viewing can be of essential importance for the instinctive value of the results. In this way science, in a process of increasing self-perfection, will strengthen its power to obtain results but debase the *value* of these results.

We observe in this context that 'mathematical viewing' is only instinctive, i.e. justified in sofar as it is directed to a world which is considered to be external; to try and direct it to inner perception is a serious error (moreover, there would never be any agreement between the results of mathematical viewing from different viewpoints). What Kant describes as 'Transcendental Analytic' can only be described as idle play.

The objectification of the world in mathematical systems by different individuals is held in mutual agreement by means of a *passionless language* which evokes in the listener a mathematical system identical to that of the speaker, although the emotional content of these systems may be completely different. The only purpose of the agreement between the mathematical systems of reality in different individuals is to enforce man's will over others out of fear or desire associated with certain elements in the system.[*]

---

[*] ⟦The *emotional content* is not the same for both; there is a correspondence; such 'sameness' would even be quite useless; only the success desired

by the speaker can connect experience and language. But sameness of *mathematical content* in speaker and listener can certainly be achieved. DJK⟧

........................

5   The creation of language is itself an example of a human uninstinctive act based on mathematical knowledge. *Common* mathematical systems are not desirable for their own sake (mathematical systems certainly lack a psychological basis for accordance), only as a means to an end. Again: the emotional representations evoked by the same word in different individuals may well be different (or rather incomparable, since comparison presupposes mathematics). This, however, does not affect the general efficiency of language. Because of the similarity of the relations in the mathematical systems, the speaker can still force the activity of the hearer with sufficient accuracy in the direction he desires, even if the representations associated with this activity are totally different in different individuals.

Perhaps the greatest merit of *mysticism* is its use of language independent of mathematical systems of human collusion, independent also of the direct animal emotions of fear and desire. If it expresses itself in such a way that these two kinds of representations cannot be detected, then the contemplative thoughts − whose mathematical restriction appears as the only live element in the mathematical system − may perhaps again come through without obscurity, since there is no mathematical system that can replace them [1]. When the mystical author joins representations of this kind into more central affections, of which they were one-sided restrictions, he can, with the most ordinary words, gradually break down the barriers round the contemplative sphere and guide us back to the 'all-embracing' which every poet seeks to approach. His language will therefore appear as meaningless to those who expect to find in words only the communication of mathematical systems or a stimulus to mathematical activity.

---

[1] For example, in the case of the word *time* the awareness of solitary weakness, of roaming, deserted after rejection of guidance, may only break through when it is no longer possible to include the independent, variable coordinate of mechanics.

........................

6   The mystical writer will even be careful to avoid anything that smacks

of mathematics or logic: weak minds might otherwise be easily made to believe and act mathematically outside the domain where this is required either by the community or their own struggle for life and end up in all kinds of follies.

We should point out that we appear to be successful in observing sequences which, to our instinct, are broadly similar (or which through mathematical induction can be brought under one single aspect with respect to approximate similarity); in particular, that 'it works' if we draw up a sequence of $\omega$ terms on which, for example, the measurement of time is based.

⟦Suggestions by K. J. Korteweg in the margin to replace the preceding by:

We notice, therefore, that we succeed in observing among the sequences given by our perception those which are approximately similar (or rather, which as to approximate similarity can be brought under one single aspect by mathematical induction); in particular, that 'it works' if we draw up a sequence of $\omega$ terms on which, for example, the measurement of time is based. DJK⟧

⟦The remainder of page 6 was not rejected by D.J. Korteweg and may be found in [1907], pp. 85-89 under the headings: 'The continuity of functions in physics' and 'Differentiability of functions in physics'. The manuscript, page 10, continues beyond the text of [1907], p.87 ED⟧:

10    Is it surprising that not only do we succeed in observing sequences which repeat themselves again and again, but that so many groups of phenomena affecting our naive senses in totally different ways can be brought together under a few general aspects which are covered by simple constructible mathematical systems? This really would be a miracle, were it not for the simple fact that the physicist concerns himself with the projections of the phenomena on his measuring instruments, all constructed by a similar process from rather similar solid bodies. It is therefore, not surprising that the phenomena are forced to record in this similar medium either similar 'laws' or no laws. For example, the laws of astronomy are no more than the laws of our measuring instruments when used to follow the course of heavenly bodies.

⟦This last sentence is crossed out ED.⟧

Science, therefore, makes sense only when man in his struggle against nature and fellow men, uses the calculations of counting and measuring; in other words, physical science has value only as a *weapon*, it does not concern life — indeed it is a disturbing and distracting factor like *everything* in any way connected with struggle.

But mathematics practised for its own sake can achieve all the harmony (i.e. an overwhelming multiplicity of different visible, simple structures within one and the same all-embracing edifice) such as can be found in architecture and music, and also yield all the illicit pleasures which ensue from the free and full development of one's faculties without external force [*]. Poincaré (*La Valeur de la Science*, p. 264) is inclined to reduce all aesthetic affections to such an affection of harmony. Perhaps his notion of aesthetic affection is simply an affection of harmony;

---

[*] 〚Mathematical systems then would have 'architectural' value and therefore a raison d'être only in sofar as they are suitable as physical hypotheses in the struggle for life. Mathematical viewing of one's own thought, or one's own language, has no no use in the struggle for life. Moreover, since systems that are applicable here are of such primitive simplicity there is no reasonable ground for the practice of this sort of mathematics. DJK〛

1) We may therefore speak of an object only relative to other phenomena.

..................

11 but even according to him it is more: he says 'outside science and aesthetics there is nothing but 'le pur néant." He therefore seems to believe that it may be this aesthetic affection which is referred to as the highest good for mankind, preserved with such great difficulty. This shows the blinding effect which the immoral, free and full development of human faculties has also on him.

One further remark: we create the mathematical systems in the exterior world as a moment in our process of externalization, i.e. of holding out against the exterior world. Since 'object' is nothing but a causal sequence, constant with respect to other variable causal sequences [1] (e.g., first of all, the rigid bodies whose group of transformations remains invariant in so many changes; but, for example, the energy of a system is as objective as its material parts), we can say we create the objective world [2] in all freedom. In this way the adept, who do not wish to externalize themselves any longer, will find that a *nothing* remains since the *objective* will have vanished. So far Poincaré is right when he defends science as 'the thought communicable to others,' and says: 'Tout ce qui n'est pas pensé est le pur néant' (*La Valeur de la Science*, p.276).

[1] We may therefore speak of an object only relative to other phenomena.
[2] Attributing to this objective world an existence independent of man himself is nothing but a habit which has grown and lived in the mutual understanding of men; as 'that which is common to all men' it is distinct from the individuals who are so strictly separated from one another.

It is remarkable that many no longer consider the phenomenal world to be objective, but still objectify their own representations of this phenomenal world, which is even more imprudent.

........................

12  What conclusion must be drawn from this as to Kant's Transcendental Aesthetic?

As far as time is concerned it remains true that the possibility of our externalization on the basis of observation of causal sequences depends on it; also, that its properties are a priori certain, independently of the nature of the observations. We can go even further and say that the creation of time as a matrix of moments is a free act of man. With this creation, however, the conditions and all elements in the construction of the whole of mathematics are given at the same time. One of these constructions is Euclidean three-dimensional geometry, a suitable schema to govern a group of phenomena within one single language and an observation, a posteriori, like any other discovery of a useful physical hypothesis.

A few years ago a work was published on the foundations of mathematics, Russell's *Essay on the Foundations of Geometry*, considered by many as a work of great merit, by Couturat (*Review de Métaphysique et de Morale* 1898) as no less than the perfection of Kant's Transcendental Aesthetic, a κτῆμα εἰς ἀεί.

〚The remainder of page 12 has not been crossed out and is found in [1907], p.94 and pp. 99 ff.
The following note appears in the manuscript on page 35:ED 〛

*Note to Chapter 2*

A danger of externalization through mathematics is that one develops great virtuosity knowing applicable mathematical systems but has forgotten to which (always very complicated) acts the applied points of the mathematical system must give rise. (One has lost the instinct which takes *all* into account; indeed if that instinct were strong, one would stubbornly resist this one-sidedness in mathematics, which always precedes the jump from end to means.)

........................

[[The following was intended by Brouwer as the final passage of Chapter 3 and of the whole of the dissertation. It follows immediately after the final sentence of the published edition ([1907], p.177). It was rejected by Korteweg, who wrote:
'I would leave it at that. It is better to give the following passage elsewhere if you so wish, and do more justice to it. It does not make sense now that parts of Chapter 2 have been dropped.' ED]]

## CHAPTER 3

33  The origin of all kinds of trivial philosophies lies in the consideration of language as something essential, and not simply as a defective instrument that should be ignored as much as possible, to be used only as a last resort; even then it will always leave behind a depressing feeling of being just half satisfied. We shall go into this only briefly and touch upon a few points which throw some light on the matter above.

First, the Kantian antinomies [2] are due to an erroneous belief that words such as 'world', 'nature', 'time', 'freedom', must be understood as representing in themselves something objective, i.e. something mathematically existent, only because they occur in some instances of pure, i.e. mathematical reasoning. Their mathematical function, however, in all such instances varies. Since they do not represent something objective, the philosopher is completely free to build behind these words any mathematical system he wishes, so long as the ordinary usage of the word, on the basis of such mathematical hypotheses, does not become absurd. Obviously there are many such hypotheses. Then perhaps it is not too bold to say that the 'unity of contradictory parts', of which some philosophers love to speak, refers to the language of the complexity of superimposed structures which all mathematical systems are. It can directly and explicitly refer to only one such structure, but immediately and through this it suggests all others. Therefore, to use their favourite expession, 'there is no end' to what language expresses directly [3].

---

[2] In the physical sciences antinomies, of course, are acceptable; each of the contradictory systems can be of some service in the human struggle. Nevertheless, one will try to rid oneself of these contradictions by a fusion of hypotheses.

[3] Even so, to enforce this principle everywhere, one tacitly makes jumps

from mathematics of a lower order to mathematics of a higher order, e.g. in the case of the word 'necessary'. This word expresses a moment in the mechanical viewing of the world. If, however, the act is itself objectified again in a higher order mathematics (i.e. in idle play) we can, of course, no longer speak of it as 'necessary'. We could then say: the reality appears as 'accidental'. But then to proclaim as some kind of philosophical wisdom that in *reality* the contradictory parts are 'verified' by one-another 'accidentally' and 'necessarily' and 'cancel out' !... [*]

---

[*] ⟦ This sentence is incomplete in the original; the exclamation mark and the context imply Brouwer's disapproval. ED⟧

........................

34  We wil still be left with the disastrous confusion caused by a *conscious* use of language and by placing it in this way above the mathematics which it accompanies, whereas mathematics demands to be placed above life which it accompanies as a simple weapon. However, *acting*, the *mystical* side of human life, can never be directed by *knowledge*, i.e. *mathematics*.

The use of language is here again the source of much evil, since we run the risk of thinking of a non-mathematical idea, clothed in words, as dependent on its mathematical synonym, which suggests the killing of the idea by the parallel intellectual thought.

In this way there exists a *causality in life*, which is the free externalization in the form of a decomposition in time of some unity into two parts still remaining linked. There also exists a causality in *science*, which is nothing but a juxtaposition of systems along a variable continuous or discontinuous coordinate which we call time. The danger is that we lose the intuition of the former as the directing force in our lives through habitual and exclusive use of the latter, either by personal acquisition or by training! A *contradiction* develops in *life* where two externalizations of the instinct clash, and either proves the other to be untenable (the free morality will choose one to escape the other, the one which appears to be the weaker of the two instincts, or to melt the two together through contemplation into a deeper harmony). There is another *contradiction* in *science* when it becomes clear that a given mathematical system cannot be fitted into another. The danger is that we lose the awareness of the first contradiction through our attempts to use the second as a guide in our lives.

## SUMMARY

Mathematics is a free creation ; it exists in developing a primordial intuition which can be called 'permanence in change' or 'the discrete in the continuous'.

Applying it to the exterior world is creating the objective world; this is characteristic of the human strategy in the general struggle for life. As such it is clearly inferior and has nothing to do with religion or with wisdom.

It is quite impossible that one day it will become empirically clear that *one* mathematical structure — be it Euclidean space or the theory of electrons — would be more true than another.

Definitions may themselves never be viewed mathematically, they can only be a means accompanying our own memory or the communication of it to others. There are elements of construction which must remain elements also in definitions, and which in the process of communication must be expressed in a single sound; they are the constructional elements which can be read off from the intuition of the continuum. Verbal signs such as 'continuous', 'discrete', 'unity', 'again', 'and-so-forth' are not derivable.

A logical construction of mathematics independent of intuition is impossible. In that way we can only construct a verbal edifice which remains absolutely distinct from real mathematics and which, like mathematics itself, needs the mathematical primordial intuition.

## Appendix 4

### Intuitieve Significa [BMS 22B]

⟦An introduction by Brouwer to a lecture given by Frederik van Eeden on 13th March 1918.⟧

De productiemethoden en rechtsverhoudingen der huidige samenleving accepteeren de menschelijke individuen als onderling volledig gescheiden centra van vrees en begeerte, elkander bestrijdend of samenwerkend ter bestrijding van derden.

In de activiteit der onheilige maatschappij kan daarom niemand goed of gelukkig zijn. En zij die daarbuiten zoeken naar geluk of heiligheid, vinden in de woorden der huidige talen, die immers in laatste instantie niets anders zijn dan de commandoseinen van het maatschappelijk arbeidsreglement, generlei impulsen tot daadkrachtige gedachten, hoogstens in hun klank en rhythme bronnen van gevolgen-arme stemmingen.

Werden nu en dan binnen beperkte groepen betere productiewijzen en rechtsverhoudingen geschapen, die een heiliger levenshouding der individuen veroorloofden, dan konden deze reeds daarom niet duurzaam zijn, omdat hunne handhaving en doorvoering op de gewone taal als verstandhoudingsmiddel was aangewezen, zoodat deze hervormers gedwongen waren, de signalen hunner nieuwe gemeenschap uit die der verworpen maatschappij te componeren, en daardoor aan een overheerschenden invloed van hersuggesties dezer laatste onderworpen bleven.

De intuitieve significa houdt zich bezig met de schepping van nieuwe woorden, vormende een code van elementaire verstandhoudingsmiddelen voor de systematische activiteit eener nieuwe en heiliger samenleving.

### Intuitive Significs (translation)

The methods of production and legal relations of present-day society accept human individuals as completely independent and separate centres of fear and desire fighting each other, or cooperating with each other in fighting a third party.

Therefore no one can be good or happy in the activity of this unholy society. Those in search of happiness or holiness outside this activity do not find any stimulus to active thought in the words of present languages, which in the last analysis are no more than signs of command in society's regulation of labour, all that their sounds and rhythm may create are inconsequential moods.

Better methods of production and better legal relations which every now and then were created within small communities and which allowed the individual a holier attitude to life, could never last very long, because for their maintenance one had to rely on ordinary language as a means of understanding. As a result, these reformers were forced to compose the signs of their new communities from those of the society they had rejected, and therefore remained subject to the dominating, re-suggestive influence of the latter.

Intuitive Significs concerns itself with the creation of new words, forming a code of elementary means of understanding in the systematic activity of a new and holier society.

## Appendix 5

**Will, Knowledge and Speech**
〚translation of [1933]〛

I
*Reflection on mathematics, science and language*

45  Knowing and speaking are forms of action through which the will to live of man and mankind are maintained and enforced. They originate in the following three anthropological phenomena: 1. mathematical viewing, 2. mathematical abstraction, 3. enforcement of will by means of signs. All three are subject to the free will both in their range and modality.

1. Mathematical viewing is an attitude which man has adopted in his struggle for existence. It comes into being in two phases, the phase of becoming-aware of time and the phase of causal attention.

The becoming-aware-of-time is the fundamental happening of the intellect: a moment in life falls apart into two qualitatively different things of which the one gives way to the other but is retained by memory. Moreover, this split moment is more or less separated from the Ego and placed in a world of its own, the perceptional world.

46  Of this temporal two-ity, born out of time-awareness, or this two-membered time sequence of phenomena, one of the elements can in turn and in the same way fall apart into two parts; in this way the temporal three-ity or three-element time sequence is born. Proceeding this process, the self-enfolding of the fundamental happening of the intellect creates the time sequence of phenomena of arbitrary multiplicity.

*Causal attention* is an act of human imagination linking and identifying different sequences of phenomena; such a phantasy is called a causal sequence. A special case of causal attention is the imagination of *objects*, i.e. of permanent (simple or complex) units of the world of perception. Through this generation of objects (which include one's own personality and one's fellow-men) the world of perception becomes itself more or less stabilized. The objects vary widely in their degree of egoicity or directedness to the Self, i.e. acceptance of the desire for their stability as the directing force of the free will.

Mathematical attention is not a necessity but a phenomenon of life subject to the free will, everyone can find this out for himself by internal experience: every human being can at will either dream-away time-awareness and the separation between the Self and the World-of-perception or by his own powers bring about this separation and call into being in the world-of-perception the condensation of separate things. Equally arbitrary is the identification of different temporal sequences of phenomena which never forces itself on us as inevitable.

The only justification of the mathematical attention lies in the expediency of the 'mathematical act', which is based on it and which is within the grasp of man because of his causal attention. It consists in a later element of a causal sequence of events, which is desired instinctively but which cannot be achieved by direct impulse (the aim), being forced into being indirectly in cool calculation: an earlier event in the sequence, perhaps not in any way desirable for its own sake (the means), is made to take place and in this way the desired event is eventually also effected as a consequence.

47    Mathematical acts and the choice of aims they serve are also subject to the free will. The choice of aims in particular reveals the order of objects as to their egoicity. The clearness of this order is even an indispensable condition for all initiatives of the will; once it vanishes 'dreaming-away' is abandoned to leave only automatic acting from sheer habit.

The causal sequence by its very nature cannot have any other existence than that of a phantasy which accompanies a tendency of the human will towards mathematical activity; there can therefore be no question of a causal coherence of the world independent of man. On the contrary, the so-called causal coherence of the world is a dark force of human thought serving a dark function of the will of mankind, which it uses like a cloud of stupefying gas, in an attempt to make the world defenseless and ready to be assaulted by its desires.

As a result of this causal attention man tries even at the lowest levels of civilization to stabilize his range of causal influence and to create an ordered domain under his power; within this sphere he first isolates the causal sequences that are useful to him, i.e. protects them from disturbing side-phenomena, and then calls into being new causal sequences by constructing new permanent objects and instruments and by subjecting in a more or less organized way the will of his fellow-creatures to his own will.

2. However, only at the highest levels of civilization does mathematical activity reach full maturity; this is achieved through the mathematical abstraction, which divests two-ity of all content leaving only its empty

form as the common substratum of all two-ities.

48  This common substratum of all two-ities forms the primordial intuition of mathematics, which through self-unfolding introduces the infinite as a perceptual form and produces first of all the collection of natural numbers, then the real numbers and finally the whole of pure mathematics or simply of mathematics. We shall not concern ourselves here further with the manner of these constructions.

The power of the mathematical abstraction is based on the human experience that many causal sequences are more easily controlled if their empty abstractions are considered to be partial systems of more extensive, but more manageable mathematical systems. In this way the relations that exist in the more elaborate systems can also be used to survey and dominate the more restricted system, which can often result in a drastic simplification of the latter. The more extended mathematical systems formed by such 'induction' are called hypotheses; constructing a dominating system of relations within the hypotheses produces the so-called scientific theories. A very essential hypothesis in the mathematical view of one's fellowmen e.g. is the supposition that there is in each of them a mathematical-scientific mechanism of viewing, acting and reflecting similar to one's own.

Some scientific theories are coined the 'theories of exact science'. They are those theories which first of all refer to causal sequences that are particularly stable, either because they are perceived as laws of nature or because they are artificially called into being as technical facts; secondly, their hypothesis achieves a very considerable simplification; thirdly, the causal sequences in them that are to be governed correspond to special values of numerical parameters whose whole domain is present in the more extended mathematical system of the hypothesis.

49  It is in these exact-scientific theories in particular that the phenomenon of the heuristic character of scientific hypotheses becomes evident; this consists in discovering behind sequences – which were originally added as hypotheses – corresponding real causal sequences in the perceptional world.

3. Like any other originally autonomic act of aggression or defense, mathematical viewing and acting can switch from serving one's own personality to that of another object, either directly by one's own free will, because of the higher degree of egoicity of that object, or by force i.e. under the influence of the will of non-egoic objects, which can also appear as the collective will of a group of human beings or of the whole of mankind. This subservience of mathematical viewing and acting to non-

egoic objects is called labour. Preparedness to labour is evoked in man either directly by suggestion, i.e. by giving fright or striking terror, by temptation, by subjugation as one does of animals, or by arousing phantasies, or indirectly by training, i.e. by influencing the experience of the person to be reduced to servitude in such a way as to lead him to a mathematical view which makes labour acceptable as a means to pleasure or avoidance of pain. One might add that such a mathematical view often disappears from consciousness once it has induced preparedness to labour, which then still remains in existence as automatic habit.

Among the mathematical views imposed on all human beings by the collective will of the whole of mankind a dominating role is played by the introduction of a hypothetical 'objective world of space and time' as the common bearer of all the temporal sequences of phenomena of all individuals, and also the exact and technical sciences as far as these in the form of industrial secrets have not been restricted to serve special interests.

50   Limited groups of people, such as political combines and professional federations impose on the individuals of which they are composed as a central mathematical view the recognition of and respect for their organization, i.e. the power-grid of will-transmission signifying within the group the special mathematical views and actions which accompany group-labour. This organization of limited groups of people has a far less stable character than the exact and technical sciences; this for two reasons: First, the organization can never control all the material circumstances on which the objectives of the group and their corresponding mathematical acts depend; it therefore has to adapt itself continuously to the unexpected changes in these material circumstances. Secondly, to be useful it not only needs to be theoretically effective, it also requires some stable feeling of loyalty and contentment in the individuals who are its subjects. The means of training at the disposal of the organization are hardly effective in fostering and maintaining such feelings. Loyalty, especially of the leaders of the organization, is endangered by the clash of personal interests and those of the group, while the contentment of the ordinary members must remain far from complete: even if they accept the necessity of authority they will obviously find it difficult to recognize the existing form and their own place in it as the only correct ones.

To secure the highest possible degree of stability in the feelings of loyalty and stability within the organized groups of people, the organization, well-aware of the limited means of training at its disposal, resorts to propaganda of moral theories, i.e. mathematical views which base the suitability and necessity of the existing organization not only on common aims and emergencies which are acceptable to the egoistic mind, but also on so-called moral values in their attitude to life, i.e. values which depart from egoistic notions.

51  Classic examples are religious commandments and notions of fatherland, property, family, solidarity, class-consciousness, honour and duty. The propaganda of the group, which in its attempts to establish the prestige of moral values has to rely almost exclusively on suggestion, would never be adequate if it were not supported by the silent operation of those mathematical considerations in separate individuals of which the final link is prevention of the undesirable success of the egoistic ambitions of others.

At the most primitive stages of civilization and in the most primitive man-to-man relationships the transmission of will to induce labour or servitude is brought about by simple gestures of all kinds especially and predominantly the emotive natural sounds of the human voice. On the other hand in a more radical organization of human society labour to be imposed is far too complex and differentiated and cannot be set in motion or maintained by such a simple cry. To be able to regulate labour by means of (spoken or written) signs of request or command even in these circumstances, the whole of laws, rules, objects and theories relating to the mathematical acts demanded of the servants is in turn subjected to mathematical view, the linguistic mathematical view. Linguistic elementary symbols are assigned to the elements of the mathematical system which is born from this mathematical view and which carries the scientific theory; with these as foundation and with the grammatical rules derived from the above scientific theory the organized languages of civilized human society are now capable of expressing the complex and differentiated will-transmissions needed. The scientific theories which form the basis of languages are by their very nature not exact-scientific theories. On the contrary, the stability and exactness which language in its grammar and vocabulary seems to possess formally, are to a great extent lost again, since far more elementary notions are needed in every-day practice than there are words and modes of connecting words in language.

52  On the other hand in every-life there is no need for stability and exactness of language, collective will and automatism-of-training have made people 'good understanders'.

Even if language originally and in the first place is a function of the activity of social man, it has some significance in the processes of reflection and mnemotechnique in the solitude of individual man, partly because of habit-automatism following the use of language, partly because of the role which science and social organization continue to play even in the thought-world of solitude.

4. What we have said so far is all rational reflection i.e. mathematical viewing, in which the content neither of the purpose nor of the objects plays any part. It is an essential hypothesis of the understanding between people that this rational reflection has a structure which is the same for all human beings. It therefore represents an eminent social value in preventing confusion when one establishes the foundations of social organization and therefore in consolidating public life.

Quite different and of more individual character is moral reflection, which tests the objects of the phenomenal world and the event of mathematics itself as to their egoicity and therefore as to their right of existence as sources forcing the direction of the free will. By concentrated listening in the border-area between dreaming and time-awareness moral reflection will try to understand the connection between the available possible purposes and one's own clear and dark destiny and origin. In that border-area causality appears only ephemerally, there is no place for mathematical action, language remains inexact, suggestive rather than adequate and 'not to be taken literally'. It will have a 'prophetic' character on the rare occasions that one receives inspiration that is transferable or collectively-active in character.

But neither is moral reflection without its social significance, because, however complete the solitude in which it is conducted, it will lead to a feeling of social justice and readiness to fight evil; also because every now and then the most useful moral theories will chrystallize from its prophetic language.

## II

### *Criticism of the attempted linguistic reform of mathematics*

Since even the smoothest transmission of will in the last analysis remains a phenomenon which is neither wholly stable nor carries the necessity of law, there cannot be any question of its complete exactness or absolute certainty. This remains the case when will-transmission is effected by means of language, and further when will-transmission refers to the construction of pure-mathematical systems. Rational reflection has therefore led us to conclude that also for pure mathematics there cannot be an infallible language, i.e. a language which in the exchange of thought excludes misunderstanding and in its mnemotechnical function provides a guarantee against error, i.e. confusing different mathematical entities. One cannot make provision against this inadequacy, as has been tried by the formalist school, by subjecting the mathematical language, i.e. the symbolic system intended by the speaker or writer to evoke pure mathematical constructions in others, in turn to mathematical viewing and through complete overhaul providing it with an exactness and stability as one finds in physical instruments or in the phenomena of exact science, and thereby making use of a 'meta-language' [Dutch: 'boventaal' Ed.], which speaks about mathematics, or a language of second order.

54 First of all, even if this meta-language may succeed with a high degree of probability (since it refers to a relatively simple, finite collection of constant objects and to a finite system of pure mathematics abstracted from it), but because of the nature of language it can never eliminate misunderstanding and error with absolute certainty. Secondly, even if this could be achieved and an exact mathematical language were to become a reality, the possibility of misunderstanding, i.e. confusing the mathematical constructions expressed in this exact mathematical language would in no way be eliminated.

There is still a wide-spread belief in the possibility of ensuring the total exactness of pure mathematics by making the language of mathematics purely mechanical. Such belief, as we have shown above, can only be due to failure to appreciate the real nature of language as a means of will-transmission; the historical explanation is to be found in a far more fundamental and radical error, much older and deep-rooted: the unjustified belief in the reliability of classical logic. The origin of this belief can be explained as follows: Already in antiquity man had at his disposal a fairly perfect (i.e. fairly successful in excluding misunderstanding) language of mathematically viewing finite systems of objects of the objective space-time world, each of these objects taken as a unit and a permanent entity. In this language there are certain ways of passing from correct affirmations (i.e. effectively indicating actual mathematical viewing) to other correct affirmations, which one has called the laws of identity, contradiction, excluded middle and syllogism, the so-called logical principles. Whenever one applied these principles purely linguistically, i.e. deduced new linguistic affirmations by means of these principles without thinking of the mathematical viewing indicated by these deductions, they apparently worked, i.e. for every new affirmation concluded in this way one could afterwards establish that it was capable of evoking in every linguistically-trained individual an actual mathematical-view which in the objective space-time world seemed to be almost universally acceptable to all linguistically-trained individuals.

55   Moreover, these logical principles seemed to be working also if one extended them to the language of science and the language of numerous other events of every-day life, at least if one restricted oneself to phenomena which were governed by the laws of nature whose unshakeable authority one had learned to trust. This led to the rash induction that one could rely on the correctness of affirmations derived by means of logical principles even in cases which apparently could not be checked directly. Such trust was placed in particular in the principle of the excluded middle, even in its more extended interpretation, which assumes that an event has actually taken place not only if another explanation of a given fact is absurd but also if it is practically not possible to find another explanation for it. On this trust are based not just the theoretical sciences such as palaeontology and cosmogeny but even state laws setting out, e.g. criminal procedures. Belief in the logical principles has become so unchallengeable that whenever they lead to obviously wrong conclusions the facts or natural laws presumed in such reasoning were questioned and changed, faith in the logical principles was never abandoned. One had completely forgotten that the established practical reliability of the logical principles was no more than the practical validity of the laws of combining affirmations of the finite-mathematical system used in describing the phenomena of the perceptional world; in other words, no more than the simple fact that human beings have succeeded in mastering an important complex of objects and mechanisms of the phenomenal world in their relation to an important complex of facts and events by considering and treating the system of the space-time world, to which these objects and mechanisms belong, as part of a finite, discrete collection whose elements are linked by a finite number of relations.

56   Instead of simply registering wonder, noting that far greater trust and satisfaction can be found in a great part of the perceptional world in its finite organization as created by man than in the organizations which man has given himself, one remained blind to such simple interpretation. The essence of the word as a means of will-transmission was completely misunderstood. Gradually the view has grown and established itself that words are names indicating entities of fetish-like character leading a permanent and immutable existence independent of language and indepen-

dent of man's causal attention, so-called 'concepts'; moreover, that between these 'concepts' there exist equally permanent and fixed relations all held together by the logical principles as some kind of a-priori law. The result was confidence that relations between concepts derived by means of the logical principles from undeniable axioms (i.e. from relations between concepts corresponding to observations of undeniable facts or laws of nature), if they correspond to testable affirmations concerning the perceptional world, will pass this test with flying colours; if not, they could with equal confidence be treated as *ideal truths*. Such ideal truths have for many centuries been deduced by philosophers, confidently and diligently. And if every now and then in this deduction as an uncomfortable incident the figure of contradiction appeared, raising doubts about the correctness of the course of reasoning, such doubts never concerned the reliability of the logical principles, but only the indisputability of the axioms on which the reasoning was based. It is therefore not surprising that in the course of time numerous axioms have again been rejected or were changed because of the contradictions in the ideal truths derived from them.

57 As to the above-mentioned attempts at linguistic reform of mathematics, their historical origin is to be found in the belief in the logical principles: In their study of pure mathematics, even in the parts dealing with infinite systems, mathematicians followed the example of philosophers and began to derive logical principles from the language of finite mathematics and apply them without reservation to new areas. In this way, especially since the discovery of set theory and the resulting expansion of the mathematics of infinite systems, for the infinite sets created by the axiom of comprehension a whole new world of 'ideal' truths was derived in which mathematicians, following the above practice, saw far more than empty word-complexes. Until, one day, there appeared contradictions also in set theory, and this time contradictions that could not be made to disappear by simply changing the axioms. This threatened the exactness of mathematics which until then one had trusted to be incontrovertible without, however, paying much attention to its origin and nature, and which one wished to continue to trust. Disregard for the real nature of language and the misconception that mathematics was based on concepts, inviolable, permanent and expressed in words, were the main reasons why attempts to support the mathematical thought-constructions had to rely on and were restricted to a programme of reviewing and renewing the architecture of the verbal edifice which correspond to these constructions. In this process of re-creating the language of mathematics, however, the

linguistic figure of the logical principle remained untouched; and yet, the real renewal that has been achieved is confined to the language which is supposed to refer to the axiomatic fundamental relations between the fundamental notions of mathematics, and to the search of concepts corresponding to newly proposed sets.

58  The final aim of these investigations, still far out of reach, in which the mathematical thought-world itself reputedly indicated by these words is completely disregarded, consists in the construction of a linguistic mechanism which, apart from a few not-too-drastic amputations, is capable of providing a verbal image of the whole of mathematics known so far, however, a linguistic mechanism that excludes the occurence of the linguistic figure of the contradiction.

### III
### *The intuitionist reconstruction of mathematics*

If on the basis of rational reflection the exactness of mathematics, in the sense of impossibility of misunderstanding and error, cannot be assured in any linguistic way, the question arises whether this assurance can be found by other means. The answer to this question must be that the languageless constructions originating in the self-unfolding of the primordial intuition, on the basis of their presence in memory alone, are exact and correct; that, however, the human power of memory, which has to oversee these constructions, by its very nature is limited and fallible. In a human mind empowered with unlimited memory therefore pure mathematics, practised in solitude and without the use of linguistic symbols would be exact. However, this exactness would again be lost in mathematical communication between individuals, even between those empowered with unlimited memory since they have to rely on language as a means of communication.

59  We may also put the question in this form: suppose there were such hypothetical human beings with unlimited power of memory, who would use words only as invariant signs for definite elements and for definite relations between elements of already constructed pure-mathematical systems, who in the reasoning accompanied by these words could make room for logical principles, i.e. rules of stringing together mathematical affirmations, or what comes down to the same thing: suppose that human beings with unlimited power of memory recorded their constructions in shortened form in a suitable language, surveyed the strings of their affir-

mations in this language and then would be able to see the occurrence of the linguistic figures of logical principles in all their mathematical modifications. Careful rational reflection would then show that as far as the principles of identity, contradiction and syllogism are concerned such an occurrence could be expected; as to the linguistic figure of the principle of the excluded middle, this would only occur if one restricted oneself to affirmations concerning parts of a definite, once and for all given, finite mathematical system. Wider applications of the latter principle would never occur since such applications to pure-mathematical affirmations usually lead to verbal complexes devoid of any mathematical sense and therefore of any sense. [[Brouwer's own complex use of language is shown in this paragraph, which in the 24-line original consists of one single sentence Ed.]]

It follows that the necessarily inadequate, limited and in-its-use-always-remaining-uncertain language of communication between human beings, whose power of memory is limited, even if it has been fashioned with all possible refinements and precision, will only then be able to play its mnemotechnical, thought-saving and communicative role in mathematical research and mathematical communication with *practically*-acceptable reliability, if in its use every application of the principle of the exluded middle is avoided unless it remains restricted to a well-defined finite system. This insight, which guides the so-called *intuitionist* practice of mathematics, has been lacking in classical mathematics, mainly as the result of the historical growth of the authority of logic. In classical mathematics through the centuries this resulted in the firm belief in a number of theorems which intuitionism has since shown to be incorrect; moreover, during the last fifty years it has become the basis of various extensive mathematical theories which, intuitionism claims, lack any mathematical meaning. In general, Intuitionism subjects mathematics to a complete overhaul, a re-construction in which unfortunately in many instances it has to lose its smooth and elegant character and take on much stiffer, more forced and complicated form. Regrettably, the world of truth is more difficult to penetrate than the world of illusion.

60    To illustrate the intuitionist reconstruction of mathematics with some simple examples, we consider in the first place the so-called fundamental theorem of algebra, which says that every higher-order equation has at least one root. All proofs of this theorem in classical mathematics have used the principle of the excluded middle. One of the best-known proofs for example, proceeds by showing that the supposition that a wholly

rational function of a complex variable in the complex plane would have a limited range, leads to absurdity. Another classical proof is based on the assumption that the discriminant of a higher-order equation is either zero or its absolute value is greater than a demonstrable rational number different from zero. Similar criticism applies to all other proofs in use until ten years ago. Intuitionist mathematics, therefore, had to reject the formulation of the Fundamental Theorem of Algebra as a meaningless combination of words until it succeeded in constructing an algorithm by means of which one can calculate a root from the coefficients at least of a considerable group of higher-order equations.

For further examples we introduce the intuitionist notions of a *fleeing property* and *the oscillatory number of a fleeing property*. A fleeing property is a property for which in the case of any given natural number either the validity can be deduced or its absurdity, while one cannot calculate a natural number that has the property nor can one prove the absurdity of the property for all natural numbers.

61     By the *critical number* $\kappa_v$ of a fleeing property $v$ we understand the (hypothetical) smallest natural number which possesses the fleeing property; by an *up-number* resp. *down-number* of $v$ we understand a natural number which is not less than, resp. less than the critical number. It is obvious that as soon as an up-number of $v$ has been found, $v$ loses its character as a fleeing property. The fleeing property is called *two-sided* if its absurdity cannot be proved neither for all even nor for all odd numbers. By the *binary oscillatory number* $s_v$ of a two-sided fleeing property $v$ we understand the limit of the fundamental sequence $a_1, a_2, \ldots$, where $a_v = (-2)^{-v}$ if $v$ is a down-number of $v$ and $a_v = (-2)^{-\kappa_v}$ if $v$ is an up-number of $v$. This limit $s_v$ will appear to be neither positive nor negative, neither equal to zero nor different from zero, neither rational nor irrational, and yet it is a real number; it therefore illustrates clearly the invalidity of the principle of the excluded middle. Further, if we define the binary oscillatory number $n_v$ of $v$ as the limit of the fundamental sequence $b_1, b_2, \ldots$, where $b_v = 2^{-v}$ if $v$ is a down-number of $v$, and $b_v = 2^{-\kappa_v}$ if $v$ is an up-number of $v$, we can then show the untenability of the following four elementary and classical properties:

1. Of every two straight lines of the projective plane one can determine at least one common point.
2. If a continuous function is negative for $x = a$ and positive for $x = b > a$ then a value of $x$ lying between $a$ and $b$ can be indicated for which the function takes the value zero.

3. If a continuous and continuously differentiable function takes the value zero for $x = a$ and for $x = b > a$ then a value of $x$ between $a$ and $b$ can be indicated for which the derivative of the function is equal to zero.
4. For two positions of a solid in Euclidean space moving about a fixed point one can determine at least one straight line through the fixed point which has the same position w.r.t. to these transformations.

To refute the first theorem we draw in the Euclidean plane fitted out with an orthogonal coordinate system a straight line $l$ through the points $(1, s_v)$ and $(-1, n_v)$. No intersection of this line $l$ with the X-axis can be found, neither in a finite point nor in a point at infinity.

To refute the second theorem we consider the function of $x$ which for the values $-2, -1, 1, 2$ of $x$ takes resp. the values $1, n_v, s_v, -1$, and which runs linearly between every two of these consecutive values. No value of $x$ can be found for which the function takes the values zero.

To refute the third theorem we consider the definite integral, measured from the ordinate $x = -3$, of the function of $x$ which for the values $-3, -2, -1, 0, 1, 2, 3$ takes the resp. values $1, n_v, s_v, -1, -n_v, -s_v, 1$ and which runs linearly between every two of these subsequent values. This is a continuous, continuously differentiable function of $x$, which takes the value zero for $x = -3$ and for $x = 3$, yet between these two there is no value of $x$ for which the value of the derivative is equal to zero.

To refute the fourth theorem we define in a Euclidean space with an orthogonal system of coordinates a fundamental sequence $r_1, r_2,...$ of rotations of a solid, movable about the origin as fixed point, all starting from the same initial position $s_o$; $r_{3v}$ as the rotation through an angle $a_{3v}$ about the X-axis, $r_{3v+1}$ as the rotation through an angle $a_{3v+1}$ about the Y-axis, and $r_{3v+2}$ as the rotation through an angle $a_{3v+2}$ about the Z-axis, where the $a_\rho$ have the meaning as defined above where we introduced the binary oscillatory number. Let $s_\rho$ be the position of the solid after the rotation $r_\rho$ from the initial position $s_o$, and let $s$ be defined as the position of the solid which is the limit of $s_1, s_2, ...$ . It is then impossible to indicate a straight line through the origin which has the same position w.r.t. $s_o$ and $s$.

Of these theorems the following parts remain intuitionistically correct:
1. For two lines $l_1$ and $l_2$ of the metric projective plane, and for every positive $\varepsilon$ a point $p_1$ of $l_1$ and a point $p_2$ of $l_2$ can be indicated, such that the distance between $p_1$ and $p_2$ is smaller than $\varepsilon$.
2. If a continuous function takes a negative value for $x = a$ and a positive

value for $x = b > a$ then for every positive $\varepsilon$ a value of $x$ between $a$ and $b$ can be indicated for which the function has an absolute value $< \varepsilon$.

3. If a continuous and continuously differentiable function takes the value zero for $x = a$ and for $x = b > a$, then for every positive $\varepsilon$ a value of $x$ between $a$ and $b$ can be indicated, for which the absolute value of the derivative is less than $\varepsilon$.

4. For every two transformations of a movable solid in Euclidean space about a fixed point and for every positive $\varepsilon$ a straight line of the solid through the fixed point can be indicated, so that the positions of this line in the two transformations form an angle which is less than $\varepsilon$.

Many similar examples can be constructed in almost every part of mathematics. That this has not simply put an end to the classical practice of mathematics is due to the fact in its favour that, even though the principle of the excluded middle is incorrect, it does not lead to contradictions as long as one restricts oneself to finite groups of properties. Intuitionism in its battle against the errors of classical mathematics, therefore, finds itself deprived from the most fashionable means of repressing errors of thought, the reductio ad absurdum, and has to rely exclusively on exhortation to reflect and reason.

## Appendix 6

**Short Retropective Notes on the Course "Intuitionist Mathematics"**
[BMS 52]
First Part, Easter Term, 1947

1  [I] In the controversies on the foundations of mathematics which were conducted a few decades ago, intuitionism intervened with two successive acts:

(a) As opposed to the formalistic point of view which sought to establish the linguistic criteria for the validity of mathematical reasoning, resulting in systems of theoretical logic, intuitionism recognized that *mathematics is an essentially languageless activity of the mind, finding its origin in the perception of time. More precisely, the perception of time may be described as the falling apart of a life moment into two distinct things, one of which recedes before the other, and is retained by memory. On divesting this perception of all its qualities, one obtains the empty form of the common substratum of all two-ities. And it is this common substratum, this empty form, this bare two-ity, which is the basic intuition of mathematics.*

The above re-orientation of the starting point of mathematics was the first intervening act of intuitionism. Inner experience reveals how, by unfolding the basic intuitions, extensive parts of mathematics can be built up. These parts include much of arithmetic and algebra, on the whole: whatever can be developed out of the theory of natural numbers (the principle of complete induction included), without essential interference of logic. In particular, a (symmetric, reflexive, transitive) relation of 'equality', and a relation of 'difference' (as the absurdity of equality) can be introduced in this domain, in a natural way.

It is true, that even these 'denumerable' branches, in their classical form, contain fundamental theorems of purely logical, thus of purely linguistic, origin. These theorems, of course, have to be verified carefully. And in this trend of thought, the following question naturally presents itself:

Suppose that, in customary mathematical laguage trying to deal with an intuitionist mathematical system, the figure of an application of one of the principles of classical logic occurs, does then this figure of language accompany an actual languageless mathematical procedure in the actual mathematical system concerned?

A careful examination reveals that the answer is, in general, in the affirmative, as far as the principles of identity, contradiction, and syllogism are concerned, if one allows for the inevitable inadequacy of language as a mode of description and communication. But with regard to the principle of the excluded third it cannot, except in special cases, be affirmative, so that *the principle of the excluded third cannot serve as a reliable instrument to discover new mathematical truths.*

In point of fact we can, within 'denumerable' mathematics already construct numbers which are neither positive nor negative, nor equal to zero, nor different from zero; moreover neither $\geq 0$ nor $\leq 0$; and not rational, even though their irrationality is absurd. To this end we make use of a *fleeing property*, i.e. property $f$, which satisfies the following three requirements:

(i) for each natural number $n$, the question whether $n$ possesses the property $f$, can be decided; (ii) no natural number $n$, which possesses $f$, is actually computable; (iii) the assumption that at least one natural number possesses $f$ is not known to lead to an abdurdity. Obviously the character of being a fleeing property is not necessarily a permanent one.

By the critical number $\kappa_f$ of the fleeing propery $f$ understand the (hypothetical) smallest natural number possessing $f$. A natural number will be called an *up-number* of $f$, if it is not smaller than $\kappa_f$ and a *down-number* of $f$, if it smaller than $\kappa_f$. Of course, $f$ would lose its character of fleeing property as soon an up-number of $f$ would be found. A fleeing property is called *two-sided with regard to parity*, if neither of the existence of an of an odd $\kappa_f$, nor of the existence of an even $\kappa_f$ the absurdity has been demonstrated. By the *binary shrinking number* $\tau_f$ of $f$ we understand the real number which is the limit of the infinite sequence $a_1, a_2, ...$, where $a_v = 2^{-v}$ if $v$ is a down-number, and $a_v = 2^{-\kappa_f}$, if $v$ is an up-number of $f$. By the *oscillatory binary shrinking number* $\sigma_f$ of $f$ we understand the real number which is the limit of the infinite sequence $b_1, b_2,...$ where $b_v = (-2)^{-v}$ if $v$ is a down-number, and $b_v = (-2)^{-\kappa_f}$ if $v$ is an up-number of $f$. Both $\tau_f$ and $\sigma_f$ are neither positive, nor negative, nor equal to zero, nor different from zero. Both $\tau_f$ and $\sigma_f$ are not rational, although their irrationality is absurd. Moreover, if $f$ is two-sided with regard to parity, $\sigma_f$ is neither $\geq 0$ nor $\leq 0$.

(b) Having pointed out this consequence of the first act of intuitionism, we remarked that in classical mathematics, the principle of complete induction is applied only to *predeterminate structures, i.e., structures in which the n-th operation has been fixed in advance for each n. It is here that* intuitionism intervenes with its second act. This act admits freely proceeding sequences as a mode of unfolding the basic intuition. Such freedom is to be understood in its widest sense: it may entail the absence of any law, or it may entail a set of restrictions, which fail to determine the sequence uniquely; these restrictions may even be allowed to change during the development of the sequence. The act also admits, at each stage of the development of mathematics, properties supposed for mathematical entities previously acquired, as new mathematical entities under the name of *species*, provided only that they satisfy the condition that, if they are realized for a mathematical entity, they are also realized for all mathematical entities which have been defined equal to it. Those mathematical entities previously acquired, for which the property proves to be realized, will be called the *elements of the species*.

Two infinite sequences of mathematical entities $a_1, a_2,...$ and $b_1, b_2,...$ are said to be *equal* or *identical*, if $a_v = b_v$ for each $v$, *different*, if the assumption of their equality has been reduced ad absurdum, and *distinct*, if for some computable $n$, $a_n$ and $b_n$ are different.

As a consequence of the second act of intuitionism the *spread* was introduced, a species of the widest import, which allows inter alia the reconstruction of the continuum and the re-creation of analysis. This species comes into being by means of a *spread law*, i.e. an instruction according to which, if *again and again an arbitrary natural number is chosen as 'index'*, each of these choices has as its effect, depending also on the preceding choices, that either a *'figure'* (viz. either nothing or a mathematical entity) is generated, or that the choice is sterilised, in which case the figures generated are destroyed and generation of any further figures is prevented, so that all following choices are sterilised likewise. The only condition to be satisfied is that after each non-sterilised initial sequence of $n - 1 > 0$ choices, one natural number at least is indicated by the spreadlaw, which, if chosen as n-th index, generates a figure.

3     The infinite columns of figures generated according to the spread law by infinitely proceeding sequences of choices are, by virtue of this genesis, together with all infinite sequences equal to one of them, the elements of a species. This species is called a *spread*.

II. Let $\sigma$ be a spread law, $s$ the corresponding spread. If $\sigma$ is such that for each natural number $n$ a natural number $k_n$ can be found with the property that each index $> k_n$ chosen at the $n$-th choice is sterilised, then $\sigma$ and $s$

are called *bounded*.

If for each two figures generated according to $\sigma$ the relation of equality either holds or is absurd, then $\sigma$ and $s$ are called *discriminate*.

Two species (in particular two spreads) are said to be *equal* or *identical*, if for each element of either of them an element of the other, equal to it, can be indicated. They are called *different* if their equality is absurd, and *congruent*, if neither can possess an element different from all elements of the other.

For spreads, apart from congruence, a wider possibility exists: we say that two spreads are *concordant*, if neither can possess an element distinct from all elements of the other.

A species is called *discrete*, if for each two of its elements the relation of equality either holds or is absurd.

If the species $M$ possesses an element which cannot possibly belong to the species $N$, or, what is the same, is different from all elements of $N$, we say that $M$ *deviates* from $N$.

The species $M$ is called a *sub-species* of the species $N$, if every element of $M$ can be proved to be equal to an element of $N$. If, in addition, $N$ deviates from $M$, then $M$ is called a *proper subspecies* of $N$.

The property of being an element of the species $M$, as well as of the species $N$, yields a species $\mathscr{D}(M, N)$ which we call the *intersection* of $M$ and $N$.

The property of being an element either of the species $M$, or of the species $N$, yields a species $\mathscr{V}(M, N)$ which we call the *union* of $M$ and $N$.

We consider a species $S$ of species $s$. The property of being an element of all species $s$, we call the *intersection* $\mathscr{D}(s)$ of the species $s$, and the property of being an element of at least one of the species $s$, the *union* $\mathscr{V}(s)$ of the species $s$.

A species which cannot possess an element is said to be *empty*. Two different species whose intersection is empty are called *disjoint*.

If $M$ and $N$ are disjoint subspecies of the species $P$ and if $\mathscr{V}(M,N)$ is congruent with $P$, then we say that $P$ is *composed* of $M$ and $N$, and that $M$ and $N$ are *conjugate subspecies* of $P$.

If $H$ and $K$ are disjoint subspecies of the species $P$ and if $\mathscr{V}(H, K)$ is identical with $P$, then we say that $P$ *splits* into $H$ and $K$ and we call $H$ and $K$ *conjugate removable subspecies* of $P$.

4    If $V$ and $W$ are conjugate subspecies of $P$, and if in addition $V$ is equal to the intersection of $P$ and $W$'s absurdity, and W equal to the intersection of $P$ and $V$'s absurdity, then we say that $P$ is *directly composed* of $V$ and

$W$, and that $V$ and $W$ are *directly conjugate subspecies* of $P$. Thus, for instance, the species $D$ and $E$, consisting of those elements of $P$, for which a certain negative property $\varphi$ is true and absurd respectively, are directly conjugate subspecies of $P$.

If the species $\delta(P)$ of the pairs of elements of $P$ is directly composed of the species of the pairs of equal elements of $P$ and the species of the pairs of different elements of $P$, i.e. if within $P$ absurdity of absurdity of equality is equivalent to equality, we say that $P$ is *semi-discrete*. Thus, for instance, each discriminate spread is semi-discrete.

Let $S$ be a discrete species of subspecies $s$, any two of which are disjoint, of the species $P$. In analogy with the above we say that $P$ is *composed* of the species $s$ (*splits* into the species $s$), and we call the species $s$ *conjugate subspecies* (*conjugate removable subspecies*) of $P$, if $\mathscr{V}(s)$ is congruent with (identical to) $P$.

If $P$ is composed of the species $s$ and if in addition each $s$ is equal to the intersection of $P$ and the simultaneous absurdity of all other $s$, then we say that $P$ is *directly composed* of the species $s$ and we call the species $s$ *directly conjugate subspecies* of $P$.

If a 1-1 correspondence has been created between two species $M$ and $N$, i.e., if to each element of $M$ an element of $N$ corresponds in such a way that to equal and only to equal elements of $M$, equal elements of $N$ correspond, while each element of $N$ corresponds to an element of $M$, we say that $M$ and $N$ are *equipotential* and possess the same *power*, or the same *cardinal number*.

A species which is equi-potential with the species of those natural numbers which precede a given natural number, is said to be *finite*. A species containing a subspecies equi-potential with the species $A$ of the natural numbers, is called *infinite*. In particular it is called *reducibly infinite*, if the subspecies concerned is removable. For instance the continuum to be introduced later on, is infinite but not reducibly infinite, because, as will be shown, it does not split. A species which is equi-potential with the species $A$, is called *denumerably infinite*. A species which is equi-potential with a subspecies of $A$, is called *denumerable*. A species which is equi-potential with a removable subspecies of $A$, is called *countable*.

After these definitions we passed to the introduction of spaces, first of Cartesian spaces, among which we treated the plane as an example, the extension of the treatment to $n$-dimensional Cartesian spaces presenting no difficulty. We started with the introduction of $\lambda^{(n)}$-squares, each consisting of a group of four pairs of binary fractions of the following form:

$$a \cdot 2^{-(n+1)}, b \cdot 2^{-(n+1)}; \ (a+2)2^{-(n+1)}, b \cdot 2^{-(n+1)};$$
$$a \cdot 2^{-(n+1)}, (b+2) \, 2^{-(n+1)}; \ (a+2)2^{-(n+1)}, (b+2)2^{-(n+1)},$$

where $a$, $b$, and $n$ are integers.

5   An infinite sequence $k_1$, $k_2$,... of $\lambda$-squares, proceeding in complete or partial freedom or without freedom, in which, for each $v$, $k_{v+1}$ is within $k_v$, was called a *point of the plane*. Two points $p'$ ($\kappa'_1, \kappa'_2,...$) *and* $p''$ ($\kappa''_1, \kappa''_2,...$) *were said to coincide,* if it is certain that $\kappa'_\mu$ and $\kappa''_v$ overlap for each $\mu$ and $v$. It is easily seen that coincidence is a symmetric, reflexive and transitive relation. The points which coincide with a point $p$, constitute a *point core* $P$ of the plane. If for the point core $P$ there is a law $\sigma(P)$ determining in advance, for each $n$, a $\lambda^{(n)}$-square $l_n$ such that each point $k_1$, $k_2$,... of $P$ has a 'tail segment' $k_v$, $k_{v+1}$,...within $l_n$, then $P$ is called a *sharp* point core of the plane. The species of the point cores of the plane is called the *full plane* or the *plane* and the species of the sharp point cores of the plane the *reduced plane*. Two points $p'$ and $p''$ of the plane are said to *deviate*, if they cannot possibly coincide, and to be *separated*, if a square of $p'$ and a square of $p''$ can be indicated which are outside one another. A point $p$ *deviates* (is *separated*) from the point core species $Q$, if it deviates (is separated) from each point of $Q$. If within every square of $p$ lies a square of $Q$, $p$ is called a *point of closure* of $Q$. If within each square of $p$ lie two mutually external squares of $Q$, $p$ is called a *point of accumulation* of $Q$.

The point core $P$ belonging to the point core species $Q$ is called an *isolated point core* of $Q$, if a square $k_n$ of $P$ can be indicated containing no two mutually external squares of $Q$ (i.e. of points of $Q$); it is called *an unapproachable point core* of $Q$, if it cannot possible be a point core of accumulation of $Q$; it is called a *covered point core of $Q$*, if it cannot possibly be an isolated point core of $Q$.

Point cores of the plane which cannot be point cores of accumulation of $Q$, are called *unapproachable* by $Q$.

Two species of points and point cores are said to *coincide* if each element of one of the species coincides with an element of the other species.

Two species of point cores are said to be *locally concordant*, or, shorter *concordant*, if neither can contain a point core separated from the other. Thus, for instance, an arbitrary point core species $Q$, the *closure* (i.e. the species of the point cores of closure) $Q'$ is concordant with the union of $Q$ and its *derivative* (i.e. the species of the point cores of accumulation) $Q''$.

Two species of point cores of the plane are said to be *locally congruent*, or shorter *congruent*, if neither can contain a point core which deviates from the other. Thus, for instance, an arbitrary point core species is congruent with the union of the two species of its isolated and of its covered point cores.

Let us denote by $V$ the plane, by $I$ the $\lambda^{(0)}$-square with centre $(\frac{1}{2}, \frac{1}{2})$ and by the *unit square U* the species of point cores possessing no square outside $I$, then $V$ coincides with a spread $s$, whose elements are the *quartering points*, i.e. the points $(k_1, k_2, ...)$, where each $k_\nu$ is a $\lambda^{(4\nu+1)}$-square, while $U$ coincides with a bounded spread $s^0$, whose elements are the quartering points of which the squares overlap $I$.

The above introduction of the plane can, without essential modifications, be carried over to the *n-dimensional Cartesian space*, which, in the case $n = 1$, is called the *continuum*, while its point cores are called *real numbers*.

6   To introduce topological spaces, we first define *a located infinite sequence* $\mathscr{F}$, into which the infinite sequence 1, 2, 3,... of the natural numbers passes if for any pair $v_1$, $v_2$ of its elements a *distance* $\rho(v_1, v_2)$ which is a non-negative real symmetric function of $v_1$ and $v_2$ satisfying $\rho(v, v) = 0$ and the usual triangle inequality, has been defined which possesses the following properties:

1. *The sequence $\mathscr{F}$ is approximated with any degree of accuracy by its successive initial segments* i.e. if the species whose elements are the natural numbers up to $n$, is denoted by $s_n$, then for each $n$ a $\mu_n$ can be indicated such that $\rho(v, s_{\mu_n}) \leq 4^{-n}$ for $v \leq \mu_n$

2. *For every degree of approximation each element of $\mathscr{F}$ is either superfluous or acceptable*, i.e. for every $n$ each element $m$ of $s_{\mu_n}$ ($\mu_n$ as above) can be characterized either as a $\beta_n$-element, or as an $\alpha_n$-element; in the former case a natural number $\mu(n,m)$ can be indicated such that $\rho(v, m) \geq \frac{5}{4} \cdot 4^{-n}$ for $v > \mu(n,m)$; in the latter case an indefinitely increasing sequence of natural numbers $\mu_1(n, m)$, $\mu_2(m, n)$, ... can be defined such that $\rho(\mu_\kappa(n, m), m) \leq \frac{3}{2} \cdot 4^{-n}$ for each $k$. The two alternatives are not mutually exclusive.

An infinite sequence of natural numbers $a_1, a_2, ...$ is called *convergent* with regard to $\mathscr{F}$, if for any $n$ a natural number $p_n$ can be indicated such that $\rho(a_\nu, a_\mu) < 2^{-n}$ for $\nu > p_n$ and $\mu > p_n$. If the sequence contains an infinite sequence of different elements, it is called a *point of $R(\mathscr{F})$*. If the sequence is predeterminate, we speak of a *sharp point of $R(\mathscr{F})$*. If two points of $R(\mathscr{F})$ can be united to a single convergent sequence, they are said to *coincide*. The species of the points of $R(\mathscr{F})$ coincident with the point $p$ of $R(\mathscr{F})$ is called a *point core* of $R(\mathscr{F})$. A point core of $R(\mathscr{F})$ is said to be *sharp*, if it contains a sharp point. The species of the point cores of $R(\mathscr{F})$ is called a *located compact topological space* and is denoted by $R(\mathscr{F})$. The species of the sharp point cores of $R(\mathscr{F})$ is called a *reduced located compact topological space*.

$R(\mathscr{F})$ coincides with a bounded spread (in a slightly widened sense), whose elements are convergent infinite sequences of elements of $\mathscr{F}$ being generated by choosing successively an $\alpha_1$-element, an $\alpha_2$-element, and so on, in such a way that after the choice of the $\alpha_n$-element $\alpha_n^0$ each $\alpha_{n+1}$-element at a distance $\leq 2 \cdot 4^{-n}$ from $\alpha_n^0$ is choosable and each $\alpha_{n+1}$-element at a distance $\geq 3 \cdot 4^{+n}$ from $\alpha_n^0$ is unchoosable.

The definitions, given above for the plane, of deviation, separation, point of closure, point of accumulation, isolated point, unapproachable point, covered point, coincidence, concordance, congruence, can, with appropriate modifications, be taken over for located compact topological spaces.

$R(\mathscr{F})$ has been called 'compact', because it is 'bounded' and because it posseses for each convergent infinite sequence of its elements an element of closure.

$R(\mathscr{F})$ has been called 'located', because its 'location' with respect to any of its points and to any element of $\mathscr{F}$ can be calculated with any degree of accuracy.

7    III. We passed to the study of the notion of order and in the first place investigated, which elements of order are implied by the 'between' which was created together with the basic intuition of the empty two-ity and which is never exhausted by the interpolation of new units, nor even by the interpolation of infinite sequences of new units. Now we had already used this 'between' as a matrix of 'point cores', and the species constituted by these point cores we had joined as 'continuum' to intuitionist mathematics. So, to find the elements of order naturally implied by the intuitive 'between', we sought a definition for a natural order of the point cores of the continuum. Starting from the classical notion of *partial order* we were led to intuitionistic definitions of *order* and of *complete order*. Then we stated that on the basis of the intuitive 'between' the continuum is partially ordered, it is true, but neither ordered nor completely ordered. On the other hand, we recognized that the natural partial order of the continuum satisfies the *axiom of virtual order*, requiring that the species of the ordered pairs *ab* of the partially ordered species concerned, is directly composed

of the subspecies forwhich $a < b$, $a = b$ and $a > b$ hold respectively. From this followed that the partial order of the continuum has a considerable degree of completeness: for each pair of point cores of the continuum the participation in the relations of order is, if not realized, at least non-contradictory; moreover, this partial order is *inextensible*, i.e. to the system $s$ of its relations every addition of a relation of order not contained in $s$, and conserving the validity of the axiom of virtual order, is impossible.

We also put the question, whether the full or the reduced continuum is susceptible of being ordered in another than the natural way, independently of the intuitive 'between', and we found that this question again must be answered in the negative.

IV Our final theme, which we only started, was the precision analysis of the continuum. We remarked that the continuum is fully determined by the denumerably infinite species $R$ of ordinal type $\eta$, over which it extends. Further, that among its elements, those defined by a *Dedekind* section, are of a very special kind. Then that the Dedekind elements are susceptible of further specializations, for instance those which, $R$ being taken as the species of rational numbers, are to be read as the species of real numbers which for any natural number $n$ can be developed into an infinite series of decreasing powers of $n$, or (still more specialized) as the species of regular continued fractions.

And we had just begun to consider those classical properties of the continuum, which are founded on the hypothesis that the continuum is completely ordered, viz. the ordinal density in itself, the ordinal separability in itself, the ordinal connectedness, the ordinal everywhere-density and the ordinal local compactness. What is to become of them in the new continuum-conception? This subject we shall now have to take up again.

October 8, 1947

## Appendix 7

## Disengagement of Mathematics from Logic [BMS 49]

1   The historical development of the mechanism of mathematical thought is naturally closely connected with the modifications which, in the course of history, have come about in the prevailing philosophical ideas, *firstly* concerning the origin of mathematical certainty, *secondly* concerning the delimitations of the object of mathematical science. And that the mental mechanism of mathematical thought during so many centuries has undergone so little fundamental change is due to the circumstance that in spite of all revolutions undergone by philosophy in general, the belief in the existence of properties of time and space, immutable and independent of language and experience, remained well-nigh intact until far into the nineteenth century. Exact knowledge of these properties was called mathematics, and was generally pursued in the following way: for some familiar regularities of (outer or inner) experience which, with any attainable degree of approximation *seemed invariable, absolute and sure invariability was postulated*. These regularities were called *axioms* and were put into language. Thereupon extensive systems of properties were developed from the linguistic substratum of the axioms by means of *reasoning* guided by experience but linguistically following and using the principles of so-called *classical logic*.

We shall call the standpoint governing this mode of thinking and working the *observational standpoint*, and the long period characterized by this standpoint the *observational period*.

During the observational period mathematics was considered functionally, if not existentially, dependent on logic, and logic itself was considered autonomous.

For space the observational standpoint became untenable when, in the course of the 19-th century and the beginning of the 20-th century, as a consequence of a series of discoveries with which the names of Lobatchefsky, Bolyai, Riemann, Cayley, Klein, Hilbert, Einstein, Levi-Civita and Hahn are associated, mathematics was gradually transformed into a mere science of numbers. Simultaneously, besides observational space, a great number of other spaces, sometimes exclusively originating from logical speculation, with properties distinct from the traditional but no less beautiful, gradually found arithmetical representation.

2    Consequently the science of classical (Euclidean) three-dimensional space had to continue its existence as a chapter without priority, on the one hand, of (exact) science of numbers, on the other hand, as applied mathematics, of (naturally only approximative) descriptive natural science.

Encouraged by the important part which, in this process of extending the domain of conceivable geometry, had been played by the *logico-linguistic method*, which, without any guidance by experience, operated on words by means of logical rules, the *Old Formalist School* (Dedekind, Cantor, Peano, Hilbert, Russell, Zermelo, Couturat) finally, for the purpose of a rigorous treatment of mathematics *and logic* (though not for the purpose of choosing the subjects of investigation of these sciences) rejected any extra-lingual element. Thus logic and mathematics were divested by this school both of their essential difference in character and their autonomy. However, the hope originally fostered by the Old Formalists that mathematical science erected according to their principles would be crowned one day with a proof of non-contradictority, was never fulfilled, and nowadays, in view of the results of certain investigations of the last few decades, has, I think, been relinquished.

Of a totally different orientation was the *Pre-intuitionist school*, led by Poincaré, Borel and Lebesgue. These thinkers seem to have maintained a modified observational standpoint for the introduction of natural numbers, of the principle of complete induction, and of all mathematical entities and theories springing from this source without the intervention of axioms of existence, hence for what might be called the 'separable' parts of arithmetic and algebra. For these parts of mathematics, even for such theorems as were deduced by means of classical logic, they postulated an existence and exactness independent of language and logic, and regarded their non-contradictority as certain, even without logical proof. For the continuum, however, they seem to have contented themselves with an ever-unfinished and ever-denumerable system of 'real numbers', generated by an ever-unfinished and ever-denumerable system of laws defining convergent infinite sequences of rational numbers.

3    In doing so they seem to have overlooked that such an ever-unfinished and ever-denumerable system of 'real numbers' is incapable of fulfilling the mathematical functions of the continuum, for the simple reason that it *cannot have a measure positively differing from zero*. On other occasions

they seem to have introduced the continuum by having recourse to some logical axiom of existence lacking sensory as well as epistemological evidence, such as the 'axiom of ordinal connectedness' or the 'axiom of completeness'. But in both cases, in their further development of mathematics, they unreservedly continued to apply classical logic, including the principle of the excluded third. They did so regardless of the fact that the non-contradictority of systems thus constructed had become very doubtful after the discovery of the logico-mathematical antinomies.

Thus, in point of fact, Pre-intuitionism re-established on the one hand the essential difference in character of logic and mathematics, and on the other hand the autonomy of logic and of a part of mathematics. On these two autonomous domains of thought the rest of mathematics remained dependent.

When the Old Formalist standpoint had been badly shaken mainly by Pre-intuitionist criticism, Hilbert founded the *New Formalist School*, which postulated existence and exactness independent of language, it is true not for mathematics proper, but for *meta-mathematics* or *mathematics of the second order*, i.e. the scientific consideration of the symbols occurring in purified mathematical language, and of the rules of manipulation of these symbols. Thus New Formalism, in contrast with Old Formalism, consciously and *in confesso*, made use of the intuition of the natural numbers and of complete induction. It is true that autonomy was postulated here for a much smaller part of mathematics than in the case of Pre-intuitionism.

But no attention was paid by New Formalism to the circumstance that, between the perfection of mathematical language and the perfection of mathematics proper, no clear connection can be seen.

The situation left by Formalism and Pre-intuitionism can be summarized as follows: for the elementary theory of natural numbers, the principle of complete induction, and more or less considerable parts of algebra and theory of numbers, exact existence, absolute reliability, and non-contradictority were universally acknowledged, independently of language and without proof. There was little concern over the existence of the continuum. Introduction of a set of predeterminate real numbers with a positive measure was attempted by logico-linguistic means, but a proof of the non-contradictory existence of such a set was lacking. For the whole of mathematics the rules of classical logic were accepted as reliable aids in the search for exact truths.

In this situation Intuitionism intervened with two acts, of which the first seems necessarily to lead to destructive and sterilizing consequences; then, however, the second yields ample possibilities for recovery and new developments. To begin with, the

*First act of intuitionism*
*completely separates mathematics from mathematical language, in particular from the phenomena of language which are described by theoretical logic, and recognizes that mathematics is a languageless activity of the mind having its origin in the basic phenomenon of the perception of a move of time, being the falling apart of a life moment into two distinct things, one of which gives way to the other, but is retained by memory. If the two-ity thus born is divested of all quality, there remains the common substratum of all two-ities, the mental creation of the empty two-ity.*

This empty two-ity and the two unities of which it is composed, constitute the basic mathematical systems. And the basic operation of mathematical construction is the mental creation of the two-ity of two mathematical systems previously acquired, and the consideration of this two-ity as a new mathematical system.

It is introspectively realized, how this basic operation, continually displaying *unaltered* retention by memory successively generates each natural number, the infinite sequence of all natural numbers, arbitrary finite sums of mathematical systems previously acquired, sums of fundamental sequences of mathematical systems previously acquired, finally a continually extending stock of mathematical systems corresponding to 'separable' systems of classical mathematics.

From the preceding follows that the first act of intuitionism affects classical mathematics in two ways: in the first place, owing to the disappearance of the logical basis for the continuum, so large a part becomes illusory that essentially only the separable parts of algebra and theory of numbers remain; in the second place even in this remaining portion, several chapters exclusively based on the belief in classical logic, have to be rejected. Under these circumstances one might fear that intuitionist mathematics must necessarily be poor and anaemic and in particular would have no place for analysis. But this fear would have presupposed that infinite sequences generated by the intuitionist self-unfolding of the basic act would, just as the infinite sequences of classical mathematics, have to be fundamental sequences, i.e. predeterminate infinite sequences in which, from the beginning, the *m*-th term is fixed for every *m*. Such, however, is not the case; on the contrary, a much wider field of development which includes an analysis, and in several places exceeds the frontiers of classical analysis, is opened by the

*Second act of intuitionism*
which recognizes the possibility of generating new mathematical entities *firstly* in the form of infinitely proceeding sequences $p_1, p_2 ,...$ whose terms are chosen more or less freely from mathematical entities previously acquired; in such a way that the freedom existing perhaps at the first choice may be irrevocably subjected, again and again, to progressive restrictions at subsequent choices, while all these restricting interventions, as well as the choices themselves, may, at any stage, be made to depend on possible future mathematical experiences of the creating subject;
*secondly*, in the form of mathematical species, i.e. properties supposable for mathematical entities previously acquired and satisfying the condition that, if they hold for a certain mathematical entity, they also hold for all mathematical entities which have been defined to be equal to it, definitions of equality having to be symmetric, reflexive and transitive, and the empty two-ity being forbidden to be equalized to an empty unity; mathematical entities previously acquired for which the property in question holds, are called elements of the corresponding species.

5   With regard to this definition of species we have to remark firstly that, during the development of intuitionist mathematics, some species will have to beconsidered as being re-defined time and again in the same way, secondly that a species can very well be an element of another species, but never an element of itself.

Two mathematical entities are called *different*, if their equality has been proved to be absurd.

Two infinitely proceeding sequences of mathematical entities $a_1, a_2,...$ and $b_1, b_2 ,...$ are called *equal* or *identical*, if $a_v = b_v$ for each $v$, and *distinct*, if a natural number $s$ can be indicated such that $a_s$ and $b_s$ are different.

In the edifice of mathematical thought based on the first and second act of intuitionism, language plays no other part than that of an efficient but never infallible or exact technique for memorizing mathematical constructions, and for suggesting them to others; so that the wording of a mathematical theorem only then has a sense, if it indicates the construction either of an actual mathematical system or of an incompatibility (e.g. the identity of the empty two-ity with an empty unity) out of some constructional condition imposed on a hypothetical mathematical system. So that mathematical language, in particular logic, by itself can never create new mathematical systems, nor deduce a mathematical state of things.

However, notwithstanding this rejection of classical logic as an instrument to discover mathematical truths, intuitionist mathematics has its general introspective theory of mathematical assertions, a theory which with some right may be called *intuitionist mathematical logic*, and to which e.g. belongs a theory of the *principle of the excluded third*.

## Appendix 8

## Notes for a Lecture [BMS 47]

⟦A two-page manuscript giving outlines of a lecture probably given in 1951. Not all sentences are completed; it starts with some cryptic references and headings Ed.⟧

'Blackbush,
Kl.Beth, pp.16,53,54,64
De Morgan, Formal Logic, London 1847, pp. 60, 118-120']

1  1. p.t.e. ongefundeerd
   2. de verschillen tusschen *niet nu* en *nooit*]

    Logica oorspronkelijk, dat er een waarheid is, waarvan een gedeelte exact en adequaat is te beschrijven, en waarvan de kennis door die beschrijving kan worden overgedragen, en aangezien in de verstandhouding alleen de beschrijving een controleerbare rol spelt, kon de waarheid met haar beschrijving min of meer worden vereenzelvigd.

    Er zijn dingen met eigenschappen, en mogelijkheden van verschijnselen, gehoorzamend aan regels, permanent voor korteren of langeren tijd; juiste en onjuiste beweringen, dat bepaalde dingen bepaalde eigenschappen bezitten, dat bepaalde verschijnselen volgens bepaalde regels verloopen deze waarheden, dus ook deze juiste beweringen zijn wellicht aan een bepaalde tijd en plaats gebonden, doch gelden daar onafh. van het subject. En iedere bewering hoort of wel of niet tot de juiste beweringen, ook al kunnen wij dit niet beslissen.

    Wiskundige waarheden golden oorspr. als eig. van een objectieve tijd en ruimte te bestudeeren in een uitwendige wereld, later als waarnemingsvormen van den menschelijken geest, te bestudeeren in een inwendige wereld. En ook van die wiskundige waarheden heeft men tot voor kort volgehouden, dat ze bestonden onafhankelijk van hun bekendheid, en dat een wiskundige bewering of juist of onjuist is, hetzij iemand dit weet of niet. Als het menschdom uitsterft, blijft er een stelsel natuurwetten bestaan en blijft de wiskunde bestaan. Gr.Beth 209.

    Wiskunde in twee stappen van de uitwendige ervaringswereld (onafh.v.h. subject gedacht) losgemaakt:

1. Kant (receptaculum van uitw. ervaring). 2. introspectie van een mentale constructie. Daarmee verliest de vraag of een significante wiskundige bewering juist of onjuist is, alle beroep op van het denkende subject onafhankelijke machten (als obj. buitenwereld, verstandhouding met anderen, of axiomas's die geen inwendige ervaringen zijn) en wordt ze een uitsluitend individueele gewetenszaak[?] van het subject en heeft de juistheid van een bewering geen andere beteekenis dan dat haar inhoud feitelijk als gewaarwording is opgetreden en hebben we 3 gevallen:

1. waar;
2. onmogelijk voor nu en altijd;
3. op het oogenblik noch waar noch onmogelijk, *a*. met, *b*. zonder bestaan van een methode, die hetzij tot 1. hetzij tot 2. moet voeren.

Het geval 3*a* is terug te voeren tot 1. and 2., zoodat we alleen onderscheiden 1., 2 en 3*b* (Het geval 3*b* komt niet voor binnen een begrensde eindige soort. Congr, 1245). Het geval 3*b* is *voorlopig*: men moet rekening houden met de mogelijkheid, dat later wel een methode wordt gevonden (eenerszijds doordat we door blijven denken, anderzijds doordat de mathematische dingen, waarover de bewering gaat, zich dikwijls in staat van wording bevinden). Daarbij blijkt soms later wel de onmogelijkheid om de onmogelijkheid aan te toonen, zonder dat evenwel de waarheid kan worden bewezen. Waarvan we straks een voorbeeld zullen zien. Het spreekt van zelf, dat zulk een voorbeeld niet alleen voor de wiskunde, doch ook voor de logica (waar b.v. de formules van de Morgan verloren gaan) een ingrijpende herziening noodig maakt, terwijl tevens de logica hiermee haar karakter van evidentie of a-prioriteit heeft verloren, dus niet meer als bron of ontdekkingsmiddel van waarheden, doch uitsluitend als een nevenverschijnsel van wiskunde en wetenschap kan worden beschouwd.

2  Aan de verarming, die de wiskunde dreigde te ondergaan door het wegvallen van het gebruik van opportunistische axioma's en van de logica, in het bijzonder het indirecte bewijs als opsporingsmiddel voor wiskundige waarheden waardoor voor die opsporing alleen directe aanschouwing (aperception, introspection) overbleef, is de *moderne wiskunde* tegemoetgekomen, door een uitgebreidere soort van oneindige reeksen in beschouwing op te nemen, dan de klassieke wiskunde had gedaan. Deze had zich namelijk beperkt tot *gepredetermineerde* oneindige reeksen, waarin van den beginne af het $n$-de element voor elke $n$ bepaald is, terwijl de moderne wiskunde ook oneindige reeksen van reeds ingevoerde wisk. entiteiten toelaat, die in volledige of gedeeltelijke vrijheid voorschreiden. Dit wordt

na het elimineeren der logica reeds noodzakelijk als men alle reëele getallen wil creëeren, waaruit het eendimensionale continuum is samengesteld. Als men n.l. hierbij alleen de gepredetermineerde convergente reeksen der klassieke wiskunde ter beschikking heeft kan men daarmee door introspectieve constructie slechts deelsoorten van een voortdurend onaffe aftelbare soort van reëele getallen tot stand brengen, welke gedoemd is, om voortdurend de maat nul te blijven bezitten. Ter invoering van een soort van reëele getallen, die het continuum kan respresenteeren en daartoe dus een positieve maat zal moeten bezitten, moet dientengevolge de klassieke wiskunde haar toevlucht nemen tot een of ander logisch proces, uitgaande van allesbehalve evidente axioma's. De meest gangbare als zoodanig gebruikte axioma's zijn opgesteld door Hilbert, en eischen dat het systeem der reëele getallen op zoodanige wijze geordend en onder geldigheid van het axioma van Archimedes aan de vier hoofdbewerkingen onderworpen kan worden, dat het systeem niet verder onder behoud der verkregen qualiteiten en onder handhaving der bestaande relaties kan worden uitgebreid (volledigheidsaxioma). Waarmee dit 'volledige' system van reëele getallen natuurlijk nog niet geschapen is, zoodat feitelijk alleen een logisch, geen wiskundig systeem werd geschapen. Op dezen grond mogen we zeggen, dat de klassieke analyse, hoe geschikt ook voor techniek en wetenschap, minder mathematische werkelijkheid bezit dan de intuitionistische analyse, die de bedoelde opbouw van het van positive maat voorziene continuum uit reëele getallen voltrekt door beschouwing van de soort der vrijelijk voortschrijdende convergente oneindige reeksen van rationale getallen, zonder daarbij de hulp van taal of logica noodig te hebben.

## Notes for a Lecture (translation)

⟦The translation follows the original closely, leaving sentences uncompleted as in the original Ed.⟧

1 Blackbush,
Kl.Beth, pp.16,53,54,64
De Morgan, Formal Logic, London 1847, pp. 60, 118-120

1. the Principle of the Excluded Middle unfounded
2. the differences between *not now* and *never*

Logic originally, that there is a truth, of which a part can be described exactly and adequately, and the knowledge of which can be transferred by means of this description. Since in mutual understanding only description plays a controllable role, truth could be more or less identified with its description. That there are things with properties, and possibilities of phenomena obeying rules, permanent for shorter or longer periods of time; that there are correct and incorrect assertions that certain things have certain properties, and that certain phenomena proceed according to certain rules; that these truths, and therefore also these correct assertions are possibly bound to a particular time and space, but nevertheless apply there independently of the subject; that every assertion either belongs or does not belong to the correct assertions, even if we cannot decide it.

Originally mathematical truths were valid as properties of an objective time and space, to be studied in an external world, later as forms of perception of the human mind to be studied in an interior world. And until recently one maintained that also these mathematical truths existed independent of them being known, and that a mathematical assertion is either corrector incorrect, whether anyone knows it or not. That, when mankind has become extinct, the system of natural laws will remain in existence and mathematics will remain in existence. (Gr. Beth 209).

Mathematics was detached from the external world of experience (i.e. considered as independent of the subject) in two steps:
1. Kant (receptaculum of external experience);
2. introspection of a mental construction. At this stage the question whether a meaningful mathematical assertion is correct or incorrect, is freed of any recourse to forces independent of the thinking subject (such as the external world, mutual understanding, or axioms which are not inner experience) and becomes exclusively a matter of individual consciousness of the subject. Correctness of an assertion then has no other meaning than that its content has in fact appeared in the consciousness of the subject. We therefore distinguish between:
1. true
2. impossible now and ever
3. at present neither true nor impossible
   *a.* either with, or
   *b.* without the existence of a method which must lead to either 1. or to 2.

The case 3*a* can be reduced to 1. and 2., so that we only need to distinguish between 1., 2. and 3b. (The case 3b does not occur in a limited finite species) Congr. 1245 [[1948C], p.1245 Ed.]]

The case 3b is *temporary*: one must take into account the possibility that one day a method may be found (because we go on thinking and because the mathematical entities underlying the assertions are often in a state of becoming). It sometimes happens that one can show the impossibility of the impossibility, although one cannot prove its truth. We shall shortly give an example of this. It stands to reason that such an example necessitates a drastic overhaul not only of mathematics but also of logic (where e.g. the formulae of dè Morgan become invalid); moreover, logic loses its evident or a-priori character and can therefore no longer be considered to be a source or a means of discovering truths, but only as a side-phenomenon of mathematics and science.

2   The impoverishment which mathematics was in danger of suffering as a result of losing the use of opportunistic axioms and of logic − in particular the use of indirect proof in the search of mathematical truths, leaving only direct intuition (aperception, intuition) − was compensated in *modern mathematics* by admitting a more extended species of infinite sequences than classical mathematics had done. Classical mathematics had restricted itself to *predetermined* infinite sequences in which from the start the $n$-th element is fixed for every $n$, whereas modern mathematics also allows infinite sequences of pre-constructed elements which proceed in total or partial freedom. After the abandonment of logic one needed this to create all the real numbers which make up the one-dimensional continuum. If only the predeterminate sequences of classical mathematics were available, one could by introspective construction only generate subspecies of an ever-unfinished countable species of real numbers which is doomed always to have the measure zero. To introduce a species of real numbers which can represent the continuum and therefore must have positive measure, classical mathematics had to resort to some logical process, starting from anything-but-evident axioms. The axioms currently most widely-used for this purpose are those defined by Hilbert; they demand that the system of real numbers can be ordered and, given the validity of Archimedes' axiom, can be subjected to the four main operations, in such a way that the system cannot be further extended without losing its acquired qualities or existing relations (axiom of completeness). Of course, this so-called complete system of real numbers has thereby not yet been created; in fact only a logical system was created, not a mathematical one. On these grounds we may say that classical analysis, however

suitable for technology and science, has less mathematical reality than intuitionist analysis, which succeeds in structuring the positively-measured continuum from real numbers by admitting the species of freely-proceeding convergent infinite sequences of rational numbers and without the need to resort to language or logic.

**Appendix 9**

**Changes in the Relation between Classical Logic and Mathematics**
[BMS 59]

⟦In Brouwer's academic estate there are a number of manuscripts of papers on the changes in classical logic caused by Intuitionism. There are indications in the text that they were used for public lectures and that Brouwer considered publication.

The first, German version: 'Wandlungen in den Beziehungen zwischen klassischer Logik und Mathematik', [BMS 41] dates probably from 1930-34. The typescript of a lecture 'The Influence of Intuitionistic Mathematics on Logic' was used, as indicated in the margin for a lecture in London on 2nd November 1951 and in Joannesburg on 29th July 1952. Under the amended title 'Changes in the relation between classical logic and mathematics' it is published as an appendix in van Dalen's edition of *Brouwer's Cambridge Lectures* (VAN DALEN[1981]). A different version with amendments is given in the hand-written manuscript [BMS59B] and is presented here.⟧

1   Classical Logic presupposed that independently of human thought there is a truth, part of which is expressible by means of sentences called 'true assertions'; mainly assigning certain properties to certain objects or stating that objects possessing certain properties exist or that certain phenomena behave according to certain laws. Furthermore, classical logic assumed the existence of general linguistic rules allowing an automatic deduction of new true assertions from old ones, so that starting from a limited stock of 'evidently' true assertions, mainly founded on experience and called axioms, an extensive supplement to existing human knowledge would theoretically be accessible by means of linguistic operations independently of experience. Finally using the term 'false' for the converse of true, classical logic assumed that by virtue of the so-called 'principle of the excluded third' each assertion, in particular each existence assertion and each assignment of a property to an object or of a behaviour to a phenomenon, is either true or false, independently of human beings knowing about this falsehood or truth, so that e.g. contradictority of falsehood implies truth, while an assertion $\alpha$ which is true as well if the assertion $\beta$

is true as if the assertion $\beta$ is false, is universally true. ⟦An unfortunate translation into English; Brouwer intended to say :'...while an assertion $\alpha$, which is true if the assertion $\beta$ is true and which is also true if the assertion $\beta$ is false, is universally true.' Ed.⟧

As long as mathematics was considered as the science of space and time, it was a beloved field of activity of this classical logic, not only in the days that space and time were believed to exist independently of human experience, but still after they had been taken for innate forms of conscious, exterior human experience. There continued to reign some conviction that a mathematical assertion is either false or true, whether we know it or not, and that after the extinction of humanity mathematical truths, just as laws of nature, will survive.

Only after intuitionism had recognized mathematics as an autonomic interior constructional activity, which although it has found extremely useful linguistic expression and can be applied to an exterior world, neither in its origin nor in the essence of its methods has anything to do with language or with an exterior world, firstly all axioms became illusory, and secondly the criterion of truth or falsehood of a mathematical assertion was confined to mathematical activity itself, without appeal either to logic or to a hypothetical omniscient being. An immediate consequence was that in mathematics no truths could be recognized which had not been experienced so that for a mathematical assertion $\alpha$ the two cases of truth and falsehood, formerly exclusively admitted were replaced by the following three:

1. $\alpha$ has been proved to be true;
2. $\alpha$ has been proved to be absurd;
3. $\alpha$ has neither been proved to be true nor to be absurd, nor do we know a finite algorithm leading to the statement either that $\alpha$ is true or that $\alpha$ is absurd.

We remark that the case that $\alpha$ has neither been proved to be true nor to be absurd, but a finite algorithm is known leading to a decision either that $\alpha$ is true or that $\alpha$ is absurd, obviously is reducible to the first and second cases. This applies in particular to assertions of possibility of some construction of bounded finite character in some finite mathematical system, because such a construction can be attempted only in a finite number of particular ways, and each attempt proves successful or abortive in a finite number of steps.

In contrast with the perpetual character of the first and second cases, an assertion being in the third case may at some time pass into one of the other ones, not only because further thinking may generate an algorithm accomplishing this passage but also because in intuitionist mathematics, as we shall see, a mathematical entity is not necessarily predeterminate, and may, in its state of free growth, at some time acquire a property which it did not possess before.

2  From the foresaid intuitionist rejection of all non-experienced truths classical mathematics is objectionable mainly in two respects:
1. for its introduction and description of the continuum;
2. for its thoughtless passage from non-contradictority to truth.

Ad 1. By means of the definition of real numbers by means of predeterminate Dedekind incisions or predeterminate convergent infinite sequences of rational numbers only an ever-unfinished and ever-denumerable system of such real numbers can be actually generated. Such an ever-unfinished and ever-denumerable system of real numbers is incapable of fulfilling the mathematical functions of the continuum for the simple reason that it can never acquire a measure positively differing from zero. Hence, to compose a continuum of positive measure out of points, classical mathematics has recourse to some logical process starting from at least one axiom. E.g. Hilbert, starting from a set of properties of order and calculation, including the archimedean property holding for the field of rational numbers, then considering successive extensions of the rational arithmetical field to larger and larger fields, each including all preceding ones and conserving the foresaid properties, finally by means of the so-called axiom of completeness, lacking sensory as well as epistemelogical evidence, postulates the existence of an ultimate such extended field admitting no further extension conserving the foresaid properties. From the intuitionist point of view a continuum created in this way has a merely linguistic, no mathematical existence. Intuitionism even proves that for the classical continuum the axiom of completeness is false.

Ad 2. To say: 'either there exists a Fermat-number $> 2$, or a Fermat-number $> 2$ cannot exist' is an assertion whose truth has not been experienced, so an unlawful application of the principle of the excluded third. Moreover, it is an assertion which is non-contradictory without being true.

More simply and more palpably, the unlawfulness even in the domain of classical mathematics of applying the principle of the excluded third and of replacing non-contradictority by truth is e.g. demonstrated by the existence of so-called 'fleeing properties'.

A property $f$ having a sense for natural numbers is called a *fleeing property* if it satisfies the following requirements:

i) For each natural number $n$, it can be decided whether or not $n$ possesses the property $f$;

ii) no way of calculating a natural number $n$ possessing $f$ is known;

iii) the assumption that at least one natural number possesses $f$ is not known to lead to an absurdity.

Obviously the fleeing nature of $f$ is not necessarily permanent, for a natural number possessing $f$ might at some time be found, or the absurdity of the existence of such a natural number might at some time be proved. Nevertheless there will always be fleeing properties, and for every existing fleeing property the assertion that $f$ either can be realized or cannot possibly be realized would only convey a truth, if this truth had been experienced, and so on account of ii) and iii) constitutes an utter falsehood.

Obviously every natural number $n$ can be proved to be either a *down-number* of $f$, meaning that no natural number $\leq n$ possesses $f$, or an *up-number* of $f$, meaning that some natural number $\leq n$ possesses $f$. In case an up-number of $f$ is found, firstly $f$ has lost its fleeing nature, secondly there exists a *smallest natural number possessing* $f$, which we shall call the *critical number of* $f$ and denote by $k_f$.

Let us consider the infinite sequence $a_1, a_2,....$, where $a_\nu = 2^{-\nu}$ if $\nu$ is a down-number, and $a_\nu = 2^{-k_f}$ *if* $\nu$ is an up-number of $f$. Obviously this infinite sequence is convergent, because beyond $a_\nu$ its field of variation is $\leq 2^{-\nu}$. So it defines a real number as its limit which we shall call the *binary shrinking number of* $f$ and denote by $b_f$. Let us assume that $f$ has not lost its fleeing nature (as was remarked above of fleeing properties not having lost their fleeing nature there will always be an abundance), then the assertion '$b_f$ is rational' is non-contradictory without being true.

3   The above objections affect classical mathematics in two ways: in the first place, owing to the disappearance of the logical basis for the continuum, so large a part becomes illusory that essentially only the 'separable' parts of algebra and theory of numbers remain; in the second place even in this remaining portion, several chapters based on the principle of the excluded third, in particular on the equivalence of non-contradictority and truth, have to be rejected. Under these circumstances one might fear that intuitionist mathematics must necessarily be poor and anaemic, and in particular would have no place for analysis. Such, however, is not the case; on the contrary, a much wider field of development which includes analysis, which in several places far exceeds the frontiers of classical math-

ematics, is opened up by what I should like to call the:

### Primordial recognitions of Intuitionism

The origin of mathematics is in the perception of a *move of time*, i.e. of the falling apart of a life moment into two distinct things, one of which gives way to the other, but is retained by memory. If the two-ity thus born is divested of all quality, there remains the common substratum of all two-ities, the mental creation of the *empty two-ity*. This empty two-ity and the two unities of which it is composed, constitute the *basic mathematical systems*. And the basic operation of mathematical construction is the *mental creation of the two-ity of two mathematical systems previously acquired,* and the consideration of this two-ity as a new mathematical system.

This basic operation, continually displaying unaltered retention by memory, generates successively each natural number, the infinite sequence of all natural numbers, arbitrary finite sequences and fundamental sequences of mathematical systems previously acquired, finally infinitely proceeding sequences whose terms *are more or less freely chosen from mathematical systems previously acquired*, in such a way that the freedom existing perhaps at the first choice may be irrevocably subjected, again and again, to progressive restrictions at subsequent choices, while all these restricting interventions, as well as the choices themselves, may at any stage, be made to depend on possible future mathematical experiences of the creating subject.

Moreover, new mathematical entities can be created in the shape of *mathematical species*, i.e., *properties supposable for mathematical entities previously acquired*, and satisfying the condition that, if they hold for a certain mathematical entity, they also hold for all mathematical entities which have been defined to be *equal* to it, equality having to be symmetric, reflexive and transitive and the empty two-ity being forbidden to be equalized to an empty unity. Mathematical entities for which the property in question holds, are called the *elements* of the corresponding species. In the finite and infinite sequences generated by intuitionism all mathematical entities previously acquired, including species and also nothing, may figure as terms.

In the edifice of mathematical thought based on the above recognitions, language plays no other part than that of an efficient, but never infallible or exact technique for memorizing mathematical constructions, and for suggesting them to others; so that the wording of a mathematical theorem has no sense unless it indicates the construction either of an actual mathematical entity or of an incompatibility (like e.g. the identity of the empty two-ity with an empty unity) out of some constructional condition imposed on a hypothetical mathematical system. So that mathematical language, in particular logic, can never by itself create new mathematical entities nor deduce a mathematical state of things.

However, notwithstanding this rejection of classical logic as an instrument to discover mathematical truths, intuitionist mathematics has its general introspective theory of mathematical assertions, a theory which with some right may be called *intuitionist mathematical logic*.

4   By virtue of the above primordial recognitions of intuitionism there is the following difference in conception of infinitely proceeding sequences between classical and intuitionist mathematics. In classical mathematics, in growing sequences of explicitly mentioned terms, this essentially free process either finishes with the indication of a final element, or with an abolition of the previously existing creative freedom, namely with the laying down of a law by which from this moment every future element is unambiguously fixed. For the intuitionist mathematician whose developments are completely autonomic, and neither committed to logical principles nor hampered by beliefs in or admissions of existences and truths exterior to his own creative activity, there is no reason whatever to accept the said alternative, so that in intuitionism an ad-infinitum freely proceeding decimal development 3,79583912...has no less mathematical importance than e.g. the development 3,14159265... in which each digit has to be calculated after a certain prescription. As a matter of course, this intuitionist freedom includes the above-mentioned right to any voluntary self-restriction, so that the freedom existing perhaps etc.

〚Could be continued as from Appendix 7, p.5. Ed.〛

## Appendix 10

### 'Second Lecture' [BMS 66]

[A manuscript of lectures given in London on 12th, 16th and 19th May 1952. It follows an introductory lecture as given in appendices 7, 8 and 9. Ed.]

1   In the preceding lecture we formulated the intuitionist conception of mathematical thinking as a process of unfolding of a basic intuition of two-ity being the substratum divested of all quality of the phenomenon of a move of time. As a consequence of this conception there do not exist mathematical systems, mathematical properties or true mathematical assertions outside human thought. From this point of view the principle of the excluded third had to be given up as an instrument for discovering mathematical truths, and in particular the so-called 'fleeing properties' and the real numbers $s_f$ generated by them showed the non-validity of the principle and of its corollary, the equivalence of truth and non-contradictority.

In particular we saw that the assertion '$s_f$ is rational' is neither true nor false, but is non-contradictory. On the other hand the assertion '$s_f = 0$' is neither true nor false nor non-contradictory. For if its contradictority were absurd, also the existence of $\kappa_f$ would have to be absurd, which is not the case as long as $f$ conserves its fleeing nature.

As further examples of the devastating consequences of the nature of the real number $s_f$ we mentioned the destruction of the complete linear order of the continuum, and the untenability of the Dedekind incision as definition of real numbers (for with respect to $s_f$ not only 0 but also each $s_{f'}$ neither belongs to the right, nor to the left subset of the continuum). But these devastating consequences are not restricted to the fundamental notions, they affect as well the more elaborate and more technical chapters, as we shall elucidate by considering the following theorem relating to plane vector fields:

If in a plane provided with a positive sense of rotation a vector field $g$ is uniformly continuous over a closed square $\mathscr{S}$, if for a certain natural number $h$ its scalar value is $> 2^{-h}$ on the periphery $\sigma$ of $\mathscr{S}$, and if for the total angle $2k\pi$ described by $g$, during a positive circuit of $\sigma$ the integer $k$, which will be called the *index* of g with respect to σ and will be denoted by $i_\sigma(g)$ is $\neq 0$, then in the interior of $\mathscr{S}$ we can construct a point in which g vanishes. The classical proof is given as follows: Let $b_\nu$ be a division of $\mathscr{S}$ into $2^{2\nu}$ congruent and homothetic partial squares $c_\nu$. Then either on the periphery of one of the $c_1$, say $c_1'$, a point bearing a vanishing value of g can be indicated, or one of the $c_1$, say $c_1''$ [a natural number $h_1$], and an integer $k_1 \neq 0$ can be indicated in such a way that the scalar value of g is $> 2^{-h_1}$ on the periphery $\sigma_1''$ of $c_1''$, while $k_1 = i_{\sigma_1''}(g)$. Dealing in the same way with $c_1''$ as we dealt with $\mathscr{S}$, and continuing in this way, we generate an infinite sequence $c_1'', c_2'',\ldots$ converging to a point $P$ of $\mathscr{S}$ in such a way that the maximum of the variation of g on $\sigma_\nu''$, so the maximum of the scalar value of g on $\sigma_\nu''$, converges to the value 0 in $P$, which process can only be stopped by the encounter of a $c_\nu'$ on whose periphery a point $Q$ bearing a vanishing value of g can be indicated.

2   This argument, after classical mathematics, would prove our theorem. But of course to this argument, applying the principle of the excluded third, intuitionism raises formal objections. There is even more: intuitionism can indicate a case in which the construction in question visibly miscarries because it is checked en route, and cannot be continued.

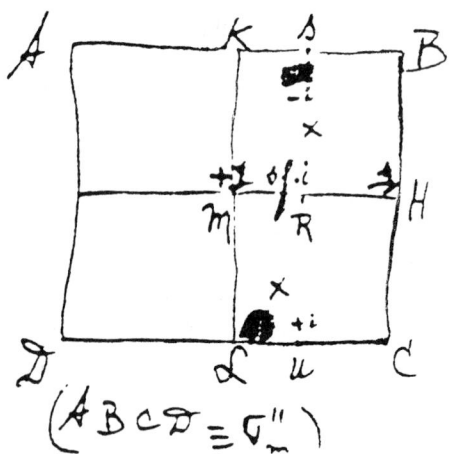

Let us suppose that $i_{\mathscr{ABCD}}(g) = i_{\mathscr{KBCL}}(g) = +1$. Let g change linearly and uniformly along $\mathscr{MKS}, \mathscr{SBH}, \mathscr{HCU}, \mathscr{ULM}, \mathscr{MR}$, and $\mathscr{RH}$. Then we have no means to decide between these possibilities of choice:
1. $\mathscr{KBHM}$ or $\mathscr{MHCL}$ as $\sigma_{m+1}'$;
2. $\mathscr{KBH}$ as $\sigma_{n+1}''$;

3. $\mathscr{MHCL}$ as $\sigma''_{m+1}$.

It may be that at some time $f$ will lose its fleeing character, and that we *may* find either a point $Q$ as mentioned above in $R$ or a point $P$ as mentioned above somewhere in the interior of $\mathscr{KBHM}$ or a point $P$ as mentioned above somewhere in $\mathscr{MHLC}$, or the process is checked anew further on, and we are again in the dark.

The invalidity of this vector theorem is of considerable purport, for in classical mathematics some theorems of central importance are usually deduced from it f.i. the fundamental theorem of algebra saying that every algebraic equation with one unknown, and the existence theorem of a fixed point for topological transformations of a square (or — which comes to the same — of a circular disk) in itself.

With regard to the algebraic equation of the $n$-th degree $\mathscr{F}(x) = 0$, classical mathematics considers the complex plane of $x$ and creates $g$ by assigning to each $x$ the vector representing the corresponding complex value of $\mathscr{F}(x)$. If then for $\mathscr{S}$ a sufficiently large square with the origin of $x$ as its centre is chosen the number $k$ will be equal to $n$, and it is the construction mentioned above which is supposed to lead to a vanishing value of $g$ which in this case means a root of $\mathscr{F}(x) = 0$. Intuitionistically rejecting this proof, we can remark that there are a considerable number of other classical proofs of the fundamental theorem by which however in general only the case of mutual indivisibility of $\mathscr{F}$ and $\mathscr{F}'$. is treated directly, while in the case of mutual divisibility of $\mathscr{F}$ and $\mathscr{F}'$ the proof detaches from $\mathscr{F}$ a factor which is in the first case. So the validity of these proofs presupposes the possibility of deciding whether $\mathscr{F}$ and $\mathscr{F}'$ have a common divisor, i.e. presupposes the possibility of deciding whether the discriminant of $\mathscr{F}$ vanishes or does not vanish.

3 And this again is denied by intuitionism. For let us consider the algebraic equation $\mathscr{F}(x) \equiv x^3 - 3x + 2\rho = 0$, in which $\rho$ is some real number. Its discriminant is easily calculated to be $108\,(\rho^2 - 1)$. Now let us put $\rho = 1 + s_f$, which is a quite normal real number, but which gives a discriminant $108\,(2s_f + s_f^2)$, which has exactly the same chances of becoming positive, zero or negative as $s_f$, so that the proofs in question of the fundamental theorem don't apply to the above equation.

Now let us turn to the mentioned fixed point theorem, and sketch its classical proof for the square $\mathscr{S}$, taking for $g$ in each point its displacement vector for the transformation in question $\tau$. Then either there is a point of $\sigma$, in which $g$ vanishes, i.e. an invariant point for $\tau$ on $\sigma$, or the suppositions of the vector lemma are fulfilled, so that from this lemma would follow the existence of an invariant point in the interior of $\mathscr{S}$. In this case not only intuitionism rejects this proof, but is even able to give a counter-example of the theorem itself. We shall give this counter-example not for a square, but for a circular disk $\mathscr{D}$ radius 2 with the origin as its centre, divided into an inner circular disk $\mathscr{I}$ with the same centre with radius 1 and a corona $\mathscr{C}$ bounded by two concentric circles. We define a topologic transformation $\tau_{2n}$ of $\mathscr{D}$ in itself as follows:

Let $d_{2n}$ be the point $Q_1 (0, \frac{1}{2})$. Let us fill $\mathscr{I}$ with circles $\gamma_1$, belonging to the pencil of circles defined by $x^2 + y^2 - 1 = 0$ and $(x - \frac{1}{2})^2 + y^2 = 0$, so containing also the straight line $x = 1\frac{1}{4}$. Then $\tau_{2n}$ transforms $\mathscr{I}$ in such a way that each radius vector through $Q_1$ is transformed into a radius vector through $Q_1$, that each of the circles $\gamma_1$ is transformed into itself, and that the circle $x^2 + y^2 - 1$ performs a rotation in itself displacing each of its points by an arc of length $2^{-2n}$. Furthermore the transformation of $\mathscr{C}$ by $\tau_{2n}$ is expressed in polar coordinates with respect to the origin as follows: $r' = r; \varphi' = \varphi + (r - 1) + (2 - r) 2^{-2n}$.

As to $\tau_{2n+1}$, let $d_{2n+1}$ be the point $Q_2 (0, -\frac{1}{2})$. Let us fill $\mathscr{I}$ with circles $\gamma_2$ belonging to the pencil defined by $x^2 + y^2 - 1 = 0$ and $(x + \frac{1}{2})^2 + y^2 = 0$. Then $\tau_{2n+1}$ transforms $\mathscr{I}$ in such a way that each radius vector through $Q_2$ remains a radius vector through $Q_2$, that each $\gamma_2$ is transformed into itself, and that the circle $x^2 + y^2 - 1$ performs a rotation in itself displacing each of its points by an arc of length $2^{-2n-1}$. And for $\mathscr{C}$, $\tau_{2n+1}$ is expressed by $r' = r, \varphi' = \varphi + (r - 1) + (2 - r) 2^{-2n-1}$.

If now $t_v$ is defined $= \tau_v$ for $v < k_f$, and $= \tau_{\kappa_f}$ for $v \geq k_f$, $t = \lim t_v$ is a topological transformation of $\mathscr{D}$ in itself, for which it will be impossible to construct an invariant point as long as $f$ will be a fleeing property.

We have seen that the first act of intuitionism (the recognition that mathematics has no other existence than as an interior activity of human mind) affects classical mathematics in two ways: in the first place it seems to have to reject a considerable part of it, sparing only the so-called 'separable' parts of arithmetic and of algebra so that in particular the continuum seems to collapse; in the second place even in the remaining parts it rejects all theorems which have been deduced by the so-called indirect method i.e. by refuting its negation. So one might fear at this stage that intuitionist mathematics will have to be poor and anaemic, and in particular will be unable to reconstruct anaysis. But this would have assumed that infinite sequences generated by the intuitionist unfolding of the basic intuition would have to be fundamental sequences, i.e. predeter-minate infinite sequences, proceeding like classical ones, in such a way that from the beginning the $n$-th term is fixed for each $n$. Such however is not the case: on the contrary, a much wider field of development, including analysis and often exceeding the frontiers of classical mathematics, is opened by what we shall call the

*Second act of intuitionism*: Admitting two ways of creating new new mathematical entities :

*firstly* in the shape of more or less freely proceeding infinite sequences of mathematical entities previously acquired so that e.g. infinite decimal fractions having neither exact values nor any guarantee of at any time acquiring exact values are admitted;

*secondly* in the shape of mathematical species, i.e. properties supposable for mathematical entities previously acquired, and satisfying the condition that if they hold for a certain mathematical entity they also hold for all mathematical entities which have been defined to be 'equal' to it, definitions of equality having to satisfy the conditions of symmetry, reflexivity, and transitivity. By the *elements* of a species we understand the mathematical entities previously acquired for which the property in question holds.

[[A footnote * appears at this point in the MS without *-reference in the text Ed.]]:

Let us surround (in a Euclidean $R_n$ a point $P$ is surrounded by a region $\mathscr{G}$ if for a suitable natural number $k$ all points of $R_n$ at a distance $< 2^{-k}$ from $P$ are contained in $\mathscr{G}$) all real numbers between 0 and 1 defined in the classical way up to 1900 by a set $\sigma_1$ of open intervals of measure $< 2^{-n-1}$, and settle that in future all such real numbers which will have been defined in the classical way up to the end of the $(19 + q)$th century will have to be surrounded by a set $\sigma_{q+1}$ of open intervals of measure $< 2^{-n-q-1}$, then the whole of classical real numbers to be defined in eternity will be surrounded by a set of open regions of measure $< 2^{-n}$. And this natural number $n$ may be chosen as great as we like. ]

5   With regard to the definition of species we have to remark firstly that during the development of intuitionist mathematics some species have to be considered as being tacitly defined again and again in the same way, secondly that a species can be an element of another species, but never an element of itself.

The second act of intuitionism authorizes us in particular to introduce the species of more or less freely proceeding but convergent infinite sequences of rational numbers, and more generally for each $n$ the species of more or less freely proceeding but convergent infinite sequences of elements of the rational grid of a Cartesian space of $n$ dimensions. These species will prove to admit of a representation which makes them in a surprising measure better surveyable and better manageable than the classical species of exact real numbers and of exact real points of $n$-dimensional space.

To arrive at this representation we have to begin with some definitions.

By a *node of order n* we understand a sequence of $n$ natural numbers ($n \geq 1$) called the *indices* of the node.

A node $p'$ of order $n+m$ ($m \geq 1$) will be called an *m-th descendant* of the node $p$ of order $n$, and $p$ will becalled the *m-th predecessor of $p'$*, if the sequence of indices of $p$ is an initial segment of the sequence of indices of $p'$. The union (conjunction) of the node $p$ and the species of its descendants will be called *a pyramid*, of which $p$ will be called the *top*.

If $m = 1$, $p'$ will also be called an *immediate descendant* of $p$ and $p$ the *immediate predecessor* of $p'$.

The immediate descendants of a node of order $n$ in their natural order (i.e. ordered according to their last index) constitute a species $Q$ of nodes. This species will be called a *row of nodes of order $n + 1$* and the *ramifying row of $p$* whilst $p$ will be called the *dominant* of $Q$.

The species of the nodes of order 1 in their natural order will be called the *row of nodes of order 1*.

A *finite* sequence of nodes consisting of a node $p_1$ of order 1, an immediate descendant $p_2$ of $p_1$, an immediate descendant $p_3$ of $p_2$, ....up to an immediate descendant $p_n$ of $p_{n-1}$, will be called a *bar of order n*.

6  An *infinite* sequence (not necessarily predeterminate) of nodes consisting of a node $p_1$ of order 1, an immediate descendant $p_2$ of $p_1$, an immediate descendant $p_3$ of $p_2$, *etc. ad infinitum*, will be called an *arrow*.

The arrow may proceed with complete freedom, i.e. in the passage from $p_v$ to $p_{v+1}$, the choice of a new index to be joined to those of $p_v$ may be completely free for each $v$, as long as the creating subject likes. But this freedom of proceeding may also at any stage be completely abolished, at the beginning, or at any $p_v$, by means of a law fixing all further nodes in advance. From this moment the arrow concerned will be called a *sharp arrow*. Finally, the freedom of proceeding, without being completely abolished, may, at some $p_v$, undergo some restriction, and later on further restrictions. All these intervening acts, as well as the choices of the $p_v$ themselves, may (at the beginning or at any later stage) be made to depend on the influence of possible future occurrences in the world of mathematical thought of the creating subject.

Let us suppose that by some law a species of nodes is defined in such a way that as nodes of order 1 only those natural numbers are admitted which do not exceed a certain natural number $m_0$, and that as immediate descendants of an arbitrary already admitted node $p$ of order $n$ only those nodes of order $n+1$ are admitted whose $(n+1)$-th index does not exceed a certain natural number $m_p$. Then this law will also indicate a species of arrows exclusively consisting of admitted nodes. This species of arrows will be called a *fan*, and the law in question a *fan key*.

Now for fans a wonderful theorem holds whose importance would justify to call it the fundamental theorem of intuitionism, but whose absolutely rigorous proof till now has not been sufficiently simplified to allow of being reported here. So I shall only give its wording which runs as follows:

*If to each arrow $\alpha$ of a fan F a natural number $v(\alpha)$ has been assigned, a natural number s can be calculated such that $v(\alpha)$ depends only on the s-th node of $\alpha$.*

Coming to the intuitionist reconstruction of the continuum, and more general, of $n$-dimensional Cartesian space, we shall consider the case of the Cartesian plane explicitly, remarking that these considerations are extensible to spaces of any finite number of dimensions with insignificant modifications.

7      In the species of the binary fractions $a \cdot 2^{-v}$ arranged in their natural order ($a$ and $v$ any integers) we consider, for given $n$, the pairs consisting of an element $a \cdot 2^{-(n+1)}$ and an element $(a+2) \cdot 2^{-(n+1)}$, and we call these pairs $\lambda^{(n)}$-intervals, understanding by a $\kappa^{(n)}$-*interval* a $\lambda^{(n)}$-interval for which $a$ is even. All $\lambda^{(n)}$-intervals for any $n$, will be called $\lambda$-*intervals*, and all $\kappa^{(n)}$-intervals for any $n$, will be called $\kappa$-*intervals*. By a $\lambda^{(n)}$-*square* we shall understand an ordered pair of $\lambda^{(n)}$-intervals, a 'former' and a 'latter' one, and by a $\kappa^{(n)}$-*square* an ordered pair of $\kappa^{(n)}$-intervals. In the following we shall characterize the relative situation of two $\lambda$-squares $a$ and $b$ deriving from the relative position of the two former and two latter intervals, by expressions being clear by their meaning when referring to squares of ordinary Euclidean plane geometry. Such expressions will be e.g. $a$ is inside (outside) $b$; $a$ touches $b$ internally (externally). In this connection we shall say that $a$ and $b$ *overlap* if their former as well as their latter intervals have a sub-interval in common; that $a$ *lies within* $b$, if $a$ is inside $b$ and does not touch $b$; that $a$ and $b$ lie *apart*, if $a$ is outside $b$ and does not touch $b$.

An infinite sequence $k_1, k_2, \ldots$ of $\lambda$-squares (predeterminate or proceeding with more or less freedom) in which, for each $v$, $k_{v+1}$ is within $k_v$ will be called a *two-dimensional Cartesian point*.

Denoting the two-dimensional Cartesian points $\kappa'_1, \kappa'_2, \ldots$ and $\kappa''_1, \kappa''_2, \ldots$ by $p'$ and $p''$ respectively, we shall say that $p'$ and $p''$ *coincide*, if it is certain that $\kappa'_\mu$ and $\kappa''_v$ overlap for each $\mu$ and $v$. It is easily seen that coincidence is a transitive relation. The species of the two-dimensional Cartesian points which coincide with the two-dimensional Cartesian point $p$ will be called a *two-dimensional Cartesian point-core*.

If, corresponding to a two-dimensional Cartesian point core $P$, there exists a *law* $\sigma(P)$ determining, for each $n$, a $\lambda^{(n)}$-square $l_n$ such that of each point $k_1, k_2, \ldots$ of $P$ a computable tail segment $k_m, k_{m+1}, \ldots$ is within $l_n$, $P$ is said to be a *sharp* two-dimensional Cartesian point core. The species of the (sharp) two-dimensional Cartesian point cores is called the *(reduced) two-dimensional Cartesian space* or *(reduced) Cartesian plane*.

The points $p'$ and $p''$ of the Cartesian plane will be said *to deviate* if they cannot possibly coincide, and to be *separated* if a square of $p'$ and a square of $p''$ can be indicated which are outside one another.

A point $p$ *deviates*, (*is separated*) from the point core species $Q$, if it deviates (is separated) from each point of $Q$. Thus, for instance, the 'origin' deviates from the species of irrational point cores, and is separated from the species of those rational point cores which do not coincide with the origin.

8   In the following we shall understand by the *filling* of a $\lambda$-square $\mathscr{S}$ the species of those point cores of the Cartesian plane which cannot possess a square outside $\mathscr{S}$. And henceforth any $\lambda$-square will be conceived in the sense of the filling of a $\lambda$-square in the sense hitherto used. Furthermore by a $k^v$ ($v > 0$) we shall understand a $\lambda^{(4v+1)}$-square contained in or overlapping the $\lambda$-square $(0 \rightarrow 1, 0 \rightarrow 1)$ which we shall call the *unity-square* and indicate by $\mathscr{L}$, and by a *standard point of $\mathscr{L}$* any point $\kappa'_{n_1}, \kappa''_{n_1 n_2}, \kappa^{(3)}_{n_1 n_2 n_3}, \ldots$ of the Cartesian plane. Obviously the species of the $k'$ as well as for each $v$ the species of the $k^{(v+1)}$ lying within one and the same $k^{(v)}$ is finite, and its elements can be distinguished in a predetermined way by indices $n'$ and $n^{(v+1)}$ respectively.

Each standard point of $\mathscr{L}$ is characterized by an arrow $n_1, n_1 n_2, n_1 n_2 n_3, \ldots$. As the species of these arrows constitutes a fan $F(\mathscr{L})$, we shall call a standard point of $\mathscr{L}$ also a *fan point* of $\mathscr{L}$.

Obviously each $k^{(v)}$ contains a $^0k^{(v)}$ and a $^1k^{(v)}$, [ED. a note in Dutch is added in the margin: '$^0k^{(v)}$ is superfluous here'] both being squares concentric and homothetic with $k^{(v)}$ and having sides of length $\frac{3}{4}$ and $\frac{7}{8}$ respectively of the side length of $k^{(v)}$. Now for an arbitrary crude point of $\mathscr{L}$, i.e. for an convergent infinite sequence $\sigma$ of points of the rational grid of $\mathscr{L}$ we can, for any $v$, indicate a $^1k^{(v)}$, say $^1k^{(v)}(\sigma)$, within which we can indicate a tail segment of $\sigma$. And although it is not certain that, for every $v$, $^1k^{(v+1)}(\sigma)$ lies within $^1k^{(v)}(\sigma)$, it is certain that, for every $v$, $k^{(v+1)}(\sigma)$ lies within $k^{(v)}(\sigma)$, so that the infinite sequence $k'(\sigma), k''(\sigma),...$ constitutes a fan point of $\mathscr{L}$, and we have proved:

*Every crude point of $\mathscr{L}$, i.e. every (not necessarily predeterminate) convergent infinite sequence of points of the rational grid of $\mathscr{L}$ 'coincides' with a fan point of $\mathscr{L}$ so that the species of crude points of $\mathscr{L}$ may be said to be represented by the species of fan points of $\mathscr{L}$.*

Now let us try to surround all crude points of $\mathscr{L}$ (i.e. surround tail segments of all crude points of $\mathscr{L}$) by a set of regions $\rho$ of the Cartesian plane. This set of regions will have to surround all fan points of $\mathscr{L}$, so for each fan point $Q$ of $\mathscr{L}$, characterized by an arrow $n_1, n_1n_2, n_1n_2n_3,...$ will have to contain a certain $k^{(v)}$ belonging to $Q$. But then by virtue of the fan theorem, a certain natural number $m$ can be calculated such that for an arbitrary $k^{(m)}$, say $\kappa^{(m)}_{n'_1...n'_m}$, a certain $k^{(v)}$ contained in $\rho$ and belonging to *each* fan point of $\mathscr{L}$ deriving from $\kappa^{(m)}_{n'_1...n'_m}$, is determined. But then this $k^{(v)}$ must also contain $\kappa^{(m)}_{n'_1...n'_m}$ itself. So all $k^{(m)}$ have to belong to $\rho$, so $\rho$ cannot have a measure $< 1$, so the measure of the species of crude points of $\mathscr{L}$ is unity. Of course it is not astonishing at all that the species of the predeterminate points with their perfectly thin and perfectly rigid future requires a less voluminous package than the species of crude points with their somewhat bulky, partly unknown and partly uncertain future.

9 We pass to another application of the fan theorem, and consider a variable intuitionist point core or 'real number' $z$ of the *unity continuum* $\mathscr{L}_1$ (i.e. the $\kappa$-interval $0 \to 1$), and a bounded intuitionist real function $f(z)$ of $z$ likewise taking only values belonging to $\mathscr{L}_1$. Then to each fan point $x$ of $\mathscr{L}_1$ a point $f(x)$ of $\mathscr{L}_1$ corresponds in such a way that to coinciding fan points $x$, coinciding points $f(x)$ are assigned. We can replace each point $f(x)$ by a fan point by replacing for each $v$ the first of its intervals of side length $\leq 2^{-4v-3}$ by the $k^v$ surrounding it as concentrically as possible with preference of right over left in cases of ambiguity. So henceforth we may suppose each $f(x)$ to be a fan point.

Then to each natural number $p$ we can assign a natural number $p'$ such that the difference of two fan points coinciding with fan points deriving from the same bar of order $p'$, is $< 2^{-p}$. Furthermore, by virtue of the fan theorem, to each natural number $p'$ we can assign a natural number $p''$ such that $\kappa^{(p')}$ of $f(x)$ is fully determined by $\kappa^{(p'')}$ of $x$, so to all fan points $x$ deriving from the same node of order $p''$ the same $k(p')$ of $\kappa^{(p')}$ corresponds, so to all fan points coinciding with fan points deriving from the same node of order $p''$ fan points coinciding with fan points deriving from *one and the same* $\kappa^{(p')}$ correspond. Again, to each natural number $p''$ we can assign a natural number $p'''$ such that of any two fan points $x$ with a difference $< 2^{-p'''}$ the $\kappa^{(p''+1)}$ overlap, so that both are lying within one and the same $\kappa^{(p'')}$, so coincide with two fan points $x$ deriving from the same node of order $p''$.

So finally to each natural number $p$ we can assign a natural number $p'''$ such that to each two fan points $x$ with a difference $< 2^{-p'''}$ two fan points of $f(x)$ correspond whose difference is $< 2^{-p}$.

## Appendix 11

### Real Functions [BMS 37]

〚English translation of 'The Table of Contents' and extracts (pp. 1-7; pp. 108-111; pp.112-114) Ed.〛


PART ONE: FOUNDATIONS OF THE THEORY OF POINT-SETS

CHAPTER 1: POINT-SPECIES AND POINT-SETS [Germ. Punktmengen]
1. Kappa-squares and Lambda-squares
2. The notion of point
3. Point-species and point-sets
4. Uniform point-species
5. Similar [German:gleichmässige] point-species
6. Union and intersection
7. Further notions

CHAPTER 2: BOUNDARY POINTS, CATALOGIZING
1. Limit-points and boundary points
2. Closure, derivation
3. Catalogizing
4. Every catalogized point-species is limitable
5. A theorem on fundamental sequences of spreads of squares
6. Domains and complements of domains
7. Inner and outer boundary species

CHAPTER 3: THE GENETIC CONCEPT OF CONTENT
1. Limited (German: limitierte) sequences, Measurability of domain and the complement of domain
2. Theorems on the content of the domain-complement
3. Uniqueness of the content of the domain-complement
4. 'Joining' (Germ. 'vereinigende') domain-complements
5. Measurability of outer boundary species
6. Uniqueness of the content of outer boundary species

7. 'Joining' outer boundary species of outer boundary species
8. Part-sets having the same content
9. Some remarks about the uniqueness of content

CHAPTER 4: THE GENERAL NOTION OF CONTENT
1. General measurability
2. Remarks about the general definition of content
3. Relation between genetic and general measurability
4. Measuring outer boundary species
5. Relations between general and genetic measurability
6. The content of measurable part-species of measurable point-species
7. The measurability of the union of a finite number of measurable point-species
8. The measurability of the intersection of a finite number of measurable point-species
9. The measurability of the union of a fundamental sequence of measurable point-species
10. The measurability of the intersection of a fundamental sequence of measurable point-species


PART TWO: FUNDAMENTAL NOTIONS OF REAL FUNCTIONS OF ONE VARIABLE

CHAPTER 1: CONTINUITY, EXTREMES
1. The concept of function
2. Continuity and uniform continuity
3. The continuous function whose domain of definition is a set (Germ.'mit Definitionsmengen')
4. Digression: natural numbers as set-functions [incl. Bar- and Fan-theorem]
5. Species of definition which coincide with sets
6. Full functions [incl. Uniform-continuity theorem]
7. 'Special' rectangle chains
8. Extremes of full functions

CHAPTER 2:
..................

# CHAPTER I
# POINT-SPECIES AND POINT-SPREADS

## §1. $\kappa$ and $\lambda$-intervals

We envisage in a plane a right-angled coordinate cross $Oxy$ and split the plane into squares $\kappa_1$ whose sides are of length 1 and whose corner-points have integral coordinates. We split each of these squares $\kappa_1$ into four congruent homothetical sub-squares $\kappa_2$ of side-length $\frac{1}{2} = 2^{1-2}$ and proceeding in a similar way, we define squares $\kappa_3, \kappa_4, \ldots$.

A square $\kappa$ or a $\kappa$-*square* then is a square $\kappa_\nu$ of arbitrary index $\nu$. The side-length of such a $\kappa_\nu$-square then is $1/2^{\nu-1} = 2^{1-\nu}$.

We widen this definition by admitting further zero and the negative integers as index $\nu$ of the square $\kappa_\nu$. A $\kappa^0$-square e.g. then has side-length 2, a $\kappa^{-1}$-square has side-length 4 etc.

According to these definitions, the corner-points of a $\kappa_\nu$-square are then given by the the binary fractions:

$$\left\{\begin{array}{l} x_1 = \dfrac{a}{2^{\nu-1}} \\ y_1 = \dfrac{b}{2^{\nu-1}} \end{array}\right. \quad \left\{\begin{array}{l} x_2 = \dfrac{a+1}{2^{\nu-1}} \\ y_2 = \dfrac{b}{2^{\nu-1}} \end{array}\right. \quad \left\{\begin{array}{l} x_3 = \dfrac{a+1}{2^{\nu-1}} \\ y_3 = \dfrac{b+1}{2^{\nu-1}} \end{array}\right. \quad \left\{\begin{array}{l} x_4 = \dfrac{a}{2^{\nu-1}} \\ y_4 = \dfrac{b+1}{2^{\nu-1}} \end{array}\right.$$

where $a$ and $b$ are integers.

Beside these $\kappa$-squares we also consider so-called $\lambda$-*squares*. They are defined as follows: every four $\kappa_{\nu+1}$-squares with one common corner-point form a $\lambda_\nu$-square. It follows that a $\lambda_\nu$-square and a $\kappa_\nu$-square (which are of the same size) either:

1. lie outside one another,
2. cover a quarter part of each other (fig.1),
3. cover half of each other (fig.2 and fig.3), or
4. cover each other completely.

The coordinates of the 4 corner-points of a $\lambda$-square then are:

$$\left\{\begin{array}{l} x_1 = \dfrac{a}{2^\nu} \\ y_1 = \dfrac{b}{2^\nu} \end{array}\right. \quad \left\{\begin{array}{l} x_2 = \dfrac{a+2}{2^\nu} \\ y_2 = \dfrac{b}{2^\nu} \end{array}\right. \quad \left\{\begin{array}{l} x_3 = \dfrac{a+2}{2^\nu} \\ y_3 = \dfrac{b+2}{2^\nu} \end{array}\right. \quad \left\{\begin{array}{l} x_4 = \dfrac{a}{2^\nu} \\ y_4 = \dfrac{b+2}{2^\nu} \end{array}\right.$$

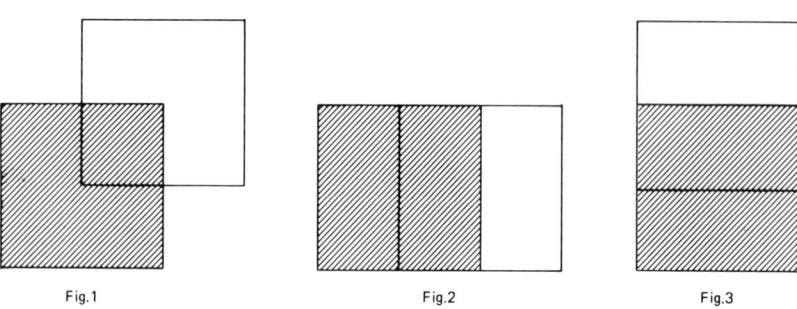

Fig.1        Fig.2        Fig.3

If two squares lie outside one-another and do not have any common side-points, we say that they *lie strictly outside* one-another. If one square lies inside another and the two squares do not have a common border-point we say that the one *lies strictly inside* the other.

The *distance* between two squares which lie strictly outside one another is defined to be the minimum distance between two points, one a border-point of one square and the other a border-point of the other square. If two squares cover each other partly or if one square lies inside the other, we say that their *distance is zero*.

It is easily ascertained that between squares the following relations apply:

1. If a $\kappa_m$-square lies strictly outside a $\kappa_n$-square and $m > n$, then their distance is greater than or equal to the side of the $\kappa_m$- square, i.e. $\geq 2^{1-m}$.
2. If a $\lambda_m$-square lies strictly outside a $\kappa_n$-square and $m \geq n$, then their distance is greater than or equal to half the side-length of the $\lambda_m$-square, i.e. $\geq 2^{-m}$.
3. If a $\kappa_m$-square lies strictly inside a $\kappa_n$-square, then $m \geq n+2$ and the shortest distance between their borders is $\geq 2^{1-m}$.
4. If a $\lambda_m$-square lies strictly inside a $\kappa_n$-square, then $m \geq n+1$ and the shortest distance between their borders is $\geq 2^{-m}$. The same applies if a $\lambda_m$-square lies strictly inside an $\lambda_n$-square.

3   What has been done so far in the $xy$-plane, can also be done on the $x$-axis and in $n$-dimensional space. We then generate $n$-dimensional $\kappa$- and $\lambda$-intervals ($n = 1,2,3...$). Two-dimensional intervals are called squares. In the following sections we shall for the sake of clarity again take $n = 2$.

## §2. *The concept 'point'*

We first take $n = 1$ and consider an indefinitely proceedable sequence *) of *'nested'* (Germ. 'ineinander geschachtelten') $\lambda$-intervals $\lambda_{v_1}, \lambda_{v_2}, \lambda_{v_3},...$ which have the property that every $\lambda_{v_{i+1}}$ *lies strictly inside* its predecessor $\lambda_{v_i}$ ($i = 1, 2, 3,...$).

According to §1 the length of the interval $\lambda_{v_{i+1}}$ at most equals half the length of $\lambda_{v_i}$, and therefore the lengths of the intervals converge to zero, i.e. given a positive integer $N$, however large, one can always indicate a minimum index $v_h$ such that the length of the interval $\lambda_{v_h}$ is smaller than $1/N$.

*We call such an indefinitely proceedable sequence of nested $\lambda$-intervals a point $P$ or a real number $P$.*

We must stress that for us the sequence the point $P$ *is* the sequence (1) $\lambda_{v_1}, \lambda_{v_2}, \lambda_{v_3},...$ *itself*; not something like 'the limiting point to which according to classical opinion the $\lambda$-intervals converge, or something to be defined as the unique accumulation point of midpoints of these intervals'.

Every one of these $\lambda$-intervals (1) is therefore *part of* the point $P$.

We must further point out that every approximating procedure can be reduced to such a nesting of $\lambda$-intervals.

---

Footnote *): An indefinitely proceedable sequence is in general not a fundamental sequence (cf.§3), since in the process of its generation free choice of elements is not excluded.

4  Let us now consider the plane. The planar point $P$ ($=$ ordered number-pair) is a sequence (1) of nested $\lambda$-squares. Each $\lambda_{v_i}$-square of (1) *is part of* the point $P$.

We further define the following:

*Two points $P' = \lambda_{\mu_1}, \lambda_{\mu_2}, ...$ and $P'' = \lambda''_{v+1}, \lambda''_{v+2},...$ coincide*, if every $\lambda'$-square strictly holds an $\lambda''$-square, and vice versa. (If in every $\lambda$-square of $P'$ there lies a square of $P''$, it follows that also in every $\lambda$-square of $P''$ there lies a square of $P'$.)

*Two points $P'$ and $P''$ are locally distinct* [Ed. lie apart] if two squares $\lambda'_{\mu_v}$ and $\lambda''_{v_\kappa}$ can be indicated which lie strictly outside one-another.

We warn that this 'coincidence' of points $P'$ and $P''$ is not what is usually understood by 'equality' of points $P'$ and $P''$. For us two points are only then equal or identical, if the square $\lambda'_i$ is identical with the square $\lambda''_i$ for every $i$, if therefore $P'$ and $P''$ are given by the same sequence of squares.

If $P'$ coincides with both $P''$ and $P'''$ then $P''$ and $P'''$ also coincide.

If $P'$ and $P''$ are locally distinct and $P''$ coincides with $P'''$, then $P'$ and $P'''$ are also locally distinct.

The two relations $(\alpha)$ = coincidence of $P'$ and $P''$ and $(\beta)$ = local distinctness of $P'$ and $P''$ are mutually exclusive, i.e. if the relation $(\alpha)$ applies then the relation $(\beta)$ cannot apply. But, if the relation $(\alpha)$ does not apply, it does *not* follow that the relation $(\beta)$ must apply: *in this case too a 'third' is not excluded.* On the other hand, if $(\beta)$ does not apply then $(\alpha)$ does, i.e. if $P'$ and $P''$ are *not* locally distinct, they must coincide.

The notion 'local distinctness' is equivalent with the classical notion of 'distinctness' of two real numbers, defined as the possibility of indicating a finite quantity $1/N$, smaller than the difference between the two numbers.

5   Finally, we particularly wish to point out that we do not consider the traditional concept of the continuum to be consistent. For us the totality of real numbers is not an ordered set; and the apparent intuitiveness of the continuous infill of the number-axes by 'real numbers' is in our system offset by the *range of choice* available when selecting $\lambda$-intervals of a *becoming* sequence of intervals.

## §3. *Point-Species and Point-Sets*

By a *point-species* $Q$ we understand a *property* which only a point can possess. If a point $A$ has this property we say that $A$ *belongs to* the point-species $Q$, or that $A$ *is a point of* $Q$, or that $A$ is an *element of* $Q$.

For example, if $Q$ is the property of coincidence with a given point $P$ (cf.§ 2), then $P$ itself is an element of $Q$. (We shall later also consider so-called 'species of second order', i.e. properties of species or species of species.)

A particular species is the *point-set*. By a *point-set* we understand a *law* by means of which with the numbers $n_1, n_2, n_3,...$ freely and in sequence selected from the sequence of natural numbers 1, 2, 3, ... either:

1) a $\lambda$-square is associated in such a way that of two subsequent $\lambda$-squares the second lies strictly within the first, or

2) *nothing* is associated with the chosen first number $n_1$, or, if $\lambda$-squares have been associated with $n_1, n_2, ..., n_{h-1}$ ($h \geq 2$), nothing is associated with the choice of the $h$-th number $n_h$ and at the same time all intervals generated so far are destroyed and the process is terminated. In the last

case, however, there must be at least one $n'_h$ different from $n_h$, with which, when chosen, the law does associate a $\lambda$-square.

6  This long-winded definition, regrettably, cannot be replaced by a shorter one. May-be the following will help and clarify the matter.

A point of a point-set comes into being as follows:

We first choose freely a number $n_1$ from 1, 2, 3,.... With this $n_1$ the law or set associates either a $\lambda$-square, $\lambda_\mu$, or nothing. In the first case we then again choose freely from 1, 2, 3, ... a number, $n_2$, (which can also be $n_1$). With this $n_2$ the set either 1) associates a $\lambda$-square, $\lambda_\nu$, which lies strictly inside $\lambda_\mu$ or 2) it results in the elimination of $\lambda_\mu$, the square already generated. In the latter case, if the second choice was just this $n_2$, no element (i.e. point) comes into being. But there is certainly at least one other $n'_2 \neq n_2$ with which the law does associate a $\lambda$-square, so that the process can be continued. We then again choose freely an $n_3$ from 1, 2, 3, ... etc.

If the law is such that we can prove that with every $n_1$ *nothing* is associated, then *no* point belongs to the set and the set is called *empty*.

Perhaps, the following diagram clarifies the definition even more:

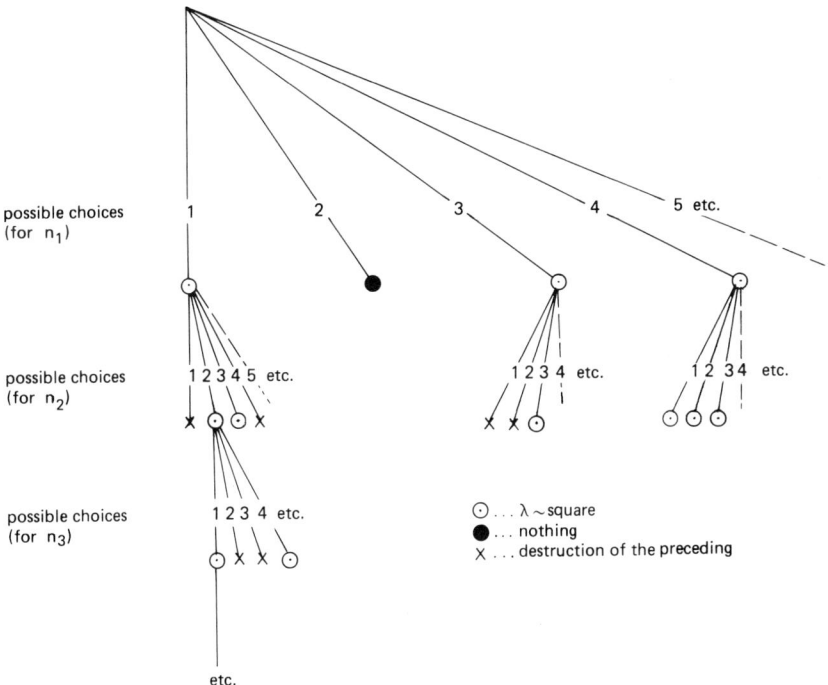

A point-set $M$ is always also a point-species, but a point-species is not necessarily a point-set. The common origin of interval sequences generated by the law $M$ is also referred-to in short as $M$.

7   The stated possibility that nothing be associated with first choice $n_1$ is recognized to be necessary to allow *empty sets*. As this gives rise to the possibility of empty 'everywhere in the set', it is necessary to have the process of 'destruction of the preceding' ($\times$ in our diagram) at one's disposal.

For us point, and therefore point of a set, is always something which is growing, and often, something for ever indeterminate; this in contrast with the classical view which considers every point to be something determinate and completed.

A *Fundamental Sequence* of points is a point-set which assigns a $\lambda$-square to every number $n_1$ selected as the first choice, and where at every subsequent choice $v$ all selected numbers $n_v$ except one lead to the 'destruction of the preceding' (as shown in the following diagram).

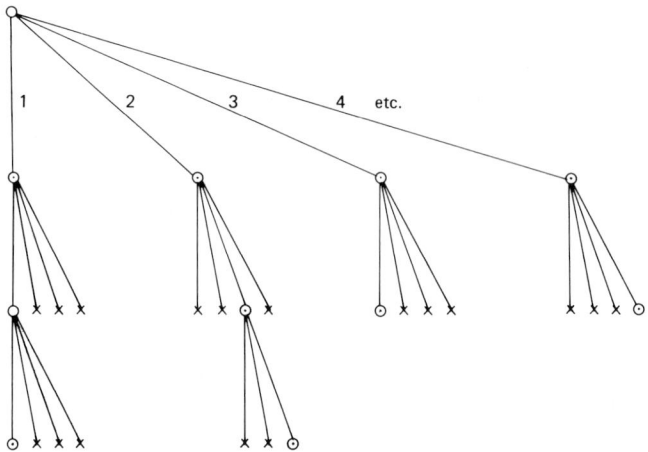

A Fundamental sequence of nested $\lambda$-intervals is a point but not every point a fundamental squence.

If, at every stage of the construction of a choice sequence only a finite number of options from the numbers 1, 2, 3, ... allow the continuation of the generating process, we speak of a *finite* point-set. In this case, for every $n_\alpha$ there exists a maximum $m_\alpha$, such that the choice of $n_\alpha > m_\alpha$ leads to the termination [Germ. 'Hemmung'] of the generating process.

..........................

## PART TWO, Chapter 1.

[[BMS37], pp.108-112, *'The Bar Theorem and The Fundamental Theorem of Finite Spreads or Fan Theorem* Ed.]

107 §4. *Digression: natural numbers as set-functions.*

Let $M$ be an arbitrary set and $\mu$ be the *countable* set of finite (arrested as well as non-arrested) choice-sequences $\mathscr{F}_{sn_1\ldots n_r}$ on which $M$ is based (where $s$ and the $n_i$ represent natural numbers ($s$ chosen at the first choice, $n_1$ at the second choice, etc.)

108  Let to each to each element of $M$ be assigned a natural number $\beta = \beta(M)$. Then there is in $\mu$ a *separable* *] countable sub-set $\mu_1$ of non-arrested finite choice-sequences such that for any one of them the same natural number $\beta$ is assigned to that (finite) choice-sequence and to all elements of $M$ issuing from it. We can therefore speak of an element of $\mu_1$ if in accordance with the algorithm of the assignment-law $\beta(M)$ the finite number $\beta$ to be assigned is completely determined and its determination is not postponed to later choices, while no finite section (initial segment) of the element has the property in question. One should by no means exclude the possibility that one might subsequently be able to indicate elements of $\mu$ which do not belong to $\mu_1$ but which have the property that one and the same natural number $\beta$ is assigned to all elements of $M$ issuing from it.

For complete determination of $\beta(M)$ one should further have a proof $h$ which shows for any non-arrested element of $\mu$ that every non-arrested infinite choice sequence which issues from it has an initial segment which is an element of $\mu_1$.

We call an element of $\mu$ '*secured*' if it is either arrested or it possesses an initial segment belonging to $\mu_1$. Then $\mu$ is split *] into a countable set $\tau$ of secured finite sequences and a countable set $\sigma$ of unsecured finite sequences. The proof $h$ should show for any arbitrary element of $\sigma$ that it is '*securable*', i.e. that each infinite non-arrested choice sequence which issues from it, possesses an initial segment which is an element of $\mu_1$.

Let $h_{sn_1\ldots n_r}$ be the 'specialized' proofs of $h$, which derive the securability of the elements $\mathscr{F}_{sn_1\ldots n_r}$ of $\sigma$. Then each of the proofs $h_{sn_1\ldots n_r}$ is based exclusively on relations which exist between the elements of $\mu$; we shall call these relations '*b*-relations'.

But these $b$ relations can all be reduced to 'elementary relations $e$', i.e. relations which exist between any two elements $\mathscr{F}_{mm_1\ldots m_g}$ and $\mathscr{F}_{mm_1\ldots m_g m_{g+1}}$, where $\mathscr{F}_{mm_1\ldots m_g m_{g+1}}$ is an immediate extension of $\mathscr{F}_{mm_1\ldots m_g}$.

109   Since any mathematical proof in which the relations used can be reduced to elementary relations, can always (although at the cost of shortness) be 'canonized' in such a way that in the canonical form only elementary relations are used, it is clear that by means of the canonical form $\kappa_{sn_1...n_r}$ of the proof $h_{sn_1...n_r}$ the securability of $\mathscr{F}_{sn_1...n_r}$ can be derived ultimately and exclusively from the elementary relations which connect $\mathscr{F}_{sn_1...n_r}$ with $\mathscr{F}_{sn_1...n_{v-1}}$ and with the $\mathscr{F}_{sn_1...n_r v}$.

For the final inference of $\kappa_{sn_1...n_r}$ one therefore uses the established securability of $\mathscr{F}_{sn_1...n_{v-1}}$ or of all $\mathscr{F}_{sn_1...n_r v}$. We refer to the elementary inference which derives the securability of $\mathscr{F}_{mm_1...m_g}$ from that of $\mathscr{F}_{mm_1...m_{g-1}}$ as a '$\zeta$-inference', and to the elementary inference which derives the securability of $\mathscr{F}_{mm_1...m_g}$ from the securability of all $\mathscr{F}_{sn_1...n_r v}$ as a '$F$-inference'.

Further, let $f_{sn_1...n_r}$ be the well-ordered species of elements of $\sigma$ whose securability is proved in order by $\kappa_{sn_1...n_r}$. Obviously, the securability of the *first* element cannot be proved by means of a $\zeta$-inference, it must therefore be proved by a $F$-inference.

We also find on the basis of transfinite induction along $f_{sn_1...n_r}$, that *at every stage of the proof* $\kappa_{sn_1...n_r}$ the extensions of all elements of $\sigma$ which have been found to be securable are also securable. Therefore, the securability of *each* element of $f_{sn_1...n_r}$ by proofs $\kappa_{sn_1...n_r}$ is derived by means of a $F$-inference.

110   If (in accordance with union-relations between part-sets of $M$) every time when by means of $\kappa_{sn_1...n_r}$ we have proved the securability of an element $\mathscr{F}_{mm_1...m_g}$ of $\sigma$, we consider this element $\mathscr{F}_{mm_1...m_g}$ as the ordered sum of the $\mathscr{F}_{mm_1...m_g v}$ (ordered by the index $v$) and consider every arrested $\mathscr{F}_{mm_1...m_g v}$ as an *empty elementary species*, it is clear on the basis of transfinite induction along $f_{sn_1...n_r}$ that $\mathscr{F}_{sn_1...n_r}$ is generated by this transfinite sequence of sum-formations as a well-ordered species $\varphi_{sn_1...n_r}$. The elementary species in the $\varphi_{sn_1...n_r}$ correspond to the 'stop-elements' of $\tau$ (Germ. 'einkehrende Elemente'), i.e. elements which are secured but which do not have secured proper segments); 'the constructive underspecies' of $\varphi_{sn_1...n_r}$, correspond to the elements of $\sigma$.

Conversely, an element of $\sigma$ or a stop-element of $\tau$ only then corresponds to a constructive underspecies, resp. to a primitive species of $\varphi_{sn_1...n_r}$, if it has an index-sequence $sn_1 ...n_r p_1 ...p_\mu$; in this case, indeed, the constructive underspecies resp. primitive species have such an index sequence: $p_1, ...p_n$.

The preceding can be summed up in the following theorem:

*If to each element of a set M a natural number β is assigned, then M is split by this assignment into a well-ordered species S of sub-sets $M_α$, each determined by a finite number of choices, and to each element of one $M_α$ the same natural number $β_α$ is assigned.*

*The species S can be constructed by means of generating operations ω of the second kind, of which each corresponds to the continuation — by a free new choice — of a certain finite initial segment of choices which is non-arrested for M. To a new choice which is arrested for M there corresponds for the operation ω an empty elementary species.*

111   If $M$ is a *finite set*, the well-ordered species $φ_{sn_1...n_r}$ is 'similar with' (Germ. ähnlich) a well-ordered species $ψ_{sn_1...n_r}$ which can be constructed without using empty elementary species and yet in parallel with the above construction of $φ_{sn_1...n_r}$ in such a way that to every ω operation used in the construction of $φ_{sn_1...n_r}$ there corresponds in the construction of $ψ_{sn_1...n_r}$ a finite number of generating operation χ of the first kind.

From this it follows that $ψ_{sn_1...n_r}$ as well as $φ_{sn_1...n_r}$ are *finite* (endlich), and in particular that for every natural number $s$ the species $φ_s$ is finite. We can therefore indicate a natural number $z$ such that any arbitrary 'stop-element' of $μ$ has at most $z$ indices.

Hence, the number $β_e$ assigned to an arbitrary alement $e$ of $M$ is completely determined by the first $z$ choices generating $e$.

We have therefore proved that:

*If to each element e of a finite set M a natural number $β_e$ is assigned then a natural number z can be indicated such that $β_e$ is completely determined by the first z choices generating e.*

---

*] If two sets $μ_1$ and $μ_2$ are disjoint and their union $\mathscr{V}(μ_1, μ_2) = μ$, $μ_1$ and $μ_2$ are called *separable* sub-sets of $μ$, and $μ$ is said to be *split* into $μ_1$ and $μ_2$.

........................

[[BMS 37], pp. 112-114, *The Uniform Continuity Theorem*]

Let the [finite] point-set $S$ be the domain-species (Germ. Definitionspecies) of a function $f(x)$. Then $f$ assigns to every nest $κ^{(μ_1)}, κ_2^{(μ_2)},...$ of $κ_ζ$ intervals of a point $K$ of $S$ uniquely a nest $λ', λ'', ...$ of $λ$-intervals on the $y$-axis.

113   Since $S$ is finite, there exists for every $ν$ an index $m_ν$ (which does not

decrease as $v$ increases) such that $\lambda^{(v)}$ depends exclusively on $\kappa_1^{(\mu_1)},\ldots,\kappa_{\mu_v}^{(\mu_{m_v})}$. Therefore, only a finite number of intervals qualify as $\lambda^{(v)}$-intervals, the longest of which is of length $b_v$. Then $\lim\limits_{v \to \infty} b_v = 0$ [1], and $\lambda^{(v+1)}$ has at most the length $\frac{1}{2}b_v$.

Let $P_1$ and $P_2$ be two points of the closure $R$ and their distance $\overline{P_1P_2} < \zeta_v$ [2], where $\zeta_v = \frac{1}{4}$ the length of a $\kappa_v$ interval. $R$ coincides with the finite point-set $S$. It will be shown that if $H_1$ and $H_2$ are the two points of $S$ which coincide resp. with $P_1$ and $P_2$ that because of [2] *the first $v$ $\kappa$-intervals of $H_1$ and $H_2$ are the same.*

Let $h_v^{(\zeta)}$ be the interval corresponding to and concentric with the interval $\kappa_v^{(\zeta)}$ and be $\frac{3}{4}$ of its length. Then an $h_v^{(\mu_v)}$ can be indicated which contains both $P_1$ and $P_2$, and therefore there exist nests of intervals $\kappa_1^{(\mu_1)}$, $\ldots, \kappa_v^{(\mu_v)}, \kappa_{v+1}^{(\sigma_1)}, \kappa_{v+2}^{(\sigma_2)}, \ldots$ and $\kappa_1^{(\mu_1)},\ldots, \kappa_v^{(\mu_v)}, \kappa_{v+1}^{(\tau_1)}, \kappa_{v+2}^{(\tau_2)},\ldots$ coinciding respectively with $P_1$ and $P_2$.

Given $\varepsilon > 0$, we chose $v_1$ so large that $b_{v_1} < \varepsilon$ (which is possible because of [1]) and we put the distance $\zeta_{m_{v_1}} = \alpha_\varepsilon$. To the points $H_1$ and $H_2$ of the finite set $S$, which have the same initial segment of intervals $\kappa_1^{(\mu_1)}, \kappa_2^{(\mu_2)},\ldots, \kappa_{m_{v_1}}^{(\mu_{m_{v_1}})}$ then are assigned the function values $f(\xi_1)$ and $f(\xi_2)$ whose absolute distance is $< b_{v_1}$ and therefore $< \varepsilon$.

Then as a result of what was shown above: to two points $P_1$ and $P_2$ of $R_1$, such that $\underline{P_1P_2} < \zeta_{m_{v_1}} = \alpha_\varepsilon$, there correspond two function values $f(\xi_1)$ and $f(\xi_2)$ such that $|f(\xi_1) - f(\xi_2)| < \varepsilon$, i.e. $f(x)$ is *uniformly continuous*.

## Appendix 12

## Intuitionist Mathematics or Die Berliner Gastvorlesungen [BMS32]

[The last part of Chapter One, p. 7 of the manuscript of the Dutch text ([BMS32A]), in which two 'versions' can be distinghuished:

1. the original, typed text;
2. the final, amended version. (see illustration).]

*The Second Act of Intuitionism*

*[Original version]*

It is clear from this that the rejection of the principle of the excluded middle, which originated in the intuitionist philosophy, deprived an important part of classical mathematics − in particular its mathematics of the continuum − of any claim to exactness. At this point one could well raise the question whether perhaps one should not reduce the significance of analysis to that of an empirically efficient means of calculation for the physical sciences without any deeper content in itself.

*A way out of this situation* is opened by *The Second Act of Intuitionism: recognizing in the self-unfolding of the Primordial Intuition not only the construction of the sequence of natural numbers but also the spread construction and the foundation of analysis on the properties of the spread.*
In spite of its simple conceptual content, the description of the spread construction has to be somewhat long-winded. It runs as follows:

First of all we determine an indefinitely proceeding sequence of symbols by a first symbol and a law which derives from each of these symbols the next one. We could e.g. choose the sequence $\zeta$ formed by the successive 'numerals' 1, 2, 3, ....

We then understand by a spread a law according to which, every time when an arbitrary numeral is chosen, each of these choices either produces a definite sign with or without terminating the process, or causes the arresting [Du. 'stuiting'] of the process and the destructionof what the process had yielded so far; moreover, for every $n > 1$, after every sequence of $n-1$ choices which is not terminated and not arrested at least one numeral can be indicated which, when chosen as the $n$-th index, will not

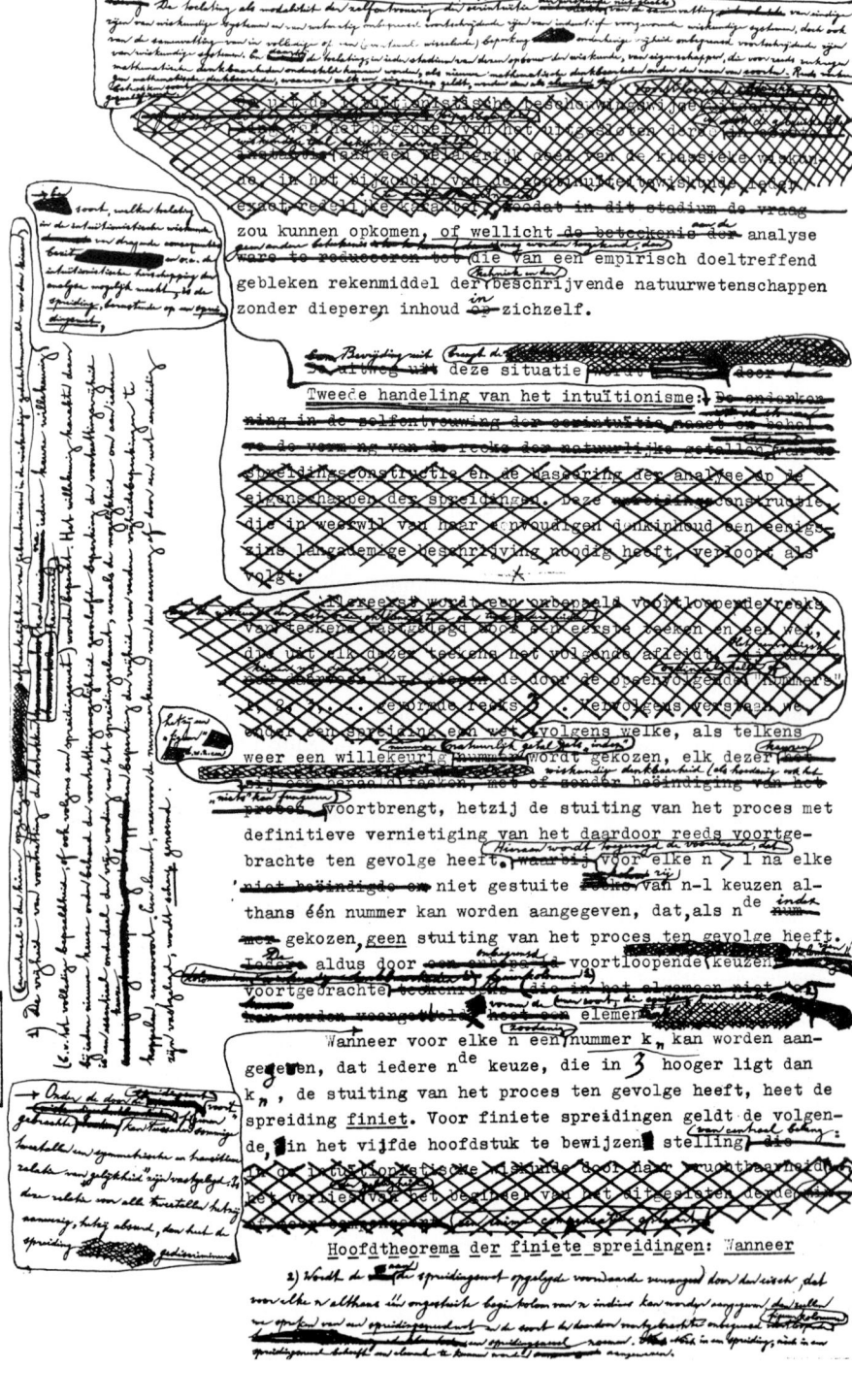

result in the process being arrested. Every sequence of symbols which in this way is generated by an indefinitely proceeding sequence of choices and which therefore in general cannot be represented as 'completed' is called an *element of the spread*.

If for every $n$ a numeral $k_n$ can be found such that every $n$-th choice which in $\zeta$ lies higher than $k_n$ results in the process being arrested, the spread is called *finite*. For finite spreads a theorem holds which is so fruitful that it more or less makes up for the loss of the principle of the excluded middle. We state it here and will prove it in Chapter Five:

*Main Theorem of Finite Spreads:*
If to every element $e$ of a finite spread $S$ a natural number $\beta_e$ is assigned, then a natural number $m$ can be indicated such that for every $e$, $\beta_e$ is completely determined by the first $m$ choices generating $e$.

....................

*[Final version]*
As long as the self-unfolding of the primordial intuition restricts itself — as is done in well-given definitions and results of classical mathematics — to configurations which are 'finished' in their entirety and in their component parts (in particular, uses only 'completely-bound' complete induction), one may consider an intuitionist re-creation of classical mathematics, but only of a large part of its number-theory and algebra while sacrificing — because of the restricted applicability of the principle of the excluded middle — many theorems which in classical mathematics were considered to be secure.

At this point one might well ask if any significance can be attributed to analysis, except perhaps as an empirically proven efficient means of calculation for mechanics and the natural sciences, without any deeper content in itself.

*Liberation from this situation* was brought about by the *Second Act of Intuitionism*: the admission as a modality of the self-unfolding of the primordial intuition of mathematics not only of the assemblage [Du. 'samenvatting'] of finite sequences of mathematical systems and of lawlike indefinitely proceeding sequences of mathematical systems pre-formed by induction, but also of the assemblage of sequences of mathematical systems proceeding indefinitely in complete freedom or in freedom subject to (possibly changing) restrictions; moreover, the admission at each stage of this construction of mathematics of properties which can be presupposed for mathematical entities [[lit. thought-constructions, Du. 'denkbaarheden'⟧ already conceived, as new mathematical entities under the name *species*. Mathematical entities which had already been conceived and to which such a property applies are called *elements* of this species.

A species whose admission in Intuitionist Mathematics has far-reaching consequences and which makes a re-creation of analysis possible is the *spread*. The spread is based on a *spread-law*, according to which every time when an arbitrary (natural) number is chosen as 'index' each of these choices either produces a 'figure', i.e. a mathematical entity (as such also 'nothing' can serve), or results in the the process being arrested together with the definite destruction of what the process had produced so far. To this must be added the condition that, for every $n > 1$, after every sequence of $n-1$ choices at least one number can be indicated which, when chosen as the $n$-th index, will *not* result in the the process being arrested. The columns of 'figures' which in this way are generated by indefinitely [Du.'onbegrensd', i.e. 'unlimitedly'] proceeding choice-sequences[1]) are the elements of a species which is called *spread*.[2])

Among the figures produced by the spread-law a symmetrical and transitive relation of 'equivalence' can have been established between some pairs. If for all pairs this relation is either present or absurd, then the spread is called *discriminated*.

If for every $n$ a number $k_n$ can be indicated such that every $n$-th choice which in $\zeta$ lies higher than $k_n$ causes the process to be arrested, then the spread is called *finite*.

For finite spread the following theorem holds, which is of central importance a proof will be given in Chapter Five:

---

1) This freedom to proceed can after every choice be arbitrarily restricted (possibly in a dependence, imposed by the choosing mathematician, on the events in his mathematical thought-world), which may lead e.g. to complete determination or to restriction by a spread-law. The arbitrary nature of this restriction at every new choice of the freedom to proceed — which is allowed provided the possibility to proceed is retained — is an essential element of the free-becoming of the element of the spread, as is the possibility to link to every choice a restriction of the freedom to make further restrictions of freedom etc. An element of which the choices of numbers have been determined uniquely in advance by a law, is said to be *sharp*.

2) If the added condition is replaced by the requirement that for every $n$ at least one initial segment of $n$ indices can be indicated which is not arrested, we shall speak of a *spread-haze law*, and call the species of sequences of figures so generated a *spread-haze*.
It is not necessary for a spread, nor for a spread-haze, that an element can be indicated.

*Main Theorem of Finite Spreads:*

If to every element e of a finite spread $S$ a natural number $\beta_e$ is assigned, then a natural number $m$ can be indicated such that for every $e$, $\beta_e$ is completely determined by the first $m$ choices generating $e$.

Professor D.J. Korteweg, 1848 – 1941

## Appendix 13

### Brouwer-Korteweg Correspondence

...ouwer and Korteweg some 146 letters... are part of the Korteweg Nachlass in ...msterdam, and some are held in the ...cted here concern Brouwer's disserta-

...dates are due to incorrect dates having ...d to the re-discoveries of letters having

7th September 1906

...me time; it is easier here to devote all my ...reading the works of others and am now ...ting a division in chapters.

...ctions, since I notice that I can still back ...ven now after all the reading I have done. ...now support them better mathematically. ...er, and to put pressure on myself, I have ...g early October. I shall come and see you ...hether you wish to see the text before it is ...e to make as many changes as I wish; of ...stricted at the proof-reading stage.

...r and I expect to see you soon.

[DJK15] 16th

Professor,

I have now divided the material I gathered for my disse[rtation into] chapters:
1. The construction of Mathematics.
2. Its genesis related to experience.
3. Its philosophical significance.
4. Its foundation on axioms.
5. Its value for society.
6. Its value for the individual.

The survey as given in chapter 1, which I sent you, serves ma[inly] and a source of reference for the following chapters; also, it c[ontains?] points at one aspect of the various recent investigations int[o founda-]tions of mathematics: their significance for constructive mat[hematics.] A few matters have been discussed in some detail, such as H[ilbert's inves-]tigation of the straight line as a minimal curve, because I [need it in] chapter 3 to refute Russell; also, the construction of the gr[oup of] operations on the continuum, because I wanted to show the [theory] of groups independent of differentiability to be essential in t[he construc-]tion of mathematics. Further, the derivation of the non-eu[clidean] element by variation, because I have not seen it done anywh[ere, and] it seems to me the only way: producing the arc-element for $n$ [dimensions] directly from what has been derived for 2 dimensions. (The u[sual proof] is based on the formulae for geodetic curves given in the inve[stigations of] Christoffel and Lipschitz in Crelle 1870 and later, and these b[ecome very] complicated in $n$ dimensions.)

I hope to be in Amsterdam to-morrow morning or the day a[fter to-mor]row, and to hear what you think.
I can then have another look at it, and perhaps the chapter c[an be] set by the end of the week.
In the meantime I am working on the next chapter and wi[ll send to] you as soon as possible.
Respectfully yours,
L.E.J. Brouwer

[DJK20]
⟦After a fateful meeting at Korteweg's house on Sunday 4th November Brouwer mounts his defense. He dispenses even with his usual, curt address 'Professor'.⟧

5th November 1906

Herewith I enclose the issue of the Göttingen Nachrichten in which Hilbert's Paris lecture 'Mathematische Probleme' was published. You will see that in Chapter One of my dissertation I have dealt in full with the First Problem ('Cantor's problem of the power of the continuum'); and this precisely by going back to the intuitive construction which is essential for all mathematics.

Problem 2 ('The non-contradictority of the axioms of arithmetic'), will be discussed in my last chapter to the extent that Hilbert's own solution given at the Heidelberger Congress is rejected; as the only solution I shall again refer back to the construction of arithmetic on the continuum, as given in my chapter one, by characterizing addition and multiplication as *the* two-parameter group.

In this way I have also solved problem 5 ('Lie's notion of continuous transformation groups independent of differentiability') for a simple case (the two-parameter linear group). Hilbert's own solution in Mathematische Annalen 56 deals with a different case (the three-parameter plane-transformation-group).

I send you the book because I felt you doubted whether the subjects in my dissertation were really worthwhile.

As to your remark that the name of Kant should not feature in a mathematical dissertation: you will find that Russell's 'Foundations' constantly refers to Kant and that Couturat's 'Les Principes Mathématiques' is complemented by an Appendix of more than 100 pages on Kant. Moreover, if you consult Kant's Transcendental Aesthetic you will see that he speaks about the very same things as do Russell and Couturat. Poincaré describes the current battle about Foundations as a continuation of the old mathematical-philosophical battle between Kant and Leibniz.

Even if the name of Kant can be avoided, his topics will still be mentioned. Or do we have to avoid his name because he is known as a philosopher? But you cannot really classify the works of Russell and Couturat as non-mathematical. Practically all mathematical journals with book-reviews have always reviewed them.

And as to my comment, which you find so absurd, that astronomy is no more than an easy way of lumping together causal sequences of the readings of our measuring instruments: Poincaré says something similar in his 'Science et Hypothèse' (although I have not copied it from him). He says: 'The earth rotates' has no other sense than 'for a convenient arrangement of some phenomena it is easiest to assume that the world rotates'. And I find that this does not strike one as absurd, on the contrary: anyone who reads it will be immediately convinced of its truth. Is not the system of heavenly bodies just a mathematical system freely built by ourselves? We human beings are so proud of it, only because it is so effective in controlling the phenomena.

Such remarks are to the point, and yet no one shall refuse 'La Science et Hypothèse' a place in the library of the faculty of mathematics and physical science. Indeed, all kinds of conference lectures of Klein, Cantor, Boltzmann and others deal with these topics.

Finally, you said on Sunday that you were not at all sure that I had studied Kant thoroughly enough to pass a judgment. Of course, I cannot supply you with that certainty, but I can *tell* you that I have read the 'Kritik der reinen Vernunft' in its entirety and many parts repeatedly and thoroughly (in particular those relevant to my dissertation).

It may be true that my work is somewhat unclear and its structure untidy, that there are signs of rushed editing, perhaps there are some inaccuracies here and there, but I emphatically reject any charge that the thoughts expressed are vague and that my preliminary studies have been superficial.

I very much hope that this is not going to turn into a haggling session between you and me about what can be left in the dissertation and what must come out. I so much wish that you would understand and recognize the fundamental ideas, the general rather than the details, what can be read between the lines, even if your own philosophy is different and you think mine is absurd, because I am a product of my time and of a different generation.

You know that when I chose my subject two years ago it was not for lack of ability to tackle a more 'ordinary' topic, but only because I felt strongly drawn towards the subject: it just sprang up and grew in me. You agreed to my choice, 'provided there was enough mathematics in it', probably suspecting it would force me into philosophy, which indeed it has done, so much so that sometimes I lost sight of mathematics altogether. But what I bring you now is exclusively concerned with the *way mathematics is rooted in life, what therefore should be the starting-point of mathematical theories*; all particular topics in my dissertation only make sense when related to this fundamental thesis. Considered in isolation, some of these parts may well retain some value (e.g. the solutions to the three problems of Hilbert mentioned above), but others, like the survey of physics, become rather trivial when wrenched from the context.

For me the essential part of the work is this general spirit; that's why I so much wanted to launch it as a dissertation, which traditionally has had the character of 'taking a stand'. Only if this spirit is valued by my supervisor, will the acceptance of my dissertation give me satisfaction.
Respectfully yours,
L.E.J. Brouwer


[Korteweg drafted a letter dated 5th November, marked 'not sent'. His more moderate reply came in his letter of 11th November (see below).Ed.]

[DJK24]
[postcard]                                                  5th November 1906


Dear Brouwer,

I wish to wait with my reply to your last letter until I have seen the whole of your work. But I can assure you now that I have never had any doubt about the thoroughness of your preliminary mathematical study.
Wishes,
Yours,
D.J. Korteweg


[DJK23]                    [Dated by Korteweg: 5 November-evening?
                                    and marked 'crossed my letter']

Professor,

Your post-card reached me in Blaricum only last night, so that I have not been able to let you have the book you asked for. I send it now together with a few others to which I have made repeated reference.

Further to our discussions on Sunday may I add that the purpose of chapter 2 is:

a. to explain *how the mathematical experience accompanies all essentially-human acting*;

b. and with reference to a.: to investigate *to what extent experience-based mathematics can be a-priori*, in particular whether space and time are *both* a-priori.

My digressions into mechanics and physics only serve as *examples* of what I have said under a). For me to put them simply on record would serve little purpose: I am not qualified to sketch in broad outline the methods of work in mechanics, astronomy and physics, and add something original. In the context of my thesis it should even be read with the understanding that perhaps it may contain something which is not wholly accurate, but which with some correction could still serve as an example of the above stated theses.

Similarly, the criticism of Russell's 'Foundations' serves as an illustration of my theses under b) and leads to my refutations of some views which for the greater part are not at all particularly Russellian.

It may well be that this structure is not very clear to the reader from the way the chapter is written, but that could be remedied.

In any case, if the chapter were to be reduced to a survey of the physical sciences and a criticism of Russell, it would lack the framework that holds it together. I'd rather scrap this chapter as well; the little bits that are needed for chapter three could be inserted there.

What I could do perhaps is to shorten the main points a) and b) even more; but in any case the survey of physics and the criticism of Russell must remain secondary.

That I have referred so exclusively to Russell's book is due to the fact that he is the only one who writes about the philosophical foundations of mathematics and (at least most of the time) uses an *exact* language with which one can battle on mathematical grounds. This is not done by Hegel, Schopenhauer, Lotze and Fechner, to name but a few. If I were to discuss them I would have to move in a purely philosophical domain; this you do not want, and I don't wish to do either.

Russell is, as far as I know, the only one who is mathematically equipped and who tackles the problem of a-priority. When criticizing him I can stay on *my own territory*, which in my own eyes is always mathematical; moreover, it gives me an opportunity to emphasize various aspects of my own point of view. And that to me is all-important, not Russell's book, which I consider to be only important because of the sort of book it is; in all other respects I consider it a complete failure.

I have been going on about these matters, on which to my great regret last Sunday you could not go along with me, and which have been constantly on my mind during these last two days. It is not that I wish to force something on you, but I am rather worried that my manuscript misrepresents my views and that also in our conversation last Sunday I have not been able to say what I really mean and have given rise to misunderstanding.

Respectfully yours,
L.E.J. Brouwer


[DJK25] 8th November 1906

Dear Brouwer,

Thank you for your letter and for sending the books. They did not come too late because, although your dissertation has priority over most other work, little things keep cropping up and demand immediate attention.

I have finished reading your chapter 2, but I still would like to check some references to Russell.

From what you mention in your letter it seems to me that there is enough left, especially since in my opinion the final part can almost entirely be retained.

I shall go through part three in the same way, and I would then like to see the first part again in its revised form. I think that this is the best way of proceeding.

Best wishes,
Yours,
D.J. Korteweg


[DJK22] 11th November 1906

Dear Brouwer,

I have now also read your third chapter. The result is very satisfactory. I think there is much in it that I find excellent. I would like a few things expressed with less bluntness: this could only cause bad feeling where it does not belong. Some statements could be put less dogmatically. For example it seems to me that you could not really object to logical structures when used outside mathematics, by themselves, as an attempt to analyse and classify the way people reason; and if you do, you are moving outside the domain of your own subject.

But that affects only a few sentences or words. I'll come back to this when we discuss the final part.

As to chapter one, 'The Construction', you know that all I would like is some clarification, and you expressed your readiness to try and do this. Therefore only chapter two remains.

After receiving your letter I have again considered whether I could accept chapter two as it stands. But honestly, Brouwer, I cannot. I find it all interwoven with a kind of pessimism and mystical attitude to life which is not mathematics and has nothing to do with the foundations of mathematics. In your mind it may well have grown together with mathematics, but that is wholly subjective. One could totally disagree with you on this point and still share your ideas about the foundations of mathematics. I am convinced that any promotor, young or old, whether he shares this philosophy of life or not, would object to it being included in a mathematical dissertation.

In my opinion yours will only improve by removing it. It now adds a rather bizarre flavour which can only damage it. It re-occurs in chapter three but only on a few pages towards the end, and therefore also needs to be taken out, by itself it would not make sense.

I have tried to indicate how these passages can be lifted out of chapter two. Just consider it at your leisure, and see if you can see your way to making something of it that *you* too think worth keeping.

I would very much regret if that proved impossible, because I have found in your treatment of Russell's book and your conclusions as to Kant's speculations on the a-priori in mathematics, much that is good and to the point.

I have a feeling that you now understand my objections somewhat better. Yet, your last letter was a great disappointment to me since it showed all kinds of misunderstanding. It grieved me the more since I had the impression last Sunday that we did understand each other.

You speak as if I were ignorant of all kinds of things which could not possibly be unknown to me, if only as the regular reviewer of the Revue de Métaphysique et de Morale. You claim to have understood that you should not have mentioned the name of Kant, even where it concerned Kant's views of mathematics. You think that I considered as absurd the view 'that astronomy is no more than an easyway of lumping together causal sequences of the readings of our measuring instruments'. No, not that view! I recognize that it is possible to look at it in this way, although in my opinion the general law of gravity has little to do with the instruments that led to its discovery except in as far as they make measuring at all possible. But the assertion that the uniformity of laws which apply in very different areas of physics would find its origin in the uniformity of the instruments used, that assertion to me seems absurd.

You further think that I suspected your preliminary studies to have been superficial. This can only be due to my asking for further explanation (and no doubt I shall go on asking for this every now and then). But the reason for this, apart from my own desire to learn, was that I would be able to say (which might be necessary) that whenever I asked you for an explanation I found deep understanding. I certainly have no doubt about this.

But that's enough. I am rather busy this week and therefore cannot see you until Saturday next, 17th November. I shall keep the whole morning free for you and would then like to ask some further questions. No doubt you will then also have completed your revision of part one.
With best wishes,
Yours,
D.J. Korteweg

[DJK26]  Blaricum, Tuesday-morning
[Korteweg added the wrong date 6.11.06;
it should be 13th November 1906. Ed.]

Professor,

I shall be coming to see you on Saturday-morning about 10 o'clock, but I don't wish to wait so long with my reply to your letter.

When I left you on Saturday 4th November I did not at all feel upset. It was only during the following days that I became more and more strongly aware of the details of the conversation and as a result increasingly more disheartened. It was mainly, I believe, remembering that paragraph which you found so absurd that you cut me off and did not let me give any further explanation. Moreover, in my own imagination the amount you wanted struck out was perhaps greater than you in fact intended. I really thought that I would not be allowed to mention Kant since, as far as I remember, you said not to feel sure that I had made adequate study of authorities on Kant.

But the first was in any case the main reason. When you considered something too absurd to even talk about it with me, I probably made the subconscious connection that you did not take it seriously and therefore doubted the sincerity and soundness of the thoughts that led to it. That's probably what made me defend myself in my last letter, which was initially intended as a brief note on the book I sent you but which spontaneously grew into what it was, due to the thoughts that dominated my mind during these days.

And I still would appreciate it if you would not consider the paragraph in question 'too absurd to talk about'. Allow me therefore to elaborate further. It matters to me more to be rehabilitated in your eyes than to keep any of it in my dissertation, provided it can be removed without damage to the whole; and I am sure that is possible.

You say (with reference to what I incorrectly remembered as the statement deemed absurd), that the general law of gravity has little to do with the instruments which led to its discovery. But aren't all laws nothing but a joining of phenomena by means of induction, a means of commanding the phenomena, and existing nowhere except in the human mind? Surely, the law of gravity as such only applies to Euclidean space, which in turn exists only by the effective but arbitrary extension of the domain of motions of solids here on earth. Without solids on earth the law of gravity would not exist, and the interaction between them is established by means of astronomical measuring instruments. The law of gravity exists in relation to astronomical phenomena as molecules do to the equation of state. Both appear effective in bringing together a group of phenomena, and as a means of predicting the future, except that the law of gravity beats molecular theory in simplicity. But again: the law of gravity is a hypothesis and so is the distance between the earth and the sun.

There is still something I wanted to add to what I said about the main question, that the similarity of laws in different physical domains is to be expected, based as they are on the similarity of instruments used. I start with the following observation:

When the electromagnetic field of a Daniell element and that of a Leclancher element are projected onto our measuring instruments no difference is shown up, and yet looking at this problem for the first one

## Appendix 13

## The Brouwer-Korteweg Correspondence

〚Of the correspondence between Brouwer and Korteweg some 146 letters have been preserved; most of them are part of the Korteweg Nachlass in the Library of the University of Amsterdam, and some are held in the Brouwer Archive. The letters selected here concern Brouwer's dissertation.

Discrepancies in numbering and dates are due to incorrect dates having been added later by Korteweg, and to the re-discoveries of letters having been made at different times. ED.〛

[DJK13]                                                                                       7th September 1906

Professor,

I have been in Blaricum for some time; it is easier here to devote all my time to my work. I have stopped reading the works of others and am now busy ordering my notes and making a division in chapters.
I feel the more firm in my convictions, since I notice that I can still back all my notes of two years ago, even now after all the reading I have done. The only difference is that I can now support them better mathematically.
I have already found a publisher, and to put pressure on myself, I have agreed that he can start printing early October. I shall come and see you some time before then to hear whether you wish to see the text before it is printed. I will then still be able to make as many changes as I wish; of course, I shall be much more restricted at the proof-reading stage.
I hope you enjoyed the summer and I expect to see you soon.
Respectfully Yours,
L.E.J. Brouwer

[DJK15] 16th October 1906

Professor,

I have now divided the material I gathered for my dissertation into 6 chapters:
1. The construction of Mathematics.
2. Its genesis related to experience.
3. Its philosophical significance.
4. Its foundation on axioms.
5. Its value for society.
6. Its value for the individual.

The survey as given in chapter 1, which I sent you, serves mainly as a basis and a source of reference for the following chapters; also, it considers and points at one aspect of the various recent investigations into the foundations of mathematics: their significance for constructive mathematics.

A few matters have been discussed in some detail, such as Hamel's investigation of the straight line as a minimal curve, because I need this in chapter 3 to refute Russell; also, the construction of the group of major operations on the continuum, because I wanted to show the construction of groups independent of differentiability to be essential in the construction of mathematics. Further, the derivation of the non-euclidean arc-element by variation, because I have not seen it done anywhere else, and it seems to me the only way: producing the arc-element for $n$ dimensions directly from what has been derived for 2 dimensions. (The usual method is based on the formulae for geodetic curves given in the investigations of Christoffel and Lipschitz in Crelle 1870 and later, and these become very complicated in $n$ dimensions.)

I hope to be in Amsterdam to-morrow morning or the day after tomorrow, and to hear what you think.

I can then have another look at it, and perhaps the chapter can be typeset by the end of the week.

In the meantime I am working on the next chapter and will send it to you as soon as possible.

Respectfully yours,
L.E.J. Brouwer

[DJK20]

⟦After a fateful meeting at Korteweg's house on Sunday 4th November Brouwer mounts his defense. He dispenses even with his usual, curt address 'Professor'.⟧

5th November 1906

Herewith I enclose the issue of the Göttingen Nachrichten in which Hilbert's Paris lecture 'Mathematische Probleme' was published. You will see that in Chapter One of my dissertation I have dealt in full with the First Problem ('Cantor's problem of the power of the continuum'); and this precisely by going back to the intuitive construction which is essential for all mathematics.

Problem 2 ('The non-contradictority of the axioms of arithmetic'), will be discussed in my last chapter to the extent that Hilbert's own solution given at the Heidelberger Congress is rejected; as the only solution I shall again refer back to the construction of arithmetic on the continuum, as given in my chapter one, by characterizing addition and multiplication as *the* two-parameter group.

In this way I have also solved problem 5 ('Lie's notion of continuous transformation groups independent of differentiability') for a simple case (the two-parameter linear group). Hilbert's own solution in Mathematische Annalen 56 deals with a different case (the three-parameter plane-transformation-group).

I send you the book because I felt you doubted whether the subjects in my dissertation were really worthwhile.

As to your remark that the name of Kant should not feature in a mathematical dissertation: you will find that Russell's 'Foundations' constantly refers to Kant and that Couturat's 'Les Principes Mathématiques' is complemented by an Appendix of more than 100 pages on Kant. Moreover, if you consult Kant's Transcendental Aesthetic you will see that he speaks about the very same things as do Russell and Couturat. Poincaré describes the current battle about Foundations as a continuation of the old mathematical-philosophical battle between Kant and Leibniz.

Even if the name of Kant can be avoided, his topics will still be mentioned. Or do we have to avoid his name because he is known as a philosopher? But you cannot really classify the works of Russell and Couturat as non-mathematical. Practically all mathematical journals with book-reviews have always reviewed them.

And as to my comment, which you find so absurd, that astronomy is no more than an easy way of lumping together causal sequences of the readings of our measuring instruments: Poincaré says something similar in his 'Science et Hypothèse' (although I have not copied it from him). He says: 'The earth rotates' has no other sense than 'for a convenient arrangement of some phenomena it is easiest to assume that the world rotates'. And I find that this does not strike one as absurd, on the contrary: anyone who reads it will be immediately convinced of its truth. Is not the system of heavenly bodies just a mathematical system freely built by ourselves? We human beings are so proud of it, only because it is so effective in controlling the phenomena.

Such remarks are to the point, and yet no one shall refuse 'La Science et Hypothèse' a place in the library of the faculty of mathematics and physical science. Indeed, all kinds of conference lectures of Klein, Cantor, Boltzmann and others deal with these topics.

Finally, you said on Sunday that you were not at all sure that I had studied Kant thoroughly enough to pass a judgment. Of course, I cannot supply you with that certainty, but I can *tell* you that I have read the 'Kritik der reinen Vernunft' in its entirety and many parts repeatedly and thoroughly (in particular those relevant to my dissertation).

It may be true that my work is somewhat unclear and its structure untidy, that there are signs of rushed editing, perhaps there are some inaccuracies here and there, but I emphatically reject any charge that the thoughts expressed are vague and that my preliminary studies have been superficial.

I very much hope that this is not going to turn into a haggling session between you and me about what can be left in the dissertation and what must come out. I so much wish that you would understand and recognize the fundamental ideas, the general rather than the details, what can be read between the lines, even if your own philosophy is different and you think mine is absurd, because I am a product of my time and of a different generation.

You know that when I chose my subject two years ago it was not for lack of ability to tackle a more 'ordinary' topic, but only because I felt strongly drawn towards the subject: it just sprang up and grew in me. You agreed to my choice, 'provided there was enough mathematics in it', probably suspecting it would force me into philosophy, which indeed it has done, so much so that sometimes I lost sight of mathematics altogether. But what I bring you now is exclusively concerned with the *way mathematics is rooted in life, what therefore should be the starting-point of mathematical theories*; all particular topics in my dissertation only make sense when related to this fundamental thesis. Considered in isolation, some of these parts may well retain some value (e.g. the solutions to the three problems of Hilbert mentioned above), but others, like the survey of physics, become rather trivial when wrenched from the context.

For me the essential part of the work is this general spirit; that's why I so much wanted to launch it as a dissertation, which traditionally has had the character of 'taking a stand'. Only if this spirit is valued by my supervisor, will the acceptance of my dissertation give me satisfaction.
Respectfully yours,
L.E.J. Brouwer

[Korteweg drafted a letter dated 5th November, marked 'not sent'. His more moderate reply came in his letter of 11th November (see below).Ed.]

[DJK24]
[postcard] 5th November 1906

Dear Brouwer,
I wish to wait with my reply to your last letter until I have seen the whole of your work. But I can assure you now that I have never had any doubt about the thoroughness of your preliminary mathematical study.
Wishes,
Yours,
D.J. Korteweg

[DJK23] [Dated by Korteweg: 5 November-evening? and marked 'crossed my letter']

Professor,

Your post-card reached me in Blaricum only last night, so that I have not been able to let you have the book you asked for. I send it now together with a few others to which I have made repeated reference.

Further to our discussions on Sunday may I add that the purpose of chapter 2 is:

a. to explain *how the mathematical experience accompanies all essentially-human acting*;

b. and with reference to a.: to investigate *to what extent experience-based mathematics can be a-priori*, in particular whether space and time are *both* a-priori.

My digressions into mechanics and physics only serve as *examples* of what I have said under a). For me to put them simply on record would serve little purpose: I am not qualified to sketch in broad outline the methods of work in mechanics, astronomy and physics, and add something original. In the context of my thesis it should even be read with the understanding that perhaps it may contain something which is not wholly accurate, but which with some correction could still serve as an example of the above stated theses.

Similarly, the criticism of Russell's 'Foundations' serves as an illustration of my theses under b) and leads to my refutations of some views which for the greater part are not at all particularly Russellian.

It may well be that this structure is not very clear to the reader from the way the chapter is written, but that could be remedied.

In any case, if the chapter were to be reduced to a survey of the physical sciences and a criticism of Russell, it would lack the framework that holds it together. I'd rather scrap this chapter as well; the little bits that are needed for chapter three could be inserted there.

What I could do perhaps is to shorten the main points a) and b) even more; but in any case the survey of physics and the criticism of Russell must remain secondary.

That I have referred so exclusively to Russell's book is due to the fact that he is the only one who writes about the philosophical foundations of mathematics and (at least most of the time) uses an *exact* language with which one can battle on mathematical grounds. This is not done by Hegel, Schopenhauer, Lotze and Fechner, to name but a few. If I were to discuss them I would have to move in a purely philosophical domain; this you do not want, and I don't wish to do either.

Russell is, as far as I know, the only one who is mathematically equipped and who tackles the problem of a-priority. When criticizing him I can stay on *my own territory*, which in my own eyes is always mathematical; moreover, it gives me an opportunity to emphasize various aspects of my own point of view. And that to me is all-important, not Russell's book, which I consider to be only important because of the sort of book it is; in all other respects I consider it a complete failure.

I have been going on about these matters, on which to my great regret last Sunday you could not go along with me, and which have been constantly on my mind during these last two days. It is not that I wish to force something on you, but I am rather worried that my manuscript misrepresents my views and that also in our conversation last Sunday I have not been able to say what I really mean and have given rise to misunderstanding.

Respectfully yours,
L.E.J. Brouwer


[DJK25]                                            8th November 1906

Dear Brouwer,

Thank you for your letter and for sending the books. They did not come too late because, although your dissertation has priority over most other work, little things keep cropping up and demand immediate attention.

I have finished reading your chapter 2, but I still would like to check some references to Russell.

From what you mention in your letter it seems to me that there is enough left, especially since in my opinion the final part can almost entirely be retained.

I shall go through part three in the same way, and I would then like to see the first part again in its revised form. I think that this is the best way of proceeding.

Best wishes,
Yours,
D.J. Korteweg


[DJK22]                                          11th November 1906

Dear Brouwer,

I have now also read your third chapter. The result is very satisfactory. I think there is much in it that I find excellent. I would like a few things expressed with less bluntness: this could only cause bad feeling where it does not belong. Some statements could be put less dogmatically. For example it seems to me that you could not really object to logical structures when used outside mathematics, by themselves, as an attempt to analyse and classify the way people reason; and if you do, you are moving outside the domain of your own subject.

But that affects only a few sentences or words. I'll come back to this when we discuss the final part.

As to chapter one, 'The Construction', you know that all I would like is some clarification, and you expressed your readiness to try and do this. Therefore only chapter two remains.

After receiving your letter I have again considered whether I could accept chapter two as it stands . But honestly, Brouwer, I cannot. I find it all interwoven with a kind of pessimism and mystical attitude to life which is not mathematics and has nothing to do with the foundations of mathematics. In your mind it may well have grown together with mathematics, but that is wholly subjective. One could totally disagree with you on this point and still share your ideas about the foundations of mathematics. I am convinced that any promotor, young or old, whether he shares this philosophy of life or not, would object to it being included in a mathematical dissertation.

In my opinion yours will only improve by removing it. It now adds a rather bizarre flavour which can only damage it. It re-occurs in chapter three but only on a few pages towards the end, and therefore also needs to be taken out, by itself it would not make sense.

I have tried to indicate how these passages can be lifted out of chapter two. Just consider it at your leisure, and see if you can see your way to making something of it that *you* too think worth keeping.

I would very much regret if that proved impossible, because I have found in your treatment of Russell's book and your conclusions as to Kant's speculations on the a-priori in mathematics, much that is good and to the point.

I have a feeling that you now understand my objections somewhat better. Yet, your last letter was a great disappointment to me since it showed all kinds of misunderstanding. It grieved me the more since I had the impression last Sunday that we did understand each other.

You speak as if I were ignorant of all kinds of things which could not possibly be unknown to me, if only as the regular reviewer of the Revue de Métaphysique et de Morale. You claim to have understood that you should not have mentioned the name of Kant, even where it concerned Kant's views of mathematics. You think that I considered as absurd the view 'that astronomy is no more than an easyway of lumping together causal sequences of the readings of our measuring instruments'. No, not that view! I recognize that it is possible to look at it in this way, although in my opinion the general law of gravity has little to do with the instruments that led to its discovery except in as far as they make measuring at all possible. But the assertion that the uniformity of laws which apply in very different areas of physics would find its origin in the uniformity of the instruments used, that assertion to me seems absurd.

You further think that I suspected your preliminary studies to have been superficial. This can only be due to my asking for further explanation (and no doubt I shall go on asking for this every now and then). But the reason for this, apart from my own desire to learn, was that I would be able to say (which might be necessary) that whenever I asked you for an explanation I found deep understanding. I certainly have no doubt about this.

But that's enough. I am rather busy this week and therefore cannot see you until Saturday next, 17th November. I shall keep the whole morning free for you and would then like to ask some further questions. No doubt you will then also have completed your revision of part one.
With best wishes,
Yours,
D.J. Korteweg

[DJK26]                          Blaricum, Tuesday-morning
[Korteweg added the wrong date 6.11.06; it should be 13th November 1906. Ed.]

Professor,

I shall be coming to see you on Saturday-morning about 10 o'clock, but I don't wish to wait so long with my reply to your letter.

When I left you on Saturday 4th November I did not at all feel upset. It was only during the following days that I became more and more strongly aware of the details of the conversation and as a result increasingly more disheartened. It was mainly, I believe, remembering that paragraph which you found so absurd that you cut me off and did not let me give any further explanation. Moreover, in my own imagination the amount you wanted struck out was perhaps greater than you in fact intended. I really thought that I would not be allowed to mention Kant since, as far as I remember, you said not to feel sure that I had made adequate study of authorities on Kant.

But the first was in any case the main reason. When you considered something too absurd to even talk about it with me, I probably made the subconscious connection that you did not take it seriously and therefore doubted the sincerity and soundness of the thoughts that led to it. That's probably what made me defend myself in my last letter, which was initially intended as a brief note on the book I sent you but which spontaneously grew into what it was, due to the thoughts that dominated my mind during these days.

And I still would appreciate it if you would not consider the paragraph in question 'too absurd to talk about'. Allow me therefore to elaborate further. It matters to me more to be rehabilitated in your eyes than to keep any of it in my dissertation, provided it can be removed without damage to the whole; and I am sure that is possible.

You say (with reference to what I incorrectly remembered as the statement deemed absurd), that the general law of gravity has little to do with the instruments which led to its discovery. But aren't all laws nothing but a joining of phenomena by means of induction, a means of commanding the phenomena, and existing nowhere except in the human mind? Surely, the law of gravity as such only applies to Euclidean space, which in turn exists only by the effective but arbitrary extension of the domain of motions of solids here on earth. Without solids on earth the law of gravity would not exist, and the interaction between them is established by means of astronomical measuring instruments. The law of gravity exists in relation to astronomical phenomena as molecules do to the equation of state. Both appear effective in bringing together a group of phenomena, and as a means of predicting the future, except that the law of gravity beats molecular theory in simplicity. But again: the law of gravity is a hypothesis and so is the distance between the earth and the sun.

There is still something I wanted to add to what I said about the main question, that the similarity of laws in different physical domains is to be expected, based as they are on the similarity of instruments used. I start with the following observation:

When the electromagnetic field of a Daniell element and that of a Leclancher element are projected onto our measuring instruments no difference is shown up, and yet looking at this problem for the first one time would expect there to be as great a difference as there is between copper-sulphate and ammonium chloride. Only our counting and measuring instinct, working with certain instruments, is affected by them in the same way and it then appears that the same mathematical system can be applied. It is only the lack of suitable instruments that has prevented us so far from finding other mathematical systems which can be applied to one but not the other. At every stage of the development of physics the measuring instruments that 'have been found to be suitable' form only a small part of the totality of measuring instruments 'which might possibly be found suitable to govern all kinds of yet unknown phenomena'. Similarly, 'the mathematical systems that have already been applied to nature' form only a small part of the whole of mathematics 'which might suitably be applied to nature if only physics had been sufficiently developed'. And if now every limited group of mathematical systems has its invariants, it is to be expected that also every limited group of natural phenomena, exactly because of this restriction, has its invariants, and this in the form of laws or principles which apply to all the phenomena of that group.

One might put the question: why should we expect invariants for the whole of currently known physics? Doesn't physics claim that it is not bound by a *particular* restriction, but draws the most heterogeneous things in an arbitrary way within the range of its study? We could answer that by saying: But there *is* a *particular* restriction, because the mathematical laws which one sees in nature in the final analysis express no more than relations between measures, taken from the rigid group; only the *influences* to which these rigid measures are exposed vary indefinitely. Other physical aspects are auxiliary, chosen to suit only certain influences of these measures; their introduction as coordinates gives the equations of motion or position a simpler form. These physical aspects themselves are never measured, only the rigid measures; the so-called link between them is purely fictitious. We therefore do not measure magnetic forces and electric currents but the twist-angle of wires, and the angular measure is taken from the rigid group. Moreover, whenever we speak of equivalent things or circumstances when certain influences are eliminated, we always mean 'relative to the readings on our measuring instruments'. The only thing that can be said to be empirically true is : 'The motion-group of solid bodies has approximately this or that property, and that property remains approximately constant in time'.

'But don't we measure things other than rigid measures, e.g. quantities of electricity? Isn't it possible to apply successively equal charges to the same conductor, by releasing twice in succession the same load, generated in the same way twice in succession, and don't we know that after the second charge the total charge in the conductor is double that in the conductor after the first charge?'

The answer is no. In what sense can we speak of *quantities of electricity*, i.e. in what sense can the effects of successive equivalent charges be added or superimposed as equivalent effects?
In as far as, for example, they produce super-imposing effects on the Coulomb torsion balance.
But in how far can we super-impose the forces which produce equal torsion on this balance?
In as far as they, for example, keep equal brass weights in equilibrium.
But in how far can we super-impose the weights of equal pieces of brass?
In as far as the accelerations which they effect on the same body (e.g. in the Atwood apparatus) can be super-imposed.
But such accelerations are only observed in relation to solid bodies, for accelerations as well as velocities are observed in relation to a rigid group. The same applies to weights of liquids, we either measure them by volume − and that is measured on the basis of the rigid group − or by translating them into forces on a solid body, e.g. a balance or a valve.

In this way every physically measurable entity is ultimately reduced to measuring in the rigid group, and *what one searches for in all kinds of different phenomena and circumstances is: the laws of these measures.* A particular, *definite* restriction on the mathematical systems which can be applied to physics is therefore to be expected, and the existence of invariant principles comes as no surprise. In the same way that an organ-pipe fails to resonate with all except certain notes, we should expect that the rigid group fails to *resonate with* phenomena other than those that satisfy the principles of energy, action and thermodynamics. The unknown, the more general, that which lies outside this sphere, could still manifest itself in the laws of physics in the form of all kinds of dielectric constants such as the unexplained weight of atoms, sound-frequences, specific gravities etc., as well as the 'accidental' fact that the laws are what they are, and not something else.

You may well find some weak points in these arguments, but in any case they show that what I said was more than vague sentiment and not just based on a pessimistic outlook on life.

Finally, I hope that you will consider what I have written here, as wel as what I wrote before, as inspired by the fear of having to give up the

feeling of being in harmony with you on the subject, and by the sincere wish to retain as much as possible of my dissertation and its detail.
Respectfully yours,
L.E.J. Brouwer


[DJK29]  10th January 1907

Dear Brouwer,

On the page that I return herewith I have suggested only some very minor alterations as you will see.

But I begin to have some doubts whether the proof on page 86 can be used for higher derivatives.

For example let $\varphi$ be the coordinate of an ordinary Weierstrass curve without differential coefficient $\varphi = C + \sum_0^\infty b^n \cos(a^n x \pi)$; $C$ can be taken such that $\varphi$ is always positive, which makes matters somewhat easier. Consider now the curves through an arbitray $A$ for which $dy/dx = \int \varphi \, dx$.

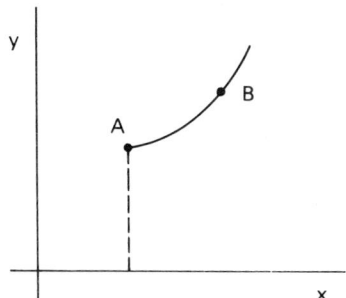

This integral does exist, at least according to Klein 'every continuous function' admits such an integral. The resulting curve certainly has a second derivative, but does it satisfy your conditions? It seems to me that it does, because the derivative increases regularly within limits which become smaller as the part of the curve under consideration becomes smaller. It seems to me that, indeed, for arguments which lie close together (such as $A$ and $B$) the value is approximately the same for all (first) derivatives since they all approach the derivatives 'continuously' as the increase gets smaller.

Maybe I have not quite understood what you mean.
Best wishes,
Yours,
D.J. Korteweg

[DJK31] 10th January 1907

Professor,

The curve you mention is indeed a case where the proof of page 86 can be given for the first but not for higher derivatives. But it does not fully satisfy the conditions I imposed.

'Approximately similar behaviour in closely neighbouring argument-points' means that all functions of the independent variable which are determined by the curve are *continuous* (and my words say this unambiguously). I further know that in the case of a continuous curve the derivatives *exist*; they are therefore functions determined by the curve, and therefore (according to my supposition) they are *continuous functions*.

The proof on p. 86 ff. shows that the first derivative *exists*. It is therefore a function determined by the orginal curve, and therefore (according to my supposition) continuous, has therefore *existing* derivatives, which again are functions determined by the original curve and (again according to my supposition) *continuous*. This leads, according to the proof of p.86, to the *existence* of a *second* derivative. We can go on in the same way: first from the *supposition* on p.86 it follows that the second derivative and its derivatives are *continuous*, and then from the *proof* of p.86 that the third derivative exists etc..

The function you mention does not satisfy my postulate, because there exist functions determined by the curve (derivatives of derivatives) which are not continuous. If these functions were continuous, then the second derivative would also exist.

Respectfully yours,
L.E.J. Brouwer

[DJK32] the morning of 11th January 1907

Professor,

On second thoughts, I feel that my letter to you last night was somewhat incomplete and might give rise to the question why I formulated my condition on page 86 in terms of 'approximately the same behaviour in closely neighbouring points', and not 'continuity of functions determined by the curve'.

The first formulation does imply the latter if one only considers functions determined by physical measures (or continuous operations applied to the results of this measuring), in other words: *observed* functions. Examples of these are derivatives, and the differential coefficients which — if they exist — can be approximated by measuring $\Delta f$, $\Delta^2 f$, $\Delta^3 f$ etc..

The second formulation, however, would also include any arbitrary *mathematical* function which my mathematical imagination might construct from the ordinate; and in this sense, of course, the conditions of the postulate will never be satisfied. But *every mathematical* curve does determine *mathematical* discontinuous functions. I thought that the word *behaviour*, i.e. of physically observed entities, expresses exactly what I meant, and does not need any further elaboration.

The more so since I wanted to show that it is only a vague feeling, something which most people have not defined for themselves in precise mathematical terms, whereas the premisses of my proof definitely are within that domain.

Of course, apart from my short remarks there is a lot more that could be said; which also applies to many other subjects that are mentioned in chapter two. The reason for this is probably that in my mind they were originally only incidental off-shoots of one fundamental idea which held them together (and which is no longer found in the dissertation), and therefore were only of secondary importance.

After their sudden appearance in the full limelight, substituting for their former leader, it was not possible to doll them up quickly in such a way that they together by themselves could save the entire performance. At least that is what I sometimes think when I look at the chapter. On the other hand I have also become more and more aware that the thoughts, as I had written them down at first, would have damaged the mathematical character of a mathematical dissertation. I have therefore tried to exlude them and to write about the relation between mathematics and experience as profoundly as possible and avoid triviality.

This letter has become longer than I intended.

The printer is again wasting time, I have not received anything during the last few days. But perhaps printing these first few sheets takes its time.
Respectfully yours,
L.E.J. Brouwer


[DJK33]                                                           18th January 1907

Professor,

Having considered your remarks on my page 128, I agree that the mathematical character is somewhat impaired, and I have therefore cut out the sentences which contain a 'value-judgment'.

I also agree that my phrase that 'theoretical logic is not aimed at the exterior world' could be misunderstood. 'Aimed at the exterior world in order to dominate it or fight against it' was meant to be: 'finding a practical application in the exterior world'; but I suppose one does not read it that way.

From your characterization of theoretical logic as a part of psychology I gather that I have not expressed myself very clearly. It was precisely my intention to show that theoretical logic, even if it is a science, can on no account be credited with any psychological significance.

I have added a few lines to page 128, and changed the last lines of page 127, resulting in an entirely new version and treatment of the topic. When I receive the proofs, I shall send you that part again together with sheet no 9 which I sent to Nijmegen yesterday. I hope to have it back tomorrow and let you have it on Sunday morning.

The topic of the differentiability of the functions of physics (sheet no 6) has now, I hope, been sufficiently enlarged.
Respectfully yours,
L.E.J. Brouwer


[DJK35]                                                        23rd January 1907

Professor,

In order not to lose any more time, and with your agreement, I shall not make any further alterations in the text. But I still would like to reply to your comment and try to move you a little further to my side on this contentious issue.

At the beginning of the chapter I show that mathematical reasoning is *not* logical reasoning, that only because of the poverty of language it makes use of the connectives of logical reasoning, and perhaps thereby will keep alive the linguistic accompaniments of logical reasoning even long after the human intellect has outgrown logical reasoning. Far from it being 'a strange race that does not think logically', I believe that it is only a sign of the inertia of language that the relevant words still exist in modern languages. One seldom meets a pure use of these words; their impure use in every-day life has led to all kinds of misunderstanding, and in mathematics to the false notions of set theory. These notions have come about not from a lack of mathematical insight but because mathematics, for lack of a pure language, had to make do with *the language of logical reasoning*, whereas its thought proceed not as logical reasoning but as mathematical reasoning, something quite different.

The theorem: 'if a triangle is isosceles it is an acute triangle' is expressed as a logical theorem: the predicate 'isosceles' in the case of triangles is considered to imply the predicate 'acute', i.e. one imagines all the triangles of a given plane represented by the pointsof an $R_6$, and one then sees that the domain of $R_6$ representing isosceles triangles is contained in the domain representing all acute triangles. This is in fact true, and logical formulation and logical language can therefore safely be applied. But the thoughts of the mathematician, who because of the poverty of his language formulates this theorem as a logical theorem, proceed in a way quite different from the above interpretation. He imagines that he is going to construct an isosceles triangle, and then finds that either at the end of the construction all angles appear to be acute or that on the postulation of a right or obtuse angle the construction cannot be executed. In other words, he thinks the construction mathematically, not in its logical interpretation.

It is precisely the main theme of Chapter 3 to show that careless use of a logical language instead of a mathematical language has led mathematics in some of its parts on a false trail.

May I just explain why I believe that logical language is out of date; in fact we already discussed this the day before yesterday. The mathematical systems which are applied to the outside world, and therefore alone should be considered for representation into language, should through their mathematical theories be able to teach us something practical. But the *mathematics of whole-and-part* does not teach us anything new for everyday life. Once this system is applied to a part of the perceptional world, even a very mediocre mind can read off directly all the consequences; he does not need any intermediary logical reasoning. One knows very well nowadays that if something about the exterior world is deduced by logical reasoning, something that was not so directly a-priori evident, it is then for this very reason wholly unreliable. The reason is, that one does not believe any more in the postulate on which it is based, the postulate that the world consists of a very large but finite number of atoms, and that every word must represent a (therefore also finite) group or group of groups of these atoms. In other words: one knows very well that the world is not a logical system, and that one cannot argue about it logically. Indeed we know very well that every debate is a load of nonsense, that the only disputes that can be settled concern mathematical problems, and in that case *not by logical reasoning* (although it sometimes appears like that because of the deficiency of language; axiomatic foundations and transfinite numbers show how false this appearance is), but *by mathematical reasoning*.

Theoretical logic teaches us nothing about the present world; anyone knows that, at least anyone with any sense. It only serves the purpose of lawyers and demagogues, not a means of teaching people but of deceiving them. They succeed because the hoi-polloi subconsciously argue like this: 'this language with its logical figures, it's there and therefore it must be useful'; and then meekly they let themselves be fooled by it. It is the argument I have heard people use when defending their gin-drinking habit, saying: 'otherwise why is there gin?' Those who still harbour illusions of making this a better world, have equally good cause to fight against the language of logical reasoning as they have for their fight against alcohol. Neither do we have any more reason to describe as 'strange people' those who do not reason logically than those who do not drink alcohol. Indeed, I believe that no abuse is more deeply-rooted than the one that has grown so intimately with the most popular parts of language.

Your question about the word 'exists' on page 141, line 5 from below is perhaps best answered by the example given 4 lines further. The conditions mentioned imply that one is searching for a *finite number*, i.e. a known mathematical entity; but there is no certainty that these conditions can be satisfied, i.e. that the mathematical entity *exists*.

Perhaps I have expressed what I meant more clearly in this rather wildly written letter than in my 'bridled' text. But maybe this letter throws a new light on my text, and that would please me enormously.

With best wishes,
Yours respectfully,
L.E.J. Brouwer

[DJK37] 16th February 1907

Professor,

Mannoury and Barrau prefer to oppose parts of the dissertation itself rather than one of the theses.
〚The formal opposition and defence of the dissertation and a number of added 'theses' is part of the graduation ceremony at Dutch universities.Ed.〛
Mannoury will oppose the intuitivity of the 'et-cetera'. He will ask me to read from page 180: 'There are elements which cannot be derived...etc.', and relate that to page 3: 'By interchanging... et-cetera. (p.4, l.2)'; also to page 142: 'We have seen...consisting of equal elements (p.143, l.1)'. He is going to maintain that *et-cetera* is no more than a relation between relations. I do not know any further details; that's all he has written to me in a letter from Helmond.

Barrau will put it to me that, admitting only constructive mathematics, without logic, I should have kept the word *continuum* completely out of my dissertation, since I operate only with discrete collections. He will ask me to read from page 8 (line 4 from below): 'Recognizing the intuition of the continuum... we can name properties of the Continuum as the matrix of points to be thought of as a whole (p.9, l.3).'and relate that to page 62: 'The continuum as a whole was given intuitively... indivualized, is inconceivable and impossible'.

He will maintain that the 'discrete by itself', i.e. the notion of 'two different points', is sufficient for the construction of the whole of mathematics, and persuade me that my own statements necessarily lead to this conclusion.

This will probably give you enough of an idea what the debate will be about. I therefore won't trouble you with the details of our further discussions, unless I hear from you.
Yours respectfully,
L.E.J. Brouwer

# BIBLIOGRAPHY

ALEXANDROFF, P. and HOPF, H.
[1935]     *Topologie I.* (Springer, Berlin)

ASHVINIKUMAR
[1966]     *Hilbert Spaces in Intuitionism.* (V.R.B., Amsterdam)

AVSITIDYSKY, S.
[1927]     Note relative au travail de MM. Barzin et Errera 'sur la logique de M.Brouwer'. *Académie Royale de Belgique, Bulletin* vol.17, pp.724-730.

BARZIN, M. and ERRERA, A.
[1927]     Sur la logique de M. Brouwer. *Académie Royale de Belgique, Bulletin* vol.13, pp. 56-71.

BERGSON, H.
[1889]     *Essai sur les données immediates de la conscience.* (Alcan, Paris)
[1897]     *Matière et Mémoire.* (Alcan, Paris)
[1903]     Introduction à la Métaphysique. *Revue de Métaphysique et de Morale* vol.11, pp.1-36.
[1907]     *L'Évolution Créatrice.* (Alcan, Paris)

BETH, E.W.
[1940]     *Inleiding tot de Wijsbegeerte der Wiskunde.* (Antwerpen, Brussel)
[1965]     *Mathematical Thought.* (Reidel, Dordrecht)
[1959A]    Remarks on Intuitionistic Logic. In: *Constructivity in Mathematics* (ed. A.Heyting). (North-Holland, Amsterdam)
[1959B]    *The Foundations of Mathematics.* (North-Holland, Amsterdam)
[1965]     *Mathematical Thought.* (Reidel, Dordrecht)

BISHOP, E.
[1967]     *Foundations of Constructive Analysis.* (McGraw-Hill, New York)
[1975]     The crisis in contemporary mathematics. *Historia mathematica* vol. 2, pp.507-517.

BOCKSTAELE, P.
[1949]     Het intuitionisme bij de Franse wiskundigen. *Verhandelingen KAW Belgie*, jaargang 11, no.32.

BOLLAND, G.J.P.J.
[1904]     Zuivere Rede, een boek voor vrienden der wijsheid. (Adriani, Leiden)

BOREL, E.
[1898]     Leçons sur la Théorie des Fonctions. (Gauthier-Villars, Paris)
[1909]     Sur les principes de la théorie des ensembles. In: *Atti del IV Congresso Internazionale dei Matematici, Roma 1908*. (Academia dei Lincei, Rome)
[1927]     À propos de la recente discussion entre M. R.Wavre et M. P.Levy. *Revue de Métaphysique et de Morale* vol.34, pp.271-276.
[1947]     Sur l'illusion des définitions asymptotiques. *Comptes Rendus* vol.224, pp.765-767.

BRIDGES, D. and BISHOP, E.
[1985]     *Constructive Analysis*. (Springer, Berlin)

CANTOR, G.
[1883]     Über unendliche, lineare Punktmannigfaltigkeiten (5). *Mathematische Annalen* vol.21, pp. 545-586.
[1895]     Beiträge zur Begründung der transfiniten Mengenlehre. *Mathematische Annalen* 46, pp. 481-512.

DE HAEN, J.I.
[1916]     *Rechtskundige significa en haar toepassing op de begrippen: 'aansprakelijk', 'verantwoordelijk' en toerekeningsvatbaar*. (Versluys, Amsterdam)

DESCARTES, R.
[1637]     *Discours de la Méthode*. (Maire, Leyden)
[1644]     *Meditationes de Prima Philosophia*. (J.Blaeu, Amsterdam)

DIJKMAN, J.G.
[1952]     *Convergentie en Divergentie in de Intuitionistische Wiskunde*. (Excelsior, 's Gravenhage)

DRESDEN, A.
[1924]     Brouwer's contribution to the foundations of mathematics. *Bulletin of the American Mathematical Association* vol.30, pp. 31-40.

DUMMETT, M.
[1977]     *Elements of Intuitionism*. (Clarendon Press, Oxford)

FRAENKEL, A.A.
[1927]     Zehn Vorlesungen über die Grundlegung der Mengenlehre. (Teubner, Berlin und Leipzig)

FRAENKEL, A.A. and BAR-HILLEL, Y.
[1958]     Foundations of Set Theory. (North-Holland, Amsterdam. Revised edition published by D. van Dalen in 1973)

FREUDENTHAL, H.
[1975]     The cradle of modern topology according to Brouwer's inedita. *Historia Mathematica* vol.2, pp. 495-502.
[1976]     *L.E.J. Brouwer Collected Works, Part 2* (an annotated edition of Brouwer's writings on geometry, analysis, topology and mechanics). (North-Holland, Amsterdam)

GIBSON, C.G.
[1967]     *The Radon Integral in Intuitionism*. (V.R.B., Groningen)

GLIVENKO, V.
[1928]     Sur la logique de M. Brouwer. *Académie Royale de Belgique, Bulletin* vol.14, pp. 225-228.
[1929]     Sur quelques points de lalogique de M. Brouwer. *Académie Royale de Belgique, Bulletin* vol.15, pp. 183-188.

GRISS, G.F.C.
[1944]     Negatielooze intuitionistische wiskunde. *Verslag KAW* vol.53, p.261. (Negationless intuitionistic mathematics. *Proceedings KAW* vol.49, p.1127.)

HALLETT, M.
[1984]     *Cantorian Set Theory and Limitation of Size*. (Clarendon Press, Oxford)

HEYTING, A.
[1925]     *Intuitionistische Axiomatiek der Projectieve Meetkunde*. (Noordhoff, Groningen)
[1930A]    Die formale Regeln der intuitionistische Logik. *Sitzungsberichte der preuszischen Akademie von Wissenschaften*, phys. math. Kl. 1930, pp. 42−56.
[1930B]    Die formale Regeln der intuitionistischen Mathematik. *Sitzungsberichte der preuszischen Akademie von Wissenschaften*, phys. math. Kl. 1930, pp. 57-71, pp. 158-169.

| | |
|---|---|
| [1930C] | Sur la logique intuitionniste. *Académie Royale de Belgique Bulletin* vol.16, pp.957-963. |
| [1934] | Mathematische Grundlagenforschung, Intuitionismus, Beweistheorie. *Ergebnisse der Mathematik und ihrer Grenzgebiete* vol.3, part 4. (Berlin) |
| [1949] | *Spanningen in de Wiskunde.* (Inaugural Address) (Noordhoff, Groningen-Batavia) |
| [1955] | *Les Fondements des mathématiques. Intuitionnisme. Théorie de la démonstration.* (Gautiers-Villars, Paris) |
| [1956A] | *Intuitionism, an Introduction.* (North-Holland, Amsterdam) |
| [1956B] | La conception intuitionniste de la logique. In: *Les Etudes Philosophiques*, no. 2 (Presses Universitaires de France) |
| [1958] | Intuitionism in mathematics. *La Philosophie au milieu du vingtième siècle* (ed. R. Klibansky), vol.1, pp.101-115. (La Nuova Italia, Firenze) |
| [1959] | Some remarks on Intuitionism. In: *Constructivity in Mathematics*, Proceedings Colloquium Amsterdam 1957. (North-Holland, Amsterdam) |
| [1960] | Remarques sur le constructivisme. *Logique et Analyse* vol.3, pp.177-182. |
| [1961] | Axiomatic Method and Intuitionism. In: *Essays on the Foundations of Mathematics dedicated to A.A.Fraenkel*, Jerusalem, 1961. |
| [1968] | Intuitionism in mathematics. In: *Contemporary Philosophy, a Survey* (ed. R. Klibansky). (La Nuova Italia, Firenze) |
| [1975] | *L.E.J. Brouwer Collected Works, Part 1* (an annotated edition of Brouwer's publications on the philosophy and foundations of mathematics). (North-Holland, Amsterdam) |
| [1981] | Continuum en keuzerij. *Nieuw Archief voor Wiskunde* (3) 29, pp.125-139. |

HILBERT, D.

| | |
|---|---|
| [1899] | *Grundlagen der Geometrie.* (Teubner, Leipzig) |
| [1900] | Mathematische Probleme. *Nachrichte der Gesellschaft der Wisschenschaften zu Göttingen*, pp.253-297. |
| [1904] | Über die Grundlagen der Logik und Arithmetik. *Verhandlungen des dritten internationalen Kongress in Heidelberg, 1904*, pp. 174-185. |
| [1922] | Neubegründung der Mathematik. *Abhandlungen aus dem mathematischen Seminar der Hamburgischen Universität* vol.1, pp. 157-177. |
| [1923] | Die logischen Grundlagen der mathematik. *Mathematische Annalen* vol.88, pp.151-165. |
| [1928] | Die Grundlagen der Mathematik. *Abhandlungen aus dem mathematischen Seminar der Hamburgischen Universität*, vol.6, pp.65-85. |

JOHNSON, D.M.
[1979]   The problem of the invariance of dimension in the growth of modern topology. *Archive for History of Exact Sciences* vol.20, number 2 1979, pp. 97-188 and vol.25, number 2/3 1981, pp.85-267.
[1987]   L.E.J. Brouwer's coming of age as a topologist. In: *Studies in the History of Mathematics*, MAA Studies vol.26 (ed. E.R. Phillips)

KANT, I.
[1783]   *Prolegomena zu einer jeden künftigen Metaphysik.* (Prolegomena to any future metaphysics; translation by P.G.Lucas, MUP, Manchester 1971)
[1787]   *Kritik der reinen Vernunft.* (Critique of Pure Reason; translation by N. Kemp-Smith, MacMillan, London 1929)

KHITCHINE, A.
[1928]   Objection à une note de MM. Barzin et Errera. *Académie Royale de Belgique, Bulletin* vol.14, pp.223-224.

KLEENE, C. and VESLEY, R.R.
[1965]   *The Foundations of Intuitionistic Mathematics.* (North-Holland, Amsterdam)

KNEEBONE, G.T.
[1963]   *Mathematical Logic and the Foundations of Mathematics.* (Van Nostrand, London)

KOLMOGOROV, A.N.
[1925]   On the Principle of the Excluded Middle. *Mame* 32, pp. 646-667; English translation in J. van Heyenoort [1967] *From Frege to Gödel.* (HUP, Cambridge Mass.)

KÖRNER, S.
[1960]   *The Philosophy of Mathematics.* (Hutchinson, London)

KORTEWEG, D.J.
[1881]   *De Wiskunde als Hulpwetenschap.* (van Heteren, Amsterdam)

KREISEL, G.
[1958]   A remark on free choice sequences and the topological completeness proofs. *Journal for Symbolic Logic* vol.23, pp.369-388.
[1968]   Lawless sequences of natural numbers. *Compositio Mathematica* vol.20, pp.222-248.

KREISEL, G. and TROELSTRA, A.S.
[1970]   Formal systems for some branches of intuitionistic analysis. *Annals of Mathematical Logic* vol.1, pp. 229-387.

LEBESGUE, H.
[1911]   Sur la non-applicabilité de deux domaines appartenant respectivement à des espaces à $n$ et $n+p$ dimensions. *Mathematische Annalen* vol.70, pp. 166-168.

LEROY, E.
[1899]   Science et Philosophie. *Revue de Métaphysique et de Morale* vol.7, pp.375-425; pp.503-562; pp.708-731.
[1905]   Sur la logique de l'invention. *Revue de Métaphysique et de Morale* vol.13, pp.193-223.
[1912]   *Une Philosophie Nouvelle: Henri Bergson.* (Alcan, Paris)

LEVY, P.
[1926A]  Sur le principe du tiers exclu et sur les théorèmes non-susceptibles de démonstration. *Revue de Métaphysique et de Morale* vol.33, pp. 253-258.
[1926B]  Critique de la logique empirique, Réponse à M. Robin Wavre. *Revue de Métaphysique et de Morale* vol.33, 545-551.
[1927]   Logique classique, Logique brouwerienne et Logique mixte. *Académie Royale de Belgique, Bulletin* vol.13, pp. 256-266.

MANNOURY, G.
[1907A]  L.E.J.Brouwer, Over de grondslagen der wiskunde. *De Beweging*, 1907, pp.241-250.
[1907B]  L.E.J.Brouwer, Over de grondslagen der wiskunde. *Nieuw Archief voor Wiskunde* series 2, vol.8, pp.175-180.
[1909]   *Methodologisches und Philosophisches zur Elementar-Mathematik.* (Visser, Haarlem)
[1924]   *Mathesis en Mystiek.* (Wereldbibliotheek, Amsterdam) (Re-published by Bohn, Scheltema en Holkema, Utrecht 1978).

MARTINO, E. and GIARETTA, P.
[1979]   Brouwer, Dummett and the bar theorem. In: *Atti del Congresso Nazionale di Logica* (ed. S. Bernini). (Bibliopolis, Napoli)

MENGER, K.
[1928]   *Dimensionstheorie.* (Springer, Leipzig und Berlin)

MYHILL, J.R.
[1967]   Notes towards an axiomatization of Intuitionistic Analysis. *Logique et Analyse* vol.35, pp. 280-297.
[1968]   Formal systems of Intuitionistic Analysis. In: *Logic, Methodology and Philosophy of Science* III (ed. B. van Rootselaar and J.F Staal). (North-Holland, Amsterdam)

PARSONS, C.
[1979]   Mathematical Intuition. *Proceedings of the Aristotelian Society* vol.80, 145-168.

POINCARÉ, H.
[1900]   Du rôle de l'intuition et de la logique en mathématique, *Compte rendu du 2e Congrès International des mathématiciens* (1900), pp.115-130.
[1902]   *La Science et l'Hypothèse*. (Flammarion, Paris)
[1903]   L'Espace et ses trois dimensions. *Revue de Métaphysique et de Morale* vol.11, pp. 281-301, pp. 407-429.
[1905]   *La Valeur de la Science*. (Flammarion, Paris)
[1906]   Les mathématiques et la logique, *Revue de Métaphysique et de Morale* vol.13, pp. 815-835, and vol.14 (1906), pp. 17-34.
[1908]   L'Invention Mathématique. *L'Enseignement Mathématique* vol.10, pp. 357-371.
[1912]   Pourquoi l'espace à trois dimensions. *Revue de Métaphysique et de Morale* vol.20, pp. 483-504.

POPPER, K.
[1968]   Epistemology without a knowing subject. In: *Logic, Methodology and Philosophy of Science III* (ed. B. van Rootselaar, J.F. Staal). (North-Holland, Amsterdam)

POSY, C.J.
[1974]   Brouwer's constructivism. *Synthese* vol.27, pp.125-159.
[1975]   Varieties of indeterminacy in the theory of general choice sequences. *Journal of Philosophical Logic* vol.5, pp.91-132.

RUSSELL, B.A.W.
[1897]   *An Essay on the Foundations of Geometry*. (CUP, Cambridge).
[1903]   Principles of Mathematics. (CUP, Cambridge).
[1918]   *Mysticism and Logic*. (Re-published by Penguin Books 1953)

RYLE, G.
[1949]    *The Concept of Mind.* (Hutchinson, London)

SCHOENFLIES, A.
[1900]    Die Entwickelung der Lehre von den Punktmannigfaltigkeiten. Bericht. Erster Teil. *Jahresbericht der D.M.V.*, vol.8, pp.1-250.
[1908]    Die Entwickelung der Lehre von den Punktmannigfaltigkeiten. Bericht. Zweiter Teil. *Ergänzungsband zu den Jahresbericht der D.M.V.* ( Leipzig)
[1913]    *Entwickelung der Mengenlehre und ihrer Anwendungen;* 1. Hälfte: Allgemeine Theorie der unendlichen Mengen und Theorie der Punktmengen. (Teubner, Leipzig und Berlin)

SCHOPENHAUER, A.
[1819]    *Die Welt als Wille und Vorstellung.* (Weichert, Berlin)
[1851]    *Parerga und Paralipomena.* (Weichert, Berlin)

TROELSTRA, A.S.
[1966]    *Intuitionistic General Topology* (V.R.B., Groningen)
[1968]    The theory of choice sequences. In: *Logic, Methodology and Philosophy of Science III* (ed. B. van Rootselaar, J. Staal). (North-Holland, Amsterdam)
[1969A]   *Principles of Intuitionism.* (Springer, Berlin)
[1969B]   Informal theory of choice sequences. *Studia Logica* vol.25, pp.31-52.
[1977A]   *The theory of Choice Sequences.* (Clarendon Press, Oxford)
[1977B]   Aspects of constructive mathematics. In: *Handbook of Constructive Mathematics* (ed. K.J. Barwise). (North-Holland, Amsterdam)
[1980]    Intuitionistic extension of the reals. *Nieuw Archief voor Wiskunde* (3) 28, pp. 63-113.
[1982]    The origin and development of Brouwer's concept of choice sequence. In: *The L.E.J. Brouwer Centenary Symposium* (ed. A.S. Troelstra and D. van Dalen). (North-Holland, Amsterdam)
[1983A]   Analysing choice sequences. *Journal of Philosophical Logic* vol.12, pp. 197-260.
[1983B]   Logic in the writings of Brouwer and Heyting. *In: Atti del Convegne Internazionale di Storia della Logica* (ed. Abrusci, Casari, Munai). (Clueb, Bologna)
[1985]    Choice sequences and informal rigour. *Synthese* vol.62, pp.217-227.

TROELSTRA, A.S. and VAN DALEN, D.
[1988]    *Constructivism in Mathematics.* (North-Holland, Amsterdam)

UNDERHILL, E.
[1911]     *Mysticism, a Study in the Nature and Development of Man's Spiritual Consciousness.* (Methuen, London)

VAN DALEN, D.
[1963]     *Extension Problems in Intuitionistic Plane Projective Geometry* (Drukkerij Holland, Amsterdam)
[1978A]   An interpretation of intuitionistic analysis. *Annals of Mathematical Logic* vol.13, pp.1-43.
          *Log* vol.13, pp.1-43.
[1978B]   *Filosofische Grondslagen van de Wiskunde* (Van Gorcum, Assen-Amsterdam)
[1981]     *Brouwer's Cambridge Lectures on Intuitionism.* (An annotated edition of [BMS51]) (CUP, Cambridge)
[1984]     *Droeve Snaar, Vriend van Mij.* (An annotated edition of the Brouwer-Adema van Scheltema correspondence) (Arbeiderspers, Amsterdam)

VAN DALEN D. and TROELSTRA, A.S.
[1970]     Projections of lawless sequences. In: *Intuitionism and Proof Theory* (ed. Myhill, Kino, Vesley). (North-Holland, Amsterdam)

VAN DANTZIG, P.
[1947]     On the principles of intuitionistic and affirmative mathematics. *Indagationes Mathematicae* vol.9, pp.429-440 and pp.506-517.

VAN EEDEN, F.
[1916]     Een machtig Brouwsel. Four articles in *De Groene Amsterdammer*, 9 September - 7 October 1916.
[1972]     *Dagboek 1878-1923* (Diary). Published by the Frederik van Eeden Genootschap. (Tjeenk Willink-Noorduyn, Culemborg)

VAN STIGT, W.P.
[1979]     The Rejected Parts of Brouwer's dissertation On the Foundations of Mathematics. *Historia Mathematica* 6, pp. 385-404.
[1981]     L.E.J. Brouwer, de Signific Interlude. In: *The L.E.J. Brouwer Centenary Symposium* (ed. A.S. Troelstra and D. van Dalen). (North-Holland, Amsterdam)
[1982]     L.E.J. Brouwer: Intuitionism and Topology. In: *Proceedings Bicentennial Congress of the Wiskundig Genootschap* part II, pp.359-374. (Mathematisch Centrum, Amsterdam).

WAVRE, R.
[1924]      Y-a-t-il une crise des mathématique? A propos de la notion d'existence et d'une application suspecte du principe du tiers exclu. *Revue de Métaphysique et de Morale* vol.31, pp.435-470.
[1926A]     Logique formelle et logique empiriste. *Revue de Métaphysique et de Morale* vol.33, pp. 65-75.
[1926B]     Sur la principe du tiers exclu. *Revue de Métaphysique et de Morale* vol.33, pp.425-430.

WELBY, V.
[1903]      *What is Meaning?*. (Macmillan, London)
[1911]      *Signifcs and Language*. (Macmillan, London)

WEYL, H.
[1917]      *Das Kontinuum. Kritische Untersuchungen über die Grundlagen der Analysis*. (Teubner, Berlin und Leipzig).
[1921]      Über die neue Grundlagenkrise der Mathematik. *Mathematische Zeitschrift* Band 10, Heft·Y812, pp. 39-79.
[1949]      *Philosophy of Mathematics and Natural Science*. (PUP, Princeton)

ZERMELO, E.
[1908]      Untersuchungen über die Grundlagen der Mengenlehre. *Mathematische Annalen* vol.65, pp.261-281.
[1909]      Sur les ensembles finis et le principe de l'induction complète. *Acta Mathematica* vol.32, pp.185-193.

# INDEX

Absurdity
   absurdity or Brouwer
      negation 239−244, 264
   absurdity as impossibility of fitting-in
      240−242
   logical absurdity 262−265
   absurdity-of-absurdity or non-
      contradictority 86, 265−270
   calculus of absurdities 87−88,
      257−260
Académie Internationale des Sciences
   107
Academy of Practical Philosophy and
   Sociology 67, 194
accompaniment (language) 201, 216,
   236
Acts of Intuitionism 97
   (see also *First* and *Second Act*)
addition 314
affirmations 227
affirmative constructivists 239
again 153
agreement of species 351−352
aid to memory (language as) 186, 201,
   205−206
Alexandroff 55, 84, 99
algorithm 168, 254, 311, 322, 357
   (see also *law*)
Amersfoort Academy 66
Amsterdam 52, 77
Amsterdam University 23, 49, 59, 65,
   68−69, 76, 84, 107, 113
Analysis
   analysis of the continuum by con-
      struction of intervals 320, 325
   analytical genesis of species 338
and-so-on 153
Annalen Affair 99−102
anthropomorphization of nature 39,
   185

apartness 348−351
applicability 54
application of mathematics 40, 60, 162,
   182−192
   (see also *science*, and *causal acting*)
   space as an application of
      mathematics 189−190
   metamathematics as an application of
      mathematics 162−163
application of logical principles 46,
   230−239
   (see also *Principle of the Ex-
      cluded Middle*)
   Intuitionist applied mathematics 191
a-priori
   Intuition of Time as a-priori of
      mathematics 150−151, 324
   Intuition of Time as a-priori of
      science 148−150, 188
   a-priori of mathematics w.r.t. science,
      language and logic 217
   synthetic a-priori judgements 161,
      166
arbitrariness
   (see *freedom*)
Aristotelian tradition (abstraction
   etc.) 171, 204
arrest
   arresting of 'fitting-in' 243
   arresting of spread-
      construction 367, 375−376
arrow 367
Ashvinikumar 275
assertions 227
assertions-of-assertions,
   see *logical principles*
attention
   (see *causal attention*)
Avsitidisky 288
*Axiomatic Foundations* 213

Axiomatics and axiomatic method  43, 214, 263, 279–285
   Hilbert's axioms  282
   Zermelo's axioms  62, 284
   Axiom of Choice  62, 75, 133, 359
   Axiom of Comprehension  82, 336

Bar Induction  92, 383
Barrau  44
Bar Theorem  92, 379, 382–384
Barzin and Errera  289, 288
basic species  338
beautiful, or good  124, 137–138, 143
becoming  156
believing  122
Bergson  112, 130–132
   Bergson and Brouwer Intuition  150–153, 324
Berlin Call  80
*Berliner Gastvorlesungen*  97–99, 298, Appendix 12
Bernstein  57
between, or 'intuitive between'  154, 156, 323–325, 327, 362
   (see also *intuitive continuum*)
Bhagavad Gîta  30, 33, 121
Bieberbach  57, 85
Bierens de Haan  47
Bishop E.  111, 274, 299, 300
Blumenthal  55, 70, 101
Boehme  30, 33, 114, 120
Bohr  101
Bolland  30, 113
Bologna Congress  101
Bolyai  233, 282
Bolzano-Weierstrass Theorem  75, 86, 241
Boole  109, 292
Borel E.  59, 129, 133, 219, 288, 319, 359
Borel, H.  66, 200
Boutroux  130
Brouwer
   Brouwer Class  338–339
   Brouwer Continuity Hypothesis (or Fundamental Hypothesis for real functions)  91, 379–381

Brouwer Logic  85, 88, 238, 257–260
International debate on Brouwer Logic  88–90, 288
Brouwer negation
   (see *absurdity*)
Brouwer's Programme  71, 81, 82, 202, 209, 270
   the demise of Brouwer's Programme  93, 270
Brouwer-Set  73ff
   (see further *Spread*)
Mrs. E. Brouwer-de Holl  26, 63, 79, 105
Buber  68, 77, 195, 196, 200
Buddhism  120
Bunch Theorem  92
   (see also *Fundamental Theorem*)
Burali-Forti paradox  62

Calculus of Absurdities  87, 257–260
Cambridge Lectures  93, 98, 299
Cantor  301, 302, 307, 337
Caratheodory  57, 99, 101–102, 162
cardinality  312–314
cardinal number  312–314
causality  31, 127, 139–143, 150
   causal (cunning or mathematical) acting  142–143
   causal attention or viewing  141–142
   causal coherence of the world  186
   isolated causal phase  137–139
   causal sequence  139, 141–142, 150
   causal thinking  141, 186
certainty  118, 217
change (time)  152
choice
   (see *free choice*)
Church  125
class (Brouwer class)  338–339
classification of constructions  165–168
coincidence  349
   coincident points or point-cores  349, 364–366

collective mind   140, 179, 186, 204
commensensical thought   226
communication
    impossibility of communication   33, 43, 104, 199−200, 207−209
    communication between Idealized Mathematicians   208
    language as means of communication   207−209
complementarity of species   352−357
complementary conjugate species   355−356
complementary law of spreads   371
Complete Principle of Judgeability or Complete Principle of the Excluded Middle   353
Complete Principle of Testability   353−354
composed   355
Compositio Mathematica   106, 109
comprehension   74, 296, 304, 333, 336, 345
Conclusiveness (Principle of Conclusiveness)   246, 263
concordance   351
congruence (of species)   351
conjugate species   355−356
conscience   124
consciousness   135−138
    phases or transitions of Consciousness   137−146
*Consciousness, Philosophy and Mathematics*   108, 138
Conseil International de Recherches   85, 99
consistency of mathematics   266
constancy   154
construction
    constructing or building   164−165, 168−169
    classification of constructions   165−168
    complex constructions   166
    elementary constructions   166
    elements of construction   166
        (see also *mathematical entities*)

construction of impossibility   242−244
    incomplete constructions   245
constructiveness   164ff
    logical construction   223−224, 230−239
    constructive thinking   168−170
    constructive interpretation of logical constants   278
continuity
    continuity arguments   91−93
    continuous function   379−380
    Brouwer continuity hypothesis   379−381
    continuity and uniform continuity of spread functions   378ff, 384−385
    apparent continuity   97
    continuous function   379−380
    Uniform Continuity Theorem   91−93, 384−385
continuum   64, 71, 318ff
    the continuous and discrete in the Primordial Intuition   153−156, 323
    continuum intuition, intuitive continuum or between   154−156, 316, 320, 323−325, 327
    made measurable continuum   328−230
    Brouwer mathematical continuum   318ff
    points on the continuum   312−313
    real-element-of-continuum   361−366
    reduced continuum   321
    unsplittability of the continuum   97, 255, 384
contradiction   227, 265
    Principle of Contradiction   246, 261
contradictority of PEM
    (see *Principle of Excluded Middle*)
convergence   322, 360, 365
correct affirmations   227
correctness   258
counterexamples refuting the PEM   252−257
Courant   101

Couturat 128
Covering Theorem (Intuitionist) 97
Creating Subject 172, 178−179
  (see further *Subject*)
crisis, new crisis in mathematics 75
cunning activity, causal or mathematical acting 142−143, 184−185

Darboux 48
decidability 248
Dedekind Cut 233
de Donder 89, 291
definable points 333
de Gruyter 95
de Haen 66, 262
Delft Lectures (Brouwer) 31
de Loor 84
Denjoy 79, 84
denumerability
  denumerability of all mathematical constructions (pre-1917) 179, 309, 332−333
  denumerably unfinished systems 332, 334
  denumerabilities 348
Descartes 129, 132, 133
destruction 367
determinacy 168, 177
deviation 348−351
de Vries 79, 85
difference 153
  absurdity of equality 240
  difference of species 348−351
Dijkman 275
directly conjugate species 356
discrete
  the discrete and continuous 153−156, 302, 323
  discrete points 330−332
  discrete systems 300−317
disjointness 348
distance 317, 328
divisibility of interval
  (see *between*)
division 314
Dummett 240, 299, 361, 383
Dyck 85

Eastern and Western thinking 200
Eckehart 30, 33, 120
Eddington 184, 185
effective computability of functions 91
Ego, Self 122, 135−138, 172
  (see also *Subject*)
Einstein 101−102
element
  elements of construction 166
    (see also *mathematical entities*)
  elements of Intuition 158, 166
  element of species 338−339
  element-species 342
  elementhood of species 341
  element of spread 366
  real-element-of-continuum 361−366
elementary constructions 166
Emotional language 209
empirical space 189
empty species 347
ending of spread process 376
epistemology 118
equipotency 313
equivalence of species 343, 348
equivalence species 342
Euclid 223−224
Euclidean geometry 189, 223−224
Euwe 110, 285
everywhere-density of rationals 316
evil 124, 143
exactness of mathematics 212, 217
(in)exactness of language 208, 212, 217
existence (mathematical) 162, 164−165, 203, 233
exodus from deepest home 126, 136
extensional definition of species relations 345
extensional interpretation of species 261, 338, 345
extensional use of spread 377
Exterior World of the Subject 139−141, 185, 186−188
externalization 39, 121, 126, 157

falling-apart, splitting 155, 302
fan
  (see *finite spread*)

Fan Theorem
  (see *Fundamental Theorem of Finite Spreads*)
Fechner 128
Fichte 112, 117
Finitary 370
Finite Set or Finite Spread (also Fan) 92−93, 370, 378
  Fundamental Theorem of Finite Spreads 92−93, 298, 379, 381−384
First Act of Intuitionism 70, 96, 199
first-order language 215
first-order mathematics 83, 161, 215
first-order species 340
fitting-in 167, 240−242, 337
Fixed-point Theorem 56
Flaubert 120, 125
Fleeing Property 253−254, 268
flesh-and-blood mathematician 169−170, 174, 215
Fokker 79
form of externality (Russell) 187
Formalism 61, 83, 100, 224, 282
  Old Formalists 279
  New Formalists 280−282
formalization 161, 281
  Formalization of Intuitionist Mathematics 279−294
  formalization of the Subject 176
  Heyting's Formalization 88, 285−292
*Foundations of Function Theory* 297
*Foundations of Mathematics* 35−44, 113
*Foundations of Set-theory* 71−75, 296
Fraenkel 82, 288
free
  freedom and will 174
  freedom and authority 194
  freedom of mathematics and Subject 158, 174, 177−179, 311
  free choice 64, 73−74, 322, 360, 363−364
  free-choice sequence 72, 178, 273, 310, 357−361
  restriction of freedom 371−372
French Intuitionist tradition 127−132
French Mathematical Intuitionism 132−135
Freudenthal 56, 99, 106, 278, 285
Fricke 57
function, continuous function 379−380
  function (definition) 380
  full function 92
  function theory 93
Fundamental Hypothesis for Real Functions 91, 379−381
fundamental relations of point species 350
fundamental sequence 310, 374
fundamental sequence of intervals 369, 374
fundamental species
  (see *species of order zero*) 365
Fundamental Theorem of Finite Spreads (also Fan or Bunch Theorem) 92−93, 97, 298, 379, 381−384
future 177

Generalization
  Principle of generalization 339
  (see also *Species Principle*)
  Brouwer's generalization of the concept 'sequence' 309−312
geometry 49
  *The Nature of Geometry* 49−51
  Euclidean geometry 189, 223−224
  geometries 49, 164, 189
  Hilbert's geometries 233
Gibson 275
Glivenko 89, 289
God 122−126
good (beautiful) 124, 137−138, 143
Göttingen 57, 70, 80, 99, 105
Griss 108
Groningen 59
Gutkind 78, 84, 200

Haalmeijer 84
Hadamard 55, 56

Hallett 302
Hamel 36, 38
haze
  (see *spread haze*)
Hecke 80
Hegel 128
Heine-Borel Covering Theorem 75, 86, 251
  Intuitionist alternative 384
Helmholtz 38, 50
Herzberg 95
Heyting 90, 109, 127, 285, 299, 300, 340
  Heyting's Intuitionism 274−279
  Heyting's Formalization of Intuitionist Logic and Mathematics 285−292
  Heyting's natural numbers 301
  Heyting constructive interpretations 278
hierarchy of species 340−345
higher-order mathematics 161, 214−215
higher-order restriction of freedom 372
higher-order species 340−345
Hilbert 36, 43, 51−53, 56, 62, 70, 80−85, 282, 283, 295
  Hilbert's testimony of Brouwer 56
  Hilbert-Brouwer controversy 99−102
  Hilbert's Paris Problems 38, 48
  Hilbert's Programme 266, 280
Hinduism 114, 136
Hölder 85
Home 120, 136
Hopf 55, 99
Hurwicz 99
'de Hut' 35
hypothesis 240, 337−338

idealism
  Brouwer's idealism 171
  the ideal nature of science 175, 185−187
idealization
  idealized character of mathematics 168−169, 174−180

Idealized Mathematician 107
Idealized Subject 172−174
  Communication between idealized mathematicians 208
immortality 136
impossibility 240−244
  (see also *absurdity*)
implication 261
inadequacy of language 43, 61, 67, 123, 217−218, 223
inclusion (of intervals) 326
incompatibility 242, 243
incomplete construction 245
  incompleteness of choice sequence 73
Indefinitely Proceeding Sequence (IPS) 74, 272, 309−312, 366−367, 371−377
independence of mathematics from language and logic 43
indeterminacy 177, 272
index 74, 312, 374
individuality 155
individuals (object) 139
Induction
  complete induction 133
  Principle of Induction 153
inexactness of mathematical language 208, 212, 217
infinite sequence 343, 346
  (see also *IPS*)
initial segment 73, 91, 246, 360−364
Inner World of the Subject
  (see *Interior World of the Subject*)
insertion of point 325
Insight
  (see *Intuition*)
instability of language 204−210, 217
instinctive 157
instruments 197
integers 315−316
integration 94
intellect 113, 141
intellectual observation 142
  (see also *causal attention*)
Interior World of Subject 160−162, 173

International Academy of Practical Philosophy and Sociology 67, 78
intersection of species 346
intersubjectivity 171–174
interval 72, 324–327, 362
  (see also *between*)
  bisection of intervals 325
  relations between intervals 326–327
  inclusion 326
  order relation between intervals 326
Intuition
  Brouwer's concept of Intuition, Insight, Introspection 34, 120–121, 124, 173
  Intuition as 'Anschauung' 34, 135
  Brouwer's Primordial Intuition of Time (the Primordial Happening, or the Primordial Phenomenon) 42, 61, 147–157
  the role of intuition in mathematics 158
  Intuition of two-oneness 149
  Intuition of the continuum 154–156
    (see also *Continuum Intuition*)
  The Primordial Intuition as the fundamental phenomenon of causality 148
  Bergson's Intuition 130–132, 134, 150–153
  Borel's Intuition 133
  Kant's Intuition 132, 151
  Poincaré's Intuition 133
  Russell's Intuition 120
Intuitionism
  French Intuitionist tradition 127–132
  French mathematical intuitionism (neo-intuitionists, pre-intuitionists) 132–135
*Intuitionism and Formalism* 61
Intuitionist Logic 238
  (see also *Brouwer Logic*)
  Intuitionist School 129
    (see also *Pre-Intuitionists*)
Intuitionist Mathematics
  Brouwer's reconstruction 77, 295ff
  (see also *Brouwer's Programme*)
Intuitionist Applied Mathematics 191
*Intuitionist Splitting of Fundamental Mathematical Concepts* 87–88, 257–260
Intuitive Continuum 154–156, 323–325, 327
  measureless intuitive continuum 328–329
Intuitive Significs 67, 200
intuitive thinking 157–159
intuitive time 152, 328
Invariance of dimension 53–55, 321
IPS
  (see *Indefinitely Proceeding Sequence*)
isolated causal phase 137

Jahnke 28
Johnson 295, 330
Jongejan 103, 110
Jordan Theorem 50, 53, 56

Kant 40, 117, 127–132, 139, 178, 180, 204
  Kant's Intuition 132, 151
Karma 135
Khintchine 89, 289
Kleene 370, 383
Kleene and Vesley 178, 299, 361
Klein 57, 59, 63, 70
Klein's Erlanger Programme 50
knowing 122
Koebe 57
Kohnstamm 47, 84, 116
Kolmogorov 88, 289
Körner 157, 180
Korteweg 24, 29, 40–44, 57–60, 69, 79, 84, 128, 162, 182
  (see also Appendix 13)
Kreisel 178, 278, 299, 361
Kreisel and Troelstra 361
Kronecker 82, 83, 300, 319

Lachelier 130
labour 67, 144, 197–198

language
  origin and purpose of language
    144−145, 196−198
  other functions of language 205
  language as accompaniment 201,
    216, 236
  language as aid to memory
    205−206
  language as means of communication
    33, 144−145, 193−198
  emotional language 209
  pure language 200−201, 212
  language levels 201, 209−210
  language of mathematics 169,
    215−220
  language of logical reasoning 43,
    220−224
  symbolic notation (Brouwer's view)
    87, 121, 204, 292
  Brouwer's use of language in his construction of mathematics 271ff
  the inadequacy of language 43, 61,
    67, 121, 123, 200, 217−218, 223
  the instability of language
    204−210, 217
  the privacy of language 203−204
  subject-predicate form of language
    212
  truth of language 212
  practical efficiency of language 208
Laren (Blaricum) 35, 60, 99
Lau van der Zee (Brouwer's pseudonym)
  30
law 74, 168, 311, 357, 363−364
  law as moral code 124
  scientific law 186
  laws of logic 227, 262, 263
  spread law 367−369, 370−371,
    374−375
  complementary law 370
Lebesque 55, 84, 129, 219, 319
Leibniz 132
Leiden 60, 68
length
  (see *distance*)
Leroy 219

Levy 89, 288
Lie groups 38, 47
*Life, Art and Mysticism* 31−35, 66, 117
linguistic engineering 214
linguistic structure 214
Lobatcheffsky 233, 282
local distinctness 349
  (see also *apartness*)
Logic
  Brouwer's view and analysis of
    logic 213ff, 224ff
  the nature of logic as a science 43,
    190, 224−230
  pure logic 225−230, 292−293
  symbolic logic 292−293
  logical reasoning 221−223
  'logical play' 225, 230, 292
  logical absurdity 262−265
  logical necessity 260−262
  logical negation 257, 262
  logical operators 278
  logical truth 260−262
  logical principles 227−229
  the application of logical
    principles 230−239
    (see also *PEM*)
  logical construction ('verbal
    edifice') 223−224, 230−239
  *The Unreliability of the Principles of
    Logic* 46, 231−235
  Intuitionist Logic 238
  the Brouwer Logic 85−90, 238,
    257−260
Logicism 132
Lorentz 60, 79
Lotze 128

Maine de Biran 130
Manifesto 67
Mannoury 26, 35, 49, 60, 69, 78, 90,
  109, 113, 200, 213
Martino and Giaretta 383
mathematical
  mathematical (cunning)
    acting 142−143

mathematical continuum 150, 318ff
  (see also *continuum*)
mathematical entity
  (construction) 165ff
mathematical entity 'previously
  acquired' 74, 167, 336, 340
mathematical existence 162 – 165
mathematical language 215 – 220
mathematical negation 239 – 244
  (see also *absurdity*)
mathematical reality 160, 162 – 165
Mathematical Significs 67
mathematical species 335
  (see also *species*)
mathematical (causal) thinking 31, 141
mathematical truth 160, 162 – 165, 204
mathematical viewing (causal
  attention) 141 – 142
Mathematical Centre 68, 69, 84, 107
mathematician
  the Idealized Mathematician, the
    Creating Subject
    (see *Subject*)
  the flesh-and-blood
    mathematician 169 – 170, 174
mathematics
  the nature of mathematics as intuitive
    thinking 158 – 161
  the idealized character of pure
    mathematics 168 – 169, 174 – 179
  the language-less nature of
    mathematics 193
  the subjective nature of
    mathematics 162, 172
  reflection on mathematics 162 – 163
  mathematics of first and higher
    order 161
    (see also *metamathematics*)
  the application of mathematics
    (see *application*)
  the reconstruction of
    mathematics 71, 76
    (see also *Brouwer's Programme*)

*Mathematics, Science and Language* 104
Mathematische Annalen 49, 55, 63, 99ff
Mauve 26, 35
meaning 204, 210
means-to-an-end 142
  (see also *cunning activity*)
measurability 152, 321
  made-measurable
    continuum 328 – 330
  measurable function 94, 97
measure 321
memory 153, 162, 165, 186
  weakness of memory 207
  (see also *aid-to-memory*)
Menger 81, 99
metamathematics 83, 100, 187, 280
  metamathematical
    consideration 161, 162 – 163, 214 – 215
Meyer 155
Mind 138 – 139
misanthropy and misogeny
  (Brouwer's) 24, 32, 39, 43, 125, 193
Mittag-Leffler 48
morality 124 – 126
  *On Morality* 193
multiplication 314
music 176, 216
Myhill 279
Mysticism 119 – 122

naive phase of Consciousness 137
natural number
  (see *number*)
*Nature of Geometry* 49 – 51
necessity
  mathematical necessity 242
    (see also *a-priori*)
  logical necessity 260 – 262
  scientific necessity 150
negation
  classical negation 238
  Brouwer mathematical
    negation 238 – 244
    (see also *absurdity*)
  other notions of negation 244 – 246

logical negation 257ff
logical absurdity 262–264
double negation 86, 246, 265
negative numbers 316
negative (essentially) properties 108, 270, 355
    negative and non-negative assertions 268
Neo-Intuitionism 61, 128, 132
nesting of intervals 72, 326, 363
Newman 99
node 367
Noether 99
non-contradictority 258, 265–270
    (see also *PEM*)
    non-contradictority of mathematical language 227
noumena 127, 129, 168
number
    genesis of number
        (see *Primordial Intuition*)
    natural (ordinal) number 300ff
    number one 155, 303, 305
    number two (two-ity, two-oneness) 155–156, 302–303
    the fundamental sequence 310, 374
    the order-type $n$ 306–308
    the order-type $\omega$ 309, 314, 332
    pure ordinal number 308
    operations on ordinal number 314
    cardinal number 312–314
    integers 315–316
    rational numbers 316–318
    real number (pre-1917) 320
    real number (post-1917) 361–366
        (see also *real point* or *real-element-of-continuum*)
    transfinite numbers 35, 43, 233
numerals 310

object 186–187, 203
object individuals 140
objectivity 180–182, 186
Old Formalists 279
one 155, 303, 305
one-to-one correspondence 313

operation
    operation
        (see *construction*)
    operation on numbers 314ff
    logical operation 88, 246
Opitz 36
order 302
    order relations 328
    order relations of intervals 326
    before-after 153
ordered pair 302
ordering the continuum 90, 97
order-type 307
order-type $\eta$ 316–318, 332
order-type $n$ 306–308
order-type $\omega$ 309, 332
ordinal number 154, 306ff
Ornstein 66

paradoxes 62
Pascal 132
past 175–176, 177
Peano 68, 78
PEM
    (see *Principle of Excluded Middle*)
perceptional experience 137
personal identity 172
pessimism (Brouwer's) 143
phenomenon 138, 144, 151
philosophy
    Brouwer's philosophy 41, 111ff
    Brouwer's view on philosophy 79, 115, 164
    Brouwer's view on philosophers and academic philosophy 115, 128
*Photogrammetry* 79
PI
    (see *Primordial Intuition*)
Picard 48
Pieri 287
Planck 85
Plato 171, 204
play and playful acting 143, 192
    logical play 225, 230, 292
plurality of mind (collective mind) 140, 173, 186, 204

Poincaré 36, 48, 57, 59, 129, 132, 134, 266, 319
   Poincaré's cut principle 330
   Poincaré's last problem 63
point
   point on the continuum (discrete point) 330–332
   sets of points on the continuum 332–333
   definable points 333–334
   possible points 50, 334
   real point of the continuum 73, 349
   point-core 349, 364–366
   order of point-cores 366
   Point-Set or Point-Spread 366–369
   *Theory of Point Sets* 75, 94
   (see also Appendix 11)
Popper 216
Posy 278, 299, 361
power 313
   power relations 348
   *On Possible Powers* 47
predicative definition 304
Pre-intuitionists 64, 82, 129, 219, 319, 321
*Preparatory Manifesto* 67, 77, 195
present (time) 175, 177
previously acquired 74, 167, 336, 338, 340
Primordial Intuition (PI), Primordial Happening 40, 42, 138, 147–157
   genesis and definition of PI 148
   discrete and continuous in PI 153–156
   PI as set construction 301–305
   PI the foundation of all mathematics 158
   PI as direct foundation of causality 148, 186
   primordial two-ity 155, 166
Principle of Conclusiveness 246, 263
Principle of Contradiction 246, 261
Principle of Generalization (Species Principle) 339–340
Principle of Reciprocity of Complementary Species 86, 256, 268, 354–355
Principle of Syllogism 261
Principle of Testability 353–354
Principle of the Excluded Middle (PEM) 46, 82, 85–86, 228, 233, 238
   Complete PEM 353
   Simple PEM 250
   unreliability of PEM 41–42, 46, 248–250
   Refutation of PEM 246ff
   counterexamples refuting PEM 252–255
   mathematics without PEM 250
   non-contradictoriety and contradictority of PEM 85, 89, 249–250, 255–257
principles (constructive) 246
principles of logic 227–229
   application of logical principles 230–239
   *The Unreliability of the Principles of Logic* 46, 231–235
   practical reliability of logical principles 237–238
privacy of language 203–204
privacy of thought 140, 173, 199
proceedability 311, 367
*Profession of Faith* 33, 111, 122–124
*Proof that every function is uniformly continuous* 92, 298
property
   (see *species*)
pseudoset 322
   (see also *species*)
pure language 217
pure logic 225–230
pure ordinal 308
pure science 191–192, 225, 228
Putsch 81–82
Putschist footnote 95

rational numbers 316–318
Ravaisson 130
real-element-of-continuum 361–366
reality (mathematical) 160, 162–165
Real Functions 93, 298

real number
  real number (pre-1917)  320
  real number (post-1917)  361−366
real point  361−366
Reason  117, 127
reciprocity
  Principle of Reciprocity of Complementary Species  86, 256, 268, 354−355
reduced continuum  321
reflection on mathematics  162
regularity in language  227, 233
regularity in Nature  39, 186
  (see also *sequence of sequences*)
Reiman  116
re-incarnation  136
relations between species  347−352
  (see also *species*)
reliability
  (see *logical principles*)
  practical reliability of language  208
  practical reliability of logic  237−238
removable  355
restriction of freedom  371−372
*The Role of the PEM in Mathematics, especially in the Theory of Functions*  85
Rome Conference 1908  47
Rosenthal  57
Russell  119, 187
  Russell distance  317, 328
  Russell's Essay  36, 41, 42, 128
  Russel's Principles  36
  Russell's mysticism  119, 120
Ryle  129

Schmidt  85, 101
Schoenflies  51−53, 62−65, 70, 295
Schogt  84
Schopenhauer  104, 112, 128, 143
Schouten  59
science
  the nature of science as applied mathematics  182−183
    (see also *causality*)
  the ideal nature of science  175, 185−187
  Brouwer's critique of scientific practice  184−188
  pure science  144, 191−192, 225, 228
  laws of science  184
  scientific necessity  150, 186
  scientific time  152, 328
  scientific truth  186, 188
  science of logic  224−230
Second Act of Intuitionism  71, 96, 296, 323, 334, 357ff
second-order mathematics  83, 100, 161, 187, 214−215
second-order species  341
Self, Ego  122−126, 135−138, 172
self-awareness  135
self-evident truths  157
semantic theory (Brouwerian)  230ff
semi-equivalence or half-identity  351
sentences  199
separable mathematics  300
separateness (separation) of species  350
separation of thought and language  199
separation mathematics and language  96, 209, 215, 219
sequence
  the sequence as fundamental concept  305ff
  fundamental sequence  310
  free-choice sequence  363−364
    (see also *free-choice*)
  Brouwer's generalization of the concept sequence  309−312, 357ff
  Indefinitely Proceeding Sequence (IPS)  309−312, 357ff
  sequences of sequences  139
    (see also *causality* and *things*)
set
  Cantor set  71, 86
  (Brouwer) mathematical set (pre-1917)  304
  the Primordial Intuition as set-construction  301
  well-constructed sets  64, 72, 249, 304, 322, 360

pseudo-sets 322−323
Brouwer's new set theory 1917 62, 64, 72ff, 344ff
  (see further *species* and *spread*)
Significs 65, 79, 195, 200−201
  Signific Circle 78
  *Signific Manifesto* 67, 77, 195
  Intuitive Significs 67, 200
    (see also Appendix 4)
  Mathematical Significs 67
Simple PEM 250
sin 124
social acting 144
social mind 144
social phase 137
society 193
Sodalitas 105
solipsism 114, 199
Sommerfeld 85
soul 118, 135−137, 141, 199
space 136
  empirical space 189
  space as an 'application of mathematics' 189−190
species
  Brouwer's development of the concept species 74, 167, 305, 323
  complementarity of species 352−357
  concept and genesis of species 335−340
  conjugate splitting species 356
  elementhood of species 341−342
  element-species 342
  equivalence species 342
  extensional interpretation of species 261, 338, 345
  higher order species 340−345
  hierarchy of species 340−345
  species principle (principle of generalization) 287, 339−340
  species of order zero or fundamental species 342, 344, 365
  theory of species 345ff
  empty species 347
  subspecies 345

intersection and union of species 346−347
relations between species 347−352
  (see also *agreement, coincidence, congruence, deviation, disjointness, difference, equivalence*)
spiritual values 200
splitting of a moment 155, 302, 306
splitting of continuum 97, 384
*Splitting of Mathematical Concepts* 87, 269
splitting species 356
spread or Brouwer set
  Brouwer's development of the concept 73−75, 323, 357ff
  point spread 366−369
  finite spread (fan or bunch) 370, 378
  Fundamental Theorem of Finite Spreads 92−93, 379, 381−384
  general spread concept 370−377
  definition of spread 373
  spread element 366, 376
  spread law 370−371, 374−375
  spread haze 375
  spread haze law 375
  spread species 341, 376−377
Springer 101
Steen 98
sterilization 367, 375−376
stillness 121, 136
*The Structure of the Continuum* 104
Struik 285
Subject, Creating Subject, Idealized Mathematician 107, 165, 170, 172−174
  communication between Idealized Mathematicians 175
  Subject's time existence 159
subject-predicate structure of language 211−212
subjectivity of mathematics 162
subspecies 345
subtraction 314
successor construction 307
supposable 337−338

syllogism 221
  Principle of Syllogism 261
symbolic logic 292−293
symbolic notation, Brouwer's view 87, 121, 204, 292

testability 252, 253
  Complete Principle of Testability 353−354
  Simple Principle of Testability 353
theosophy 114
things 139, 185
three 306, 307
time 136, 150−153, 178, 324
  intuitive and scientific time 152, 328
  time existence of Subject and mathematics 175−176
tools, instruments 197
Topological Activity 49−55
*Towards a Foundation of Intuitionist Mathematics* 93
transfinite numbers 43, 233
tree 366−367
Troelstra 178, 275, 278, 290, 299, 300, 361, 372, 385
Troelstra and Van Dalen 288, 299, 361
*Troostgronden* 30
truth 107, 125, 191
  mathematical truth 160, 204, 248
  logical truth 260−262
  scientific truth 188
  truth of language 211−212
truth-predicates 267
two, two-ity, two-oneness 149−150, 302, 305−306
two-ity as ordered pair 302

Underhill 120
Uniform continuity 379
Uniform Continuity Theorem 92−93, 97, 298, 379, 384−385
union of species 346

*Unreliability of Logical Principles* 46, 231−235
unsolvable mathematical problems 254−255
Urysohn 84

Van Dalen 98, 275, 278, 285, 299, 361
Van Dalen and Troelstra 278, 361
Van Dantzig 108
Van der Waerden 84
Van Eeden 35, 65, 78
Van Ginneken 200
Van Rootselaar 275, 299
Veblen 53
Vienna Lectures 104
Vietoris 99

Wavre 89, 289
Weitzenböck 85
Welby, Lady Victoria 65
well-constructed set 64, 72, 249, 304, 322, 360
Well-ordering Theorem 75
Weyl 57, 75−77, 80, 287, 319, 339−340, 379−380
Whitehead 185, 287
Wijdenes 84
will 112, 143
  freedom and will 174
  will-transmission 104, 144−145, 197−198
*Will, Knowledge and Speech* 106−107, 201
  (see also Appendix 5)
Wilson, 99
Wiskundig Genootschap 45, 69
Wittgenstein 119
women 32, 125
words 211

Zeeman 79
Zermelo's axioms 61, 284, 296